SpringerWienNewYork

# CISM COURSES AND LECTURES



The series presents lecture notes, monographs, edited works and proceedings in the field of Mechanics, Engineering, Computer Science and Applied Mathematics.
Purpose of the series is to make known in the international scientific and technical community results obtained in some of the activities organized by CISM, the International Centre for Mechanical Sciences.

INTERNATIONAL CENTRE FOR MECHANICAL SCIENCES

COURSES AND LECTURES - No. 481

# SURFACE WAVES IN GEOMECHANICS: DIRECT AND INVERSE MODELLING FOR SOILS AND ROCKS

EDITED BY
CARLO G. LAI
EUCENTRE, PAVIA, ITALY

KRZYSZTOF WILMAŃSKI
WEIERSTRASS INSTITUT-WIAS, BERLIN, GERMANY

SpringerWienNewYork

The publication of this volume was co-sponsored and co-financed by the UNESCO Venice Office - Regional Bureau for Science in Europe (ROSTE) and its content corresponds to a CISM Advanced Course supported by the same UNESCO Regional Bureau.

This volume contains 163 illustrations


Printed in Italy
SPIN 11526254

In order to make this volume available as economically and as rapidly as possible the authors' typescripts have been reproduced in their original forms. This method unfortunately has its typographical limitations but it is hoped that they in no way distract the reader.

**Additional material to this book can be downloaded from http://extras.springer.com.**
ISBN-10 3-211-27740-4 SpringerWienNewYork
ISBN-13 978-3-211-27740-9 SpringerWienNewYork

## *PREFACE*

*Theories of surface waves develop since the end of XIX century and many fundamental problems like existence, phase and group velocities, attenuation (quality factor), mode conversion, etc. have been, in part successfully, solved within the framework of such simple models as ideal fluids*[1] *or linear elasticity. However, a sufficiently complete presentation of this subject, particularly for solids, is still missing in the literature. The sole exception is the book of I. A. Viktorov*[2] *which contains an extensive discussion of fundamental properties of surface waves in homogeneous and stratified linear elastic solids with particular emphasis on contributions of Russian scientists. Unfortunately, the book has never been translated to English and its Russian version is also hardly available.*

*Practical applications of surface waves develop intensively since a much shorter period of time than theories even though the motivation of discoverers of surface waves such as Lord Rayleigh stems from their appearance in geophysics and seismology. Nowadays the growing interest in practical applications of surface waves stem from the following two main factors:*

- *surface waves are ideal for developing relatively cheap and convenient methods of nondestructive testing of various systems spanning from nanomaterials (e.g. surface coating) through semiconductors (e.g. detection of lattice defects), metals and composites (e.g. detection of surface cracks), near-surface geophysical and geotechnical characterization of geomaterials (e.g. determination of the speeds of propagation of longitudinal and transversal waves, porosity, etc.) to seismological structures (e.g. investigation of the Earth crust and of its upper mantle);*
- *the reliability of surface wave-based techniques increases rapidly through the improvement of testing equipment (e.g. laser techniques, "intelligent" geophones with remote connections, etc.) and interpretation software (e.g. simulated annealing, waveform and fractal inversion, enumerative algorithms, etc.).*

*The aim of the CISM course entitled "Surface Waves in Geomechanics: Direct and Inverse Modelling for Soils and Rocks" held in Udine on September 6 – 10, 2004, was to cover both theoretical and experimental aspects of surface waves with special emphasis on applications to geomaterials for near-surface, non-invasive characterization of sites. Although the frequency range used in these field applications varies typically in the band from 0.5-1 Hz to 100 Hz, in geophysical laboratory measurements of soil specimens are applied much higher frequencies. For this reason,*

---

[1] *e.g. John V. Wehhausen and Edmund V. Laitone; Surface Waves, in: S. Flügge, C. Truesdell (eds.), Encyclopedia of Physics, vol. IX, Fluid Dynamics III, Springer, Berlin, 1960. See also the online edition, Regents of the University of California, 2002 (www.coe.berkeley.edu/SurfaceWaves/)*

[2] *I. A. Viktorov; Zvukovye povierkhnostnye volny v tverdykh telakh (Acoustic surface waves in solids, in Russian), Nauka, Moscow, 1981.*

*theoretical considerations of these notes are not limited to low frequency approximations that are typical of near-surface geophysics.*

*The articles which we present in this volume can be divided into four groups. The first group of two contributions (Rix and Foti) contains a thorough description of surface wave methods for geotechnical site characterization. Chapter 1 in particular illustrates in some details the field equipment used in surface wave testing as well as an updated overview of the signal processing techniques used to construct the experimental dispersion curve(s) in both active and passive testing. The chapter ends with an introduction to the inverse problem associated with surface wave data and an illustration of the strategies used to overcome its inherent ill-posedness. Chapter 2 is also dedicated to the applications of surface wave testing for geotechnical characterization but here the focus is more on enlighting the limitations and drawbacks of the various techniques used in surface wave analysis and to illustrate the advantages of a joint inversion with seismic refraction and electric data.*

*The second group (Lai, Kausel) is devoted to the theoretical analysis of surface waves in vertically inhomogeneous, viscoelastic solids. Two main issues are addressed in these contributions: an illustration of the main characteristics of surface wave propagation in linear viscoelastic, vertically inhomogeneous continua in both forward and inverse modelling and an introduction to computational methods for generalized surface and interface waves in stratified media. Chapter 3 gives a detailed account of the theory of surface waves in linear, one-constituent, vertically inhomogeneous, dissipative continua. Although the basic theory is developed for both Love and Rayleigh waves, most of the focus is dedicated to Rayleigh waves because they are more relevant in the applications. The chapter begins with an illustration of standard results of the theory of surface wave propagation in elastic and weakly dissipative media (including solution of the Lamb's problem) using the formalism of variational calculus. Then the full theory of Rayleigh wave propagation is developed for arbitrarily dissipative linear viscoelastic materials. The chapter final section illustrates the main properties of the inverse problem associated with surface wave motion focusing in particular on the joint inversion of dispersion and attenuation data. Chapter 4 is an introduction to computational methods in geomechanics. It begins with a brief review of normal modes in simple systems which are then used to illustrate the fundamental concepts of wave propagation, including complex wave spectra. These concepts are then generalized to the case of horizontally stratified media through the Stiffness Matrix Method (SMM) and its discrete counterpart the Thin Layer Method (TLM) which constitutes a powerful tool to obtain the normal modes for both propagating and evanescent generalized surface and interface waves.*

*The third group (Wilmanski, Albers) presents a review of theoretical results for various interface waves of single- and two-component elastic materials. Apart from classical Rayleigh waves, Love, Stoneley, various leaky waves, etc. are presented. Chapter 5 contains the theory of the most common classical surface and interfacial waves. In particular, it is shown how the analysis of the dispersion relation yields seismograms, some comments are made on pseudosurface (leaky) Rayleigh and Love waves. It ends with the introduction to the analysis of surface waves in poroelastic saturated materials. This problem is the subject of the detailed numerical analysis in Chapter 6. This part contains also some unpublished results on attenuation of leaky surface waves in such materials.*

*Finally, the fourth group (Maugin) is devoted to theoretical foundations of nonlinear surface waves (solitons). Chapter 7 contains the presentation of the most popular nonlinear differential equations, such as Korteweg-de Vries, Sine-Gordon, nonlinear Schrödinger, Zakharov system, which lead to the existence of soliton solutions. The analysis of the nonlinear surface waves without and with dispersion presented in this contribution is based on the second order approximation of nonlinear elasticity. Some nonlocal effects for solitary Rayleigh waves are also indicated.*

*This volume contains two appendixes. The first is a short essay (Lancellotta) devoted to an illustration of certain features of experimental soil behaviour and of its testing by wave propagation techniques. a series of remarks are made in relation to the dependence of soil mechanical response on current mean effective pressure (barotropy) and current porosity (pycnotropy) and by the induced anisotropy.*

*The second appendix to the book is a CD (Kausel, Rix and Lai) containing freeware computer software for wave propagation problems in stratified media (Kausel) and surface wave analysis (Rix and Lai). See the CD content for a detailed description of the programs together with the instructions for their installation.*

*Carlo G. Lai*
*Krzysztof Wilmanski*

# Near-Surface Site Characterization Using Surface Waves

Glenn J. Rix
School of Civil and Environmental Engineering
Georgia Institute of Technology
Atlanta, Georgia, USA

**Abstract** Modern surface wave testing methods rely on advanced signal processing and inversion algorithms to extract information about the shear wave velocity profile from observations of Rayleigh wave propagation at the free surface. Extensive use is made of temporal and spatial Fourier transforms to measure Rayleigh wave dispersion. In this context, it is important to understand the influence of finite temporal and spatial sampling on frequency and wavenumber resolution and aliasing. The aperture smoothing function is shown to be a valuable tool to understand these characteristics for a particular receiver array. Four commonly used methods for calculating dispersion curves — Spectral Analysis of Surface Waves (SASW), Multi-Offset Phase Analysis (MOPA), Spatial Autocorrelation (SPAC), and conventional frequency-domain beamforming — are presented and discussed for both active and passive tests.

Surface wave inversion, like most geophysical inverse problems, is ill posed due to finite, uncertain data and modelling errors. To overcome these difficulties, it is possible employ one or more of several strategies that conceptually involve introducing additional information about the inverse problem. Candidates include introducing information about uncertainty in the experimental dispersion data, a priori values of shear wave velocity, and/or desired characteristics of the shear wave velocity profile such as smoothness. Several least squares inversion algorithms are presented that allow the user to incorporate this type of information.

## 1 Introduction

Profiles of small-strain dynamic material properties including the shear wave velocity $V_s$, initial tangent shear modulus $G_{max}$, material damping ratio $D_{min}$, and/or quality factor $Q$ are useful for a variety of applications in geotechnical engineering and engineering seismology. Recently, there has been renewed interest in methods used to measure $V_s$ profiles in situ due to the adoption of building codes that utilize the average shear wave velocity in the upper 30 m to assign a site class for estimating local site effects (Borcherdt, 1994). Surface wave tests are one method that may be used to measure $V_s$

profiles for this purpose. They belong to a class of methods called *non-invasive* methods in which the source and receivers are located on the ground surface, avoiding the need for boreholes or probes. Other non-invasive methods include seismic refraction and seismic reflection. *Invasive* methods include crosshole, downhole (e.g., seismic cone penetration test), and borehole logging (e.g., suspension P-S logger) techniques. In this chapter, we will focus exclusively on the use of surface wave methods to measure the shear wave velocity profile.

The dispersive nature of Rayleigh waves propagating in a vertically heterogeneous medium forms the basis of surface wave testing. Consider the example shown in Figure 1. The left side of the figure depicts a horizontally layered medium consisting of two layers overlying a half space. In the center of the figure, the vertical particle motion associated with a high-frequency (i.e., short wavelength) Rayleigh wave is shown. Most of the particle motion is confined to within approximately one wavelength of the free surface of the medium. In this case the particle motion occurs almost exclusively in Layer 1. As such, the material properties of Layer 1 will control the velocity of the Rayleigh wave. The right side of the figure illustrates the vertical particle motion associated with a low-frequency (i.e., long wavelength) Rayleigh wave. In this case the particle motion extends to greater depth and there is significant particle motion in Layers 1 and 2 and less in Layer 3. The velocity of this low-frequency Rayleigh wave will be controlled by some combination of the material properties of all three layers, perhaps in rough proportion to the relative amount of particle motion occurring within each layer.

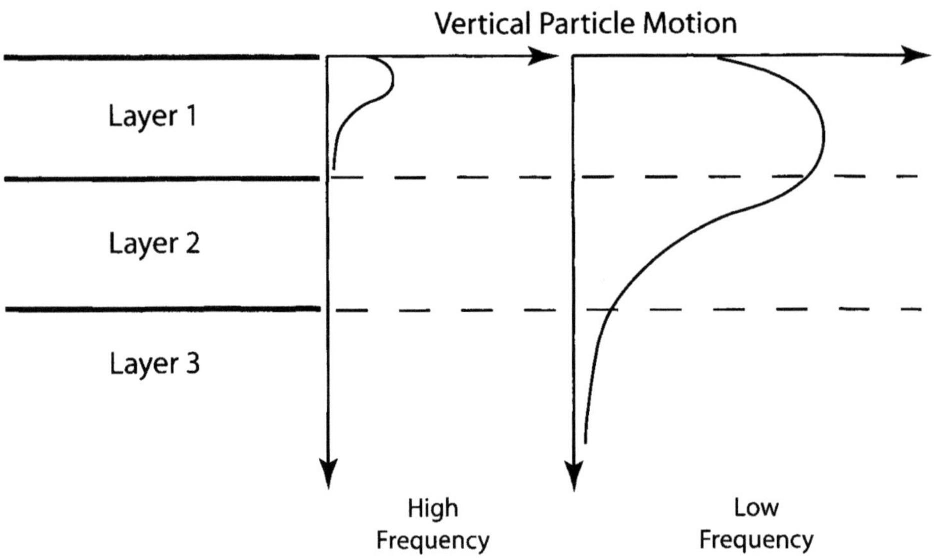


**Figure 1.** Principle of Geometric Dispersion

This qualitative description of Rayleigh wave propagation in a layered medium suggests a procedure that may be used to determine the $V_s$ profile. First, measurements of

Rayleigh wave velocity are made over a broad range of frequencies, resulting in a *dispersion* curve for the site. An example of a dispersion curve is shown in Figure 2. The characteristics of the curve are typical of those measured at sites where the stiffness of the material generally increases with depth. Low-frequency waves with longer wavelengths penetrate into deeper, stiffer layers and thus have larger phase velocities. The second phase of the procedure is to perform an *inversion* to determine the shear wave velocity of the each layer within the profile. Generally, this inversion is performed by a procedure in which a theoretically calculated dispersion curve for an assumed $V_s$ profile is compared to the experimentally measured dispersion curve. The $V_s$ profile is iteratively updated until the match between the theoretical and experimental dispersion curves is satisfactory.

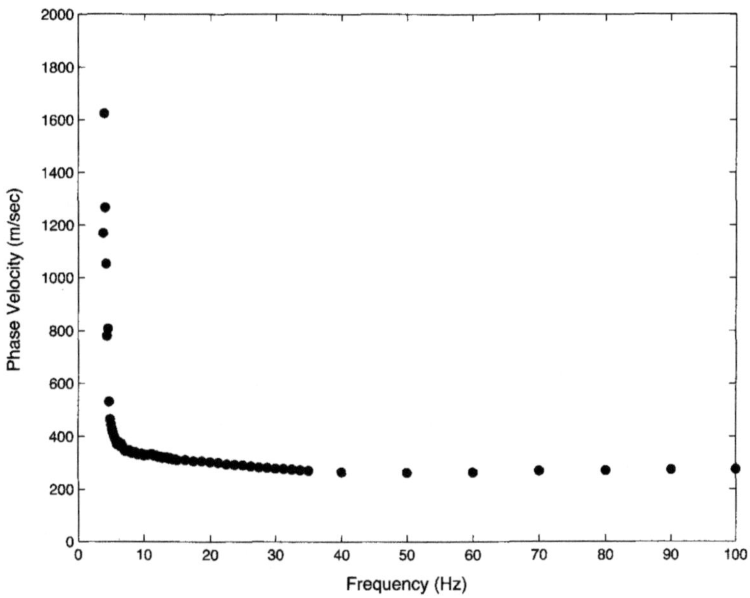


**Figure 2.** Example of a Rayleigh Wave Dispersion Curve

The successful use of surface wave tests for near-surface site characterization depends on having a basic knowledge of three topics: (1) wave propagation theory, (2) signal processing, and (3) inverse theory. Aspects of wave propagation theory pertinent to surface wave testing are presented in other contributions to this volume. This chapter addresses the signal processing and inversion techniques that are used in modern surface wave methods.

## 2 Field Equipment

Surface wave tests are performed using (1) a source, (2) one or more receivers, and (3) instrumentation to record and process the data. Sources may be either *active* or *passive* and may produce either *transient* or *continuous* surface wave energy. Common examples of active transient sources include hammers and large dropped weights such as those shown in Figures 3a and 3b, respectively. The advantages of transient sources are their low cost and portability (in most cases). Frequently, however, the user has poor control over the frequency content of the transient signal and the signal may not be repeatable. Active continuous sources usually produce pseudo-random or harmonic signals and include small electro-mechanical shakers or larger field vibrators such as those depicted in Figures 3c and 3d, respectively. Although they are more expensive and cumbersome to deploy, continuous sources offer the user a more controlled, repeatable signal.

**Figure 3.** Examples of Active Sources Used for Surface Wave Tests

Many active sources are limited to higher frequencies because of their relatively small mass and may not produce sufficiently long wavelengths to develop a $V_s$ profile to the desired depth. In these cases, passive sources such as microtremors and cultural noise (Asten, 1978) generally produce energy at lower frequencies than most active sources and thus allow deeper $V_s$ profiling. However, passive tests may be impractical at some sites, especially those located in rural areas, because of insufficient passive energy. An important assumption associated with passive testing is that the measured particle motions are those of Rayleigh waves (Aki, 1957; Toksöz, 1964; Liaw and McEvilly, 1979).

Receivers may be either geophones or seismic accelerometers. An example is shown in Figure 4. It is important that the receivers have a frequency response that corresponds to the range of frequencies that will be used in the test and that all of the receivers be calibrated properly or, more simply, have identical frequency response characteristics. Care should be taken in coupling the receivers to the ground surface to ensure that the particle motion is accurately measured, and the location of each receiver should be recorded.

**Figure 4.** Example of Receivers Used for Surface Wave Tests

Modern surface wave tests were enabled by advances in field instrumentation and the availability of high-speed computers for calculating theoretical dispersion curves rapidly. In the early 1970s, the availability of portable Fast Fourier Transform (FFT) analyzers contributed significantly to the development of the Spectral Analysis of Surface Waves (SASW) method (Nazarian, 1984). These FFT analyzers permitted the user to examine the data in the frequency domain in real time. More recently, the availability of multi-channel data acquisition systems has enhanced surface wave testing. Today, the most frequently used types of instruments include (1) multi-channel seismographs, (2) multi-channel modular (e.g., VXI-based) systems, and (3) PC-based systems. The system used by the author is shown in Figure 5. For all of these systems, it is important to consider the frequency response and dynamic range to ensure they are appropriate for surface wave tests.

**Figure 5.** Example of Equipment Used for Surface Wave Tests

## 3 Signals in Time and Space

We begin our analysis of dispersed Rayleigh waves by assuming that the vertical particle motion at a point on the ground surface may be represented by the spatiotemporal signal $s(x, y, t)$ where $x$ and $y$ are the Cartesian coordinates of the point and $t$ is time. We can abbreviate the notation by introducing the position vector $\mathbf{x} = (x, y)$ to form $s(\mathbf{x}, t)$.

Fourier analysis is a powerful tool in the analysis of dispersed surface waves. We can express the *temporal* Fourier transform of $s(\mathbf{x}, t)$ as:

$$S(\mathbf{x}, \omega) = \int_{-\infty}^{\infty} s(\mathbf{x}, t)e^{-i\omega t}dt \tag{3.1}$$

where $\omega = 2\pi f$ is the circular frequency. The inverse Fourier transform allows us to recover the time domain signal from its frequency domain representation:

$$s(\mathbf{x}, t) = \frac{1}{2\pi}\int_{-\infty}^{\infty} S(\mathbf{x}, \omega)e^{i\omega t}d\omega \tag{3.2}$$

Equation 3.1 allows us to represent the signal in the temporal frequency domain. We can extend this concept to represent the signal in the spatial frequency domain as well:

$$S(k_x, k_y, t) = \int_{-\infty}^{\infty} \int_{-\infty}^{\infty} s(x, y, t) e^{i(k_x x + k_y y)} dx dy \tag{3.3a}$$

$$S(\mathbf{k}, t) = \int_{-\infty}^{\infty} s(\mathbf{x}, t) e^{i\mathbf{k}\cdot\mathbf{x}} d\mathbf{x} \tag{3.3b}$$

where $k_x$ and $k_y$ are the wavenumbers in the $x$ and $y$ directions, respectively, and $\mathbf{k} = (k_x, k_y)$ is the vector wavenumber. The inverse spatial Fourier transform is given by:

$$s(\mathbf{x}, t) = \frac{1}{(2\pi)^2} \int_{-\infty}^{\infty} S(\mathbf{k}, t) e^{-i\mathbf{k}\cdot\mathbf{x}} d\mathbf{k} \tag{3.4}$$

Finally, we may combine both temporal and spatial Fourier transforms to calculate the *frequency-wavenumber spectrum* of the signal:

$$S(k_x, k_y, \omega) = \int_{-\infty}^{\infty} \int_{-\infty}^{\infty} \int_{-\infty}^{\infty} s(x, y, t) e^{-i(\omega t - k_x x - k_y y)} dx\, dy\, dt \tag{3.5a}$$

$$S(\mathbf{k}, \omega) = \int_{-\infty}^{\infty} \int_{-\infty}^{\infty} s(\mathbf{x}, t) e^{-i(\omega t - \mathbf{k}\cdot\mathbf{x})} d\mathbf{x}\, dt \tag{3.5b}$$

and we may recover the spatiotemporal signal via the inverse transform:

$$s(\mathbf{x}, t) = \frac{1}{(2\pi)^3} \int_{-\infty}^{\infty} \int_{-\infty}^{\infty} S(\mathbf{k}, \omega) e^{i(\omega t - \mathbf{k}\cdot\mathbf{x})} d\mathbf{k}\, d\omega \tag{3.6}$$

As an example, consider the example of a unit amplitude plane wave given by:

$$s(\mathbf{x}, t) = e^{i(\omega_0 t - \mathbf{k}_0\cdot\mathbf{x})} \tag{3.7}$$

where $\omega_0$ and $\mathbf{k}_0$ are the circular frequency and wavenumber of the signal, respectively. The propagation velocity of the wave is:

$$c = \frac{\omega_0}{|\mathbf{k}_0|} \tag{3.8}$$

and the wavelength is:

$$\lambda = \frac{2\pi}{|\mathbf{k}_0|} \tag{3.9}$$

The frequency-wavenumber spectrum of the signal is:

$$S(\mathbf{k}, \omega) = \delta^2(\mathbf{k} - \mathbf{k}_0)\, \delta(\omega - \omega_0) \tag{3.10}$$

Figure 6 shows a graphical representation of the frequency-wavenumber spectrum. The signal is represented by a single point in frequency-wavenumber space.

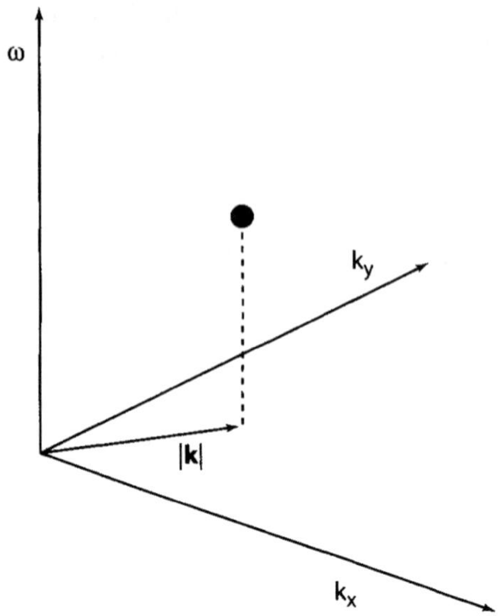


**Figure 6.** Frequency-Wavenumber Spectrum of a Harmonic Plane Wave

### 3.1 Effects of Finite Sampling in Time and Space

In Equation 3.5, the infinite integrals over time and space imply perfect sampling of the signal. In practice, of course, this is not possible and we sample the signal at a finite number of times and locations. This limitation impacts our ability to determine the characteristics $\omega_0$ and $\mathbf{k}_0$ of the signal.

We begin by examining the effects of finite sampling in the time domain. Let us assume that the signal is sampled for a finite period of time $T$ sec. This can be viewed as the multiplying the signal by a window $w(t)$:

$$y(\mathbf{x}, t) = w(t)\ s(\mathbf{x}, t) \tag{3.11}$$

where $w(t)$ is defined by:

$$w(t) = \begin{cases} 1 & 0 \leq t \leq T \\ 0 & \text{otherwise} \end{cases} \tag{3.12}$$

The Fourier transform of $y(\mathbf{x}, t)$ can be expressed as the convolution of the temporal Fourier transforms of $w(t)$ and $s(\mathbf{x}, t)$:

$$Y(\mathbf{x}, \omega) = \int_{-\infty}^{\infty} W(\omega - \zeta)\ S(\mathbf{x}, \zeta)\ d\zeta \tag{3.13}$$

where $\zeta$ is a dummy variable. For the $w(t)$ defined in Equation 3.12, the amplitude of the Fourier transform is:

$$|W(\omega)| = T\left|\frac{\sin\frac{\omega T}{2}}{\frac{\omega T}{2}}\right| \tag{3.14}$$

Figure 7 shows a plot of $|W(\omega)|$. The result of the finite length time window is apparent; the spectrum is spread over a broad range of frequencies (i.e., *leakage*) and frequency resolution is decreased. The frequency resolution defined according to the Rayleigh criterion is equal to one half of the width of the main lobe, which is $f_{Rayleigh} = 1/T$ for the window used here. See Bendat and Piersol (1986) for additional information on the use of alternative windowing functions to control side-lobe leakage and frequency resolution.

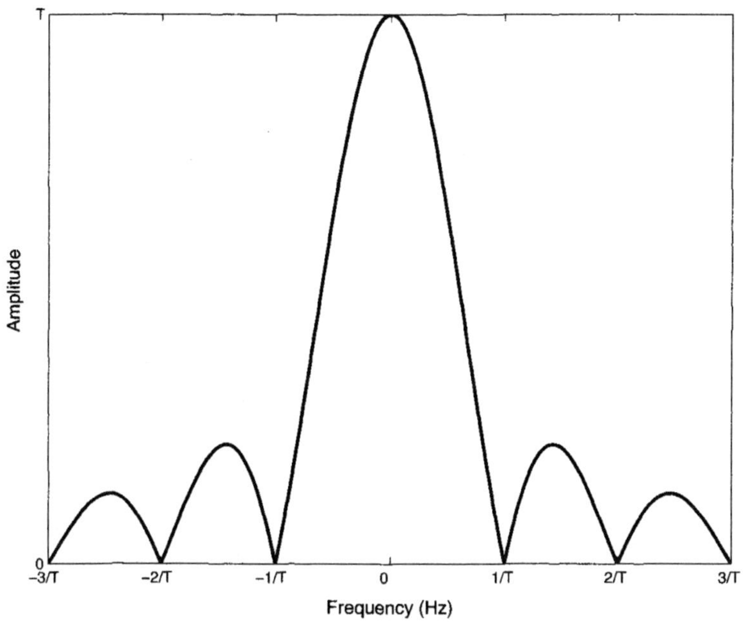


**Figure 7.** Fourier Transform of $w(t)$

Another consequence of finite sampling is *aliasing.* Consider a signal that is regularly sampled at an interval of $\Delta t$ (i.e., the sampling rate is $1/\Delta t$ samples per second). The maximum frequency that may be accurately measured using this sampling interval is called the Nyquist frequency:

$$f_{Nyquist} = \frac{1}{2\Delta t} \tag{3.15}$$

Frequency components of the signal that exceed the Nyquist frequency will be aliased into the frequency range $0 \leq f \leq f_{Nyquist}$. Aliasing may be avoided by using analog low-pass filters to remove frequency components of the signal that exceed $f_{Nyquist}$ prior to sampling the signal.

Similar issues arise due to finite sampling of the signal in the spatial domain. For simplicity, let's consider a one-dimensional signal propagating along the $x$ axis. Assume that the signal is sampled within a *continuous aperture* of length $D$ m. The spatially sampled signal may be viewed as the product of the (original) signal and a window $w(x)$:

$$z(x,t) = w(x)\ s(x,t) \tag{3.16}$$

where $w(x)$ is defined by:

$$w(x) = \begin{cases} 1 & 0 \leq x \leq D \\ 0 & \text{otherwise} \end{cases} \tag{3.17}$$

The spatial Fourier transform of $z(x,t)$ can be expressed as the convolution of the spatial Fourier transforms of $w(x)$ and $s(x,t)$:

$$Z(k_x,t) = \int_{-\infty}^{\infty} W(k_x - \zeta)\ S(\zeta,t)\ d\zeta \tag{3.18}$$

For the $w(x)$ defined in Equation 3.17, the amplitude of the Fourier transform is:

$$|W(k_x)| = D\left|\frac{\sin\frac{k_x D}{2}}{\frac{k_x D}{2}}\right| \tag{3.19}$$

Figure 8 shows a plot of $|W(k_x)|$ called the *aperture smoothing function.* The result of the finite spatial aperture is analogous to the finite time window; the wavenumber spectrum is spread over a broad range of wavenumbers and wavenumber resolution is decreased. The wavenumber resolution defined according to the Rayleigh criterion is equal to one half of the width of the main lobe, which is $k_{Rayleigh} = 2\pi/D$ for the aperture used here. Clearly, increasing the aperture of the array improves the spatial resolution.

Figure 9 illustrates the effects of limited wavenumber resolution. In Figure 9a, the signal is composed of two components with wavenumbers separated by a sufficient amount to allow them to be resolved easily. In Figure 9b, the wavenumbers differ by less, but the individual peaks may still be resolved. In Figure 9c, the wavenumbers are too similar to be resolved independently, and the apparent wavenumber corresponding to the peak of the combined signal lies between the two individual components. Figure 9d illustrates that by doubling the spatial aperture, the wavenumber resolution is improved and the two individual components can once again be resolved.

This concept may be easily extended to determine the resolution associated with multidimensional arrays of discrete sensors. For this case the spatial window is defined as:

$$w(\mathbf{x}) = \sum_{m=1}^{M} w_m\ \delta(\mathbf{x} - \mathbf{x}_m) \tag{3.20}$$

where $M$ is the number of receiver in the array and $w_m$ is the *shading weight* assigned to each receiver. Uniform weighting is taken to mean $w_m = 1,\ m = 1, \ldots, M$. The aperture smoothing function for this general case is given by:

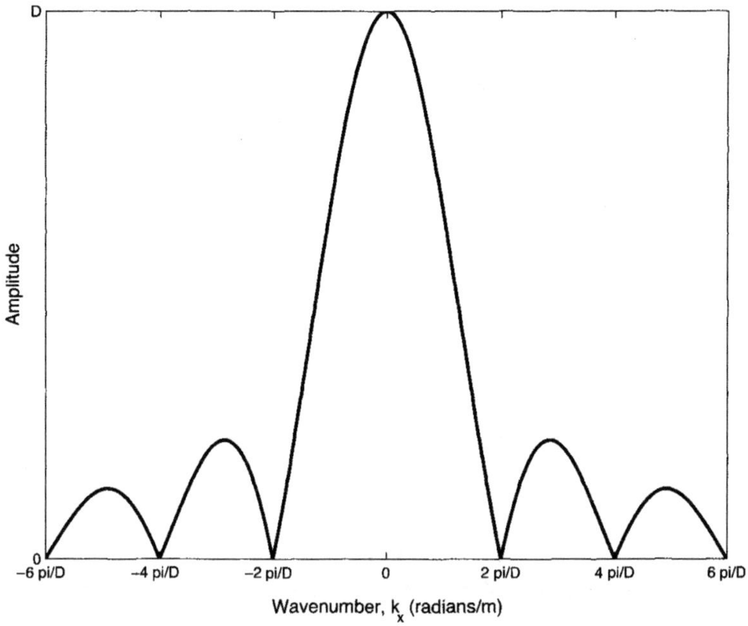


**Figure 8.** Fourier Transform of $w(x)$

$$W(\mathbf{k}) = \sum_{m=1}^{M} w_m \, e^{i\mathbf{k}\cdot\mathbf{x}_m} \tag{3.21}$$

Sampling the signal at discrete spatial locations also gives rise to *spatial aliasing* in a manner analogous to temporal aliasing. The Nyquist wavenumber is related to the minimum distance between adjacent receivers:

$$k_{Nyquist} = \frac{\pi}{d_{min}} \tag{3.22}$$

A comparison of the expressions for $k_{Rayleigh}$ and $k_{Nyquist}$ reveals that achieving good wavenumber resolution by increasing the spatial aperture of the array comes at the expense of spatial aliasing for a fixed number of receivers $M$ and vice versa. The use of an array with irregular spacing helps to balance these two competing objectives. For active surface wave tests, the author uses an irregular linear array of $M = 15$ receivers with the spacing shown in Figure 10a. The aperture smoothing function for this array is shown in Figure 10b. It represents a reasonable compromise between wavenumber resolution as governed by the width of the main lobe and spatial aliasing as governed by the maximum wavenumber.

For passive tests, a two-dimensional array is needed because the exact location(s) of the passive source(s) of energy is often unknown, and the array must be able to accurately measure waves propagating from any direction. Several two-dimensional configurations

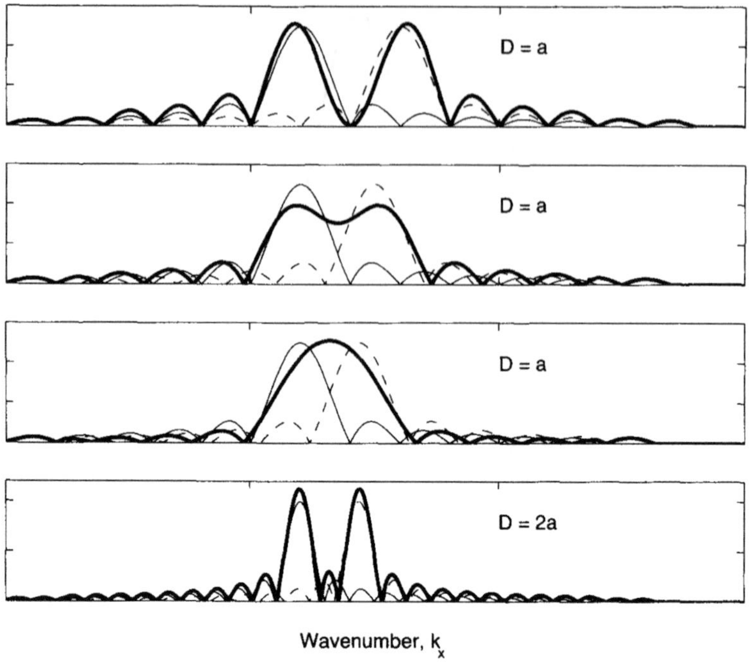


**Figure 9.** Example Illustrating Effects of Limited Wavenumber Resolution

are commonly used. Figure 11a shows the geometry a circular array of $M = 16$ receivers with a radius of 30 m. Figure 11b shows a contour plot of the two-dimensional aperture smoothing function for this array. The main lobe is located at the origin of the plot; side lobes are arranged in a symmetric pattern about the origin. For some sites it may only be feasible to arrange the receivers in a cross-shape pattern such as that shown in Figure 12a. The aperture smoothing function is shown in Figure 12b. Of course, there is no requirement that the geometry of the array be determined by a regular pattern such as the circle or cross illustrated herein. See Johnson and Dudgeon (1993) for a summary of the characteristics of *random arrays* in which the sensor locations are selected randomly according to a particular probability density function.

In summary, aperture smoothing functions are the best means available to understand the spatial sampling characteristics including wavenumber resolution and spatial aliasing for a particular receiver configuration.

## 4 Dispersion Calculations

Conceptually, our task is straightforward: determine the frequency-wavenumber spectrum of Rayleigh waves from measurements of particle motions obtained during testing subject to the limitations of finite sampling in time and space. The dispersion curve (i.e., Rayleigh phase velocity vs. frequency) is obtained via Equation 3.8. In this section, we

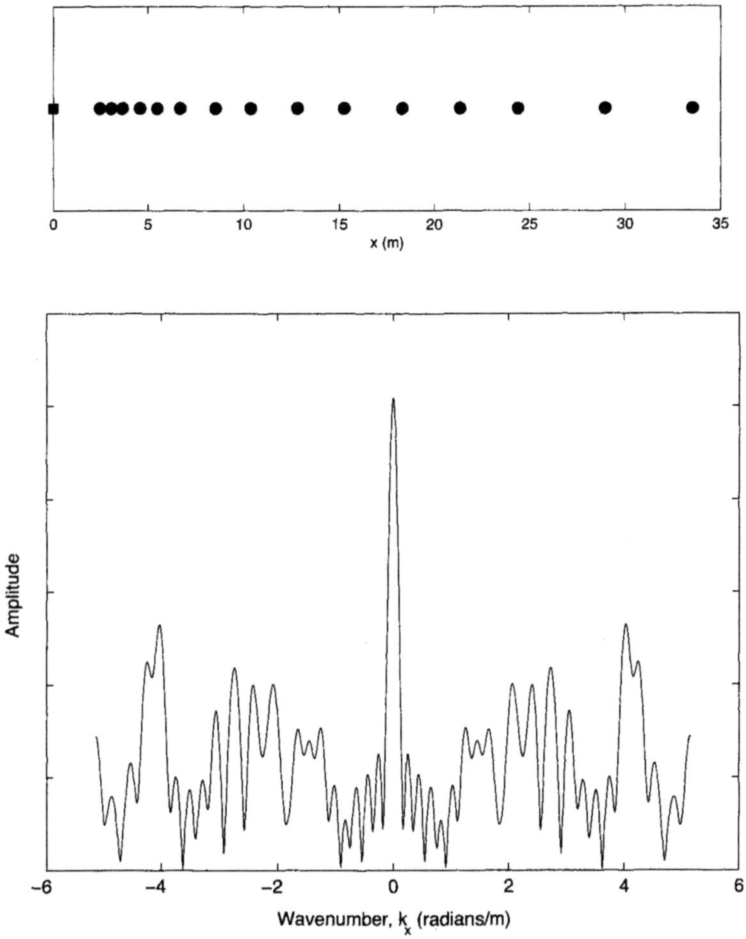


**Figure 10.** (a) Receiver Configuration and (b) aperture Smoothing Function for Irregular Linear Array with $M = 15$ Receivers Used for Active Surface Wave Tests

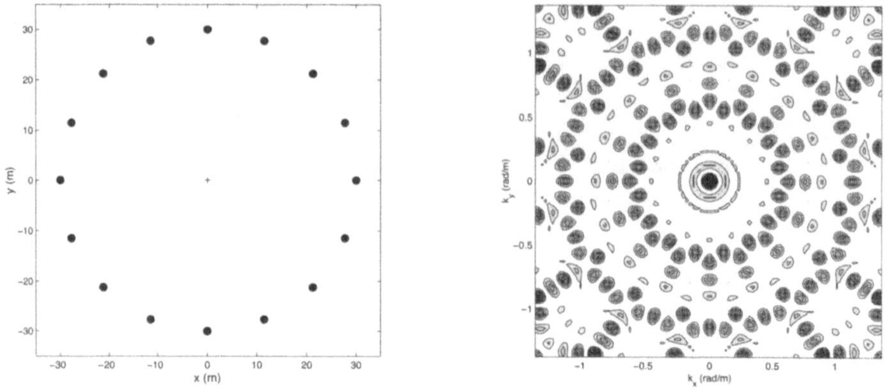


**Figure 11.** (a) Receiver Configuration and (b) aperture Smoothing Function for a 30-m Circular Array with $M = 16$ Receivers

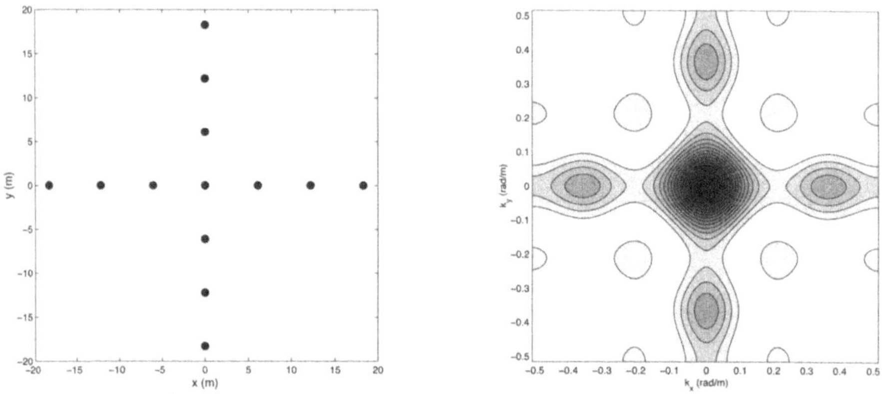


**Figure 12.** (a) Receiver Configuration and (b) aperture Smoothing Function for a Cross-Shaped Array with $M = 13$ Receivers

will focus on four methods that are widely used in engineering surface wave testing: (1) Spectral Analysis of Surface Waves (SASW), (2) multi-offset phase analysis (MOPA), (3) spatial autocorrelation (SPAC), and (4) conventional frequency-domain beamforming.

### 4.1 Spectral Analysis of Surface Waves

Spectral Analysis of Surface Waves (SASW) tests are performed with $M = 2$ receivers that are co-linear with an active source as shown in Figure 13. Typically, the receivers are placed such that $x_2 = 2x_1$ (Sánchez-Salinero, 1987).

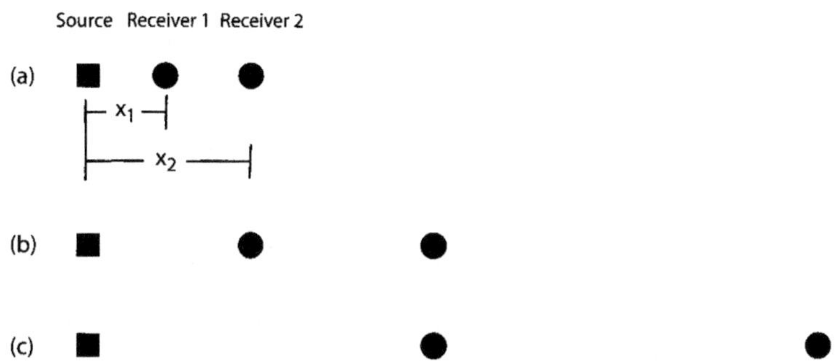


**Figure 13.** Source-Receiver Configuration Used for SASW Test

The time domain signal recorded at each receiver is given by $s(x_m, t)$, $m = 1, 2$. Taking the temporal Fourier transform of each signal yields:

$$S(x_m, \omega) = |S(x_m, \omega)| e^{i(\phi(\omega)+k(\omega)x_m)} \tag{4.1}$$

where $\phi(\omega)$ is the arbitrary phase of the source. The cross-power spectrum between the two signals is:

$$S_{12}(\omega) = S^*(x_1, \omega)\, S(x_2, \omega) \tag{4.2a}$$

$$= |S(x_1, \omega)||S(x_2, \omega)| e^{-i(\phi(\omega)+k(\omega)x_1)} e^{i(\phi(\omega)+k(\omega)x_2)} \tag{4.2b}$$

$$= |S(x_1, \omega)||S(x_2, \omega)| e^{ik(\omega)(x_2 - x_1)} \tag{4.2c}$$

where the asterisk denotes the complex conjugate of a quantity. In practice, it is desirable to use ensemble averaging to reduce the variance of the measured cross-power spectra (Bendat and Piersol, 1986):

$$\widehat{S_{ij}}(\omega) = \frac{1}{n_d} \sum_{k=1}^{n_d} S_k^*(x_i, \omega)\, S_k(x_j, \omega) \tag{4.3}$$

where $n_d$ is the number of averages. The phase spectrum equals:

$$\theta_{12}(\omega) = \arg(\widehat{S_{12}}(\omega)) = k(\omega)(x_2 - x_1) \tag{4.4}$$

Thus we may determine the dispersion curve by rearranging Equation 4.4 as follows[1]:

$$c(\omega) = \frac{\omega(x_2 - x_1)}{\theta_{12}(\omega)} \tag{4.5}$$

Figure 14 shows an example of a phase spectrum measured during an SASW test using a transient source (a sledge hammer) with a pair of receivers separated by 12.8 m. This example is typical of phase spectra measured in this manner. The large uncertainty in the phase spectrum at low and high frequencies is likely due to the inability of the source to generate sufficient energy within those ranges. The plot is presented as *wrapped phase* with values of $\theta_{12}(\omega)$ ranging from $\pi$ to $-\pi$ (i.e., modulo $2\pi$). To apply Equation 4.5, the user must unwrap the phase by adding integer multiples of $2\pi$ in the following manner:

$$\theta_{unwrap}(\omega) = \theta_{wrap}(\omega) \pm 2n\pi, \; n = 0, 1, 2, \ldots \tag{4.6}$$

Choosing the proper values of $n$ can be a source of considerable uncertainty in SASW tests. This issue will be discussed in more detail subsequently.

The dispersion curve corresponding to this phase spectrum is shown in Figure 15. Notice that only frequencies between 15 Hz and 100 Hz have been included due to the poor quality of the phase spectrum outside of that range. If data at other frequencies is required, the test may be repeated using different receiver spacings such as those shown in Figures 13b and c[2]. Dispersion curves from individual receiver spacings are combined to form a composite dispersion curve for the site. The source may also be moved to the opposite side of the array to determine the dispersion curve in the "reverse" direction in an effort to average any differences in phase velocity due to lateral inhomogeneity at the site.

The SASW technique has several important limitations that arise from the use of only $M = 2$ receivers. First, spatial aliasing complicates interpretation of the test results. To avoid spatial aliasing, Equation 3.22 must be satisfied. In the context of the SASW test:

$$k(\omega) \le k_{Nyquist} \tag{4.7a}$$

$$\frac{\theta_{12}(\omega)}{(x_2 - x_1)} \le \frac{\pi}{(x_2 - x_1)} \tag{4.7b}$$

$$\theta_{12}(\omega) \le \pi \tag{4.7c}$$

[1] Negative values of $\theta$ are often used to represent a phase *lag* due to propagation delay. It may be necessary to multiply Equation 4.5 by -1.

[2] As a matter of practice, the receiver spacing is often doubled.

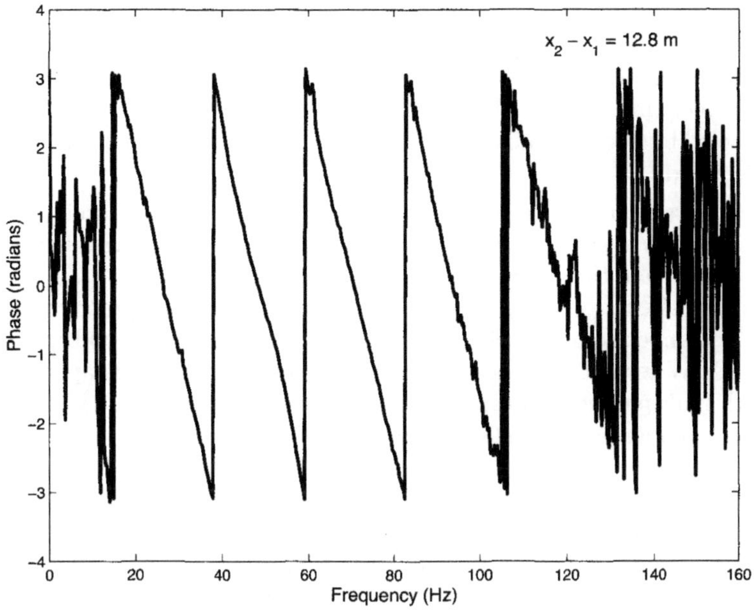


**Figure 14.** Example of Phase Spectrum from an SASW Test

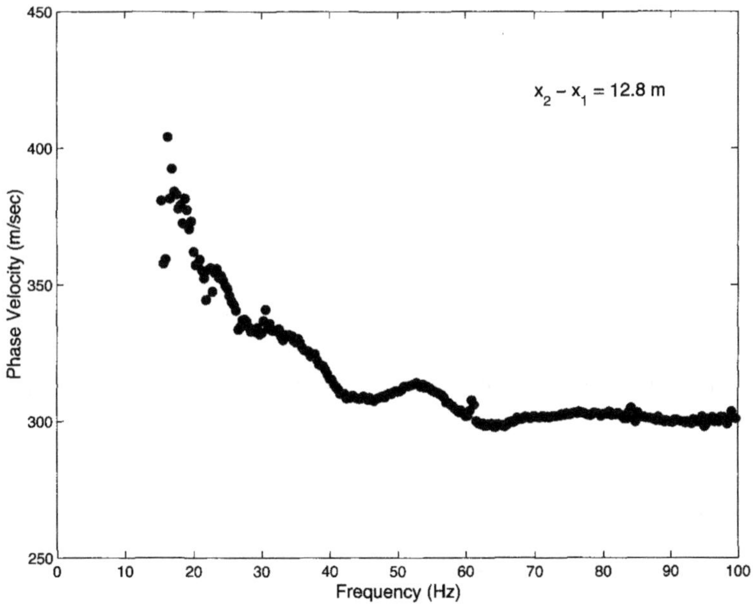


**Figure 15.** Example of Dispersion Curve from SASW Test for a Single Receiver Spacing

Thus, the "jumps" in the phase spectrum and, more specifically, the need to choose $n$ in Equation 4.6 are the result of spatial aliasing. As noted above, this aspect of SASW test interpretation often results in considerable uncertainty. Although "automatic" phase unwrapping algorithms are available (e.g., the `unwrap` function in Matlab), they are frequently fooled by poorer quality data and manual interpretation of the phase spectrum is usually required.

Second, the wavenumber resolution is generally very poor for receiver spacings normally used in SASW tests. Figure 16 shows the aperture smoothing function for $x_2 - x_1$ = 12.8 m. Note that the main lobe is wide and spans the entire range of wavenumbers defined by $-k_{Nyquist} \leq k \leq k_{Nyquist}$. The most important practical implication of the poor wavenumber resolution is that it is impossible to resolve multiple surface wave modes in a SASW test. The smoothing produced by convolving the aperture smoothing function with multiple modes propagating at different wavenumbers yields only a single apparent wavenumber as illustrated in Figure 9c.

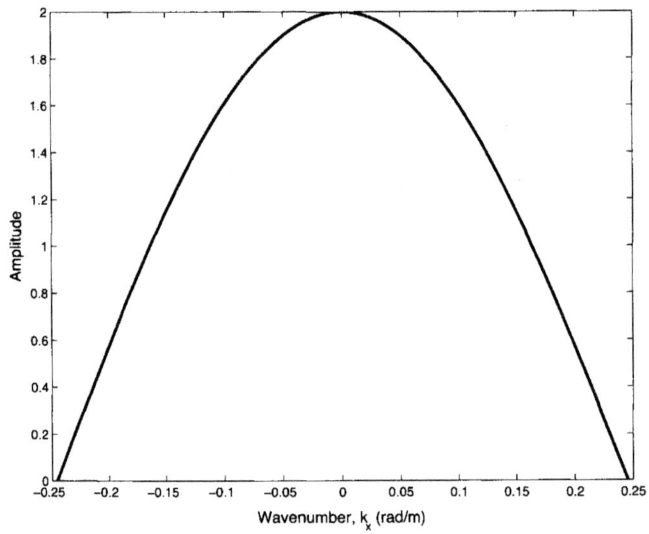


**Figure 16.** aperture Smoothing Function for an SASW Test with $x_2 - x_1 = 12.8$ m

## 4.2 Multi-Offset Phase Analysis

Multi-offset phase analysis (MOPA) is an extension of the SASW method in which the phase differences between multiple receivers ($M > 2$) arranged in a linear array is used to determine the phase velocity (Rix et al., 2001; Lai et al., 2002; Strobbia and Foti, 2004). Consider the cross-power spectra between the first receiver located nearest the source and the remaining receivers in the array:

$$S_{1m}(\omega) = S^*(x_1,\omega)\, S(x_m,\omega) \tag{4.8a}$$

$$= |S(x_1,\omega)||S(x_m,\omega)|e^{-i(\phi(\omega)+k(\omega)x_1)}e^{i(\phi(\omega)+k(\omega)x_m)} \tag{4.8b}$$

$$= |S(x_1,\omega)||S(x_m,\omega)|e^{ik(\omega)(x_m-x_1)} \tag{4.8c}$$

where $m = 1, \ldots, M$ and other terms have been defined previously. Ensemble averaging may be used to reduce the variance of the estimated cross-power spectra. The phase spectrum between each pair of receivers equals:

$$\theta_{1m}(\omega) = \arg(\widehat{S_{1m}}(\omega)) = k(\omega)(x_m - x_1) \tag{4.9}$$

Thus the dispersion curve may be determined by finding the value of wavenumber that best satisfies Equation 4.9 in a least squares sense at each frequency. Figure 17 shows examples of this process for an active surface wave test using the receiver array depicted in Figure 10a. The dispersion curve resulting from similar analyses applied to phase data over a broad range of frequencies is shown in Figure 18.

Calculating the best wavenumber in a least squares sense from multiple receivers provides for a more robust estimate of the dispersion curve compared to using only two receivers in the SASW test. Despite the fact that, however, that the aperture smoothing function for the multi-receiver array is superior to the two-receiver SASW array, the MOPA technique can only be used to determine the (apparent) phase velocity of a single mode of propagation.

This technique may also be easily adapted to simultaneously determine the dispersion and attenuation curves by using both the magnitude and phase of the cross-power spectrum to determine a complex-valued wavenumber at each frequency (Rix et al., 2001; Lai et al., 2002):

$$k^*(\omega) = \frac{\omega}{c^*(\omega)} = \frac{\omega}{c(\omega)} - i\alpha(\omega) \tag{4.10}$$

where $\alpha$ is the spatial attenuation coefficient of the Rayleigh waves. The attenuation curve is subsequently inverted in either an uncoupled (Rix et al., 2000) or coupled (Lai et al., 2002) manner to obtain the $D_{min}$ (or $Q$) profile.

### 4.3 Spatial Autocorrelation

The spatial autocorrelation (SPAC) method was developed by Aki (1957, 1965) to calculate dispersion curves using microtremor data. The original implementation of the method utilized a circular receiver array with an additional receiver located at the center of the circle. We will take a more general approach similar to Ohori et al. (2002) that does not impose such restrictions on the geometry of the array.

The basis of the method is the observation that the spatial autocorrelation function may be obtained via an inverse spatial Fourier transform of the wavenumber spectrum:

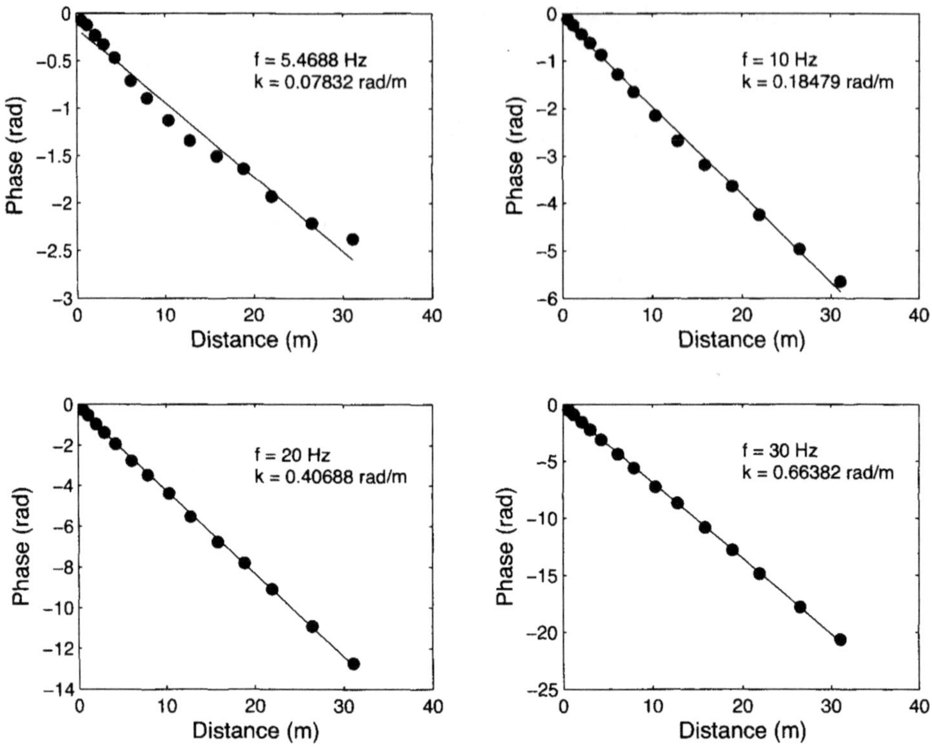


**Figure 17.** Examples of Linear Regression Analyses to Determine Wavenumber from Multi-Offset Phase Data

$$R(\chi) = E[s(\mathbf{x}, t)\ s(\mathbf{x} + \chi, t)] \tag{4.11a}$$

$$= \frac{1}{2\pi} \int_{-\infty}^{\infty} S(\mathbf{k}) e^{-i\mathbf{k}\cdot\chi}\ d\mathbf{k} \tag{4.11b}$$

where $\chi$ is the spatial lag and $E[\cdot]$ denotes the expected value of a quantity.

Consider a uniform plane wave described by:

$$s(\mathbf{x}, t) = e^{i(\omega_0 t - \mathbf{k}_0 \cdot \mathbf{x})} \tag{4.12}$$

The wavenumber spectrum of this signal is:

$$S(\mathbf{k}) = \delta^2(\mathbf{k} - \mathbf{k}_0) \tag{4.13}$$

Substituting Equation 4.13 into Equation 4.11b yields:

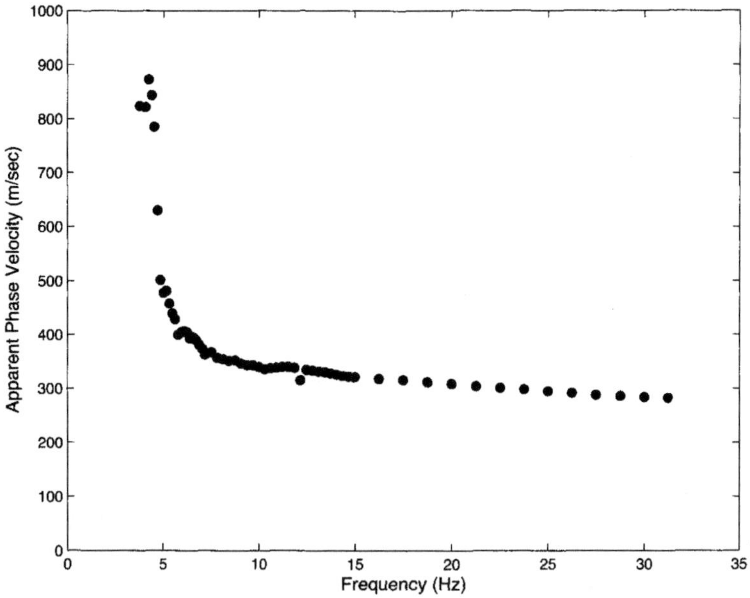


**Figure 18.** Dispersion Curve from Multi-Offset Analysis of Phase Data

$$R(\boldsymbol{\chi}) = \frac{1}{2\pi}\int_{-\infty}^{\infty} \delta^2(\mathbf{k}-\mathbf{k}_0)e^{-i\mathbf{k}\cdot\mathbf{x}}\,d\mathbf{k} \tag{4.14a}$$

$$= \frac{1}{2\pi}e^{-i\mathbf{k}_0\cdot\boldsymbol{\chi}} \tag{4.14b}$$

$$= \frac{1}{2\pi}\left[\cos(\mathbf{k}_0\cdot\boldsymbol{\chi}) - i\sin(\mathbf{k}_0\cdot\boldsymbol{\chi})\right] \tag{4.14c}$$

Because the autocorrelation function is an even function of the spatial lag, we can limit our interest to the real part of the solution. Thus:

$$R(\boldsymbol{\chi}) = \frac{1}{2\pi}\cos(\mathbf{k}_0\cdot\boldsymbol{\chi}) \tag{4.15}$$

Finally, the autocorrelation function may be normalized to obtain the autocorrelation coefficient:

$$\rho(\boldsymbol{\chi}) = \frac{R(\boldsymbol{\chi})}{R(0)} = \cos(\mathbf{k}_0\cdot\boldsymbol{\chi}) \tag{4.16}$$

Equation 4.16 may be used to estimate the wavenumber (and thus phase velocity) from observations of the spatial autocorrelation coefficient of the wavefield. The most efficient means to calculate experimental spatial autocorrelation coefficients is via the normalized cross-power spectra between signals at two receiver positions:

$$\rho(\chi) = \frac{Re[\widehat{S_{ij}}(\omega)]}{\sqrt{\widehat{S_{ii}}(\omega)\widehat{S_{jj}}(\omega)}} \tag{4.17}$$

where $\widehat{S_{ij}}(\omega)$ is the average cross-power spectrum between receivers $i$ and $j$ calculated using Equation 4.3.

Figure 19 shows examples of the use of the spatial autocorrelation method to determine the dispersion curve from an active surface wave test using a linear array of receivers. In this case the wavenumber and spatial lag may be represented by scalar quantities. The resulting dispersion curve is shown in Figure 20.

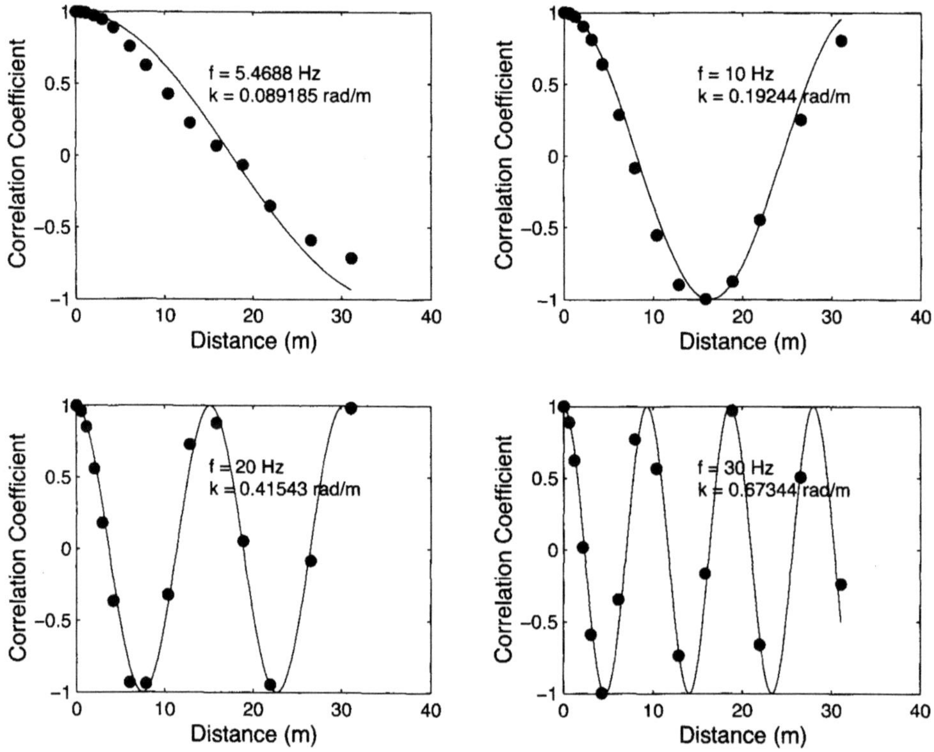


**Figure 19.** Examples of Regression Analyses to Determine Wavenumber from Spatial Autocorrelation Function - Active Test

It is of interest to consider an *isotropic wavefield*, which is comprised of random waves propagating in all possible directions with equal probability. This ideal situation may be reasonably assumed to represent a passive surface wave test with multiple, spatially distributed sources. This problem naturally lends itself to polar coordinates where:

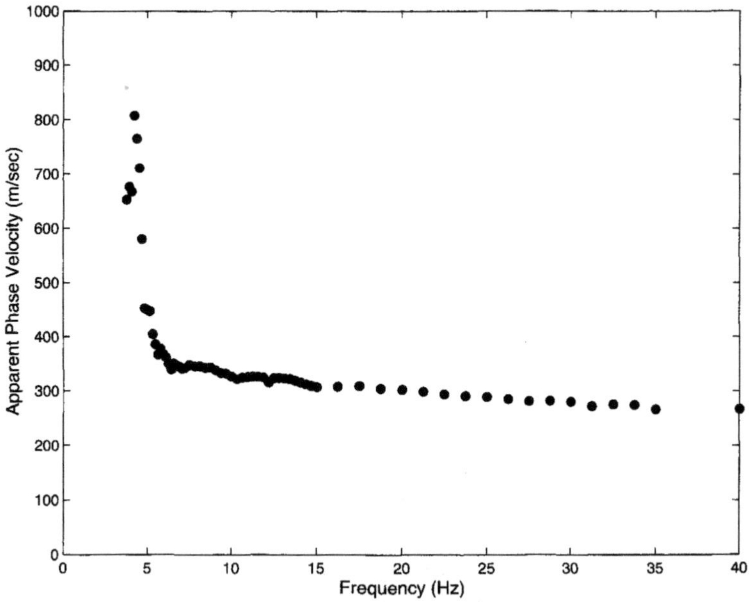


**Figure 20.** Dispersion Curve from Spatial Autocorrelation Method - Active Test

$$x = r\cos(\theta) \tag{4.18a}$$

$$y = r\sin(\theta) \tag{4.18b}$$

$$k_x = k\cos(\phi) \tag{4.18c}$$

$$k_y = k\sin(\phi) \tag{4.18d}$$

and the signal may be represeted by $s(r,t)$. The spatial autocorrelation function for this radially symmetric problem becomes (Bracewell, 1978):

$$R(r) = E[s(0,t)\ s(r,t)] \tag{4.19a}$$

$$= \int_0^\infty S(k)J_0(kr)kdk \tag{4.19b}$$

which is an inverse Hankel transform where $J_0$ is the zeroth order Bessel function of the first kind.

Consider a unit-amplitude isotropic wavefield described by:

$$s(r,t) = e^{i(\omega_0 t - k_0 r)} \tag{4.20}$$

with a wavenumber spectrum given by:

$$S(k) = \delta(k - k_0) \tag{4.21}$$

Substituting Equation 4.21 into Equation 4.19b yields:

$$R(r) = \int_0^\infty \delta(k - k_0) J_0(kr) k dk \tag{4.22a}$$

$$= J_0(k_0 r) k_0 \tag{4.22b}$$

As noted earlier, the autocorrelation function may be normalized to obtain the autocorrelation coefficient:

$$\rho(r) = \frac{R(r)}{R(0)} = J_0(k_0 r) \tag{4.23}$$

To comply with the assumption of radial symmetry, it is necessary to calculate the experimental autocorrelation coefficient using one receiver as a reference (i.e., the receiver at the "origin"). Thus:

$$\rho(r) = \frac{Re[\widehat{S_{0j}}(\omega)]}{\sqrt{\widehat{S_{00}}(\omega)\widehat{S_{jj}}(\omega)}} \tag{4.24}$$

where the subscript 0 denotes the reference receiver.

Figure 21 shows examples at four frequencies of the spatial autocorrelation method applied to a passive surface wave test using the cross-shaped array depicted in Figure 12a. The center receiver was used as the reference receiver. The value of the wavenumber derived from a least-squares fit of Equation 4.23 to the data is also shown. The dispersion curve resulting from similar calculations over a range of frequencies from 4 to 16 Hz is provided in Figure 22.

In his original implementation of the method, Aki (1957, 1965) obtained Equation 4.23 by taking the azimuthal average of the spatial autocorrelation function, although this required the use of a circular array (i.e., constant $r$) as noted previously. The current implementation has no such requirement; Equations 4.16 and 4.23 may be used to determine the wavenumber by minimizing the least-squares error between the theoretical and experimental autocorrelation coefficients as a function of $x$ and $r$, respectively.

A limitation of the SPAC method is that it may be used only for unidirectional (Equation 4.16) or isotropic (Equation 4.23) wavefields (Horike, 1985). For active tests, this poses no problem because the user creates unidirectional waves via a controlled source. For passive tests, however, the wavefield consists, in general, of multiple sources and may not conform to either of the conditions above. Furthermore, the user must have some a priori knowledge of the source characteristics to properly choose between Equations 4.16 and 4.23.

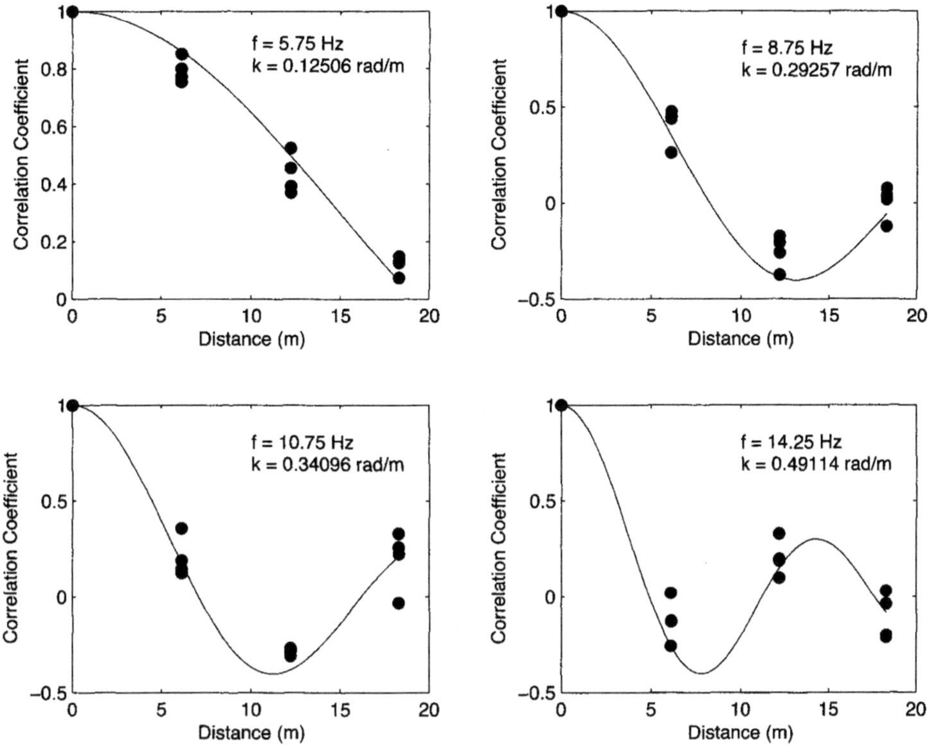


**Figure 21.** Examples of Regression Analyses to Determine Wavenumber from Spatial Autocorrelation Function - Passive Test

### 4.4 Conventional Frequency-Domain Beamforming

The frequency-wavenumber spectrum may be estimated via a process called *frequency domain beamforming* (LaCoss et al., 1969; Johnson and Dudgeon, 1993). The method is similar to other techniques including that of McMechan and Yedlin (1981) that seek to transform the signals from the time-space domain ($t - x$) to the frequency-wavenumber ($\omega - k$) or frequency-slowness ($\omega - p$) domain to identify and characterize dispersive Rayleigh waves.

Let $s(x_m, t),\ m = 1, \ldots, M$ be the signal observed at the $m^{th}$ receiver of an array. The column vector $\mathbf{S}(\omega) = [S(x_1, \omega), \ldots, S(x_M, \omega)]^T$ contains the temporal Fourier transform of the signal for each receiver and $\mathbf{e}(\mathbf{k})$ denotes a *steering vector* defined as:

$$\mathbf{e}(\mathbf{k}) = [e^{-i\mathbf{k}\cdot\mathbf{x}_1}, \ldots, e^{-i\mathbf{k}\cdot\mathbf{x}_M}]^T \tag{4.25}$$

where $T$ denotes the transpose of the vector. The output of the conventional frequency-domain beamformer is given by (Johnson and Dudgeon, 1993):

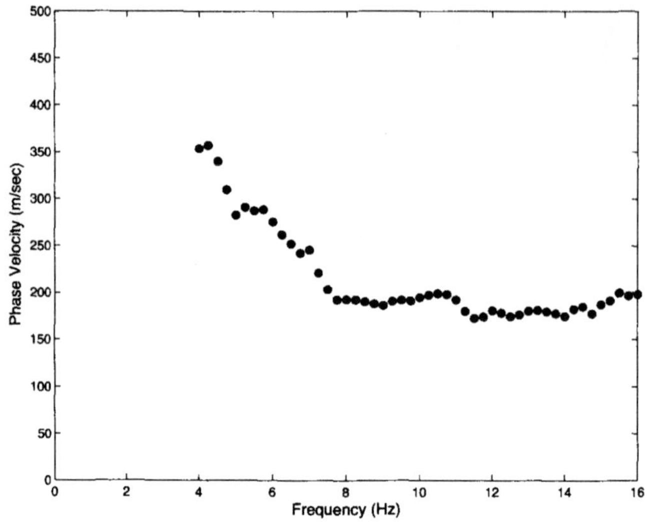


**Figure 22.** Dispersion Curve from Spatial Autocorrelation Method - Passive Test

$$Z(\mathbf{k},\omega) = \sum_{m=1}^{M} w_m S(x_m,\omega) e^{i\mathbf{k}\cdot\mathbf{x}_m} \tag{4.26a}$$

$$= \mathbf{e}^H \mathbf{W}\mathbf{S} \tag{4.26b}$$

where $\mathbf{W} = \text{diag}[w_1, \ldots, w_M]$ is a diagonal matrix containing *shading weights* for each receiver and $H$ denotes the Hermitian transpose of the vector. The power in the beamformer output is given by the *steered response power spectrum*:

$$P(\mathbf{k},\omega) = ZZ^H \tag{4.27a}$$

$$= \mathbf{e}^H \mathbf{W}\mathbf{S}\mathbf{S}^H \mathbf{W}^H \mathbf{e} \tag{4.27b}$$

$$= \mathbf{e}^H \mathbf{W}\mathbf{R}\mathbf{W}^H \mathbf{e} \tag{4.27c}$$

where $\mathbf{R}$ is the spectral correlation matrix:

$$\mathbf{R}(\omega) = \mathbf{S}\mathbf{S}^H = \begin{bmatrix} S_{11}(\omega) & S_{12}(\omega) & \cdots & S_{1M}(\omega) \\ S_{21}(\omega) & S_{22}(\omega) & \cdots & S_{2M}(\omega) \\ \vdots & \vdots & \ddots & \vdots \\ S_{M1}(\omega) & S_{M2}(\omega) & \cdots & S_{MM}(\omega) \end{bmatrix} \tag{4.28}$$

As noted earlier, it is desirable to use ensemble averaging via Equation 4.3 to reduce the variance of the measured cross-power spectra (LaCoss et al., 1969; Bendat and Piersol, 1986). Equation 4.27 thus becomes:

$$P(\mathbf{k}, \omega) = \mathbf{e}^H \mathbf{W} \widehat{\mathbf{R}} \mathbf{W}^H \mathbf{e} \tag{4.29}$$

where $\widehat{\mathbf{R}}$ denotes a spectral correlation matrix with terms that are the average cross-power spectra between pairs of receivers. Equation 4.29 is in quadratic form and may also be expressed as follows:

$$P(\mathbf{k}, \omega) = \sum_{i,j=1}^{M} w_i w_j \widehat{S_{ij}}(\omega) e^{i\mathbf{k}\cdot(\mathbf{x}_i - \mathbf{x}_j)} \tag{4.30a}$$

$$= \sum_{i,j=1}^{M} w_i w_j \widehat{S_{ij}}(\omega) e^{i\mathbf{k}\cdot\boldsymbol{\chi}_{ij}} \tag{4.30b}$$

where $w_i$ and $w_j$ are the shading weights of receivers $i$ and $j$, respectively, and $\chi_{ij}$ is the spatial lag between the two receivers.

Peaks in the steered response power spectrum correspond to the wavenumbers of propagating surface waves and may be used to calculate surface wave phase velocities using:

$$c = \frac{\omega}{|\mathbf{k}_{peak}|} \tag{4.31}$$

**Beamforming - Active Test with Transient Source** Beamforming is a powerful technique that is easily adapted to a variety of specific surface wave test procedures. As a first example, consider an active surface wave test performed using a transient source. The receivers are configured in the irregular linear array depicted in Figure 10a. Because the location of the source and the direction of wave propagation are known for active surface wave tests, the vector wavenumber $\mathbf{k}$ and receiver position $\mathbf{x}_m$ can be simplified to scalar values $k_x$ and $x_m$. For active surface wave measurements the shading weights are specified as follows:

$$w_m = \sqrt{x_m} \tag{4.32}$$

to approximately account for the geometric spreading of Rayleigh waves. Zywicki (1999) demonstrated that Equation 4.32 results in better wavenumber resolution than other forms of signal normalization. Figure 23 shows an example of the transient signals recorded at each receiver location in two formats: true amplitude and normalized amplitude.

The spectral correlation matrix is represented graphically in Figure 24. Note that along the diagonal of the "matrix" the phase spectrum is equal to zero at all frequencies because that term of the matrix is the auto-power spectrum of the receiver corresponding

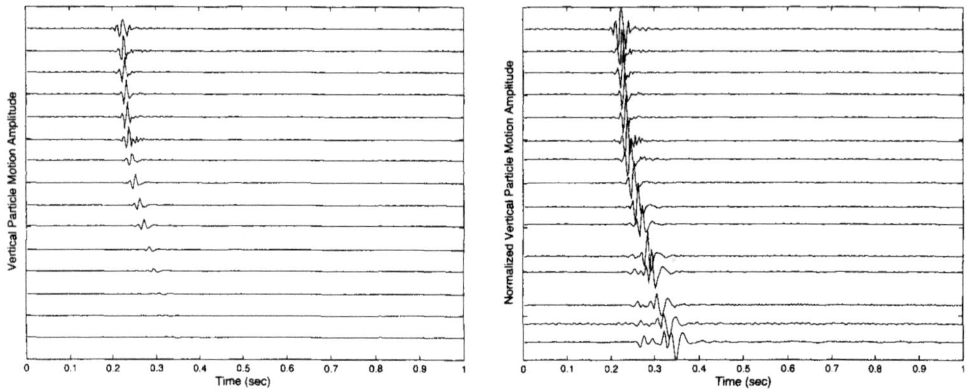


**Figure 23.** Transient Signals Recorded in an Active Surface Wave Test (a) True Amplitude and (b) Normalized Amplitude

to that position in the matrix. Each term in the spectral correlation matrix is the linear, frequency-domain average of 10 transient impacts.

The steered response power spectrum resulting from the application of Equation 4.29 to the data is shown in Figure 25a. A slice through this plot at constant frequency yields the wavenumber spectrum at that frequency. An example is shown in Figure 25b. This plot may be easily converted to a plot of phase velocity vs. frequency to visualize the dispersion curve as shown in Figure 26a. Finally, the peak value in the steered response power spectrum (or phase velocity vs. frequency) plot is selected to define the specific $(f, c)$ pairs to be used in a subsequent inversion procedure to obtain the shear wave velocity profile. If one chooses only the largest peak at each frequency, the resulting dispersion curve may be regarded as containing only the dominant mode (or mode combination) at that frequency. Multiple modes may be selected by identifying secondary peaks in the steered response power spectrum at each frequency (assuming there is sufficient wavenumber resolution). In this example, no attempt was made to identify secondary peaks and the resulting dispersion curve is shown in Figure 26b.

The Multichannel Analysis of Surface Waves (MASW) method (Park et al., 1999b,a) is also an active test that uses conventional frequency-domain beamforming to determine the dispersion curve. In the MASW technique the shading weights are chosen in the following manner:

$$w_m = \frac{1}{|S(x_m, \omega)|} \tag{4.33}$$

so that the normalized amplitude of each receiver is 1.0.

**Beamforming - Active Test with Harmonic Source** Beamforming may also be used with continuous sources such as those pictured in Figures 3c and 3d. The small electro-mechanical shaker pictured in Figure 3c was used for the following example. The source was programmed to oscillate vertically in stepped sine mode at approximately 70

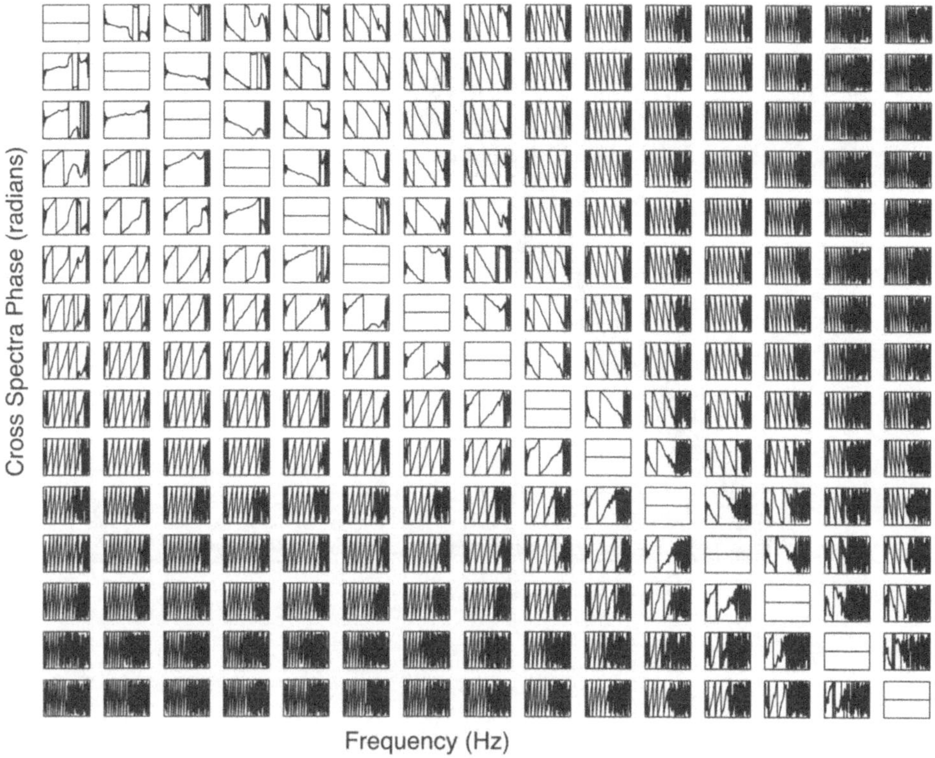


**Figure 24.** Graphical Representation of the Spectral Correlation Matrix

excitation frequencies between 3.75 and 100 Hz. For each source excitation frequency $\omega_0$, vertical particle motions were recorded at each receiver position for 32 sec at a sampling frequency of 320 Hz. An example of the time histories recorded for a source frequency of 5.3125 Hz is shown in Figure 27. Each time history was subsequently divided into $n_d = 10$ blocks that were 3.2 sec long.

The cross-power spectra between pairs of receivers is calculated for each source frequency to form the spectral correlation matrix $\widehat{\mathbf{R}}$ and the steered response power spectrum $P(k_x, \omega)$ is calculated as described previously. The latter is shown in Figure 28. Selecting the dominant wavenumber (i.e., mode) at each frequency yields the dispersion curve shown in Figure 29.

For the data presented here, both transient and harmonic sources produce reasonably good dispersion curves. This is not always the case, and, in general, the author prefers using the harmonic source because it affords much greater control over the frequency content of the source. In short, the author has found that the combination of a stepped-sine (i.e., harmonic) source and beamforming methods yields robust dispersion curves that make subsequent processing and interpretation easier.

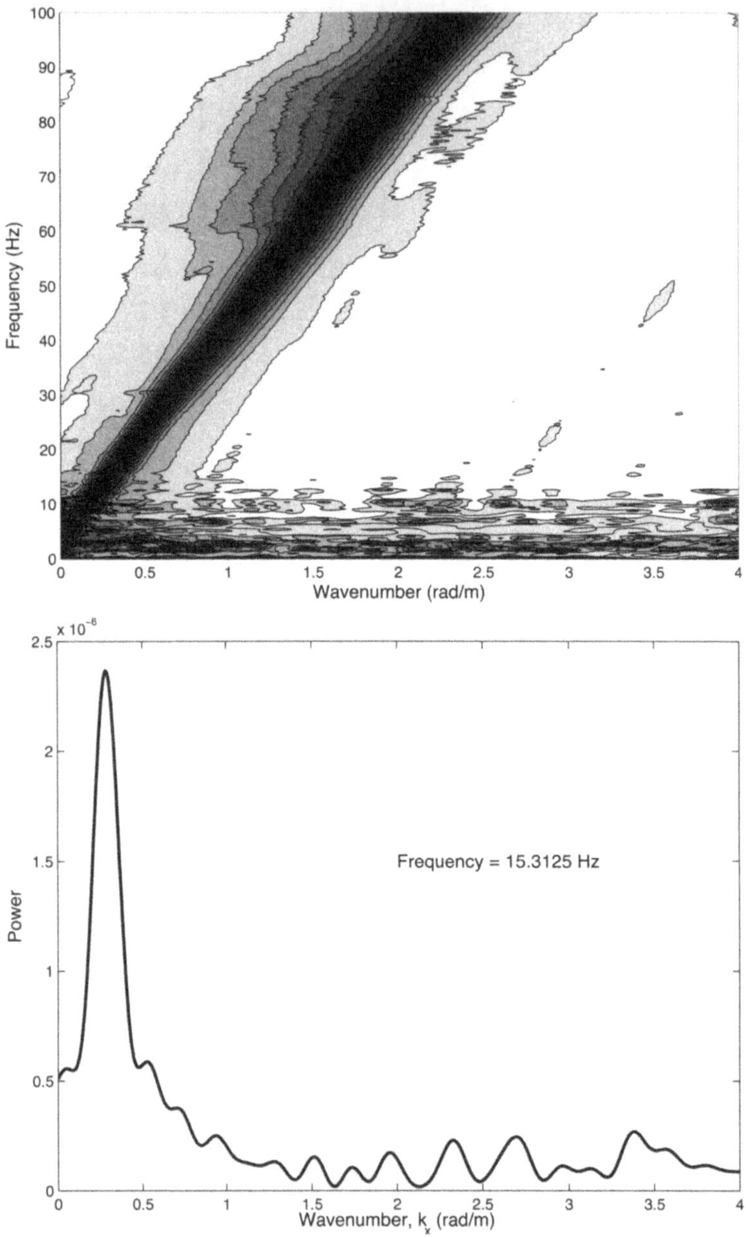


**Figure 25.** Frequency Domain Beamformer Results for an Active Test Using a Transient Source: (a) Steered Response Power Spectrum and (b) Individual Wavenumber Spectrum

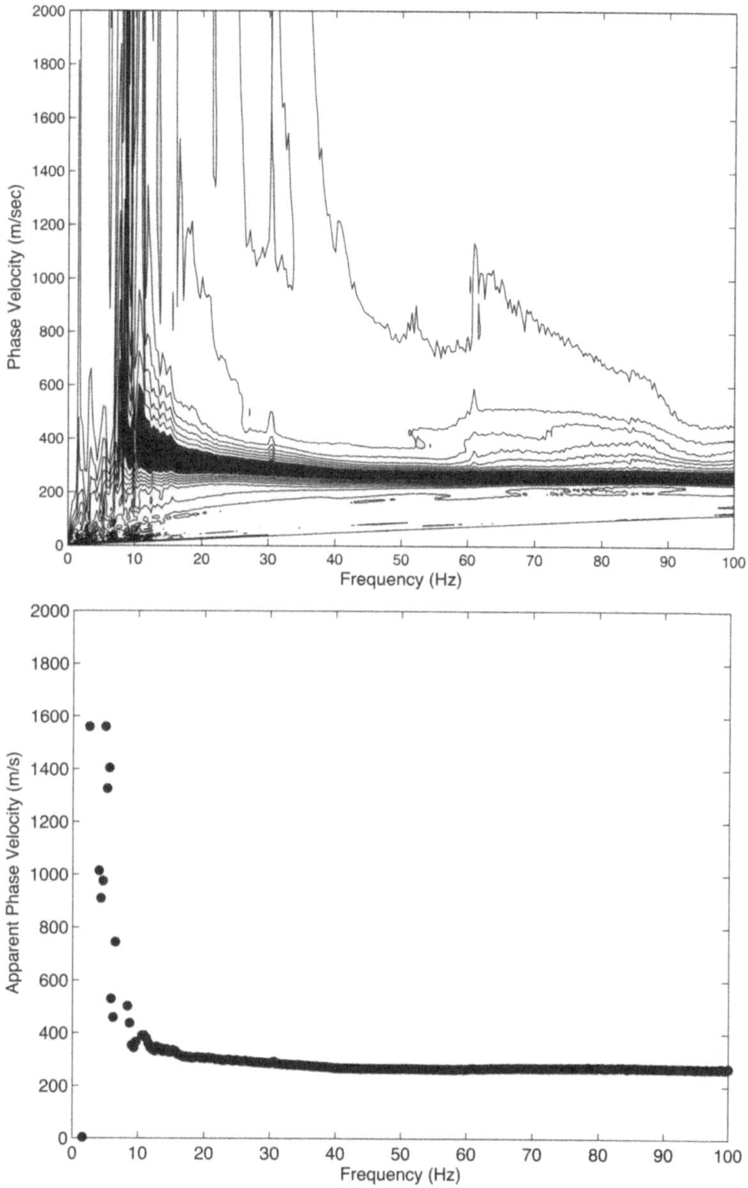


**Figure 26.** Dispersion Curves from Frequency Domain Beamforming - Active Test Using a Transient Source: (a) Contour Plot and (b) $(f, c)$ Pairs

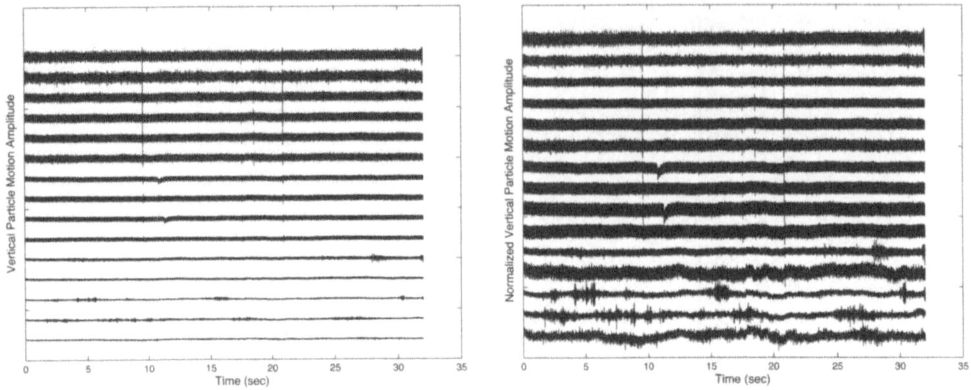


**Figure 27.** Harmonic Signals Recorded in an Active Surface Wave Test (a) True Amplitude and (b) Normalized Amplitude

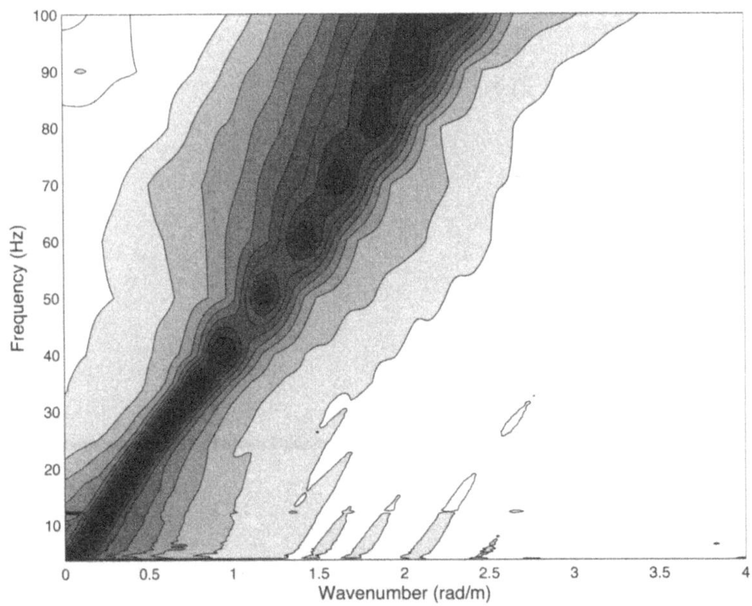


**Figure 28.** Steered Response Power Spectrum - Active Test Using a Harmonic Source

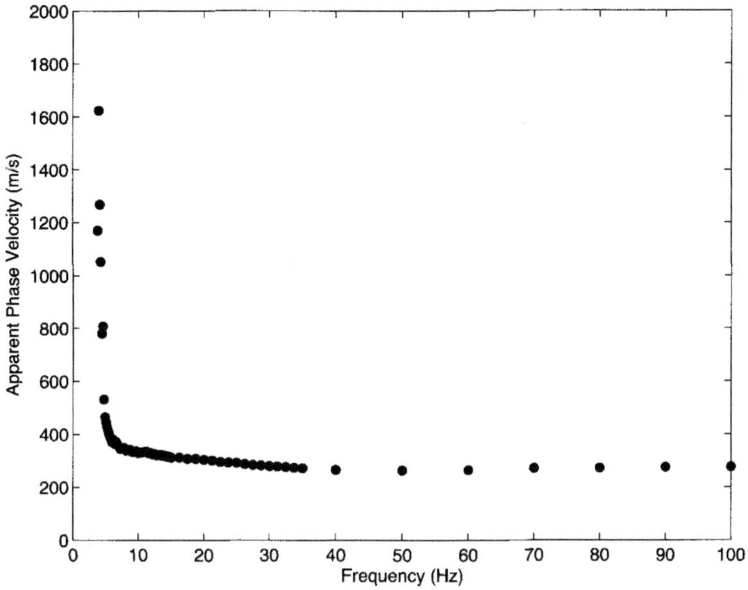


**Figure 29.** Dispersion Curve from Frequency Domain Beamforming - Active Test a Harmonic Source

**Beamforming - Passive Test** As noted previously, typical passive sources such as microtremors and cultural noise generally produce energy at lower frequencies than most active sources and thus allow deeper $V_s$ profiling. The exact location of the passive energy source is often not known, and it is necessary to use a two-dimensional array of receivers to enable the measurement of seismic waves originating from any direction. Thus, $\mathbf{x}_m$ and $\mathbf{k}$ remain vector quantities. The array geometry used in this example consisted of the cross-shaped array of 13 receivers as shown in Figure 12a. Particle motion time histories were recorded for 256 sec at a sampling frequency of 256 Hz[3] and were subsequently divided into $n_d = 32$ equal blocks for ensemble averaging. The receivers were uniformly weighted (i.e., $w_m = 1.0$) since surface waves from passive sources are assumed to be plane.

Figure 30 shows examples of steered response power spectra at four individual frequencies. In Figures 30a and 30c, there is a single dominant peak, but artifacts associated with the aperture smoothing function (see Figure 12b) are clearly visible. The circle is the locus of points with constant $|\mathbf{k}|$. In Figure 30b, it appears that energy may be propagating from two or more sources. Note that the two dominant peaks are closely spaced, and the limited wavenumber resolution is beginning to become apparent. In Figure 30d, there are also multiple peaks, but they are more widely separated and bet-

[3] A high sampling frequency was chosen to provide maximum flexibility during post-processing of the data. In practice, a much lower sampling frequency could be selected to reflect the emphasis on low-frequency data during passive testing.

ter resolved. Figure 31 shows the dispersion curve resulting from similar analyses at frequencies between 6 and 16 Hz.

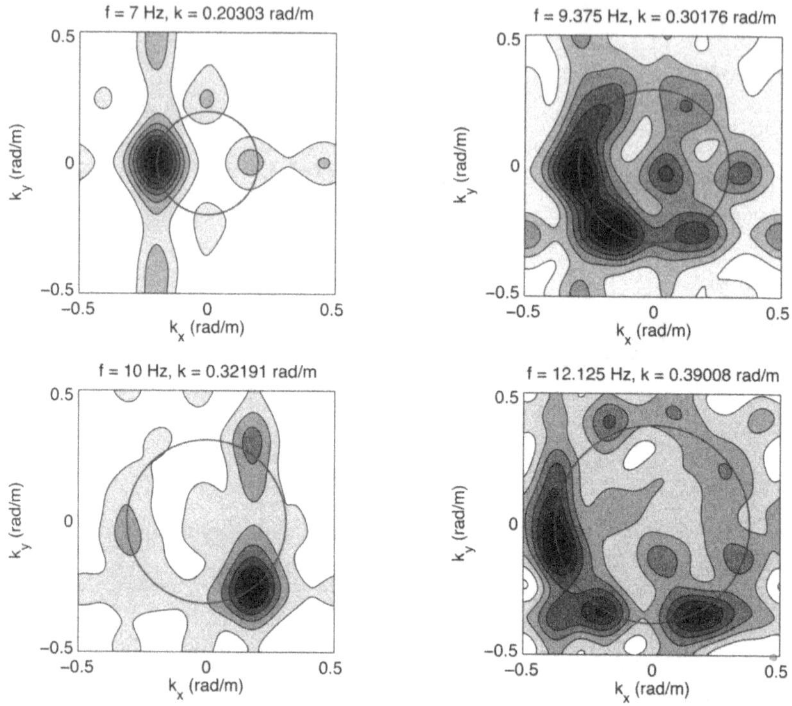


**Figure 30.** Examples of Steered Response Power Spectra - Passive Test

## 4.5 Adaptive Array Processing Methods

A variety of more advanced signal processing techniques are available that may generally be called *adaptive* array processing methods (Johnson and Dudgeon, 1993). Examples include minimum variance, linear prediction, Pisarenko harmonic decomposition, eigenvector, and multiple signal classification methods. A detailed presentation of these methods is beyond the scope of this chapter; interested readers are referred to Johnson and Dudgeon (1993) for a discussion of these advanced methods and to Zywicki (1999) for examples of the application of these methods to engineering surface wave analyses.

## 4.6 Combined Active-Passive Tests

Active and passive methods may be combined to enable measurements of the shear wave velocity profile to greater depths without sacrificing the near-surface resolution offered by active methods (Rix et al., 2002; Yoon and Rix, 2004). Figure 32 shows an example of this approach. The active measurements were performed using a harmonic

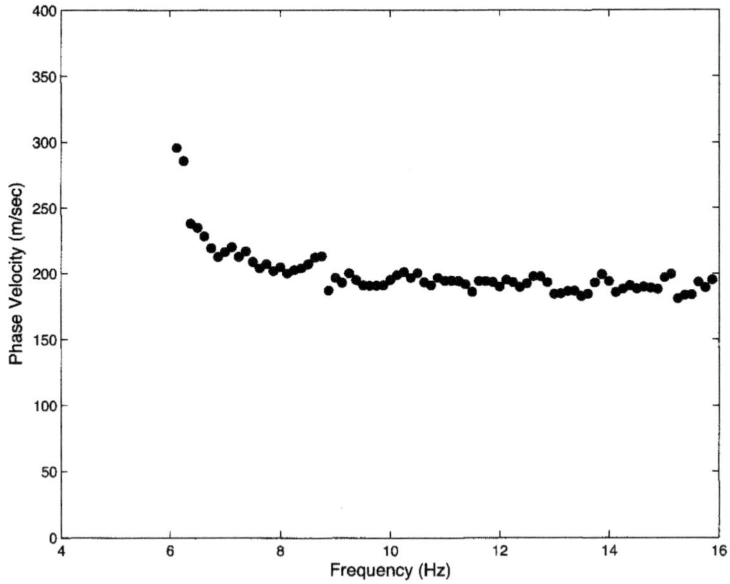


**Figure 31.** Dispersion Curve from Frequency Domain Beamforming - Passive Test

source and the resulting dispersion curve ranges from approximately 5 to 100 Hz. There was sufficient passive energy at this site to generate a dispersion curve from 1.875 to 10 Hz. Adding the passive data decreases the minimum frequency from 5.0 to 1.875 Hz and increases the maximum measured wavelength from approximately 60 to 250 m. The active and passive measurements overlap in the frequency range of approximately 5 to 10 Hz as shown in the figure insert, and the values of phase velocity from the active test are generally less than from the passive test. One possible strategy is to simply average the active and passive values in the range where they overlap, but this is an issue deserving additional study.

## 5 Inversion

The experimental dispersion curve is used to determine the shear wave velocity profile of the site via a process called *inversion.* We begin our discussion of surface wave inversion with a formal statement of the *forward problem,* which allows us to calculate a theoretical dispersion curve corresponding to an assumed shear wave velocity profile:

$$\mathbf{d} = \mathbf{g}(\mathbf{m}) \tag{5.1}$$

where **d** is a data vector, **m** is a model vector, and **g** is an operator that maps the model vector to the data vector. The most common subsurface model that is used in surface

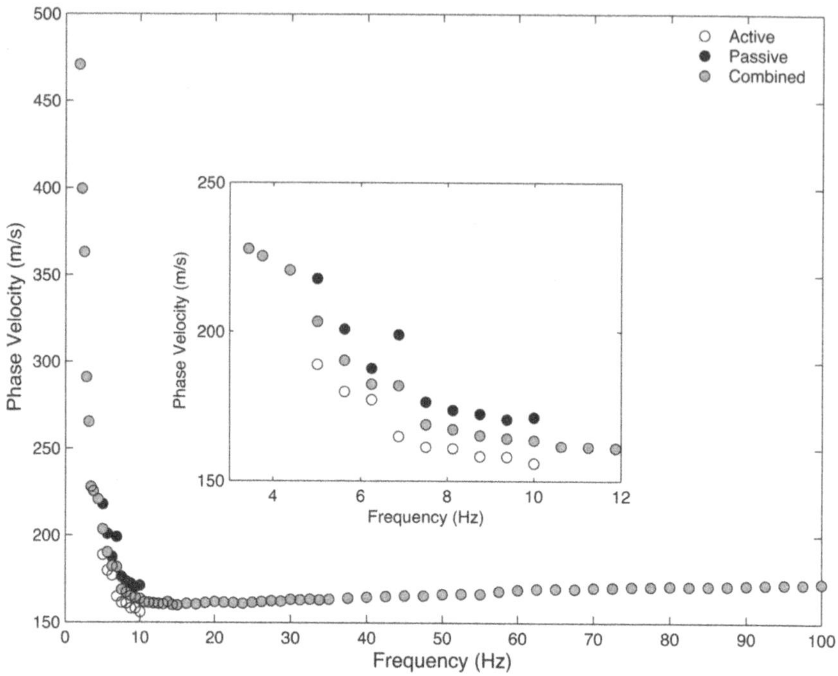


**Figure 32.** Combined Active-Passive Dispersion Curve

wave inversion consists of a stack of uniform layers over a half space as shown in Figure 33. In the context of surface wave inversion, **d** is a vector of surface wave phase velocities and **m** is the vector of shear wave velocities:

$$\mathbf{d} = \begin{pmatrix} c_1 \\ c_2 \\ \vdots \\ c_{nf} \end{pmatrix} \quad \mathbf{m} = \begin{pmatrix} V_{s,1} \\ V_{s,2} \\ \vdots \\ V_{s,nl} \end{pmatrix} \tag{5.2}$$

where $nf$ is the number of frequencies at which the dispersion curve is defined and $nl$ is the number of layers including the half space in the subsurface profile. The other parameters of each layer—the compression wave velocity, mass density, and thickness—are usually held constant[4]. The operator **g** that maps the shear wave velocities to surface wave phase velocities is nonlinear.

Following Tarantola (1987), if the posterior probability density function of the model space may be reasonably assumed to be Gaussian, the *maximum likelihood* solution may be found by minimizing the following objective function:

[4]In many inversion algorithms, Poisson's ratio is used in place of the compression wave velocity.

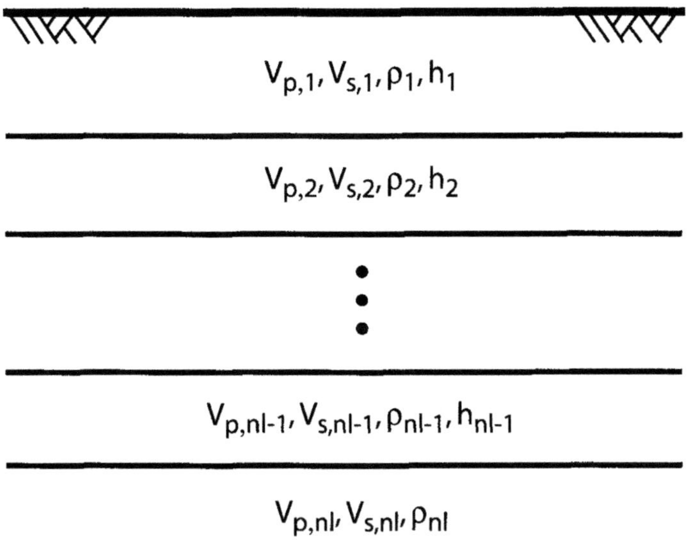


**Figure 33.** Subsurface Model Commonly Used for Surface Wave Inversion

$$S(\mathbf{m}) = \frac{1}{2}\left[(\mathbf{d}_{obs} - \mathbf{g}(\mathbf{m}))^T \, (\mathbf{d}_{obs} - \mathbf{g}(\mathbf{m}))\right] \tag{5.3}$$

where $\mathbf{d}_{obs}$ is the $nf \times 1$ vector of observed (i.e., experimental) phase velocities. The minimum of the objective function may be found by setting its derivative equal to zero:

$$\frac{\partial S(\mathbf{m})}{\partial \mathbf{m}} = \mathbf{G}^T(\mathbf{d}_{obs} - \mathbf{g}(\mathbf{m})) = 0 \tag{5.4}$$

where $\mathbf{G}$ is a matrix containing the partial derivatives of the data with respect to the model parameters:

$$G_{ij} = \frac{\partial g_i(\mathbf{m})}{\partial m_j} = \frac{\partial c_i}{\partial V_{s,j}} \quad i = 1, \ldots, nf; \; j = 1, \ldots, nl \tag{5.5}$$

Frequently, the partial derivatives may be obtained using the variational principle of Rayleigh waves (Aki and Richards, 1980; Lai, 1998), which leads to closed-form expressions.

One approach to solving Equation 5.4 is to linearize the solution of the forward problem about a particular model:

$$\mathbf{g}(\mathbf{m}_{n+1}) \approx \mathbf{g}(\mathbf{m}_n) + \mathbf{G}_n(\mathbf{m}_{n+1} - \mathbf{m}_n) \tag{5.6}$$

where $n$ is an iteration index. Substituting Equation 5.6 into Equation 5.4 and solving for $\mathbf{m}_{n+1}$ yields:

$$\mathbf{m}_{n+1} = \left(\mathbf{G}_n^T \mathbf{G}_n\right)^{-1} \mathbf{G}_n^T (\mathbf{d}_{obs} - \mathbf{g}(\mathbf{m}_n) + \mathbf{G}_n(\mathbf{m}_n)) \tag{5.7a}$$

$$= \mathbf{m}_n + \left(\mathbf{G}_n^T \mathbf{G}_n\right)^{-1} \mathbf{G}_n^T (\mathbf{d}_{obs} - \mathbf{g}(\mathbf{m}_n)) \tag{5.7b}$$

Equation 5.7a or 5.7b may be used iteratively to calculate the solution (i.e., the shear wave velocity profile) to the surface wave inverse problem starting from an initial estimate $\mathbf{m}_0$ of the profile and continuing until $\|\mathbf{m}_{n+1} - \mathbf{m}_n\|$ is less than some threshold.

This inverse problem, like most geophysical inverse problems, is ill posed due to finite, uncertain data and modelling errors. To overcome these difficulties, it is possible employ one or more of several strategies that conceptually involve introducing additional information about the inverse problem (Santamarina and Fratta, 1998). One possible strategy is to introduce information about the experimental data and errors associated with the use of a particular forward model. This information may be incorporated via the $nf \times nf$ matrix $\mathbf{C}_D$ defined as (Tarantola, 1987):

$$\mathbf{C}_D = \mathbf{C}_d + \mathbf{C}_T \tag{5.8}$$

where $\mathbf{C}_d$ contains the covariances of the observed data $\mathbf{d}_{obs}$ and $\mathbf{C}_T$ contains the covariances of the theoretical modeling errors. Several investigators have evaluated the uncertainty of observed dispersion data. Marosi and Hiltunen (2004) found that the data was normally distributed with a coefficient of variation of 2 percent over a range of frequencies from 20 to 150 Hz based on 30 replicate SASW tests at two sites. Lai et al. (2004) observed coefficients of variation that differed with frequency using beamforming measurements at two Italian sites. At frequencies less than about 10 to 12 Hz, the coefficients of variation ranged from 1.1 to 13.9 percent; most were less than 5 percent. At higher frequencies, smaller coefficients of variation ranging from 0.1 to 1.25 percent were observed. Lai et al. (2004) also found that the phase velocity measurements were normally distributed. Uncertainties due to modeling errors are likely larger and may be especially important for surface wave inversion because the subsurface model shown in Figure 33 is often an idealized representation of actual subsurface conditions. However, the magnitude of these modeling errors remains largely unknown. In light of these issues, the author typically uses a simple form of $\mathbf{C}_D$ consisting only of a diagonal matrix:

$$\mathbf{C}_D = \operatorname{diag}\left(\sigma_1^2, \sigma_2^2, \ldots, \sigma_{nf}^2\right) \tag{5.9}$$

where $\sigma_i^2$ is the total variance associated with $c_i$. The author often assumes a coefficient of variation of 5 percent (i.e., $\sigma_i = 0.05 \cdot c_i$) to account for data and modeling errors. The treatment of uncertainties in surface wave testing is a topic requiring further study.

Given a set of uncertainties represented by $\mathbf{C}_D$, the objective function may be reformulated as follows:

$$S(\mathbf{m}) = \frac{1}{2}\left[(\mathbf{d}_{obs} - \mathbf{g}(\mathbf{m}))^T \mathbf{C}_D^{-1} (\mathbf{d}_{obs} - \mathbf{g}(\mathbf{m}))\right] \tag{5.10}$$

Proceeding as before yields the following solution to the inverse problem:

$$\mathbf{m}_{n+1} = (\mathbf{G}_n^T \mathbf{C}_D^{-1} \mathbf{G}_n)^{-1}\ \mathbf{G}_n^T \mathbf{C}_D^{-1} (\mathbf{d}_{obs} - \mathbf{g}(\mathbf{m}_n) + \mathbf{G}_n(\mathbf{m}_n)) \tag{5.11a}$$

$$= \mathbf{m}_n + (\mathbf{G}_n^T \mathbf{C}_D^{-1} \mathbf{G}_n)^{-1}\ \mathbf{G}_n^T \mathbf{C}_D^{-1} (\mathbf{d}_{obs} - \mathbf{g}(\mathbf{m}_n)) \tag{5.11b}$$

Another approach is to introduce information about the anticipated solution. Suppose that an initial estimate of the solution $\mathbf{m}_{prior}$ is available using independent information. We may expect that the shear wave velocity profile derived from the surface wave inversion will not differ too much from this initial estimate. Mathematically, this expectation can be expressed by the following objective function:

$$S(\mathbf{m}) = \frac{1}{2}\Big[(\mathbf{d}_{obs} - \mathbf{g}(\mathbf{m}))^T\ \mathbf{C}_D^{-1}\ (\mathbf{d}_{obs} - \mathbf{g}(\mathbf{m})) + (\mathbf{m} - \mathbf{m}_{prior})^T\ \mathbf{C}_M^{-1}\ (\mathbf{m} - \mathbf{m}_{prior})\Big] \tag{5.12}$$

where $\mathbf{C}_M$ is the $nl \times nl$ matrix that contains the covariances of the a priori model parameters. Proceeding as before yields the following solution:

$$\begin{aligned}\mathbf{m}_{n+1} &= (\mathbf{G}_n^T \mathbf{C}_D^{-1} \mathbf{G}_n + C_M^{-1})^{-1} \\ &\Big[\mathbf{G}_n^T \mathbf{C}_D^{-1} (\mathbf{d}_{obs} - \mathbf{g}(\mathbf{m}_n) + \mathbf{G}_n(\mathbf{m}_n)) + \mathbf{C}_M^{-1}\ (\mathbf{m}_n - \mathbf{m}_{prior})\Big] \\ &= \mathbf{m}_n + (\mathbf{G}_n^T \mathbf{C}_D^{-1} \mathbf{G}_n + C_M^{-1})^{-1} \\ &\Big[\mathbf{G}_n^T \mathbf{C}_D^{-1} (\mathbf{d}_{obs} - \mathbf{g}(\mathbf{m}_n)) + \mathbf{C}_M^{-1}\ (\mathbf{m}_n - \mathbf{m}_{prior})\Big]\end{aligned} \tag{5.13}$$

If there is no a priori information on the model parameters (i.e., the shear wave velocity profile), the covariance matrix $\mathbf{C}_M = \infty\mathbf{I}$ and Equation 5.13 simplifies to Equation 5.11.

Finally, we may introduce information about particular features we desire in the solution via *regularization*. One example is a constrained least squares inversion algorithm proposed by Constable et al. (1987) that seeks to find the smoothest shear wave velocity profile that fits the experimental dispersion curve with a specified tolerance. In the words of Constable et al. (1987), the profile should "depart from the simplest case only as far as is necessary to fit the data." Their approach is motivated by the observation that inversions performed with unconstrained least squares techniques often lead to physically unreasonable profiles of model parameters. The algorithm requires a measure of smoothness or its converse, roughness, for a candidate shear wave velocity profile. In a layered medium, roughness may be defined by:

$$R_1 = (\boldsymbol{\delta}\mathbf{m})^T (\boldsymbol{\delta}\mathbf{m}) \tag{5.14}$$

where $\boldsymbol{\delta}$ is an $nl \times nl$ two-point, backward difference operator:

$$\boldsymbol{\delta} = \begin{bmatrix} 0 & & & & \\ -1 & 1 & & 0 & \\ & -1 & 1 & & \\ & & & \cdots & \\ & 0 & & -1 & 1 \end{bmatrix} \tag{5.15}$$

Roughness defined in this manner is thus the sum of squared differences between the values of $V_s$ of adjacent layers in the profile.

The solution of the nonlinear inverse problem consists of finding a soluton that minimizes $R_1$ with the constraint that the squared error between predicted and observed values of the phase velocity be equal to a specified value. The method of Lagrange multipliers may be employed to solve this constrained minimization problem. The corresponding unconstrained objective function is:

$$S(\mathbf{m}) = (\boldsymbol{\delta}\mathbf{m})^T(\boldsymbol{\delta}\mathbf{m}) + \mu^{-1}\left[(\mathbf{d}_{obs} - \mathbf{g}(\mathbf{m}))^T \, \mathbf{C}_D^{-1} \, (\mathbf{d}_{obs} - \mathbf{g}(\mathbf{m})) - \eta^2\right] \tag{5.16}$$

where $\mu$ is the Lagrange multiplier, which may be interpreted as a smoothing parameter, and $\eta^2$ is the value of the squared error that is thought to be reasonable. Constable et al. (1987) make a compelling argument that an appropriate value for $\eta^2$ is $nf$, the number of points in the experimental dispersion curve. This criterion corresponds to a root-mean-square (rms) error of 1.0.

The solution may be calculated iteratively using:

$$\mathbf{m}_{n+1}(\mu) = \left[\mu\boldsymbol{\delta}^T\boldsymbol{\delta} + \mathbf{G}_n^T\mathbf{C}_D^{-1}\mathbf{G}_n\right]^{-1} \mathbf{G}_n^T\mathbf{C}_D^{-1}(\mathbf{d}_{obs} - \mathbf{g}(\mathbf{m}_n) + \mathbf{G}_n(\mathbf{m}_n)) \tag{5.17}$$

For each iteration, the "smoothing parameter" $\mu$ is chosen using the ad-hoc procedure recommended by Constable et al. (1987) to achieve the proper balance between minimizing the data misfit and roughness of the shear wave velocity profile. Figure 34a shows an example of the variation of the rms error as a function of the smoothing parameter for each iteration, and Figure 34b shows the rms error as a function of the iteration number.

The shear wave velocity profiles corresponding to each iteration are shown in Figure 35. The initial estimate was a uniform half space with a shear wave velocity of 1500 m/sec. Note the rapid convergence to the final profile. Figure 36 shows the final profile with the corresponding estimates of the uncertainty in the shear wave velocity profile of each layer using the method described by Lai et al. (2004).

As a check on the inversion process, the theoretical dispersion curve corresponding to the final shear wave velocity profile may be compared to the experimental dispersion curve. An example is shown in Figure 37. Satisfactory agreement between the experimental and theoretical dispersion curves is a necessary, but not sufficient, condition for a successful inversion.

It is important to note that for a given frequency, the solution of the forward problem given by Equation 5.1 may yield dispersion curves corresponding to several modes of propagation. Usually the theoretical phase velocities and the Jacobian matrix used in

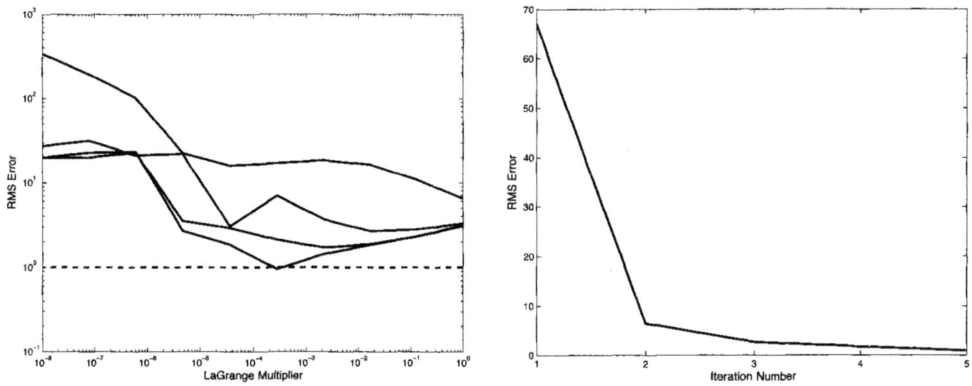


**Figure 34.** Variation of RMS Error with (a) LaGrange Multiplier and (b) Iteration Number

the inversion correspond to the fundamental mode, but the inversion algorithms described herein do not preclude the use of higher modes or the effective phase velocity (Lai, 1998) in the inversion for sites where the fundamental mode may not dominate the response (Gucunski and Woods, 1992; Roësset et al., 1991; Tokimatsu, 1995).

This algorithm is one of many including trial and error forward modeling, Monte Carlo methods, iterative methods, singular value decomposition, gradient methods, artificial neural networks, genetic algorithms, and others that may be used for surface wave inversion. Nonetheless, the author has found that it yields reasonable, if not unique, profiles for a wide range of conditions.

## 6 Acknowledgments

I had the privilege of working for Professor Kenneth H. Stokoe, II as a graduate research assistant at the University of Texas at Austin where I first became interested in surface wave methods. While on the faculty at the Georgia Institute of Technology, I have had the good fortune of having numerous talented graduate research assistants work on projects related to surface wave testing including Christopher Haker, Gregory L. Hebeler, Carlo G. Lai, Elizabeth A. Leipski, M. Catalina Orozco, A. Wesley Spang, Jr., Sungsoo Yoon, and Daren J. Zywicki. Sungsoo Yoon deserves special thanks for processing all of the data included in this chapter. Two visiting Ph.D. students from Poliecnico di Torino, Sebastiano Foti and Vitantonio Roma, have also contributed to my understanding of surface wave methods.

I have also enjoyed my association with numerous colleagues around the world that share an interest in various aspects of surface wave testing and have made important contributions that have helped me continue to learn including James Bay, David M. Boore, Nenad Gucunski, Robert B. Herrmann, Dennis Hiltunen, Eduardo Kausel, Barbara Luke, Soheil Nazarian, José M. Röesset, Ignacio Sánchez-Salinero, Kohji Tokimatsu, and Richard D. Woods.

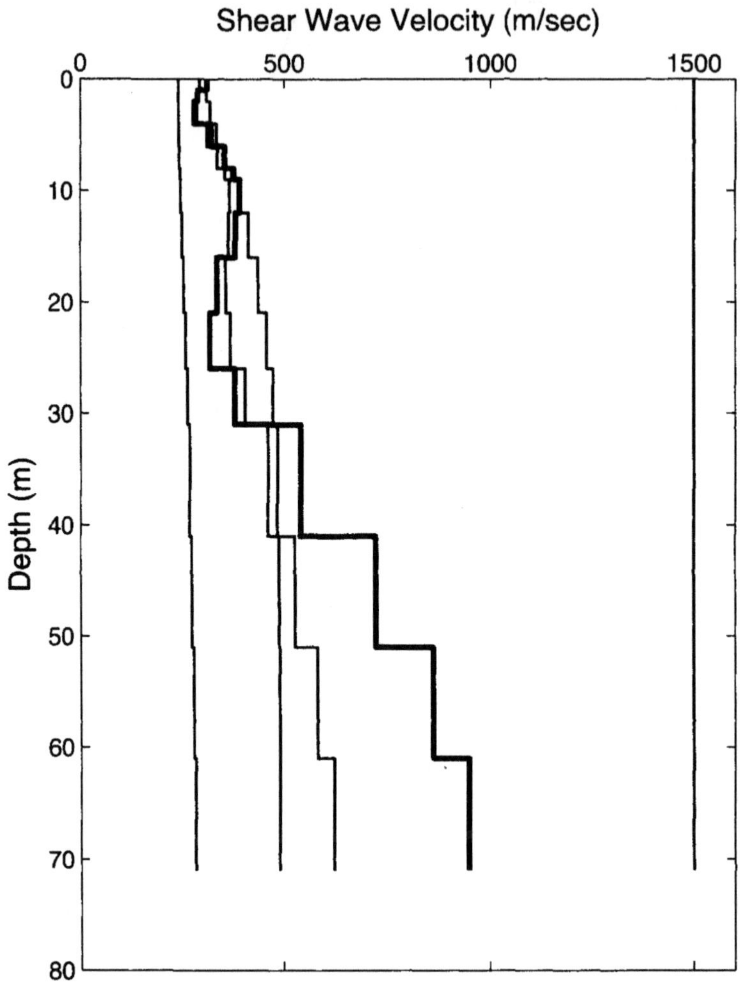


**Figure 35.** Shear Wave Velocity Profile for Each Iteration and the Final Profile

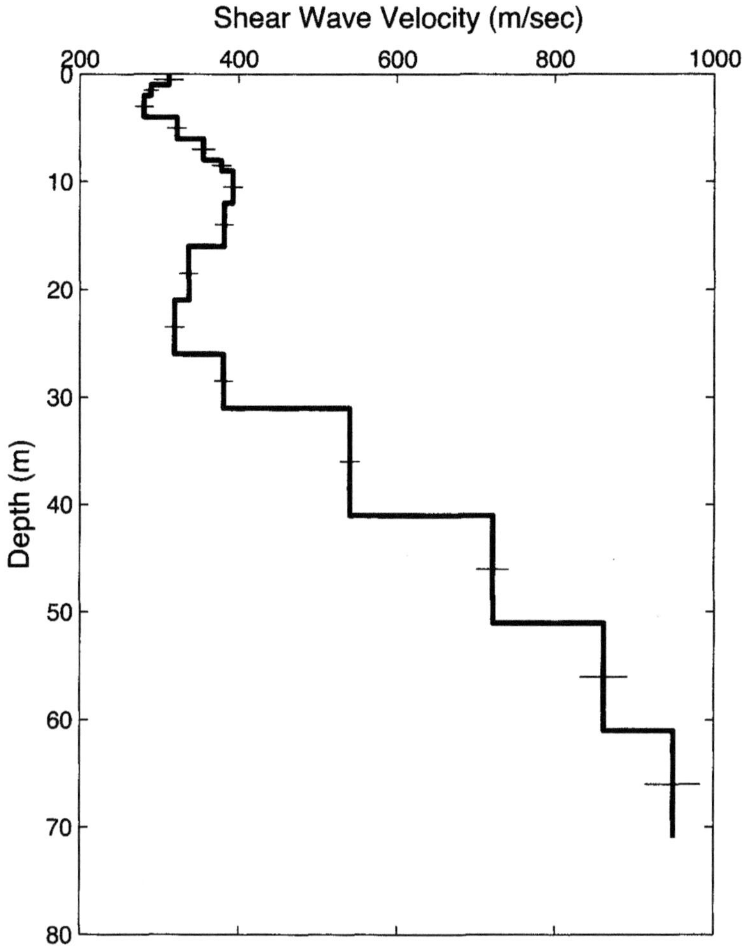


**Figure 36.** Final Shear Wave Velocity Profile and Estimates of Uncertainty

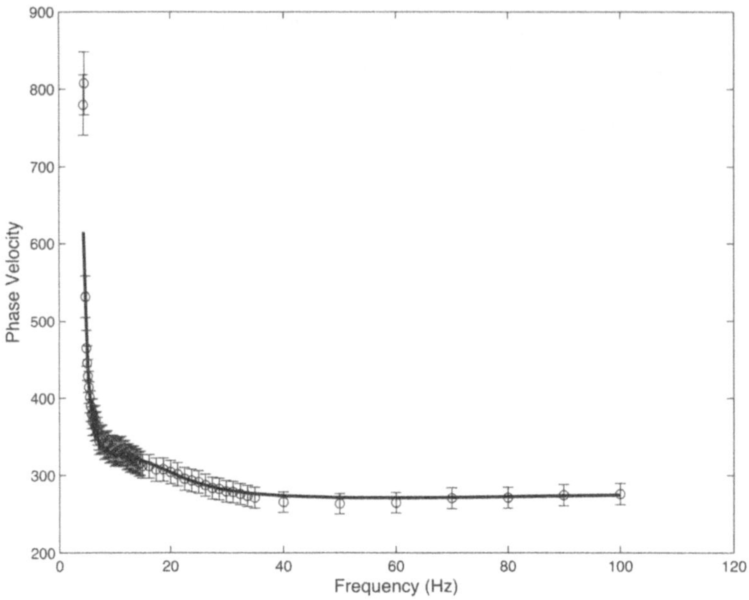


**Figure 37.** Comparison Between Experimental and Theoretical Dispersion Curves for the Final Iteration

This material is partially based upon work supported by the National Science Foundation under Grant Nos. CMS-9402358 and CMS-9616013, the U.S. Geological Survey under Award No. 1434-95-G-2634, and the Mid-America Earthquake Center under a grant from the Earthquake Engineering Research Centers program of the National Science Foundation under Award No. EEC-9701785. The author gratefully acknowledges the support of these organizations.

## Bibliography

K. Aki. Space and time spectra of stationary stochastic waves with special reference to microtremors. *Bulletin of the Earthquake Research Institute*, 35:415–456, 1957.

K. Aki. A note on using microseisms in determining the shallow structures of the Earth's crust. *Geophysics*, 30(4):665–666, August 1965.

K. Aki and P. G. Richards. *Quantitative Seismology: Theory and Methods.* W.H. Freeman and Company, New York, San Francisco 1980.

M. Asten. Control on 3-component spectra of Rayleigh-wave microseisms. *Bulletin of the Seismological Society of America*, 68:1623–1636, 1978.

J. S. Bendat and A. G. Piersol. *Random Data: Analysis and Measurement Procedures.* John Wiley & Sons, New York, $2^{nd}$ edition, 1986.

R.D. Borcherdt. Estimates of site-dependent response spectra for design (methodology and justification). *Earthquake Spectra*, 10:617–653, 1994.

R.N. Bracewell. *The Fourier Transform and Its Applications.* McGraw-Hill, New York, 2nd edition, 1978.

S.C. Constable, R.L. Parker, and G.G. Constable. Occam's inversion: A practial algorithm for generating smooth models from electromagnetic sounding data. *Geophysics*, 52:289–300, 1987.

N. Gucunski and R. D. Woods. Numerical simulation of the SASW test. *Soil Dynamics and Earthquake Engineering*, 11:213–227, 1992.

M. Horike. Inversion of phase velocity of long-period microtremors to the S-wave-velocity structure down to the basement in urbanized areas. *Journal of the Physics of the Earth*, 33:59–96, 1985.

D. H. Johnson and D. E. Dudgeon. *Array Signal Processing: Concepts and Techniques.* PTR Prentice-Hall, Inc., Upper Saddle River, New Jersey, 1993.

R.T. LaCoss, E.J. Kelly, and M. N. Toksöz. Estimation of seismic noise structure using arrays. *Geophysics*, 34(1):21–38, February 1969.

C. G. Lai. *Simultaneous Inversion of Rayleigh Phase Velocity and Attenuation for Near-Surface Site Characterization.* PhD thesis, Georgia Institute of Technology, 1998.

C. G. Lai, G. J. Rix, S. Foti, and V. Roma. Simultaneous measurement and inversion of surface wave dispersion and attenuation curves. *Soil Dynamics and Earthquake Engineering*, 22:923–930, 2002.

C.G. Lai, S. Foti, and G.J. Rix. Propagation of data uncertainty in surface wave inversion. *submitted to Journal of Environmental and Engineering Geophysics*, 2004.

L.J. Liaw and T.V. McEvilly. Microseisms in geothermal exploration - studies in Grass Valley, Nevada. *Geophysics*, 44:1097–1115, 1979.

K.T. Marosi and D.R. Hiltunen. Characterization of SASW phase angle and phase velocity measurement uncertainty. *Geotechnical Testing Journal*, 27(2):205–213, 2004.

G.A. McMechan and M.J. Yedlin. Analysis of dispersive waves by wave field transformation. *Geophysics*, 46:869–874, 1981.

S. Nazarian. *In Situ Determination of Elastic Moduli of Soil Deposits and Pavement Systems by Spectral Analysis of Surface Waves Method.* PhD thesis, University of Texas at Austin, Austin, Texas, 1984.

M. Ohori, A. Nobata, and K. Wakamatsu. A comparison of ESAC and FK methods of estimating phase velocity using arbitrarily shaped microtremor arrays. *Bulletin of the Seismological Society of America*, 92(6):2323–2332, August 2002.

C.B. Park, R.D. Miller, and J. Xia. Multichannel analysis of surface waves (MASW). *Geophysics*, 64:800–808, 1999a.

C.B. Park, R.D. Miller, and J. Xia. Multimodal analysis of high frequency surface waves. In M.H. Powers, L. Cramer, and R.S. Bell, editors, *Symposium on the Applications of Geophysics to Engineering and Environmental Problems*, pages 115–121, Denver, CO, 1999b. Environmental and Engineering Geophysical Society.

G. J. Rix, G.L. Hebeler, and M.C. Orozco. Near-surface $v_s$ profiling in the New Madrid Seismic Zone using surface-wave methods. *Seismological Research Letters*, 73(3):380–392, May/June 2002.

G.J. Rix, C.G. Lai, and S. Foti. Simultaneous measurement of surface wave dispersion and attenuation curves. *Geotechnical Testing Journal*, 24(4):350–358, 2001.

G.J. Rix, C.G. Lai, and A.W. Spang. In situ measurement of damping ratio using surface waves. *Journal of Geotechnical and Geoenvironmental Engineering*, 126(5):472–480, May 2000.

J. M. Roësset, D.W. Chang, and K. H. Stokoe, II. Comparison of 2-D and 3-D models for analysis of surface wave tests. In *5th International Conference on Soil Dynamics and Earthquake Engineering*, pages 111–126, Karlsruhe, Germany, 1991.

I. Sánchez-Salinero. *Analytical Investigation of Seismic Methods Used for Engineering Applications*. PhD thesis, University of Texas at Austin, 1987.

J.C. Santamarina and D. Fratta. *Introduction to Discrete Signals and Inverse Problems in Civil Engineering*. ASCE Press, Reston, Virginia, 1998.

C. Strobbia and S. Foti. Multi-offset phase analysis of surface wave data. *submitted to Journal of Applied Geophysics*, 2004.

A. Tarantola. *Inverse Problem Theory*. Elsevier, Amsterdam, 1987.

K. Tokimatsu. Geotechnical site characterization using surface waves. In Kenji Ishihara, editor, *Earthquake Geotechnical Engineering*, pages 1333–1368, Rotterdam, 1995. Balkema.

M. N. Toksöz. Microseisms and an attempted application. *Geophysics*, 39:154–177, 1964.

S. Yoon and G. J. Rix. Combined active-passive surface wave measurements for near-surface site characterization. In *Symposium on the Applications of Geophysics to Engineering and Environmental Problems*, pages 1556–1564. Environmental and Engineering Geophysical Society, 2004.

D. J. Zywicki. *Advanced Signal Processing Methods Applied to Engineering Analysis of Seismic Surface Waves*. PhD thesis, Georgia Institute of Technology, 1999.

# Surface Wave Testing for Geotechnical Characterization

Sebastiano Foti

Dept of Structural and Geotechnical Eng., Politecnico di Torino, Torino, Italy

**Abstract.** This chapter focuses on execution and interpretation of surface wave tests for soil characterization. The different steps (acquisition, processing and inversion) will be considered specifying standard procedures and possible alternatives. Limitations and pitfalls will be also considered, showing with experimental examples the difference between theory and practice in surface wave testing.

## 1 Introduction

Surface waves have been studied in seismology for the characterization of Earth's interior since the 1920s, but their widespread use started during the 1950s and 1960s thanks to the increased possibilities of numerical analysis and to the improvements of instrumentation for recording seismic events associated to earthquakes (Aki and Richards, 1980; Ben-Menhaem, 1995). Geophysical applications at regional scale for the characterization of geological basins make use of seismic signals from explosions (Malagnini et al., 1995) and microtremors (Horike, 1985). Engineering applications started in the 1950s with the Steady State Rayleigh Method (Jones, 1958), but their diffusion arrived only in the last two decades, initially with the introduction of the SASW (Spectral Analysis of Surface Waves) method (Stokoe et al., 1994) and then with the spreading of multistation methods (Foti, 2000). The recent interest in surface waves methods in shallow geophysics is witnessed by workshops and sessions in International Conferences (e.g. Foti and Butcher, 2004) and by dedicated issues of International Journals (EAGE – Near Surface Geophysics, November 2004; Journal of Engineering and Environmental Geophysics, in preparation, expected June 2005).

Despite the different scales, the aforementioned applications rely on the same basic principles. They are founded on the geometrical dispersion, which makes Rayleigh waves velocity of propagation frequency dependent in vertically heterogeneous media. High frequency (short wavelength) Rayleigh waves propagate in shallow zones close to the free surface and are informative about their mechanical properties, whereas low frequency (long wavelength) components involve deeper layers. Surface wave methods use this property to characterize materials in a very wide range of scales, from microns to kilometers. The essential difference between applications is given by the frequency range of interest and on spatial sampling, as it will be detailed in next sections. The present paper is focused on applications for soil characterization for engineering application, where the depth of interest is typically confined to few tens of meters.

Surface wave tests are typically devoted to the determination of a small strain stiffness profile for the site under investigation, although other applications have been proposed (e.g. void detection; see Nasseri-Moghaddam et al., 2004). The basics of surface waves for site characterization have been addressed in another chapter in the present volume (Rix, 2005). The different steps of data acquisition and processing in surface wave test will be separately analyzed in the present contribution focusing on a restricted number of specific aspects which need to be taken in account when performing a test.

Figure 1 reports the standard procedure for surface wave tests, which can be subdivided in three main steps:

1) acquisition of experimental data
2) signal processing to obtain the experimental dispersion curve
3) inversion process to estimate site properties

It is very important to recognize that the above steps are strongly interconnected and their interaction must be adequately accounted for during the whole interpretation process.

Appealing alternatives for the interpretation of surface wave data are the inversion of field data based on full waveform simulations and the inversion of the Fourier frequency spectra of observed ground motion (Szelwis and Behle, 1987), but these strategies are rarely used because of their complexity. Moreover the experimental dispersion curve is informative about trends to be expected by the final solution, so that its visual inspection is important for the qualitative validation of the results. Indeed engineering judgment plays a certain role in test interpretation. Since the site and the acquisition are never "ideal", also the results of fully automated interpretation procedures must be carefully examined, with special attention to intermediate results during each step of the interpretation process. A deep knowledge of theoretical aspects and experience are hence essential.

Surface wave data can also be used to characterize the dissipative behaviour of soils. Indeed the spatial attenuation of surface waves is associated to the internal dissipation of energy. Using a procedure substantially analogous to the one outlined in Figure 1, it is possible to extract from field data the experimental attenuation curve, i.e. the coefficient of attenuation of surface wave as a function of frequency, and then use this piece of information in an inversion process aimed at estimating the damping ratio profile for the site.

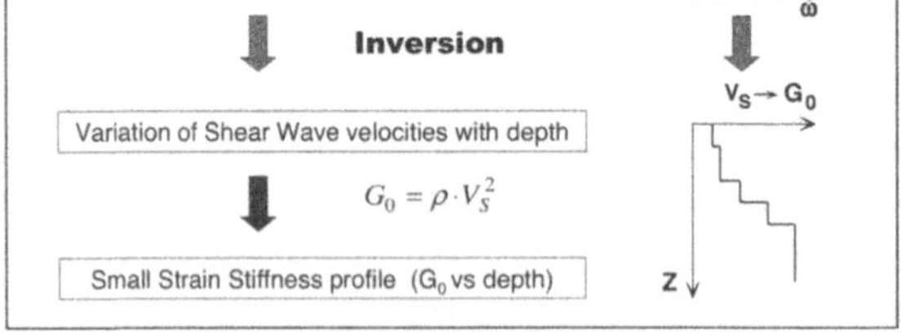


**Figure 1.** Flow chart of surface wave tests

## 2 Acquisition

The choice of equipment and testing configuration is closely linked to the scope of the test and to the technique to be used in the interpretation. In the following the attention will be focused mainly on surface wave test for shallow surveys in soils, considering the typical depth of interest of geotechnical engineering (<30m). Indeed, although the principles and the interpretation steps outlined in the introduction are essentially the same with respect to other applications related to characterization of pavement systems or deep explorations, the different scales require different frequency ranges and hence different equipment.

Proper design of the test is of great importance and some rules of thumbs can be useful in choosing the parameters for the acquisition (both in time and space). Typically the length of the survey line should be at least double than the desired depth for the investigation. It is always important to remember specific characteristics of Rayleigh waves, such as that they travel in a zone of about one wavelength from the free surface and hence they are informative about this zone. Moreover the link between frequency, wavelength and phase velocity makes the frequency range of interest closely linked to the materials to be investigated, so that for example deep penetration in soft soils requires low frequency components. Preliminary information related to the site under investigation can be very helpful at this stage.

Independently from subsequent processing and interpretation, signals are detected in time and space. The information related to wave propagation is embedded in the signals and can be extracted using a variety of techniques. A good practice is always to store the original unprocessed signals in time-offset domain, so that they can be reinterpreted with different processing techniques.

### 2.1 Testing Equipment

Provided that the general requirements in terms of frequency range and quality of the seismic signals are fulfilled, there is no restriction on the type of equipment. The frequency range of interest is a function of the desired depth of penetration and resolution for shallow layers. The basic testing equipment is composed of receivers (geophones or accelerometers) connected to an acquisition device, which digitizes and stores seismic signals. In active test, the wave is artificially generated using a seismic source with a trigger device connected to the acquisition device, in order to set the starting time of the acquisition.

**Receivers.** Type and specification of the receivers are usually chosen in relationship with the desired depth of penetration. Typically for soil testing the frequency range of interest is between few Hz and 100Hz and the measurements are taken using velocity transducers (geophones). Geophones measure particle velocities using a magnet, which is tied to the geophone case, and an electric coil system mounted on a suspended mass (Figure 2a). The relative movement between the case and the suspended mass, due to a movement at the base of the geophone and to inertia forces on the mass, produces a voltage in the electric coil proportional to the velocity of motion. The shunt resistor absorbs the unwanted vibrational energy in order to damp the mass movement and to stop its resonant vibration that would swamp the records. The damping introduced is such that the instrument response is flat enough over the frequency range of interest; usual values for the damping range between 0.6 and 0.7 (Figure 2b).

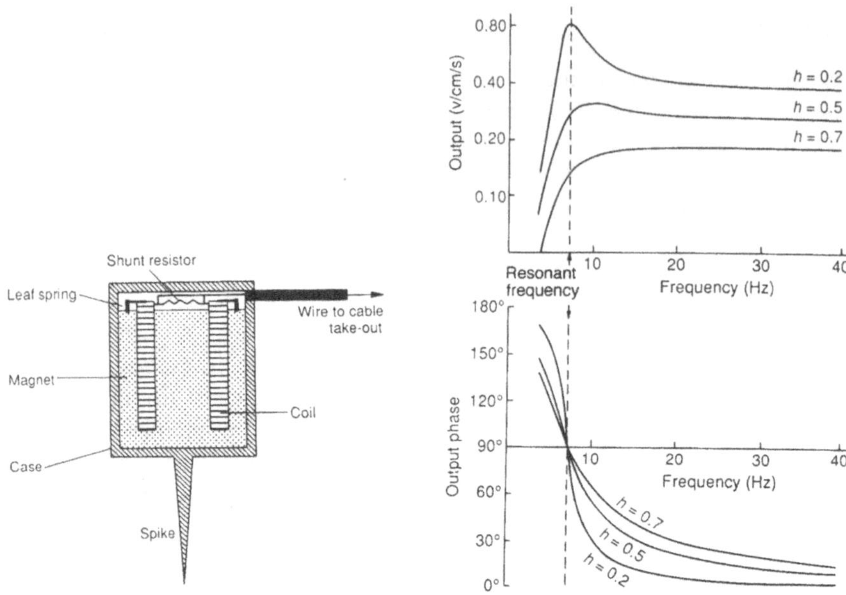


Figure 2. Geophones: (a) construction scheme; (b) instrument response (Doyle, 1995)

Geophones are classified on the basis of their resonance frequency or natural frequency, below which the geophone response to the external motion rapidly decreases to zero (Figure 2b). The natural period of vibration $T_0$ of the instrument can be evaluated with the classical formula:

$$T_0 = 2\pi\sqrt{\frac{m}{k}} \quad (1.)$$

where $k$ is the spring constant and $m$ the suspended mass. Low frequency geophones need a quite large mass, so that the sensitivity must be higher and they are largely more expensive than other ones. Moreover the weight of the suspended mass can easily damage the spring during setup operations.

In order to have information on deeper layers, low frequency geophones must be used. 1Hz or 2Hz geophones (also called seismometers) are sometimes used. Nevertheless it must be considered that they are quite expensive and very delicate to be used in the field. A good compromise is to use 4.5Hz geophones that are less expensive and do not need particular care to be used on site. Indicatively 2Hz geophones cost more or less four times 4.5Hz geophones, while 1Hz geophones cost twice 2Hz geophones. Usually the geophones used for seismic refraction or reflection have higher natural frequency (10Hz or more), so that they are not the best choice for surface wave tests unless the depth of interest is confined to few meters below the ground surface.

For pavement testing the frequency range of interest is considerably higher, indeed the materials to be tested are considerably faster (stiffer) and the depth of interest is restricted to few decimeters. The receivers used in these cases are typically accelerometers, which can reach operative frequencies in the kHz range.

**Recording devices.** The recording equipment can be of different types. Its main function is to digitalize and record analog electrical signals generated by the receivers.

Digital signal analyzers allow real-time processing of signals, so that signal quality assessment and preliminary interpretation can be performed directly on site. They have been extensively used for 2-station procedures (Stokoe et al., 1994). These systems are usually designed to be used in laboratory and hence they are not enough robust to be used on site and require considerable setup efforts in the field. Moreover the cost per channel is considerably high, so that they are less used for multistation tests.

Seismographs are designed for more classical geophysical tests (e.g. seismic refraction and reflection). They can be used for surface wave testing provided their cut-off frequency in the low frequency band does not involve useful frequencies for the scope of the test. Typically they are very robust, waterproof and resistant to dust because they are designed to be used in the field. One typical limitation is their acquisition software, which is designed for seismic refraction and reflection, so that the only possible preliminary check on site is the visual inspection of signals in time domain or at most in frequency domain. New generation seismographs are composed by scalable acquisition blocks to be used in connection with field computers, hence allowing preliminary processing of data on site.

All the above instruments are quite expensive. A possible option to limit the budget required for SW tests is the use of acquisition boards connected to PCs or laptops. High quality boards for acquisition on multiple channels are available on the market, but they are typically to be used with desktop computers or racks for laboratory testing. An example of how they can be used to built a portable system for on site testing is reported by Foti and Fahey (2003). A great advantage is that the acquisition software can be realized on purpose and hence accounting for surface wave testing requirements and allowing on site processing of data. Simpler options are PCMCIA cards for laptop computers, but they are limited in the number of channels, so that specific acquisition procedures must be used to process the data with multistation techniques (Ryden et al., 2002).

**Sources.** In active measurements the wave is generated using a seismic source. Several types of sources can be used ranging from ordinary hammers to expensive controlled sources. Impact sources are often preferred for economic reasons. Heavy sources are needed to input energy in the low frequency range, while lighter sources can be used for shallow characterization. As an example Figure 4 reports a comparison of energy spectra associated to different impact sources.

More sophisticated controlled sources can be useful for advanced applications. Indeed they can be driven by a signal generator allowing the generation of monochromatic or user-specific input such as sweep-sine used for deep explorations. Using controlled sources it is possible to get high quality signals with high signal to noise ratios. Controlled sources range from portable electromechanical shakers to large track-mounted vibroseis. Figure 3 reports plots of vertical versus horizontal particle velocities obtained with a sinusoidal input to an electromechanical shaker, showing the typical ellipse-shaped particle motion of surface waves. Note that the ellipse has not necessarily its principal axis vertical and the motion is not necessarily prograde on the free surface has it would be for a Rayleigh wave in a linear elastic homogeneous medium. Similar results are obtained in numerical simulations of Rayleigh waves for vertically heterogenous linear visco-elastic media (Lai, 2005).

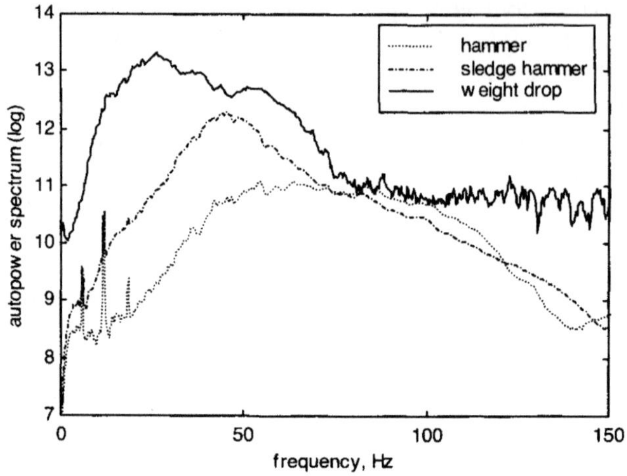


**Figure 4.** Comparison of energy spectra for different sources (receiver offset 3m) (Foti, 2000)

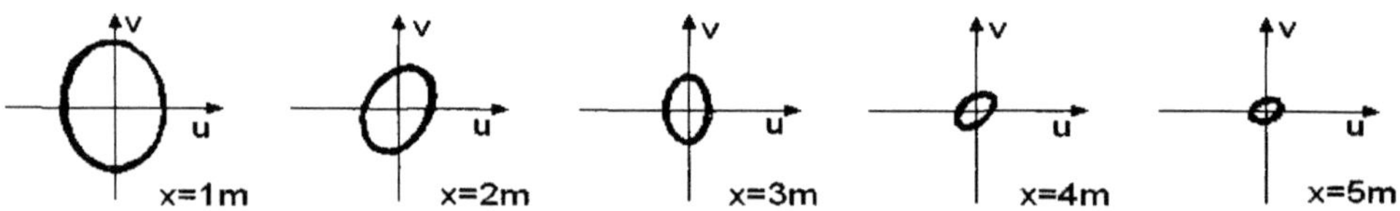


**Figure 3.** Experimental evidence of elliptical particle motion associated to surface wave propagation obtained with an electromechanical shaker working at 50Hz

## 2.2 Testing Configuration

The choice of sampling parameters both in time and space has significant effects. Sampling parameters in time are set on the acquisition device; sampling parameters in space are determined by the layout geometry and are typically subjected to a number of restrictions due to available receivers and testing space.

**Layout geometry.** The typical arrangement for receivers in surface wave tests is to have multiple receivers along a straight line and the seismic source acting on one end of the array. The use of constant receiver spacing simplifies the processing and can be considered the standard of practice, nevertheless considering the trade-off between depth of penetration and resolution variable receiver spacing acquisition setup have also been proposed (Rix, 2005). Indeed the number of receivers is fixed in relationship to the available equipment so that the total length of the array is determined by receiver spacing. Wide spacing allow the detection of long wavelength and hence a

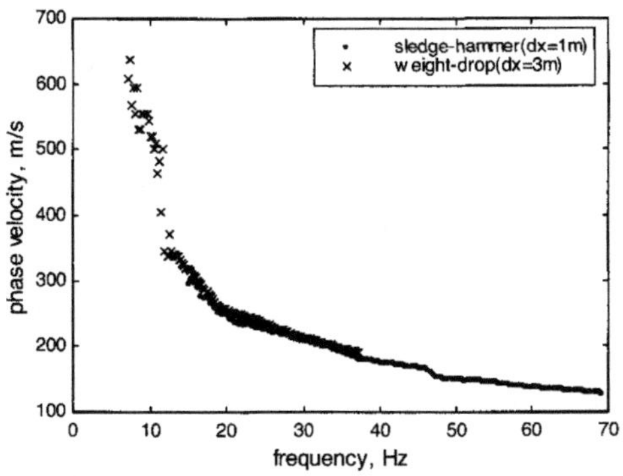


**Figure 5.** Branches of the experimental dispersion curve (Foti, 2000)

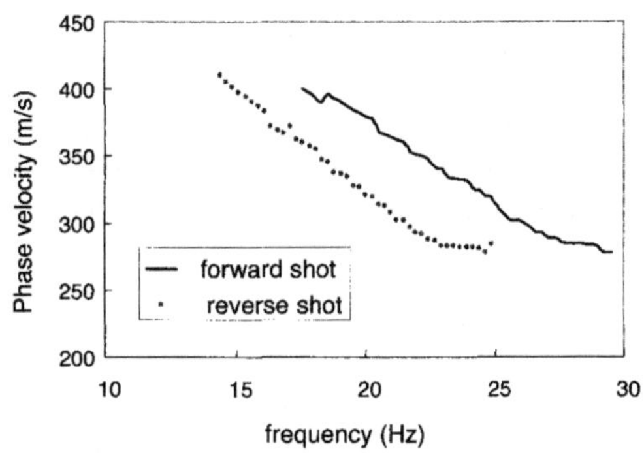


**Figure 6.** Evidence of lateral variations in experimental dispersion curves

characterization of deeper layers, but the attenuation of high frequency components causes loss of information for the high frequency range and hence of resolution for shallow layers.

Several strategies can be adopted to overcome these limitations. For example using 24 channels a couple of testing setup (e.g. spacing 1m and 3m) is likely to give information over a reasonable frequency range covering the need of soil testing for engineering purposes (Figure 5). Variable spacing with shorter distances close to the source can be useful to get a wide spectrum of information from a single testing array, but they require processing procedures more complicate (Rix, 2005; Strobbia and Foti, 2004). One aspect to be considered is that using constant spacing it is easier to repeat the test with forward and reverse shot position, by moving the seismic source on the opposite side of the receiver array. This procedure has to be considered as part of good common practice in surface wave test, indeed the comparison of experimental dispersion curve obtained for the same receiver array with forward and reverse shots can give clear indication about the presence of lateral variations (Figure 6), which violates the basic assumption of horizontally stratified medium on which the entire interpretation process is based (see also Section 3.2).

Passive tests, in which a seismic source is not used, require 2D arrangements of receivers on the ground surface (typical choices are circles, triangles or L-shaped arrays). Indeed in this case the location of the source position is not specified and special processing techniques are required in order to get the correct estimate of phase velocity (Zwicki, 1999; Okada, 2003; Rix, 2005). Sometimes passive surface wave tests are performed using linear arrays assuming that the energy is arriving simultaneously from all the possible directions (Louie, 2001). Data processing in this case is much more cumbersome and aleatory; indeed noise from passive sources not in line with the receiver array will give higher apparent velocities than the actual Rayleigh phase velocity for the site, so that it is not possible an automatic pick of the absolute maxima in *fk* or *fp* panels and the operator needs to select manually relative peaks.

**Sampling parameters.** The choice of sampling parameters is related to the frequency range of interest (typically 5-100Hz for engineering applications). A sampling ratio of 2ms (corresponding to a Nyquist frequency of 250Hz) is typically sufficient for surface wave testing because resolution in time domain is not important. The total length in time of the signal is related to the frequency interval of the transformed signal in the frequency domain, hence it is important to have at least an acquisition window of at least 1s (frequency resolution 1Hz). In any case it is of primary importance that the whole event is recorded. A pre-trigger time is typically used in order to have signals centered in the time window, allowing the use of hanning or hamming windows to avoid leakage (Santamarina & Fratta, 1998) without altering energy contents in the signals.

## 2.3 Practical Issues

Preliminary analysis on site can be very useful to assess the need for more data or for a change in the testing configuration. Anyway good practice is always to store original signals in the time-space domain also if part of the processing is performed directly on site. This allows subsequent interpretation and the use of alternative approaches on the same dataset.

A great concern when acquiring seismic data is ambient noise that can mask the actual signals and causes difficulties in the interpretation especially in time domain. In this respect surface wave tests are less sensitive to external noise than other techniques because their interpretation in totally performed in the frequency domain. In the Author's experience surface wave tests are successful also in very noisy environments such as urbanized environments or industrial facilities.

Records of ambient noise at the site can help during the processing of field data, especially to identify less reliable portions of the experimental dispersion curve. Figure 8a reports the spectral analysis of the noise recorded at a site clearly showing a peak around 12 Hz, probably caused by machinery or other facilities close by. The uncertainty of the experimental dispersion curve (Figure 8b) expressed by the coefficient of variation of phase velocity at each frequency (Figure 8b) obtained with multiple repetition of the tests, clearly shows the greater uncertainty caused by ambient noise. Nevertheless it must be noted that few data points in the dispersion curve have a limited influence on the inversion process and hence on the final result of the test.

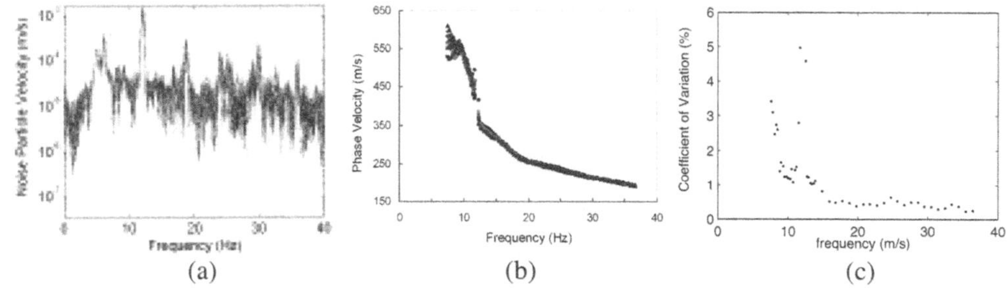


**Figure 7.** Influence of ambient noise on the experimental dispersion curve (Lai et al., 2004)

## 3 Processing

Several approaches can be used to process field data in order to get the experimental dispersion curve. The phase velocity can be evaluated on the basis of the phase difference between a couple of receivers or from the simultaneous processing of several traces.

For further details on the two station method, often referred as SASW method in the literature (Stokoe et al., 1994), the reader is referred to the article by Rix (2005) in the present volume. It is anyway worthwhile to mention that the evaluation of phase velocity from the phase difference at two receivers (i.e. using the cross-power spectrum) is equivalent to any of the multistation procedure that will be presented in the following, if they are applied on a couple of receivers. Hence the two station method is simply a particular case of multistation methods. Even if the equipment is less expensive, testing procedures for the two-station method are much more involved and processing is considerably more complex, moreover the results are very sensible to errors at a single receiver. As a consequence multistation methods are nowadays the standard (Foti & Butcher, 2004).

### 3.1 Transform-Based Methods

These methods are undoubtedly the most widely used. They are based on the transformation of experimental data from the time-offset domain to different domains, where the dispersion curve is easily extracted from spectral maxima. For example applying a double Fourier transform to field data the dispersion curve can be identified as the maxima in the frequency-wavenumber panel (Figure 8). Other methods use different transforms obtaining similar results, e.g. the $\omega$-$p$ (frequency-slowness) panel obtained with the slant-slack transform (McMechan & Yedlin, 1981) or the MASW method (Park et al., 1999). The formal equivalence of these approaches can be proved considering the mathematical properties of the different transforms (Santamarina & Fratta, 1998) and there is practically no difference in the obtained dispersion curves (Foti, 2000).

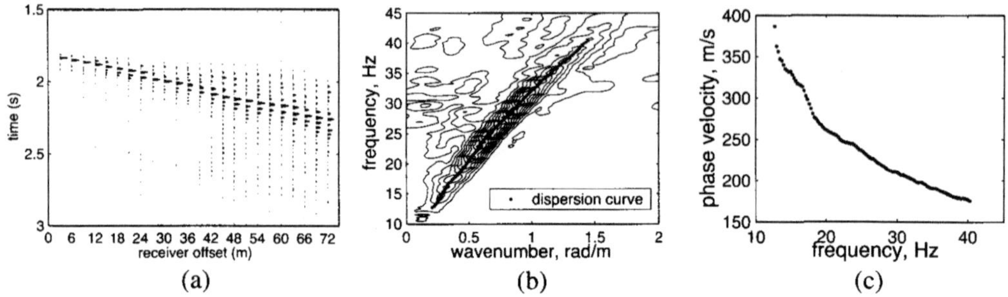


**Figure 8.** fk analysis: (a) field data; (b) fk panel; (c) dispersion curve

In theory transform-based methods allow the identification of several distinct Rayleigh modes. Tselentis and Delis (1998) showed that the *fk* spectrum for surface waves in layered media can be written as the following sum of modal contributions:

$$F(f,k)=\sum_{m} S_m(f)\cdot\left[\sum_{n=1}^{N} e^{-\alpha_m(f)\cdot x_n}\cdot e^{i(k-k_m(f))\cdot x_n}\right] \tag{2.}$$

where $S_m$ is a source function, $x_n$ is the distance from the source of the $n^{th}$ receiver, $\alpha_m$ and $k_m$ are respectively attenuation and wavenumber for the $m^{th}$ mode. Observing the quantity in square bracket it is evident that, if material attenuation is neglected, the maxima of the energy spectrum are obtained for $k = k_m(f)$. Furthermore it can be shown that also if the above differentiation is conducted without neglecting the material attenuation the conclusion is the same, i.e. the accuracy is not conditioned by material attenuation (Tselentis and Delis 1998).

Once the modal wavenumbers have been estimated for each frequency, they can be used to evaluate the dispersion curve recalling that phase velocity is given by the ratio between frequency and wavenumber.

Using a very large number of signals (256) Gabriels et al. (1987) were able to identify six experimental Rayleigh modes for a site and then they used these modes for the inversion process. The possibility of using modal dispersion curves is a great advantage with respect to methods giving only a single dispersion curve (as the two-station method) because having more information means a better constrained inversion. Nevertheless it has to be considered that in standard practice the number of receivers for engineering applications is typically small and reduced spatial sampling strongly affects the resolution of surface wave test. Receiver spacing influences aliasing in the wavenumber domain, so that if high frequency components are to be sought spacing must be small. On the other side, the total length of the receiver array influences the resolution in the wavenumber domain. Obviously using a finite number of receivers this aspect generates a trade-off similar to the one existing between resolution in time and in frequency. Indeed the resolution in the wavenumber domain is inversely proportional to the total length of the acquisition array. Using a simple 2D Fourier Transform on the original dataset to obtain the experimental fk panel would lead to a spatial resolution not sufficient to obtain a reliable estimate of the dispersion curve. The use of zero padding or the use of advanced spectral analysis techniques such as beamforming

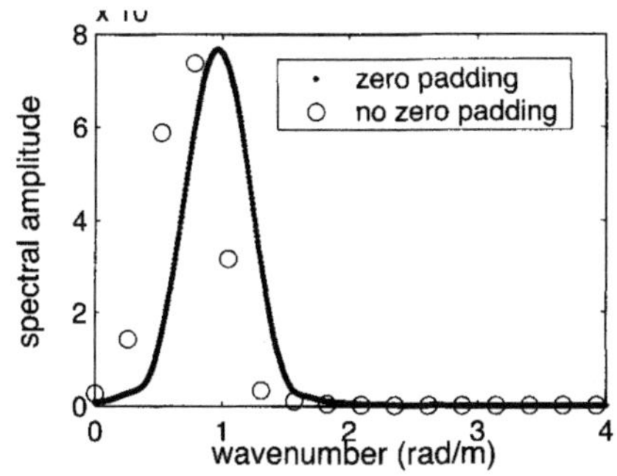


**Figure 10.** Effect of zero padding on resolution in the wavenumber domain: slice of the fk panel for a given frequency.

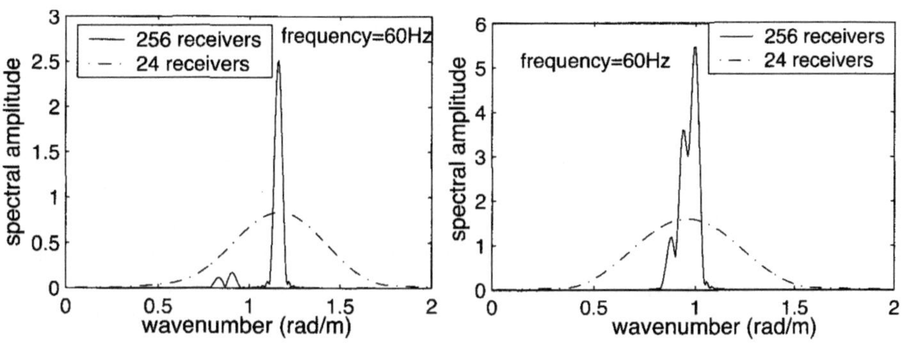


**Figure 9.** Influence of the effective wavenumber resolution on the dispersion curve (a) dominant fundamental mode (b) relevant higher modes

(Zwicki, 1999; Rix, 2005) apparently makes it possible to locate the correct position of the maxima in the fk panel (**Figure 10**).

Unluckily these strategies do not improve the real resolution, hence it is not possible to separate modal contributions when more than a single mode plays a relevant role in the propagation (Foti et al., 2002). This aspect is exemplified in **Figure 9** where slices of the fk spectrum for a given frequency are reported for two different synthetic dataset. If a large number of receivers is used to get evaluate the fk spectrum, the resolution is very high and the energy peaks are well defined, but if the number of receivers is low the resolution is very poor. With poor resolution it is only possible to locale a single peak in the fk panel, which in principle is not associated to a single mode but to several superposed modes. The concept of apparent phase velocity has been introduced to denote the velocity of propagation corresponding to this single peak representing several modes (Tokimatsu, 1995). In the example of **Figure 9**a the fundamental mode is the dominant mode in the propagation, meaning that almost all energy is associated to this mode. In this situation the apparent phase velocity is the phase velocity associated to the fundamental mode and the inversion process can be simplified inverting the apparent dispersion curve as a fundamental mode. This situation is usual in soil deposit with stiffness always increasing with depth with no

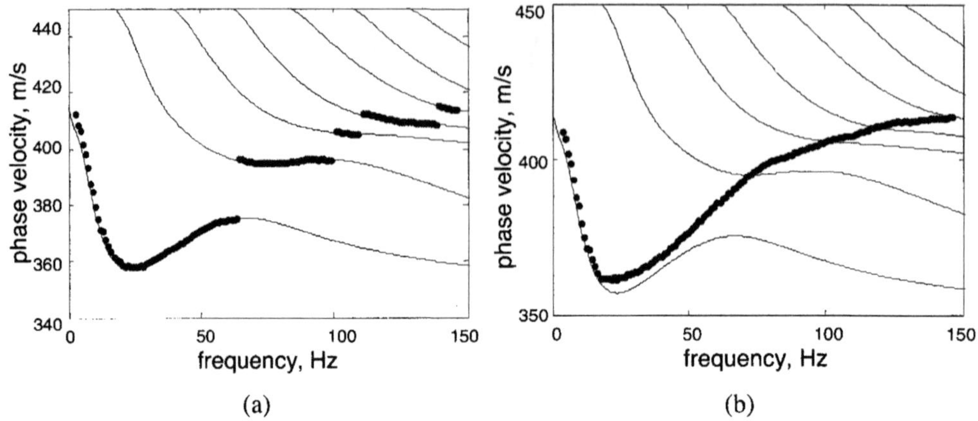


**Figure 11.** Influence of wavenumber resolution on the dispersion curve from synthetic data: (a) 256 receivers; (b) 24 receivers (Foti, 2000)

marked impedance jumps between different layers. On the contrary in the example of **Figure 9**b the fundamental mode is still the one carrying more energy, but it is not anymore dominant, meaning that higher mode plays a relevant role in the propagation. If few receivers are used a single peak will be observed and a single value of phase velocity will be obtained. This value is not necessarily the phase velocity of one of the modes involved in the propagation, but it is rather a sort of average value, often referred as apparent phase velocity or effective phase velocity (Tokimatsu, 1995). In this case it is not anymore possible to use inversion processes based on the fundamental mode or on modal dispersion, but it is necessary to use algorithm that can account for mode superposition effects (Lai, 1998) and for the actual testing configuration (O'Neill, 2004). This situation is usual when strong impedance contrasts are present in the soil profile or in inversely dispersive profiles, i.e. profiles in which soft layers underlie stiff layers.

The dispersion curves obtained with fk analysis on the synthetic dataset used for **Figure 9**b are reported in **Figure 11**. As explained above if a sufficiently high number of receivers is used, it will be possible to obtain the modal dispersion curves (**Figure 11**a), whereas with the number of receivers used in standard practice a single apparent dispersion curve will be obtained (**Figure 11**b).

As mentioned above, the apparent dispersion curve is dependent on the spatial array so that if higher modes are relevant for a given site the inversion process will be cumbersome. On the other side if the fundamental mode is dominant, the inversion process can be noticeably simplified. Unluckily, it is not always clear from the simple inspection of the experimental dispersion curve if higher modes are involved. In this respect the MOPA method, presented in the next section, can be a convenient tool to highlight the influence of higher modes.

## 3.2 Regression-Based Methods

Regression methods are founded on theoretical formulas of surface wave propagation in layered (visco-)elastic media. The experimental variation of particle motion with distance from the source can be used in connection with the correspondent mathematical equation in order to infer the relevant parameters of the propagation. Although these methods suffer some intrinsic limitations, they are worth to be mentioned because they can give additional insight into the phenomenon of surface wave propagation.

**MOPA (Multi-offset phase analysis).** The motion associated to surface wave propagation in a layered visco-elastic medium can be written as the sum of modal contributions (Aki and Richards, 1980):

$$s(\omega, x) = \sum_m A_m(\omega, x) e^{i(\omega t - k_m(\omega) x)} \tag{3.}$$

where $x$ it the source to receiver distance, $\omega$ is the circular frequency, $k_m(\omega)$ and $A_m(\omega, x)$ are respectively the wavenumber and the amplitude associated to the $m^{th}$ mode. The amplitude can be expressed as a function of the source and of the path so that Eq. 3 can be rearranged separating the space dependent and frequency dependent components of the amplitude:

$$s(\omega, x) = \sum_m I(\omega) R_m(\omega) \frac{e^{-\alpha_m(\omega) x}}{\sqrt{x}} e^{i(\omega t - k_m(\omega) x + \varphi_0(\omega))} \tag{4.}$$

where $I(\omega)$ and $\varphi_0(\omega)$ are respectively the amplitude and the phase of the source, while $R_m(\omega)$ and $\alpha_m$ are the site response and the intrinsic attenuation for the $m^{th}$ mode. The amplitude has a geometric attenuation inversely proportional to the square root of the source to receiver distance because the propagation is characterized by a cylindrical wavefront (Richart et al., 1970).

If a single mode of propagation is dominant, the complex displacement of Eq. 4 can be decomposed in modal amplitude and modal phase, which can be used for two separate regression processes in order to infer modal attenuation and modal phase velocity. This procedure is strictly valid only if a single mode is dominating the actual propagation phenomenon.

The theoretical modal phase is a straight line in space for a given frequency:

$$\varphi(\omega, x) = -k(\omega) \cdot x + \varphi_0(\omega) \tag{5.}$$

Eq. 5 can be used in a regression process to estimate the wavenumber under the assumption that a single mode is dominant in the propagation. This hypothesis is reasonable for many practical cases. The influence of higher modes will be discussed later on.

The procedure can be outlined as follows. The experimental phase as function of the offset and of the frequency can be obtained with a Fourier Transform of multi-offset data, transforming the seismogram from time-offset domain to frequency-offset domain and considering only the phase of the transformed gather. Then for any given frequency in the range of interest, the best fitting line is sought and its slope represents the dominant wavenumber for that frequency, from which the phase velocity can be determined. The regression process is then repeated for the other frequencies. This analysis can be straightforwardly used to analyse data from non evenly spaced receiver arrays (see section 2.2).

The above procedure can be considered an extension of the two-station method to the use of multiple receivers. From this point of view it would be an improvement in terms of robustness of the estimate of the experimental dispersion curve.

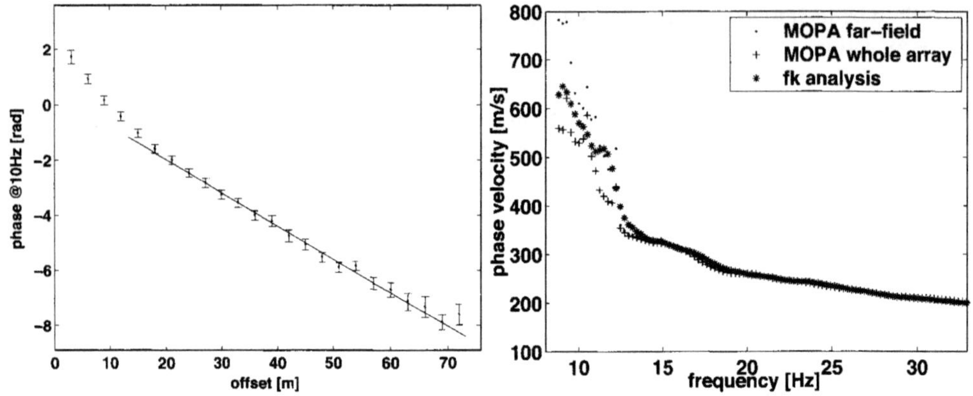


**Figure 12.** Near field effects: (a) experimental evidence; (b) influence on the experimental dispersion curve (Strobbia and Foti, 2004)

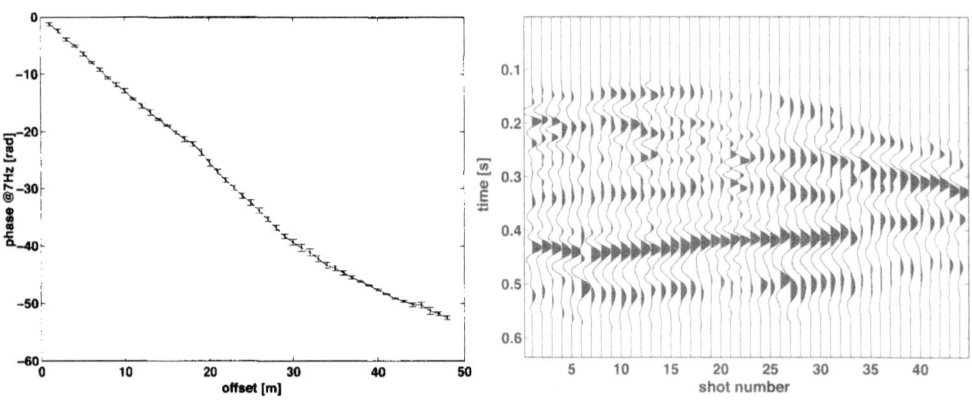


**Figure 13.** Lateral variations: (a) experimental evidence in surface wave tests; (b) SH seismic reflection at the same site (Strobbia and Foti, 2004)

Compared to transform based method the procedure is undoubtedly more involved and has the disadvantage that it is not possible at all to obtain separate modes. The interest in this approach is related to the possibility of detecting anomalies. Indeed any deviation from a straight line of the phase-offset data is associated to a violation of underlying assumption. Phillips et al. (2003) proposed the use of this approach in order to detect the presence of voids. Strobbia and Foti (2004) developed a statistical test to automatically recognise deviations from the basic assumptions of surface wave testing. The MOPA statistical test is passed if the experimental behaviour is close enough to what is expected from theory accounting also for data uncertainty. Departures from the

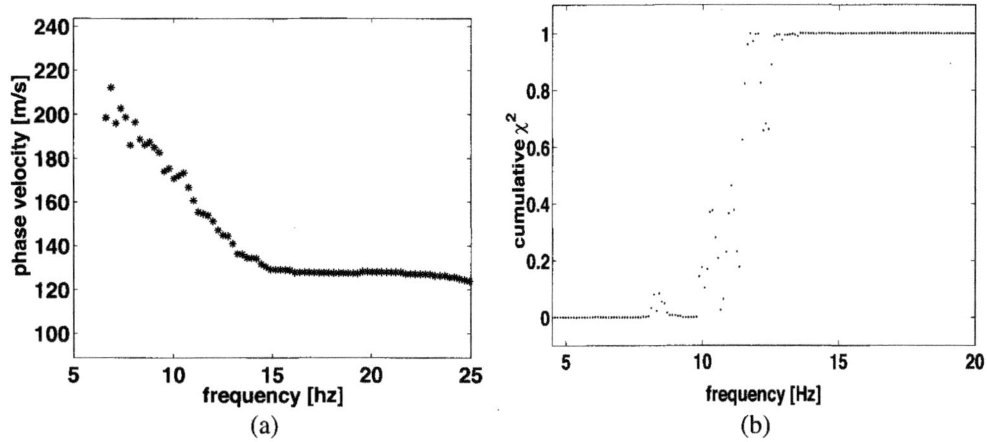


**Figure 14.** Hidden higher mode influence: (a) experimental dispersion curve from fk analysis; (b) MOPA statistical test (Strobbia and Foti, 2004)

expected linearity of the phase-offset data can be caused by the following: near-field effects, lateral variations, modal superposition.

Near field effects are effects of proximity of the source that tends to modify the experimental Rayleigh dispersion curve. They are mainly due to body waves generate by the source and to the plane wavefront approximation used in the analysis. Their influence can be recognised in multistation data, if part of the array is outside the near field. The apparent phase velocity is indeed different in the near field and in the far field zones, leading to different slopes in the phase-offset plot (Figure 12a). Filtering out the traces in the vicinity of the source it is possible to eliminate near field effects from the dispersion curve. For the example of Figure 12a, if the regression is performed filtering the data, so that the closest 4 receivers are not considered in the analysis, the behaviour is characterised by a straight line and the MOPA statistical test is passed. Figure 12b reports for the results obtained with MOPA procedure, with or without filtering the data in space. The comparison shows that near field effects tend to reduce the observed phase velocity in the low frequency range, leading to an underestimation of soil properties especially for deep layers. This result is in agreement with numerical simulations reported in the literature (Sanchez-Salinero, 1987).

Similarly it is also possible to recognise the presence of lateral variations. Redundancy of data allows discriminating between the two: for instance comparing data acquired with shot in different positions, since the near field depends on the shot position while the lateral variations do not. If lateral variations are identified, it is not possible to invert the data with a 1D model. As an example in Figure 13 experimental data from a site characterised by strong lateral variation are reported. In this case phase-offset plots (e.g. Figure 13a) show marked deviations from linearity with several changes of average slope. Lateral variations have been subsequently evidenced by other geophysical tests performed along the same alignment of surface wave tests. For example Figure 13b shows the seismogram of a SH reflection test. It is worthwhile to recall that if lateral variations are not properly recognised during the interpretation, the final results can be quite misleading. Procedures for 2D interpretations of surface wave data have been proposed in the literature to deal with lateral variations, but typically the 2D solution is obtained as a sequence of 1D inversion processes rather than being based on a true 2D model (Xia et al., 2002; Hayashi and Hikima, 2003).

Finally, the statistical test can be useful to discriminate situations in which apparently the fundamental mode is dominating, while in reality the influence of higher modes is relevant, at least in a frequency range. The example reported in Figure 14 is related to one of these cases, indeed the experimental curve obtained with fk analysis (Figure 14a) apparently can be interpreted as a fundamental mode dispersion curve. The application of the MOPA statistical test shows that this assumption is correct only for frequencies up to about 12Hz. Above this value phase-offset data show marked deviations from linearity, indeed the statistical test (Figure 14b) is not passed (values higher than the conventional threshold 0.95). The inversion of the experimental curve as a fundamental mode would lead in this case to an overestimation of soil properties for shallow layers.

**Transfer function method.** The transfer function or frequency response of a system is defined as the ratio of the output and the input in frequency domain (Santamarina & Fratta, 1998). It can be used as a reference quantity to characterize the system (in this case the soil deposit).

In its original formulation, the method was implemented with a controlled source, so that the transfer function could be obtained from the ratio of the response at the receivers over the measured acceleration of the source (Rix et al., 2001). Later the method has been modified (Foti, 2003) defining the transfer function with respect to a reference receiver rather than to the actual source, using the concept of deconvolution (Dziewonski and Hales, 1972). The modified transfer function method has some relevant advantages: the influence of the source-soil coupling is avoided and the method can be used with ordinary seismic data obtained using impact sources.

Considering modal superposition, the vertical displacement $U_z(r,\omega)$ induced by a harmonic source $R_z \cdot e^{i\omega t}$ acting on the ground surface can be expressed as:

$$\mathrm{U}_z(r,\omega) = R_z \cdot \mathrm{G}(r,\omega) \cdot e^{i[\omega t - \Psi(r,\omega)]} \tag{6.}$$

where $\Psi(r,\omega)$ is the complex-valued phase angle and $\mathrm{G}(r,\omega)$ is the geometrical spread function (Lai, 1998). Hence, if the response of the receiver placed at $r=r_1$ is used as the reference trace, the transfer function can be written as:

$$\tilde{\mathrm{F}}(r,\omega) = \frac{\mathrm{U}_z(r,\omega)}{\mathrm{U}_z(r_1,\omega)} = \frac{\mathrm{G}(r,\omega) \cdot e^{-i\Psi(r,\omega)}}{\mathrm{G}(r_1,\omega) \cdot e^{-i\Psi(r_1,\omega)}} \tag{7.}$$

Assuming $\Psi(r,\omega) = K(\omega) \cdot r$ the implicit dependence of the complex-valued phase angle on the source-to-receiver distance is eliminated and Eq. 7 becomes:

$$\tilde{\mathrm{F}}(r,\omega) = \frac{\mathrm{G}(r,\omega)}{\mathrm{G}(r_1,\omega)} \cdot e^{-i \cdot K(\omega) \cdot (r - r_1)} \tag{8.}$$

where $K(\omega) = \left[ \frac{\omega}{V_R(\omega)} + i\alpha_R(\omega) \right]$ is a complex wavenumber with $V_R(\omega)$ phase velocity and $\alpha_R(\omega)$ attenuation coefficient of Rayleigh waves. The assumption $\Psi(r,\omega) = K(\omega) \cdot r$ is equivalent to consider the phase angle $\Psi(r,\omega)$ to be the result of a single mode of propagation (Rix et al., 2001).

Eq. 8 is used in a non-linear regression analysis to estimate the complex-valued wave number $K(\omega)$ from the experimental values of the complex-valued transfer function. If higher mode contributions are relevant for the site under investigation, the result will be an apparent quantity.

Figure 15 reports an example of results obtained with the transfer function method. The original data in space-time domain are transferred to the space-frequency domain with a FFT, then the experimental transfer function is evaluated as the ratio of the response of a given receiver over the response of the closest receiver to the source. For any given frequency the optimal complex-valued wavenumber $K(\omega)$ is sought with a regression process (Figure 15a). Repeating the process over the frequency range of interest it is then possible to reconstruct the dispersion and attenuation curves (Figure 15b), which can subsequently be used for a simultaneous inversion process aimed at estimating both the small stiffness and damping ratio profiles for the site (Lai et al., 2002; Foti, 2003).

The Transfer Function procedure has the following advantages if compared to the regression of phase-offset data (MOPA):

- It is implicitly a coupled regression of phase and amplitude, hence it is more robust from a mathematical point of view

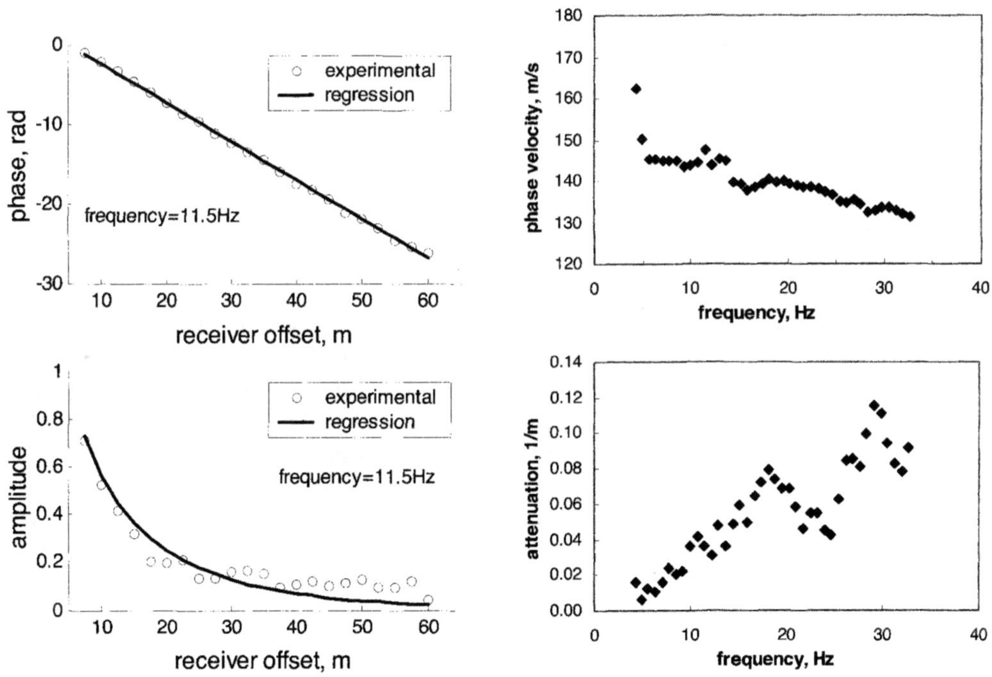


(a) (b)

**Figure 15.** Transfer Function Method: (a) regression of experimental data (b) experimental dispersion and attenuation curves (Foti, 2003).

- It gives the possibility of estimating also the experimental attenuation curve that can be used to estimate dissipative properties of soils
- It is easier to implement because it uses directly the complex values obtained with the FFT, with no need for phase unwrapping, which is a delicate step in MOPA

## 4 Inversion

The last step of surface wave analysis is the inversion process needed to estimate soil parameters on the basis of the experimental dispersion curve. The usual practice is to use local search algorithms starting from a tentative solution (Herrmann, 1994; Rix and Lai, 2000; O'Neill, 2004), nevertheless global search methods have been applied and they seem very promising for future applications (Williams and Gucunski, 1995; Hunaidi, 1998; Beaty et al., 2002). The trial and error method (Menke, 1989), although very rough, is still sometimes used especially when higher mode contributions make the interpretation more cumbersome. A detailed discussion of inversion methods is outside the scope of the present article; in the following some specific topics of surface wave inversion will be examined.

A peculiar aspect is related to the way in which the forward problem is solved. The easiest strategy is to use a fundamental mode approach, assuming that the experimental dispersion curve is representative of the fundamental mode at the site. This approach is substantially correct only if the fundamental Rayleigh mode is effectively dominating the propagation phenomenon. Nevertheless it must be considered that in some situations higher modes plays a relevant role and the experimental dispersion curve is given by an apparent phase velocity given by modal superposition and influenced by the testing setup (see section 3). Typical examples are inversely dispersive profiles (i.e. stiff shallow layers) and profile with abrupt variations of soil impedance. In these situations the forward problem solution needs to account for mode superposition and for the actual configuration of the receivers. An example of inversion algorithm based on a full waveform simulation is reported by O'Neill (2004). Mode superposition can be simulated directly in frequency domain applying a spatial Fourier transform on the particle displacements associated to harmonic sources, hence saving computational time (Strobbia, 2002).

Typically, the model assumed for the interpretation of SW tests is a stack of homogeneous linear elastic layers over a halfspace. For a model with $n$ layers, including the halfspace, model parameters are $4n$-1 (two elastic parameters, soil density and thickness for each layer, except for the halfspace, which thickness is not defined). Considering that inversion problems are ill-posed from a mathematical point of view, it is necessary to adopt adequate strategies in order to reduce the number of unknowns and possibly to introduce constrains in order to limit equivalence problems (non-uniqueness of the solution). The usual strategy consist in assuming a priori values of density and Poisson ratio, taking advantage of the reduced sensitivity of the dispersion curve with respect to these parameters (Nazarian, 1984).

A priori information or constrains given by other tests performed at the site are of primary importance and will be discussed in the following.

## 4.1 Borehole Logs

Often in order to further reduce the number of unknowns, a priori layer thicknesses are used in the inversion (e.g. Rix and Lai, 2000). The corresponding values can be arbitrary chosen, typically with thicknesses increasing with depth, in order to comply with the decrease of resolution with depth, which is an implicit limitation of SW tests. Obviously this reduces the ability of the test in locating layer interfaces.

Stratigraphic information from borehole logs can help in the choice of more appropriate layer thickness greatly improving the reliability of the final solution.

## 4.2 P-wave Refraction

Testing configuration for multistation surface wave test and seismic P-wave refraction are basically the same, so that with a minimum effort it is possible to obtain addition information, which can be very useful during the interpretation (Foti et al., 2003). Two typical situations in which refraction data are particularly valuable for the inversion of surface wave data are the presence of bedrock and the presence of shallow water table. Moreover P-wave refraction can be useful to recognize the presence of dipping layers or other lateral variations that cannot be resolved with the 1D model used for the interpretation of surface wave data.

In the example reported in Figure 16, the bedrock position evaluated with seismic refraction has been used as a constrain in the inversion of surface wave data and the final result is in good agreement with subsequent borehole logs at the same site.

The role that water table position plays in surface wave inversion is a more subtle one. Indeed although water table has not a great influence on the values of shear wave velocity that are the final objective of SW tests, it has a great relevance for the velocity of P-waves and hence on the values of Poisson ratio. Also if the sensitivity of the dispersion curve to Poisson ratio is reduced

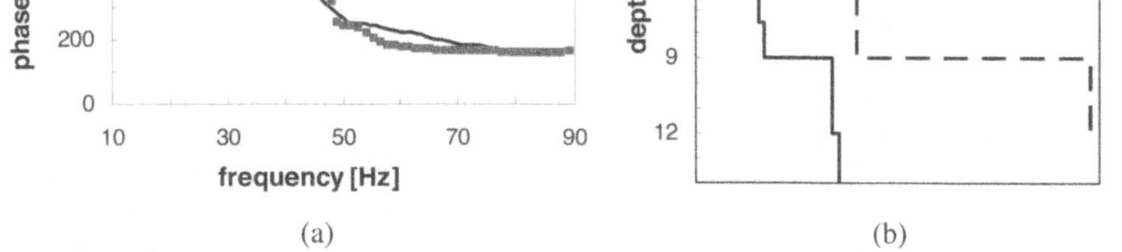


**Figure 16.** Inversion of SW data with constrains from P-wave refraction (a) dispersion curves; (b) seismic profiles (Foti et al., 2003)

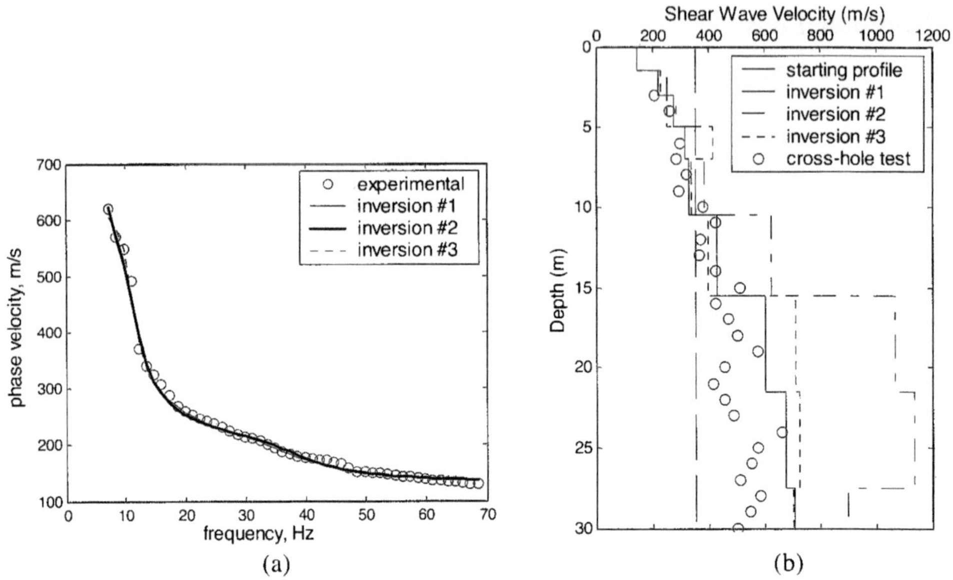


**Figure 17.** Influence of water table position in SW inversion: (a) dispersion curves; (b) shear wave velocity profile (Foti and Strobbia, 2002)

with respect to the sensitivity to shear stiffness and layer thicknesses, the role played by this parameter during the inversion is not always negligible (Foti and Strobbia, 2002). The transition from dry to wet soil, which is sharp for coarse grained soils, is associated to an abrupt change of Poisson ratio and soil density. If this transition is not properly recognized it can lead to misleading results, as shown in the example reported in Figure 17. The inversion process has been repeated with different assumptions for the water table position, associating to the dry soil a Poisson ratio of 0.2 and density $1.8 Mg/cm^3$ and to the wet soil a Poisson ratio of 0.49 and density $2.1 Mg/cm^3$. The obtained shear wave velocity profiles (Figure 17b) are equivalent with respect to the propagation of surface waves, in the sense that the correspondent numerical dispersion curves are practical identical and show a good agreement with the experimental dispersion curve (Figure 17a). Inversion #1 corresponds with the position of the water table obtained from P-wave refraction (3m below the ground surface) and it is the one that gives the best result, if compared to the results of a Cross-Hole test at the same site. Inversion #2 assumes no water table (dry soil) and gives results totally inconsistent with the Cross-Hole test. Inversion #3 assumes a biased position for the water table (7m below the ground surface).

The above example shows the importance of a-priori information on the water table position when interpreting surface wave data. Although this information can be obtained in various ways (electrical surveys, borehole or piezometer data), P-wave refraction appears a convenient option. When no information is available at all it is necessary to account for the uncertainty related to the unknown water table position. A strategy in this respect can be given by some trials with reasonable guesses of the water table position, then considering the final results of the different inversions in the view of engineering judgment.

## 4.3 Vertical Electrical Sounding

Synergies of surface wave tests with other geophysical tests can be useful to improve the reliability of the final result and to obtain a more consistent soil model. Hayashi et al. (2004) report a joint analysis of surface wave and microgravity data aimed at detect the presence of underground voids. Another very appealing possibility is given by the joint inversion of SW and VES (Vertical Electrical Sounding) data (Hering et al. 1995, Misiek et al. 1997, Comina et al., 2002). Indeed the interpretation of these tests present strong similarities. SEV are based on the inversion of field data passing through an intermediate quantity, the apparent resistivity curve, which can be considered the analogous of the dispersion curve for surface wave data. Moreover the inversion is based on a layered model similar to the one used for surface wave tests but in this case the unknown parameters are the electrical resistivities of the layers.

Assuming that layer interfaces are the same for the seismic and the electrical model, it is possible to use a joint inversion strategy, simultaneously inverting the experimental dispersion and apparent resistivity curves to obtain thickness, shear wave velocity and resistivity of each layer. In practice 3n-1 unknowns are determined from the joint inversion of the two experimental dataset whereas each uncoupled inversion has to determine 2n-1 parameters from a single dataset. From a mathematical point of view the problem is hence better conditioned.

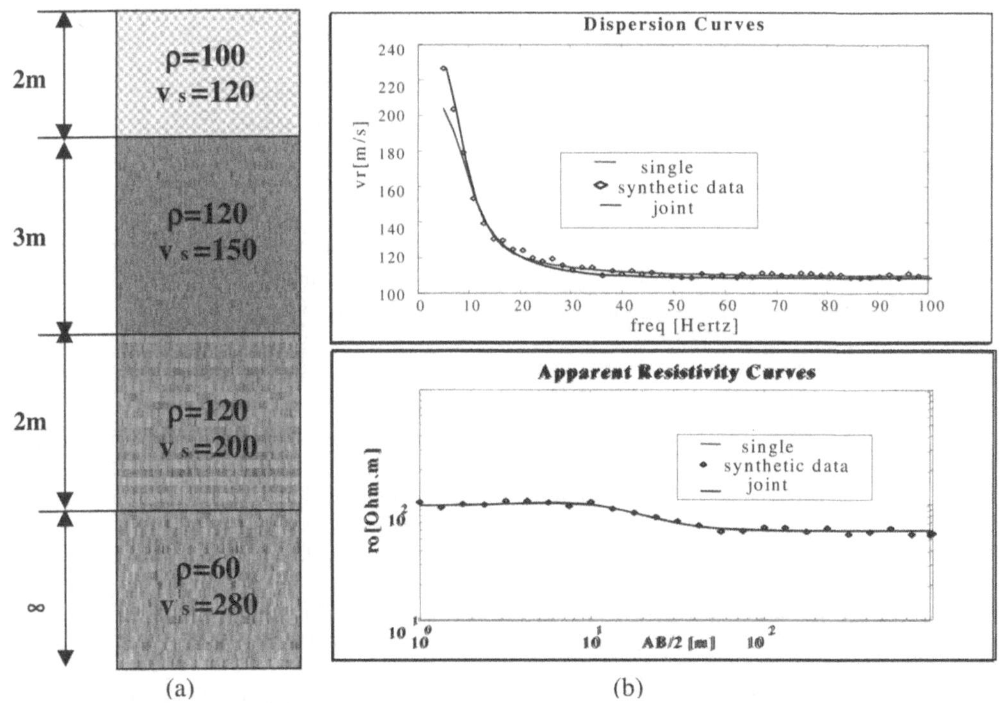


**Figure 18.** Joint inversion of SW test and SEV: (a) synthetic seismic and electrical profile; (b) dispersion and apparent resistivity curves (Comina et al., 2002).

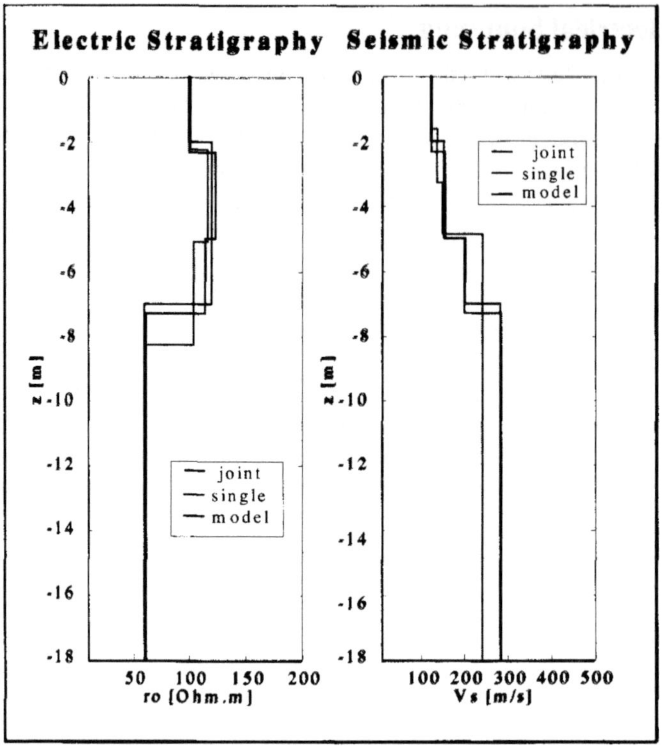


**Figure 19.** Joint inversion of SW test and SEV: electrical and seismic profiles for the example reported in Figure 18 (Comina et al., 2002).

Numerical simulations (Figure 18 and Figure **19**) and experimental results obtained by Comina et al. (2002) using this approach show the effectiveness of joint inversion, especially for the localization of layer interfaces, which is typically one weak point of surface wave tests.

## References


Aki K., Richards P.G. (1980). *Quantitative seismology: theory and methods* - 2 vol. Freeman. S. Francisco

Beaty K.S., Schmitt D.R., Sacchi M. (2002). Simulated annealing inversion of multimode Rayleigh wave dispersion curves for geological structure. *Geophys. J. Internat.*, 151, 622-631

Ben-Menhaem A. (1995). A concise history of mainstream seismology: Origins, legacy and perspectives. *Bull. Seism. Soc. of A.*, 85 (4). 1202-1225

Comina C., Foti S., Sambuelli L., Socco L.V., Strobbia C. (2002) Joint inversion of VES and surface wave data. *Proc. of SAGEEP 2002*, Las Vegas, USA, February 10-14, CD-Rom

Doyle H. (1995). *Seismology*. J. Wiley & sons, Chichester

Dziewonki A.M., Hales A.L. (1972). Numerical Analysis of Dispersed Seismic Waves. In *Methods in Computational Physics* vol.11 Seismology: Surface waves and Earth Oscillations. B.A. Bolt Ed. Academic Press. New York: 39-85

Foti S. (2000). *Multistation Methods for Geotechnical Characterization using Surface Waves.* PhD dissertation, Politecnico di Torino, Italy

Foti S. (2003). Small Strain Stiffness and Damping Ratio of Pisa Clay from Surface Wave Tests. *Geotechnique*, 53(5): 455-461

Foti S., Butcher A.P. (2004) General Report: Geophysical methods applied to geotechnical engineering, *Proc. ISC-2 on Geotechnical and Geophysical Site Characterization*, Viana da Fonseca & Mayne (eds.), Millpress, Rotterdam, 409-418

Foti S., Fahey M. (2003) Applications of multistation surface wave testing, *Deformation Characteristics of Geomaterials*, Di Benedetto H., Doanh T., Geoffroy H. & Sauzéat C. Eds, vol.1, Balkema, Rotterdam, 13-20

Foti S., Sambuelli L., Socco L.V., Strobbia C. (2002). Spatial sampling issues in fk analysis of surface waves. *Proc. of SAGEEP2002*. Las Vegas. USA. February 10-14. CD-Rom

Foti S., Sambuelli L., Socco L.V., Strobbia C. (2003) Experiments of joint acquisition of seismic refraction and surface wave data, *Near Surface Geophysics*, EAGE, 119-129

Foti S., Strobbia C. (2002) Some notes on model parameters for surface wave data inversion, *Proc. of SAGEEP 2002*, Las Vegas, USA, February 10-14, CD-Rom

Gabriels P., Snieder R., Nolet G. (1987). In situ measurements of shear-wave velocity in sediments with higher-mode Rayleigh waves. *Geophys. Prospect.* 35: 187-196

Hayashi K., Hikima K. (2003). CMP analysis of multi-channel surface wave data and its application to near-surface S-wave velocity delineation. *Proc. of SAGEEP2003*. San Antonio. USA. April 6-10. CD-Rom

Hayashi K., Okada A., Matsuoka T., Hatakeyama H. (2004). Joint Analysis of a surface-wave method and a Micro-Gravity survey. *Proc. of SAGEEP 2004*, Colorado Springs, USA, CD-Rom

Hering A., Misiek R., Gyulai A., Ormos T., Dobroka M., Dresen L. (1995). A joint inversion algorithm to process geoelectric and surface wave sismic data. Part I: basic ideas, *Geophysical Prospecting*, 43, 135-156.

Herrmann R.B. (1994). *Computer programs in seismology.* User's Manual. S.Louis University, Missouri (USA)

Horike M. (1985) Inversion of phase velocity of long-period microtremors to the S-wave-velocity structure down to the basement in urbanized areas. *J. Phys. Earth*, 33, 59-96

Hunaidi O. (1998). Evolution-based genetic algorithms for analysis of non-destructive surface wave tests on pavements. *NDT&E Int.*, 31, no.4, 273-280

Jones R.B. (1958). In-situ measurement of the dynamic properties of soil by vibration methods. *Geotechnique*, 8 (1), 1-21

Lai C.G. (1998). *Simultaneous inversion of Rayleigh phase velocity and attenuation for near-surface site characterization.* PhD Diss., Georgia Inst. of Techn., Atlanta (Georgia, USA)

Lai C.G. (2005). Surface waves in dissipative media: forward and inverse modelling. *In this CISM volume*, Springer. Wien

Lai C.G., Foti S., Rix G.J. (2004). Propagation of data uncertainty in surface wave inversion. Submitted for publication to *J. of Eng. and Envir. Geophysics.* EEGS

Lai C.G., Rix G.J., Foti S., Roma V. (2002). Simultaneous Measurement and Inversion of Surface Wave Dispersion and Attenuation Curves. *Soil Dynamics and Earthquake Engineering*, 22 (9-12), 923-930

Louie J.N. (2001). Faster, better: shear-wave velocity to 100 meters depth from refraction microtremor arrays. *Bull. Seismol. Soc. Amer.*, 91, n.2 , 347-364.

Malagnini L., Herrmann R.B., Biella G., de Frando R. (1995). Rayleigh waves in quaternary alluvium from explosive sources: determination of shear-wave velocity and Q structure. *Bull. of Seism. Soc. of A.*, 85, pp. 900-922

McMechan G.A., Yedlin M.J. (1981). Analysis of dispersive waves by wave field transformation. *Geophysics*. 46: 869-874

Menke, W. (1989). *Geophysical data analysis: discrete inverse theory*. Academic Press. San Diego. 260 pp.

Misiek R., Liebig A., Gyulai A., Ormos T., Dobroka M., Dresen L. (1997). A joint inversion algorithm to process geoelectric and surface wave sismic data. Part II: applications, *Geophysical Prospecting*, 45, 65-85

Nasseri-Moghaddam A., Cascante G., Phillips C., Hutchinson J. (2004). A new quantitative procedure to determine the location and embeddement depth of a void with surface waves. *Proc. of SAGEEP 2004*, Colorado Springs, USA, CD-Rom

Nazarian S. (1984). *In situ determination of elastic moduli of soil deposits and pavement systems by Spectral-Analysis-of-Surface waves method*, PhD Diss., Un. of Texas at Austin

O'Neill A. (2004) Full waveform reflectivity for inversion of surface wave dispersion in shallow site investigations, *Proc. ISC-2 on Geotechnical and Geophysical Site Characterization*, Viana da Fonseca & Mayne (eds.), Millpress, Rotterdam, 547-554

Okada H. (2003). *The microtremor survey method.* Geophysical monograph series, number 12, SEG, Tulsa, USA

Park C.B., Miller R.D., Xia J. (1999). Multichannel analysis of surface waves. *Geophysics* 64: 800-808

Phillips C., Moghaddam A.N., Moore T., Cascante G., Hutchinson D.J. (2003). A simple automated method of SASW analysis using multiple receivers. *Proc. of SAGEEP2003*. San Antonio. USA. April 6-10. CD-Rom

Richart F.E. Jr, Wood R.D., Hall J.R. Jr (1970). *Vibration of soils and foundations*. Prentice-Hall, New Jersey

Rix G.J. (2005). Surface wave testing for near surface site characterization. *In this CISM volume*, Springer. Wien

Rix G.J., Lai C.G., Foti S. (2001). Simultaneous measurement of surface wave dispersion and attenuation curves. *Geotechn. Testing J.* ASTM: 350-358

Rix, G.J. and Lai, C.G. (2000). *Software tools for surface wave analysis*. Available at the WEB site: http://www.ce.gatech.edu/~grix/surface_wave.html#Software.

Ryden N.,Ulriksen P.,Park C., Miller R. (2002). Portable Seismic Acquisition System (PSAS) for pavement MASW. *Proc. of SAGEEP 2002*, Las Vegas, USA, February 10-14, CD-Rom

Sànchez-Salinero I. (1987). Analytical investigation of seismic methods used for engineering applications. PhD Diss. Un. of Texas at Austin

Santamarina J.C., Fratta D. (1998). Introduction to discrete signals and inverse problems in civil engineering. Asce Press. Reston

Stokoe K.H. II, Wright S.G., J.A. Bay, J.M. Roesset. (1994). Characterization of geotechnical sites by SASW method. *Geophysical Characterization of Sites*. R.D. Woods Ed.: 15-25

Strobbia C. (2002). *Surface Wave Methods: Acquisition, Processing and Inversion*. PhD Diss. Politecnico di Torino. Italy

Strobbia C., Foti S. (2004) Multi-Offset Phase Analysis of Surface Wave Data (MOPA), submitted to *J. Applied Geophysics*

Szelwis R., Behle A. (1987). Shallow shear-wave velocity estimation from multimodal Rayleigh waves, in Danbom, S. and Domenico, S. N., Ed., *Shear-wave exploration*: Soc. Expl. Geophys., pp.214-226

Tokimatsu K. (1995) Geotechnical Site Characterisation using Surface Waves. Proc. IS Tokyo 1995, Balkema, 1333-1368

Tselentis G-A., Delis G. (1998). Rapid assessment of S-wave profiles from the inversion of multichannel surface wave dispersion data. *Annali di Geofisica*. 41. 1-15

Williams T.P., Gucunski N. (1995). Neural networks for backcalculation of moduli from SASW test, *J. of Computing in Civil Eng.*, 9 (1), ASCE, pp. 1-8

Xia J., Li P.H., Lewis M.J., Miller R.D., Park C.B. (2002). Using surface wave method to define a sinkhole impact area in a noisy environment. *Proc. of SAGEEP2002*. Las Vegas. USA. February 10-14. CD-Rom

Zwicki D.J. (1999). *Advanced signal processing methods applied to engineering analysis of seismic surface waves*. PhD dissertation, Georgia Institute of Technology, Atlanta, USA

# Surface Waves in Dissipative Media: Forward and Inverse Modelling

Carlo G. Lai[1,2]

[1] European Centre for Training and Research in Earthquake Engineering, EUCENTRE, Pavia, Italy
[2] European School for Advanced Studies in Reduction of Seismic Risk, ROSE School, Pavia, Italy

## TABLE OF CONTENTS

1 Introductory Notes
2 Linear Surface Waves in Vertically Inhomogeneous Elastic Continua
    2.1 Overview
    2.2 Eigenvalue Problem: Free Surface Waves
        2.2.1 Numerical solutions
        2.2.2 A notable degenerate case: surface Rayleigh waves in homogeneous continua
    2.3 Variational Principles of Love and Rayleigh Waves
        2.3.1 Introduction
        2.3.2 Hamilton's principle applied to surface waves
        2.3.3 Analytical Jacobian of Rayleigh phase velocity
    2.4 Inhomogeneous Boundary Value Problems of Elastodynamics
        2.4.1 Introduction
        2.4.2 Lamb problem for time-harmonic point source
        2.4.3 Green's functions
        2.4.4 Modal and apparent phase velocity
3 Linear Surface Waves in Vertically Inhomogeneous Inelastic Continua
    3.1 Introduction
    3.2 Constitutive Modeling of Linear Dissipative Materials
    3.3 Viscoelastic Waves in Unbounded Homogeneous Continua
    3.4 Surface Rayleigh Waves in Weakly Dissipative Media
    3.5 Surface Rayleigh Waves in Strongly Dissipative Media
        3.5.1. A solution based on Cauchy's theorem of complex analysis
4 Inverse Problem Associated with Surface Wave Motion
    4.1 Introduction
    4.2 Stability, Regularization and a-Priori Information
    4.3 Inversion Strategies, LS and GS Methods
    4.4 Joint-Causal Inversion of Rayleigh Dispersion and Attenuation Curves
5 References

# Surface Waves in Dissipative Media: Forward and Inverse Modelling

Carlo G. Lai[1,2]

[1] European Centre for Training and Research in Earthquake Engineering, EUCENTRE, Pavia, Italy
[2] European School for Advanced Studies in Reduction of Seismic Risk, ROSE School, Pavia, Italy

**Abstract.** Scope of this chapter is to review the theory of surface wave propagation in linear, one-constituent, vertically inhomogeneous, dissipative continua. Subjects of discussion will include both forward and inverse modelling. Although the basic theory will be applicable to both Love and Rayleigh waves, most of the attention will be dedicated to Rayleigh waves due to their major relevance in the applications. After reviewing standard results of the theory of surface wave propagation in elastic and weakly dissipative media (including solution of the Lamb's problem) using the formalism of variational calculus, the full theory of Rayleigh wave propagation is developed for arbitrarily dissipative linear viscoelastic materials. Emphasis is placed on illustrating a technique for the solution of the Rayleigh eigenproblem based on the application of Cauchy's theorem of complex variable theory. The last section of the chapter is dedicated to illustrate the inverse problem associated with surface wave motion and its properties. The previous results obtained for the solution of the coupled forward problem are used to develop an algorithm for the joint inversion of Rayleigh dispersion and attenuation curves to determine the transversal phase velocity and quality factor of viscoelastic, layered systems. Application of the inversion algorithm is shown through examples using synthetic and real surface wave data.

## 1 Introductory Notes

Since the time Lord Rayleigh predicted their existence in 1887 (Rayleigh, 1887) surface waves have attracted the interest of a constantly increasing number of researchers embracing disciplines as diverse as solid-state physics, microwave engineering, geotechnical engineering, non-destructive testing, seismology, geophysics, material science, ultrasonic acoustics, and others. Despite their marked differences these disciplines share the goal of exploiting surface waves propagating along the boundary of a domain for obtaining some type of information about the interior of that domain, usually expressed in terms of one or more scalar fields.

Surface waves are appealing because they are ideal for the development of non-invasive techniques for material characterization at a small scale (e.g. ultrasonic surface waves can be used to identify material defects, see for instance Viktorov, 1967; Blake and Bond, 1990, 1992; Pecorari, 2001), large scale (e.g. seismologists have long employed surface waves to investigate the structure of the Earth crust and of its upper mantle, see for instance Ewing et al, 1957; Dorman et al, 1960; Kovach 1978; Lee and Solomon, 1979; Keilis-Borok, 1989; Aki and Richards, 2002) and intermediate scale (e.g. geophysicists and geotechnical engineers use surface waves for shallow characterization of geomaterials, see for instance Jones, 1962; Al-Eqabi and Herrmann, 1993;

Stokoe et al., 1994; Tokimatsu, 1995; Malagnini et al., 1997; Xia et al., 1999; Glangeaud et al., 1999; Strobbia, 2003). A detailed illustration of the use of surface waves for near-surface site characterization in geophysical-geotechnical applications can be found in the chapters by Rix (2005) and Foti (2005) of this volume.

Some of the properties of surface waves make them particularly suitable for material characterization. They originate from the condition of vanishing stress at a boundary of a domain and their radiation pattern (see Figure 1) is essentially two-dimensional thereby characterized by a much lesser rate of geometrical attenuation than body (or bulk) waves whose energy spreads in both horizontal and vertical directions. As an example in an homogeneous, elastic half-space Rayleigh surface waves generated by a line source do not suffer any geometric attenuation whereas for a point load the rate of spatial decay is proportional to the inverse of the square root of the distance from the source. By contrast for the same geometry of the source the geometric attenuation factor of body waves propagating along the boundary of an elastic half-space is proportional to the inverse of the square of the distance (Ewing et al., 1957). Thus, at distances on the order of one to two wavelengths from the source, the contribution of body waves becomes negligible and the wave field is dominated by Rayleigh waves. Lamb (1904) was among the first to recognize this fundamental property of surface waves and its implications in the transmission of earthquake energy at large distances.

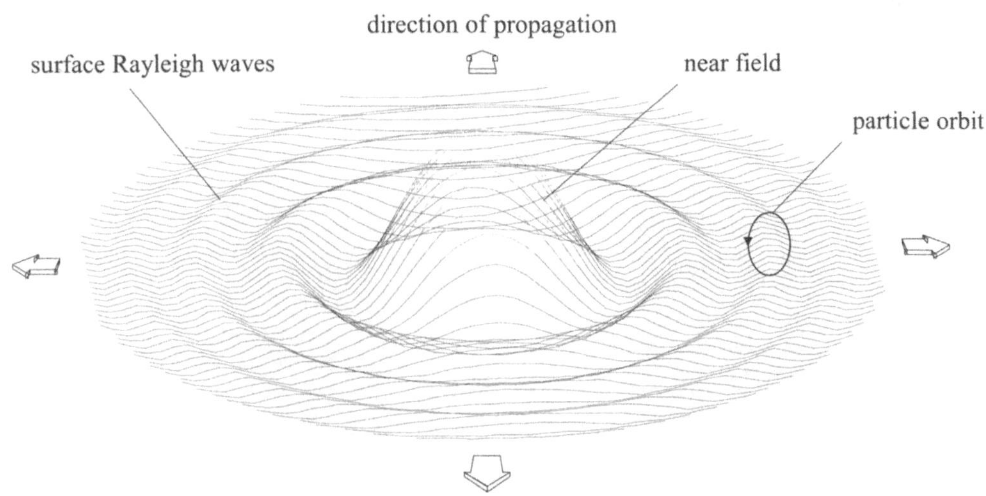


**Figure 1.** Two-dimensional radiation pattern of surface Rayleigh waves generated by a vertical source.

In the direction orthogonal to the direction of propagation the displacement field generated by a surface wave decays exponentially[1] as no energy is propagated in the interior of the half-space. It can be shown (Achenbach, 1984) that most of the strain energy associated with surface wave

[1] Several books on wave propagation and theoretical seismology use this property as a definition of *surface waves* (e.g. Achenbach, 1984; Bullen and Bolt, 1985; Ben-Menahem and Singh, 2000).

motion is confined within a depth of about a wavelength from the free boundary. One means of describing this property is through the concept of *skin depth*, the depth at which the amplitude decreases by a factor of 1/e. For Rayleigh waves in a homogeneous medium, the skin depth is approximately 0.94λ. Hence Rayleigh waves of larger wavelengths penetrate deeper into the interior of a medium. As shown in Figure 2, the opposite is true for shorter wavelengths.

Since the speed of propagation of surface waves is influenced by the mechanical properties up to a depth of about a wavelength, if the medium is inhomogeneous surface waves of different wavelengths will travel at different speeds (i.e. they are *dispersive*) because they will sample a medium whose properties vary with depth (see Figure 2). Exploitation of the dispersive characteristics of surface waves is the principle upon which are founded surface wave-based methods in geophysical prospecting for near-surface site characterization and in seismology for the identification of the Earth structure.

In these methods as well as in the techniques used in non-destructive testing the objective is to determine the location, shape and constitution of the interior of a medium from measurements of a two-dimensional surface wave field at its free boundary. Mathematical formalization of this objective leads to the theory of *inverse problems*, specifically to a particular branch of this theory known as *system* or *parameter-identification problems* (Engl, 1993). Solution and properties of the solution of such problems depend strongly upon the constitutive model used to describe the thermo-mechanical-hydraulic response of a deformable body to the propagation of disturbances, as well as on the properties of its boundaries.

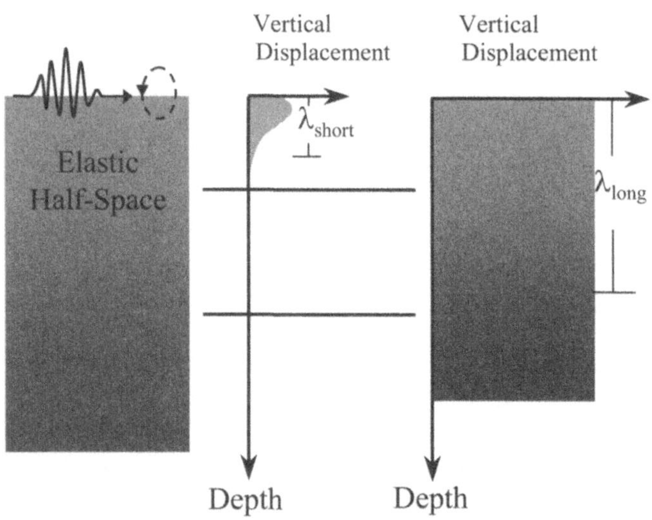


**Figure 2.** Dependence of *skin depth* from wavelength in a Rayleigh surface wave.

A constitutive model often used in the applications is one-constituent, linear elasticity, mainly because of its simplicity but also for its ability to predict many fundamental features of surface wave propagation observed experimentally. However one-constituent, linear elasticity is not capable to account for important phenomena occurring when a mechanical disturbance propagates through real materials. One such phenomenon is energy dissipation which in most materials occurs even at very low strain levels when their response is still linear though inelastic. Another example is represented by geomaterials like soils and sedimentary rocks, that are composed by a solid phase interacting with one or more fluids.

Such situations require the employment of more advanced constitutive models able to adequately describe complex phenomena such as material dispersion, multimode wave propagation, existence of additional kind of surface waves (in addition to classical Rayleigh waves, pseudo-Rayleigh waves with attenuation, etc.), not to mention nonlinear effects such as *solitary waves* in which a surface wave is a traveling package of nonlinear disturbances, and so on (Maugin, 2005).

This chapter is dedicated to illustrate the theory of surface wave propagation in linear, one-constituent, vertically inhomogeneous dissipative continua in both forward and inverse modelling. Although the basic theory will be developed for both Love and Rayleigh waves, the main focus of the chapter is devoted to Rayleigh waves because of their greater importance in the applications. The chapter is organized in four main sections.

After these introductory notes, Section 2 reviews the theory of surface wave propagation in linear elastic, vertically inhomogeneous continua. Well-known results are re-derived using *Hamilton's principle* and the formalism of variational calculus which includes an analytical expression for the *Jacobian* of Rayleigh phase velocity with respect to medium parameters. An explicit formula is derived for the *apparent* Rayleigh phase velocity resulting from mode superposition. The classical *Lamb's problem* is briefly revisited when discussing the solution of the inhomogeneous boundary value problem associated with the propagation of surface waves (i.e. the so-called *source problem*).

Section 3 is devoted to illustrate the theory of surface wave propagation in linear, vertically inhomogeneous dissipative continua. After a discussion on constitutive modelling of dissipative materials and of main properties of viscoelastic waves, the attention is focused on the solution of the Rayleigh eigenproblem in *weakly* dissipative media. This result forms the basis of the most common procedures used by seismologists and geophysicists to solve surface waves propagation problems in inelastic media. The general theory of surface wave propagation in arbitrarily dissipative, linear viscoelastic continua is then presented which include an elegant technique for the solution of the Rayleigh eigenproblem using the Cauchy's residue theorem of complex analysis.

The chapter ends with a section dedicated to the *inverse problem* associated with surface wave motion and its solution. After a general introduction to the theory of inverse problems in geophysical applications, an algorithm for the joint inversion of Rayleigh dispersion and attenuation data is presented to determine the transversal phase velocity and quality factor (or damping ratio) of viscoelastic, layered systems. The algorithm is based on using the *holomorphic* properties of Rayleigh phase velocity viewed as an *holomorphic mapping* of the complex-valued speed of propagation of transversal waves characteristic of the medium. An application of the algorithm is illustrated via a joint-causal inversion of synthetic and real surface wave data.

Elastic modelling of surface waves in multi-component systems as well as the nonlinear theory of surface waves are not treated as they will be the subjects of other contributions of this volume namely those by Albers (2005), Maugin (2005), and Wilmanski (2005).

## 2 Linear Surface Waves in Vertically Inhomogeneous Elastic Continua

### 2.1 Overview

The purpose of this section is to describe the most relevant features of surface wave propagation in vertically inhomogeneous, linear elastic continua delimited by a plane boundary. The special case of Rayleigh surface waves in homogeneous media will not be discussed in detail since it has been already treated in this volume in the chapter by Wilmanski (2005). However a summary of the most relevant properties of Rayleigh wave propagation in homogeneous half-spaces will be briefly reviewed.

The first subject to be discussed is the eigenvalue problem associated with the propagation of *free* surface waves from which fundamental results such as the Green's function and the apparent phase velocity can be easily derived. Solution of the eigenvalue problem also leads to the important concept of *geometric dispersion*, a phenomenon by which, in inhomogeneous continua, the phase velocity of surface waves is a multi-valued function of the frequency of excitation.

Geometric dispersion must be distinguished from *material dispersion* which arises in dissipative media as a necessary condition to satisfy the physical *principle of causality* (see Section 3). The phenomenon of geometric dispersion is caused by effects of constructive interference occurring in media that are either bounded (e.g. rods, plates, and other types of waveguides) or inhomogeneous. It is responsible for the existence of several modes of propagation each traveling at a different phase and group velocity which are named *modal* velocities. Another effect produced by geometric dispersion is to alter the geometric spreading law governing the spatial attenuation of surface waves in elastic inhomogeneous media.

Some peculiar properties of surface wave propagation in inhomogeneous media can be easily derived from the application of certain variational principles whose conditions of applicability are rather general. Among the results obtained from this principle is a closed-form expression for the *Jacobian* of Rayleigh phase velocity with respect to medium parameters which is of fundamental importance in the solution of the inverse surface-wave problem (see Section 4), and an approximate solution of the Rayleigh eigenproblem in weakly dissipative media (see Section 3).

For surface waves generated by harmonic sources applied at the boundary or in the interior of a vertically inhomogeneous half space, the various modes of propagation of Rayleigh waves are superimposed like in a spatial Fourier series. The phase velocity of the resulting waveform can be obtained from an apposite superposition of modal quantities (phase and group velocities, eigenfunctions, etc.). This kinematical quantity is given the name of *apparent* (or effective) phase velocity and it will be shown to be a local quantity in the sense that its magnitude depends on the spatial position where it is measured. The notion of apparent phase velocity is particularly relevant in applications of surface waves for near-surface site characterization (see chapters by Rix, 2005 and Foti, 2005). A discussion on source problems should also include the solution to the *Lamb's problem*, a classical in elastodynamics which will be here briefly revisited for its implications with the surface wave radiation pattern induced by a point source in the far-field.

The solutions presented in this section are all obtained in the *frequency* domain. The mathematics of wave propagation problems is often fairly involved, and explicit non-integral solutions can rarely be obtained. One of the few exceptions are boundary value problems where the boundary conditions and body forces are specified as harmonic functions of time. However working in the frequency domain does not necessarily imply a loss of generality since the availability of harmonic solutions is often a sufficient condition to obtain far more general results using the Fourier integral theorem. In the sequel, the properties of the medium are assumed to be arbitrary (hence not necessarily continuous) functions of depth *y*. Explicit solutions, however, are presented only for the case of a finite number of homogeneous strata overlaying a homogeneous half-space (i.e. a horizontally stratified medium).

## 2.2 Eigenvalue Problem: Free Surface Waves

The dynamic behavior of a continuous system can be described by a scalar function, called the *Lagrangian* density and denoted by $\mathscr{L}$, of a certain number N of *generalized coordinates* $q_i$ $(i=1,N)$ and their spatial and temporal derivatives $q_{i,j}$ and $\dot{q}_i$ $(j=1,3)$ (Goldstein, 1980). The equations governing the motion of such a system can be derived from *Hamilton's principle* and are called *Lagrange's equations*. When the system is *conservative*, the Lagrangian density $\mathscr{L}(q_i, \dot{q}_i, q_{i,j})$ is equal to $\mathscr{L} = T - U$ where *T* and *U* are the kinetic and the potential energy of the system, respectively. For a conservative system Lagrange's equations are (Achenbach, 1984):

$$\frac{\partial}{\partial t}\left(\frac{\partial \mathscr{L}}{\partial \dot{q}_i}\right)+\sum_{j=1}^{3}\frac{\partial}{\partial x_j}\left[\frac{\partial \mathscr{L}}{\partial (q_{i,j})}\right]-\frac{\partial \mathscr{L}}{\partial q_i}=0 \tag{1}$$

where $x_j$ are the components of the position vector $\mathbf{x}$ in Cartesian coordinates. For a linear elastic material the potential energy identifies with the elastic strain energy, and thus the Lagrangian density is given by:

$$\mathscr{L}\left(\dot{u}_i, u_{i,j}\right)=\frac{1}{2}\rho \dot{u}_i \dot{u}_i-\frac{1}{2}\sigma_{ij}\varepsilon_{ij} \tag{2}$$

and $u_i$ $(i=1,3)$ are the components of the displacement vector $\mathbf{u}(\mathbf{x},t)$, $\sigma_{ij}$ and $\varepsilon_{ij}$ are the components of Cauchy stress and strain tensors respectively, and $\rho$ is the mass density which is assumed constant with time. It is seen from Eq. (2) that in a linear elastic body the generalized coordinates $q_i$ identify with the components of displacement vector $\mathbf{u}(\mathbf{x},t)$.

Considering *Hooke's law* $\sigma_{ij} = \lambda\varepsilon_{kk}\delta_{ij} + 2G\varepsilon_{ij}$ where $\varepsilon_{ij} = \frac{1}{2}\left(u_{i,j} + u_{j,i}\right)$, $\delta_{ij}$ is the Kronecker symbol, and $\lambda$ and G are *Lamè's* elastic moduli, Eq. (2) can be rewritten as:

$$\mathscr{L}(\dot{u}_i, u_{i,j})=\frac{1}{2}\rho \dot{u}_i \dot{u}_i-\left[\frac{1}{2}\lambda\left(\varepsilon_{kk}\right)^2+G\varepsilon_{ij}\varepsilon_{ij}\right] \tag{3}$$

In general, Lamè's parameters and the mass density are functions of the coordinates, namely $\lambda=\lambda(x_j)$, $G=G(x_j)$, and $\rho=\rho(x_j)$. However, in this chapter the elastic medium is assumed vertically heterogeneous hence $\lambda=\lambda(x_2)$, $G=G(x_2)$, and $\rho=\rho(x_2)$ with $x_2=y$. Using the definition of Lagrangian density given by Eq. (3), Lagrange's equations of motion (1) yield the following result in vector notation (Ben-Menahem and Singh, 2000):

$$G\nabla^2\mathbf{u}+(\lambda+G)\mathrm{grad}(\mathrm{div}\mathbf{u})+\mathbf{e}_y\frac{d\lambda}{dy}\mathrm{div}\mathbf{u}+\frac{dG}{dy}\left(\mathbf{e}_y\times\mathrm{curl}\mathbf{u}+2\frac{\partial \mathbf{u}}{\partial y}\right)=\rho\frac{\partial^2\mathbf{u}}{\partial t^2} \tag{4}$$

Equations (4) are the *Navier's* equations of motion for vertically inhomogeneous continua in absence of body forces. Obviously Eq. (4) is written in Cartesian coordinates specified by a set of basis vectors $\{\mathbf{e}_j\}$ and a set of coordinate axes $\{x_j\}$. For convenience the base vector $\mathbf{e}_2$ has been denoted by $\mathbf{e}_y$. Finally, the symbol $(\cdot)\times(\cdot)$ is used to indicate the vector product.

To seek solutions of Eq.(4) representing surface waves, the displacement field $\mathbf{u}(\mathbf{x},t)$ must be assumed of the form:

For *Love waves*: $$\mathbf{u}:\{u_1=0,\quad u_2=l_1(y,k,\omega)\cdot e^{i(\omega t-kx)},\quad u_3=0\} \tag{5a}$$

For *Rayleigh waves*: $$\mathbf{u}:\{u_1=r_1(y,k,\omega)\cdot e^{i(\omega t-kx)},\quad u_2=0,\quad u_3=i\cdot r_2(y,k,\omega)\cdot e^{i(\omega t-kx)}\} \tag{5b}$$

In fact whereas *Love waves* are horizontally polarized transversal SH waves (see Figure 4), *Rayleigh wave* particle motion is *elliptical* since the horizontal and vertical components of the displacement field are $\pi/2$ radians out of phase, with the minor axis of the ellipse parallel to the free boundary. In a homogeneous, linear elastic continuum the particle orbit is *retrograde* at the free-surface then at a depth of about of $0.2\cdot\lambda$ it becomes *prograde* while the amplitude of particle motion attenuate exponentially as the depth increases (see Figure 3).

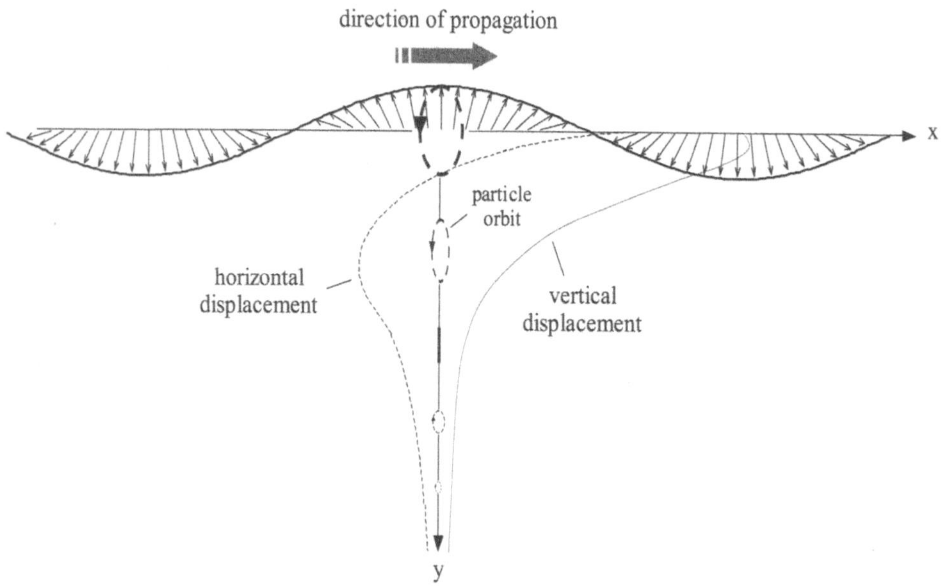


**Figure 3.** Elliptical polarization of particle motion in a Rayleigh wave in a homogeneous half-space. At the free boundary the particle orbit is *retrograde* then at a depth of ~$0.2\cdot\lambda$ it becomes *prograde*.

Equation (5a) and (5b) represent a one and two-dimensional (plane) strain field respectively. This assumption does not imply any loss of generality in the theory since it can be shown (Aki and Richards, 2002) that for instance cylindrical Rayleigh waves share the same *y*-dependence described by Eq. (5b). In Eqs.(5a) and (5b) the term $k = k(\omega)$ denotes the real wavenumber which, in general, is a multi-valued function of the frequency of excitation $\omega$. Finally $x$ is used to indicate the direction of propagation $x_1$ and $i = \sqrt{-1}$. Figure (4) illustrates the sign convention assumed for the Cartesian coordinate axes $\{x_1=x,\ x_2=y,\ x_3=z\}$. The medium occupies the half-space $y > 0$.

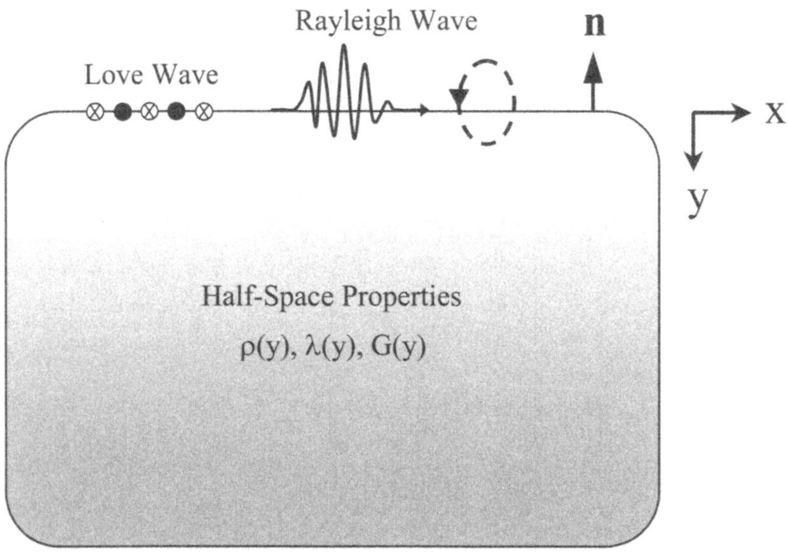


**Figure 4.** *Love* and *Rayleigh* surface waves in a vertically inhomogeneous continuum.

To represent surface waves Eqs.(5a) and (5b) must be supplemented with appropriate boundary conditions which are vanishing of the stress field at the free surface of the half-space, and vanishing of stress and displacement fields as $y \to \infty$. The latter is often called the *Sommerfeld* or *radiation* or *source free* condition. Formally these boundary conditions can be written as follows:

$$\begin{aligned} &\boldsymbol{\sigma}(\mathbf{x}, y)\cdot\mathbf{n} = \mathbf{0} \qquad \text{at } y = 0 \\ &\mathbf{u}(\mathbf{x}, y) \to \mathbf{0}, \qquad \boldsymbol{\sigma}(\mathbf{x}, y) \to \mathbf{0} \quad \text{as } y \to \infty \end{aligned} \tag{6}$$

where the notation $\boldsymbol{\sigma}(\mathbf{x}, y)$ denotes the Cauchy's stress tensor and $\mathbf{n}$ is a unit normal vector (see Figure 4). In vertically inhomogeneous media where the material properties ($\lambda$, G, and $\rho$) have *jump* discontinuities, the stress and displacement fields must be continuous at each layer interface:

$$\begin{aligned} &\mathbf{u}(\mathbf{x}, y^{+}) = \mathbf{u}(\mathbf{x}, y^{-}) \\ &\boldsymbol{\sigma}(\mathbf{x}, y^{+})\cdot\mathbf{n} = \boldsymbol{\sigma}(\mathbf{x}, y^{-})\cdot\mathbf{n} \end{aligned} \tag{7}$$

otherwise the stress and displacement discontinuities would correspond to the existence of seismic sources[2]. From the displacement field defined by Eqs.(5a) and (5b) the stress field can be easily computed using Hooke's law $\sigma_{ij} = \lambda\varepsilon_{kk}\delta_{ij} + 2G\varepsilon_{ij}$ thereby obtaining:

For *Love waves*:
$$\begin{cases} \tau_{xx} = \tau_{zz} = \tau_{yy} = \tau_{yx} = 0 \\ \tau_{zy} = G\dfrac{dl_1}{dy} \cdot e^{i(\omega t - kx)} \\ \tau_{xz} = -ikGl_1 \cdot e^{i(\omega t - kx)} \end{cases} \quad (8a)$$

For *Rayleigh waves*:
$$\begin{cases} \tau_{zy} = \tau_{xz} = 0 \\ \tau_{xx} = i\left[\lambda\dfrac{dr_2}{dy} - k(\lambda + 2G) \cdot r_1\right] \cdot e^{i(\omega t - kx)} \\ \tau_{zz} = i\left(\lambda\dfrac{dr_2}{dy} - k\lambda r_1\right) \cdot e^{i(\omega t - kx)} \\ \tau_{yy} = i\left[(\lambda + 2G) \cdot \dfrac{dr_2}{dy} - k\lambda r_1\right] \cdot e^{i(\omega t - kx)} \\ \tau_{yx} = G\left(\dfrac{dr_1}{dy} + kr_2\right) \cdot e^{i(\omega t - kx)} \end{cases} \quad (8b)$$

For Eq.(7) the stress components $\tau_{zy}$ for Love waves and $\tau_{yx}$, $\tau_{yy}$ for Rayleigh waves are continuous in y. It is convenient to re-write them as follows:

For *Love waves*:
$$\tau_{zy} = l_2(y,k,\omega) \cdot e^{i(\omega t - kx)} \quad (9a)$$

For *Rayleigh waves*:
$$\begin{cases} \tau_{yx} = r_3(y,k,\omega) \cdot e^{i(\omega t - kx)} \\ \tau_{yy} = i \cdot r_4(y,k,\omega) \cdot e^{i(\omega t - kx)} \end{cases} \quad (9b)$$

By substituting Eqs.(5a) and (5b) into *Navier's* equations of motion (4) and rearranging Eqs.(8a) and (8b) in light of Eqs.(9a) and (9b), one obtains (Aki and Richards, 2002):

For *Love waves*:
$$\frac{d}{dy}\begin{bmatrix} l_1 \\ l_2 \end{bmatrix} = \begin{bmatrix} 0 & G(y)^{-1} \\ k^2 \cdot G(y) - \omega^2 \cdot \rho(y) & 0 \end{bmatrix} \cdot \begin{bmatrix} l_1 \\ l_2 \end{bmatrix} \quad (10a)$$

and

[2] A discontinuity in the displacement field across a layer interface would be equivalent to a *dislocation*.

$$\frac{d}{dy}\begin{bmatrix} r_1 \\ r_2 \\ r_3 \\ r_4 \end{bmatrix} = \begin{bmatrix} 0 & k(\omega) & G^{-1}(y) & 0 \\ -k(\omega)\lambda(y)\cdot[\lambda(y)+2G(y)]^{-1} & 0 & 0 & [\lambda(y)+2G(y)]^{-1} \\ [k(\omega)]^2\zeta(y)-\omega^2\rho(y) & 0 & 0 & k(\omega)\lambda(y)[\lambda(y)+2G(y)]^{-1} \\ 0 & -\omega^2\rho(y) & -k(\omega) & 0 \end{bmatrix} \cdot \begin{bmatrix} r_1 \\ r_2 \\ r_3 \\ r_4 \end{bmatrix} \tag{10b}$$

for *Rayleigh waves*. The functions $l_2$ for Love waves and $r_3$, $r_4$ for Rayleigh waves can be expressed in terms of the eigenfunctions $l_1$ and $r_1$, $r_2$ respectively defined in Eqs.(5a) and (5b):

For *Love waves*:

$$l_2(y,k,\omega) = G(y)\cdot\frac{dl_1}{dy} \tag{11a}$$

For *Rayleigh waves*:

$$\begin{cases} r_3(y,k,\omega) = G(y)\cdot\left(\dfrac{dr_1}{dy} - kr_2\right) \\ r_4(y,k,\omega) = \left[(\lambda(y)+2G(y))\dfrac{dr_2}{dy} + k\lambda(y)r_1\right] \end{cases} \tag{11b}$$

The function $\zeta(y)$ in Eq.(10b) is defined in terms of *Lamè's* parameters[3] $\lambda(y)$ and $G(y)$:

$$\zeta(y) = 4G(y)\cdot\frac{[\lambda(y)+G(y)]}{[\lambda(y)+2G(y)]} \tag{12}$$

Now by defining vectors $\mathbf{g}(y) = [l_1 \quad l_2]^T$ and $\mathbf{f}(y) = [r_1 \quad r_2 \quad r_3 \quad r_4]^T$, and matrices $\mathbf{B}(y)$ and $\mathbf{A}(y)$ denoting respectively the 2×2 and 4×4 arrays of Eqs.(10a) and (10b), these equations can be re-written in the following compact form:

For *Love waves*:

$$\frac{d\mathbf{g}(y)}{dy} = \mathbf{B}(y)\cdot\mathbf{g}(y) \tag{13a}$$

For *Rayleigh waves*:

$$\frac{d\mathbf{f}(y)}{dy} = \mathbf{A}(y)\cdot\mathbf{f}(y) \tag{13b}$$

The two sets of first-order, ordinary differential equations (13a) and (13b) define two *eigenvalue problems* with *linear operator d/dy*. The displacement and stress eigenfunctions are respectively $l_1(y,k,\omega)$ and $l_2(y,k,\omega)$ for *Love waves*, whereas for *Rayleigh waves* they are the pair $[r_1(y,k,\omega), r_2(y,k,\omega)]$ and $[r_3(y,k,\omega), r_4(y,k,\omega)]$ respectively.

The boundary conditions associated with the eigenproblems (13a) and (13b) can be immediately deduced from Eq.(6) and Eqs.(9a) and (9b):

[3] It would be more appropriate to identify *Lamè parameters* $\lambda$ and G as *Lamé functions* in light of their dependence on depth y.

For *Love waves*:
$$\begin{cases} l_2(y,k,\omega)=0 & \text{at } y=0 \\ l_1(y,k,\omega)\rightarrow 0 & \text{as } y\rightarrow\infty \end{cases} \tag{14a}$$

For *Rayleigh waves*:
$$\begin{cases} r_3(y,k,\omega)=r_4(y,k,\omega)=0 & \text{at } y=0 \\ r_1(y,k,\omega)\rightarrow 0,\ r_1(y,k,\omega)\rightarrow 0 & \text{as } y\rightarrow\infty \end{cases} \tag{14b}$$

For a given frequency $\omega$, non-trivial solutions of the two eigenproblems (13a) and (13b) subjected to boundary conditions (14a) and (14b) exist only for special values of the wavenumber $k_m=k_m(\omega)$, m=1,M[4]. These particular values $k_m$ are called the *eigenvalues* of a given eigenproblem and the corresponding solutions $l_i(y,k,\omega)$, $r_n(y,k,\omega)$, i=1,2; n=1,4 are the *eigenfunctions*. It can be shown (Keilis-Borok, 1989) that, for a given frequency $\omega$, the set of eigenfunctions satisfy to a series of orthogonality conditions with appropriate weighting functions.

The systems of linear, first-order, ordinary differential equations (13a) and (13b) have variable coefficients since the Lamé parameters $\lambda$ and G and the mass density $\rho$ are functions of y. It is precisely this characteristic that make their solution not trivial. In will be shown in the next section that if the medium is homogeneous (i.e. $\lambda$, G and $\rho$ are independent of y) the solution to Eqs.(13a) and (13b) can be found in closed-form.

In the most general case of inhomogeneous media the relation $k_m=k_m(\omega)$ is only known implicitly and maybe written in the form $\mathscr{F}_{L/R}[\lambda(y),G(y),\rho(y),k_m,\omega]=0$ where $\mathscr{F}_{L/R}[\cdot]=0$ is a highly non-linear, transcendental function of Lamè's parameters $\lambda$ and G, mass density $\rho$, wavenumber $k_m$ and frequency of excitation $\omega$. This relation $\mathscr{F}_{L/R}[\cdot]=0$ is called either the *Love* or the *Rayleigh dispersion equation* depending on whether it is obtained from Eq.(13a) and (14a) or (13b) and (14b). It states that in vertically inhomogeneous media the velocity of propagation of surface Love and Rayleigh waves is, in general, a *multi-valued* function of frequency.

Each set $\{\{k_m, l_i(y, k_m, \omega)\}$ or $\{k_m, r_n(y, k_m, \omega)\}$defines a mode of propagation and, in general, there are $M^4$ *normal* modes of propagation at any given frequency. The number *M* can be finite or infinite, depending upon the *y*-dependence of the medium properties and on the frequency of excitation. Furthermore, the distribution of the modes, namely the *mode spectrum*, can be continuous or discrete, and in some cases both (Keilis-Borok, 1989). It can be shown that in a medium composed of a finite number of homogeneous layers overlaying a homogeneous half-space, the total number of surface wave modes of propagation is always finite (Ewing et al., 1957).

From a physical point of view, the existence of different modes of propagation at a given frequency can be explained by the constructive interference occurring among waves. In continuously varying heterogeneous media the ray paths are curved (as a result of *Fermat's principle*), and hence they interfere with each other. In stratified media the ray paths are rectilinear, and interference phenomena occur among waves undergoing multiple reflections at the layer interfaces. In either case, the dispersion equation is the mathematical statement of this condition of *constructive interference* (Achenbach, 1984).

---

[4] The fact that to simplify the notation it has been used the same symbol *M* to denote the number of modes of propagation resulting from the solution of either the *Love* or the *Rayleigh* eigenproblem does not necessarily imply that they are equal.

**Numerical solutions.** This section will briefly review some of the most common techniques used to solve the Love and Rayleigh eigenproblems Eqs.(13a) and (13b) with their associate boundary conditions Eqs.(14a) and (14b). A detailed description of the methods will not be given since they are the subject of the chapter by Kausel (2005) in this volume.

Although several techniques could be used to solve a linear system of first-order, ordinary differential equations with variable coefficients including numerical integration, finite difference, finite element, boundary element, and spectral element methods, some of the most common and frequently used techniques (especially by seismologists and geophysicists) are those belonging to the class of *propagator-matrix methods* (Kennett, 1983). The Thomson-Haskell algorithm (Thomson, 1950; Haskell, 1953), also called the *transfer matrix method*, is probably the most notorious of this class of methods because of its simplicity and ease of computer implementation. The application of this algorithm, however, is limited to vertically inhomogeneous media that can be idealized by a stack of homogeneous strata overlying a homogeneous half-space.

In the Thomson-Haskell algorithm the non-trivial solutions of the surface wave eigenproblem are found from the roots of the dispersion relations $\mathscr{F}_{L/R}[\lambda(y),G(y),\rho(y),k_m,\omega] = 0$. The latter are constructed through a sequence of matrix multiplications using the closed-form solution of surface wave eigenproblems in homogeneous media, which is valid within a given layer and imposing continuity of stress and displacement fields at the layer interfaces. Each matrix is formed by elements which are transcendental functions of the material properties of a specific layer and for this reason is named *layer matrix*. The roots of the dispersion equations are the wavenumbers corresponding to the modes of propagation of surface waves at different frequencies. Once solutions of the dispersion equations are found, it is easy to determine the eigenfunctions for each mode of propagation by simple algebraic manipulations. The eigenfunctions give the depth-dependence of stress and displacement fields.

An important contribution to the transfer matrix method was given by Haskell (1953) who developed asymptotic expressions for the dispersion equations for the important limiting cases of short and long wavelengths. Because the original formulation of the Thomson-Haskell algorithm suffered at high frequencies from problems of numerical instability (Knopoff, 1964), this method has been modified and improved throughout the years by numerous researchers (Schwab and Knopoff, 1970; Abo-Zena, 1979; Harvey, 1981). It is also worth mentioning that the matrix propagator method has been recently generalized to layered porous media (Jocker et al., 2004).

Turning the attention to different techniques, Kausel and Roësset (1981) derived a finite element formulation of the Thomson-Haskell algorithm, which was then called the *dynamic stiffness matrix method*. The main feature of this method is the replacement of the Thomson-Haskell transfer matrices with layer stiffness matrices that are similar to conventional stiffness matrices used in structural analysis. The advantage of this formulation is the ability to use standard structural analysis techniques such as condensation and substructuring to solve both the Love and Rayleigh eigenproblems and the inhomogeneous elastodynamic problem of layered media subjected to dynamic loads (Kausel, 1981). The first attempts of using finite element techniques to solve wave propagation problems in seismology and earthquake engineering date back to the early 70's with the works of Lysmer and Waas (1972) and Lysmer and Drake (1972).

Another important class of algorithms for solving the surface waves eigenvalue problems is the *method of reflection and transmission coefficients* developed by Kennett and his co-workers (Kennett, 1974; Kennett and Kerry, 1979; Kennett, 1983), and modified and/or improved by others researchers (Luco and Apsel, 1983; Chen, 1993; Hisada, 1994; Hisada, 1995). Also this method, like the Thomson-Haskell algorithm, is only suitable for applications in multi-layered media. It is based on the use of reflection and transmission coefficients to construct reflection and transmission matrices for a stratified medium. The result is a very efficient iterative algorithm for constructing the dispersion relations $\mathcal{F}_{L/R}[\lambda(y),G(y),\rho(y),k_m,\omega] = 0$. The method of reflection and transmission coefficients also offers an interesting physical interpretation because it explicitly models the constructive interference that leads to formation of the surface waves modes (Kennett, 1983). Earlier versions of this algorithm were numerically unstable at high frequencies because of the presence of certain frequency-dependent terms that have been eliminated in more recent formulations (Chen, 1993). Most of the computational efforts of the algorithms used for the solution of the surface wave eigenproblems are devoted to the following two tasks:

- construction of the Love and Rayleigh *secular functions* $\mathcal{F}_{L/R}[\lambda(y),G(y),\rho(y),k_m,\omega]$;
- computation of the roots of $\mathcal{F}_{L/R}[\cdot]$ as a function of frequency or wavenumber $k_m=k_m(\omega)$[5].

For an elastic medium the use of complex arithmetic in constructing the Love and Rayleigh secular functions $\mathcal{F}_{L/R}[\cdot]$ can be avoided (Haskell, 1953; Schwab and Knopoff, 1971), and the roots of the dispersion relations can in general be obtained by means of root-bracketing techniques combined with bisection (Hisada, 1995). The use of these slow converging root-finding techniques is recommended by the rapidly oscillating behavior of the Love and Rayleigh secular functions particularly at high frequencies, which requires the use of *robust* methods (Press et al., 1992).

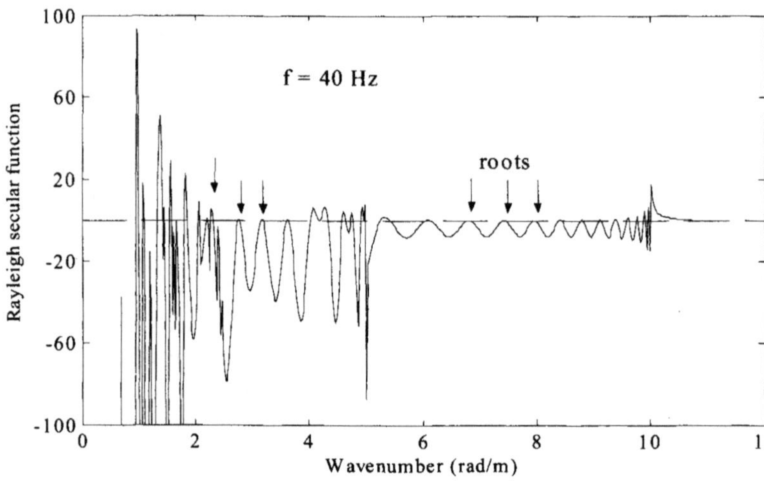


**Figure 5.** Irregular, oscillating behavior of the Rayleigh secular function. The plot shows $\mathcal{F}_R[\cdot]$ constructed using the Thomson-Haskell algorithm as a function of wavenumber at a given frequency.

[5] Computation of the relation $k_m=k_m(\omega)$ or of its inverse $\omega_m=\omega_m(k)$ is not *equivalent*. One way of proceeding may reveal unexpected difficulties (not only numerical) and results. See at this purpose the chapter of this volume by Wilmanski (2005).

Figure 5 above shows the (real-valued ) Rayleigh secular function plotted against the wavenumber for a prescribed frequency of excitation. The function $\mathcal{F}_R[\cdot]$ has been constructed using the Thomson-Haskell algorithm modified by Schwab and Knopoff (1972). The material properties of the layered medium considered for the analysis are illustrated in Table 1. The symbols α and β in Table 1 denote respectively the speed of propagation of longitudinal and transversal waves. They are functions of the Lamè parameters λ, G and mass density ρ through the relations[6] $\alpha = \sqrt{\lambda + 2G/\rho}$ and $\beta = \sqrt{G/\rho}$ .

In Figure 5 are shown the closely-spaced roots of the Rayleigh secular function identifiable from the zero-crossing of $\mathcal{F}_R[\cdot]$ with the wavenumber axis. The number of roots, at a given frequency, define the eigenvalues of the eigenproblem (13b) and identify the number of Rayleigh modes that can exist for a particular layered medium. Their precise and reliable computation is not an easy task, especially at high frequencies (Strobbia, 2003).

**Table 1.** Medium properties used to construct the Rayleigh secular function plotted in Figure 5.

| Stratum | Thickness (m) | α (m/s) | β (m/s) | ρ ($t/m^3$) |
|---|---|---|---|---|
| 1 | 5 | 100 | 50 | 1.8 |
| 2 | 10 | 300 | 200 | 1.8 |
| 3 | 20 | 800 | 500 | 1.8 |
| half-space | ∞ | 1400 | 800 | 1.8 |

As will be discussed in Section 3, determination of the roots of the Rayleigh (or Love) secular function in viscoelastic media is even more difficult because in this case $\mathcal{F}_{L/R}[\cdot]$ is a complex-valued mapping of the complex-valued wavenumber.

Figure 6 shows a graph of the roots of the Rayleigh dispersion equation $\mathcal{F}_R[\lambda(y), G(y), \rho(y), k_m, \omega] = 0$ for the layered medium of Table 1 where phase velocity, rather than wavenumber, has been plotted against frequency. Each dispersion curve is associated with a particular mode of propagation. In general, there are several modes of propagation at a given frequency with the higher modes characterized by a higher speed of propagation. Another important feature of multi-mode surface wave propagation is readily apparent from Figure 6: as the frequency increases the number of modes associated with that frequency also increases and the modes become more closely spaced. As $\omega \to \infty$ all the modes tend to a common limit which is the Rayleigh phase velocity of the thin layer neighboring the free-boundary of the vertically inhomogeneous half-space and that corresponds to a 5 m thick stratum with β = 50 m/s. Conversely, as the frequency decreases the higher modes cease to exist below their *cut-off* frequencies and the dispersion curve associated with the fundamental mode tend to approach the Rayleigh phase velocity of the half-space with β = 800 m/s.

Once the roots of the Rayleigh secular function have been obtained, computation of the eigenfunctions is a straightforward task which requires only algebraic manipulations. Figures 7 and 8 illustrate the mode shapes of respectively the *displacement* and *stress* eigenfunctions (components

[6] See the chapter by Wilmanski (2005) of this volume for a derivation of these relations.

$\tau_{yx}$ and $\tau_{yy}$) for the layered medium of Table 1 at $f$=15 Hz. The eigenfunctions have been normalized with respect to the maximum of the absolute value of the vertical displacement (see Figure 7) and maximum value of the stress component $\tau_{yy}$ (see Figure 8). A common feature of multi-mode Rayleigh wave propagation is that higher modes have a greater penetration depth than lower modes. This property, which is clearly shown in Figures 7 and 8, is very important in the solution of the inverse problem because the ability to resolve deeper layers can be directly related to the ability to explicitly recognize higher modes of propagation.

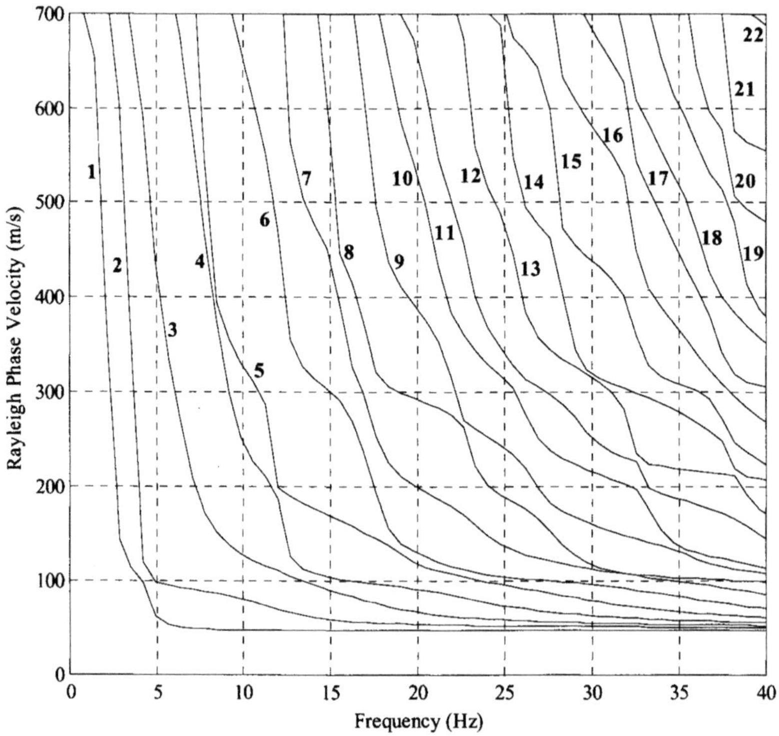


**Figure 6.** Rayleigh dispersion curves for the layered medium of Table 1. Each curve denotes a single mode of propagation and corresponds to an eigenvalue of Eq.(13b) for different values of frequencies.

As mentioned at the beginning of this section, other numerical techniques can be used to solve the systems of linear, first-order, ordinary differential equations (13a) and (13b) including the finite difference method (Boore, 1972), numerical integration (Takeuchi and Saito, 1972), the boundary element method (Manolis and Beskos, 1988), and the spectral element method (Faccioli et al., 1996; Komatitsch and Vilotte, 1998). Although these techniques are less popular than the propagator matrix methods briefly described in this section, they have several advantages. Numerical integration and spectral element method, for example, can be used in vertically inhomogeneous media where the medium properties vary *continuously* with depth, and therefore they are more general than techniques belonging to the class of propagator matrix methods. Boundary element methods are best suited for modeling unbounded or semi-infinite media because they require only boundary discretization. As a result, they reduce the dimension of the problem by one. Moreover, boundary element methods eliminate the need, required by finite element based methods, of using fictitious or non-reflecting boundaries to simulate the *Sommerfeld* condition at infinity.

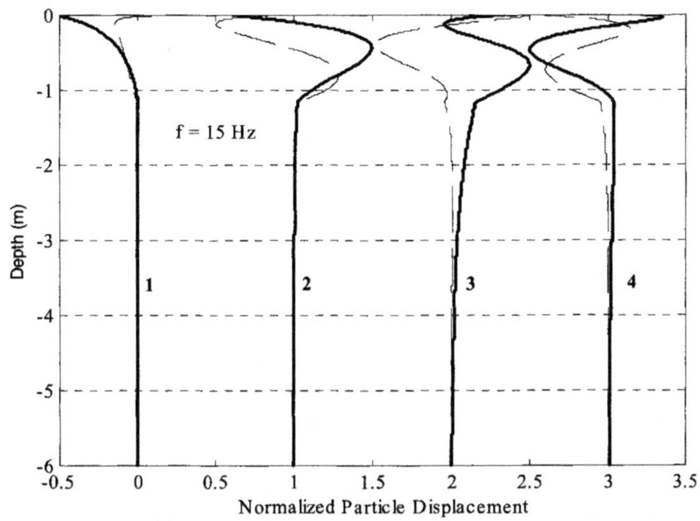


**Figure 7.** Horizontal (dashed lines) and vertical (bolded lines) components of Rayleigh *displacement* eigenfunctions for the layered medium of Table 1 at $f$=15 Hz. Each pair of curves denotes a single mode.

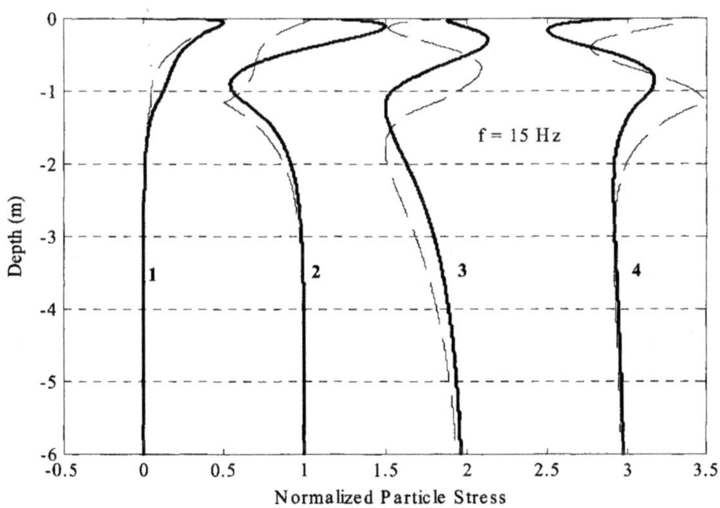


**Figure 8.** Components $\tau_{yx}$ (dashed lines) and $\tau_{yy}$ (bolded lines) of Rayleigh *stress* eigenfunctions for the layered medium of Table 1 at $f$=15 Hz. Each pair of curves denotes a single mode.

**A notable degenerate case: surface Rayleigh waves in homogeneous continua.** The systems of linear, first-order, ordinary differential equations (13a) and (13b) and their associated boundary conditions (14a) and (14b) govern the propagation of *free*, monochromatic Love and Rayleigh surface waves in general, linear elastic vertically inhomogeneous continua. Mathematically, the main feature of equations (13a) and (13b) is to have variable coefficients since Lamé parameters $\lambda$ and $G$ and mass density $\rho$ are all functions of depth y, and it is precisely this attribute that makes their solution non trivial as it was briefly discussed in the previous section.

However in homogeneous media matrices **B** and **A** of equations (13a) and (13b) become independent of y and their solution can immediately be found in closed-form. A well known result of the theory of linear, first-order, ordinary differential equations (Birkhoff and Rota, 1978) is that under appropriate conditions (which are rather weak) the most general solution of a system like $\mathbf{df}(y)/dy = \mathbf{A}\cdot\mathbf{f}(y)$ with **A** being an $n \times n$ matrix of constant coefficients and **f**(y) an $n$-dimensional vector is simply given by a linear combination of the type:

$$\mathbf{f}(y) = \mathbf{V}(y)\cdot\mathbf{\Lambda}(y)\cdot\mathbf{c} \tag{15}$$

where **V**(y) is an $n \times n$ matrix formed by $n$ linearly independent eigenvectors $\mathbf{v}^{(i)}(y)$, $i$=1,$n$ of matrix **A**, $\Lambda(y)$ is a diagonal matrix whose elements are terms like $\exp(\lambda^{(i)}\cdot y)$ where $\lambda^{(i)}$ is the $i^{th}$ eigenvalue of matrix **A** associated with the eigenvector $\mathbf{v}^{(i)}(y)$. Finally the symbol **c** in Eq.(15) denotes an $n$-dimensional vector of constants.

Now if the matrix **A** of equation $\mathbf{df}(y)/dy = \mathbf{A}\cdot\mathbf{f}(y)$ is identified with matrix **A** of Eq.(13b), Eq. (15) formally represents the most general solution **f**(y) of the Rayleigh eigenproblem in linear elastic, homogeneous media. The four eigenvalues $\lambda^{(i)}$, $i$=1,4 are given in this case by the relations:

$$\begin{cases} \lambda^{(1,2)} = \pm\sqrt{\dfrac{k^2-\omega^2}{\alpha^2}} \\ \lambda^{(3,4)} = \pm\sqrt{\dfrac{k^2-\omega^2}{\beta^2}} \end{cases} \tag{16}$$

whereas the non-zero elements of the 4 × 4 factored matrix $\mathbf{W}(y)=\mathbf{V}(y)\cdot\Lambda(y)$ are given (Aki and Richards, 2002) by Eq.(17).

To truly represent surface Rayleigh waves, the general solution given by Eqs.(15), (16) and (17) must satisfy the boundary conditions (14b). Application of relations (14b) to Eq.(15) for $\mathbf{f}(y) = [r_1 \quad r_2 \quad r_3 \quad r_4]^T$ yields four homogeneous algebraic equations in the four unknown constants represented by the vector **c**. A non trivial solution of this system of equations exists if the *determinant* of the coefficients vanishes leading to the well-known *Rayleigh dispersion equation*:

$$\left(\frac{V_R}{\beta}\right)^6 - 8\cdot\left(\frac{V_R}{\beta}\right)^4 + 8\cdot\left(\frac{V_R}{\beta}\right)^2\cdot\left[1+2\cdot\left(1-\frac{\beta^2}{\alpha^2}\right)\right] - 16\cdot\left(1-\frac{\beta^2}{\alpha^2}\right) = 0 \tag{18}$$

where $V_R$ is the phase velocity of Rayleigh waves in homogeneous, linear elastic half-spaces.

$$\begin{cases} w_{11}=w_{33}=1+\dfrac{2G}{\omega^2\rho}\cdot\left[2k^2\sinh^2\dfrac{\lambda^{(1)}y}{2}-[k^2+(\lambda^{(3)})^2]\cdot\sinh^2\dfrac{\lambda^{(3)}y}{2}\right] \\ w_{12}=-w_{43}=\dfrac{kG}{\omega^2\rho}\cdot\left[[k^2+(\lambda^{(3)})^2]\cdot\dfrac{\sinh\lambda^{(1)}y}{\lambda^{(1)}}-2\lambda^{(3)}\sinh\lambda^{(3)}y\right] \\ w_{13}=\dfrac{1}{\omega^2\rho}\cdot\left[k^2\cdot\dfrac{\sinh\lambda^{(1)}y}{\lambda^{(1)}}-\lambda^{(3)}\sinh\lambda^{(3)}y\right] \\ w_{14}=-w_{23}=\dfrac{2k}{\omega^2\rho}\cdot\left[\sinh^2\dfrac{\lambda^{(1)}y}{2}-\sinh^2\dfrac{\lambda^{(3)}y}{2}\right] \\ w_{21}=-w_{34}=\dfrac{kG}{\omega^2\rho}\cdot\left[[k^2+(\lambda^{(3)})^2]\cdot\dfrac{\sinh\lambda^{(3)}y}{\lambda^{(3)}}-2\lambda^{(1)}\sinh\lambda^{(1)}y\right] \\ w_{22}=w_{44}=1+\dfrac{2G}{\omega^2\rho}\cdot\left[2k^2\sinh^2\dfrac{\lambda^{(3)}y}{2}-[k^2+(\lambda^{(3)})^2]\cdot\sinh^2\dfrac{\lambda^{(1)}y}{2}\right] \\ w_{24}=\dfrac{1}{\omega^2\rho}\cdot\left[k^2\cdot\dfrac{\sinh\lambda^{(3)}y}{\lambda^{(3)}}-\lambda^{(1)}\sinh\lambda^{(1)}y\right] \\ w_{31}=\dfrac{G^2}{\omega^2\rho}\cdot\left[4k^2\lambda^{(1)}\sinh\lambda^{(1)}y-[k^2+(\lambda^{(3)})^2]^2\cdot\dfrac{\sinh\lambda^{(3)}y}{\lambda^{(3)}}\right] \\ w_{32}=-w_{41}=2G^2\cdot[k^2+(\lambda^{(3)})^2]\cdot w_{14} \\ w_{42}=\dfrac{G^2}{\omega^2\rho}\cdot\left[4k^2\lambda^{(3)}\sinh\lambda^{(3)}y-[k^2+(\lambda^{(3)})^2]^2\cdot\dfrac{\sinh\lambda^{(1)}y}{\lambda^{(1)}}\right] \end{cases} \tag{17}$$

Equation (18) shows that $V_R$ depend solely on $\alpha$ and $\beta$ which are intrinsic properties of the medium[6] and does not exhibit any dependence upon the wavenumber $k$ proving the assertion that Rayleigh waves at the free-boundary of a linear elastic, homogeneous medium are *non-dispersive* that is their speed of propagation is independent of frequency. This happens because a homogeneous half-space does not possess an intrinsic length scale (Aki and Richards, 2002).

It is easy to show that Eq.(18) can be re-written with the ratio $V_R/\beta$ expressed solely as a function of the *Poisson ratio* $\nu$ of the medium[7]. Figure 9 shows a plot of the variation of $V_R/\beta$ with Poisson ratio. The dependence upon $\nu$ exhibited by the ratio $V_R/\beta$ is rather weak: as $\nu$ varies from 0 to 0.5 $V_R/\beta$ changes from 0.862 to 0.955. In the same plot it is also shown the variation of $\alpha/\beta$ with Poisson ratio which is, as expected, much stronger: in fact as $\nu \to 0.5$ the ratio $\alpha/\beta \to \infty$ because for $\nu = 0.5$ the deformation of the medium becomes *isochoric*.

Equation (18) is an algebraic equation of the six degree in $V_R/\beta$ and a question arises upon the number of its roots along the real axis and their physical meaning. The problem is non-trivial and a careful analysis is beyond the scope of this section. The matter has been thoroughly investigated by several authors[8] using the elegant method based on the *principle of the argument* of complex variable theory (Remmert, 1998). The conclusion is that Eq.(18) may only have two *real* roots of which only the positive one is physically meaningful for it correspond to $V_R$: the speed of propagation of Rayleigh waves in linear elastic, homogeneous media.

[7] Since $\nu$ can be expressed only as a function of $\alpha$ and $\beta$ via the relation $\nu = 0.5\cdot[(\alpha^2-2\cdot\beta^2)/(\alpha^2-\beta^2)]$.

[8] See for instance Achenbach (1984).

After computing $V_R$ from Eq.(18), the last step to completely resolve the Rayleigh eigenproblem is the computation of displacement and stress eigenfunctions a task which is accomplished by using Eqs.(15), (16) and (17).

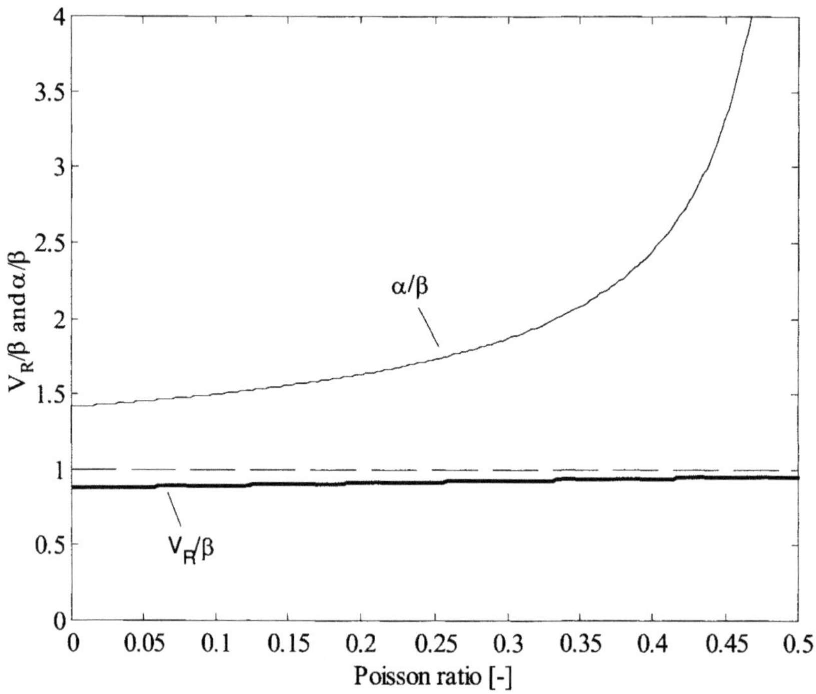


**Figure 9.** Variation of the ratios $V_R/\beta$ and $\alpha/\beta$ with Poisson ratio. For $\nu$ varying from 0 to 0.5 the ratio $V_R/\beta$ increases from 0.862 to 0.955. For $\alpha/\beta$ the variation is much greater with $\alpha/\beta \rightarrow \infty$ as $\nu \rightarrow 0.5$.

It is found that the particle orbit is elliptical retrograde at the free boundary (see Figure 3) with a ratio of vertical to horizontal displacement of about 1.5. The particle motion become prograde elliptical at a depth of about $0.2 \cdot \lambda$.

## 2.3 Variational Principles of Love and Rayleigh Waves

**Introduction.** This section will derive some rather interesting results that can be obtained from the application of *Hamilton's principle* to a vertically inhomogeneous, linear elastic half-space deformed by the passage of free, plane Love and Rayleigh waves. The most significant of these results is the derivation of closed-form expressions for the *Jacobian* of phase velocity of surface waves with respect to medium parameters $\alpha$ and $\beta$. As already mentioned in Section 2.1, these expressions are fundamental in the solution of the inverse problem, a topic that will be discussed in Section 4. The Love and Rayleigh variational principles can be used to compute the Jacobian of both modal and apparent surface wave phase velocity with respect to $\alpha$ and $\beta$ (see Section 2.4 and Lai, 1998).

These principles can also be used to obtain important results for the propagation of surface waves in weakly attenuating media. In fact, it will be shown in Section 3 that the solution of the Love and Rayleigh eigenproblems in linear elastic media can be used as a basis to compute an approximation of the attenuation coefficients $\alpha_{L/R}(\omega)$ which are the parameters that characterize the response of dissipative media to the propagation of surface waves.

**Hamilton's principle applied to surface waves.** Hamilton's principle applied to a continuous, conservative system of volume V, states that among all the possible paths of motion between two instants in time $t_1$ and $t_2$ the actual path is such that the integral:

$$\mathscr{I} = \int_{t_1}^{t_2}\int_V \mathscr{L}(q_i, \dot{q}_i, q_{i,j})\, dV\, dt \tag{19}$$

where $\mathscr{L}$= T-U has a stationary value (Goldstein, 1980). The function $\mathscr{L}=(\cdot)$ already introduced in Section 2.2, is called the *Lagrangian density*, whereas the variables $q_i$, i=1,N describing the behavior of the system are the *N* generalized coordinates. Finally *T* and *U* are the kinetic and the potential energy of the system, respectively. It is well known from Calculus of Variations (Logan, 1997) that a necessary condition for the integral of Eq.(19) to have a stationary value it is required that:

$$\delta(\mathscr{I}) = \delta \int_{t_1}^{t_2}\int_V \mathscr{L}(q_i, \dot{q}_i, q_{i,j})\, dV\, dt = 0 \tag{20}$$

that is the first variation of the integral $\mathscr{I}$ vanishes for arbitrary changes $\delta q_i$ which vanish at the boundary of the volume V and at times $t_1$ and $t_2$. Practical implementation of Eq.(20) yields the *Lagrange's equations of motion* of the system and their associated natural boundary conditions. In Section 2.2. it was shown that in a continuous linear elastic material the Lagrangian density is given by Eq.(3) which is reported here for convenience:

$$\mathscr{L}(\dot{u}_i, u_{i,j}) = \frac{1}{2}\rho \dot{u}_i \dot{u}_i - \left[\frac{1}{2}\lambda(\varepsilon_{kk})^2 + G\varepsilon_{ij}\varepsilon_{ij}\right] \tag{3bis}$$

where $\lambda = \lambda(y)$, $G = G(y)$, and $\rho = \rho(y)$. If the elastic body of volume V is identified with a vertically inhomogeneous, linear elastic half-space with no body forces and surface tractions, and if the displacement field $\mathbf{u}(\mathbf{x}, t)$ is specified according to Eq.(5a) for Love waves and Eq.(5b) for Rayleigh waves, the expression for the Lagrangian density becomes:

For *Love waves*:
$$\langle \mathscr{L} \rangle = \frac{1}{2}\rho\omega^2 l_1^2 - \frac{1}{4}G \cdot \left[k^2 l_1^2 + \left(\frac{dl_1}{dy}\right)^2\right] \tag{21a}$$

For *Rayleigh waves*:
$$\langle \mathscr{L} \rangle = \frac{1}{4}\rho\omega^2\left(r_1^2 + r_2^2\right) - \frac{1}{4} \cdot \left[\lambda\left(kr_1 + \frac{dr_2}{dy}\right)^2 + G\left(\frac{dr_1}{dy} - kr_2\right)^2 + 2G\left(k^2r_1^2 + \left(\frac{dr_2}{dy}\right)^2\right)\right] \tag{21b}$$

where the symbol $\langle\cdot\rangle$ denotes the *average value* of the quantity inside the brackets which has been used to eliminate the time dependence from the definition of the harmonic Lagrangian density.

Equations (21a) and (21b) defines respectively the average Lagrangian density $\langle \mathscr{L} \rangle$ for the Love and Rayleigh eigenproblem. Application of Hamilton's principle with the Lagrangian density given by Eqs. (21a) and (21b) for any perturbation of the eigenfunctions $l_1(y,k,\omega)$ for Love waves and $r_1(y,k,\omega)$, $r_2(y,k,\omega)$ for Rayleigh waves satisfying the boundary conditions (14a) and (14b) yields:

For *Love waves*:
$$\delta(\mathscr{I}) = \delta\int_0^\infty \langle \mathscr{L} \rangle dy = \delta\left[\omega^2 J_1 - k^2 J_2 - J_3\right] = 0 \tag{22a}$$

For *Rayleigh waves*:
$$\delta(\mathscr{I}) = \delta\int_0^\infty \langle \mathscr{L} \rangle dy = \delta\left[\omega^2 I_1 - k^2 I_2 - kI_3 - I_4\right] = 0 \tag{22b}$$

where J1, J2, J3, and I1, I2, I3, I4 are respectively the Love and Rayleigh energy integrals, and are defined as follows (Ben-Menahem and Singh, 2000; Aki and Richards, 2002):

For *Love waves*:
$$\begin{cases} J_1 = \dfrac{1}{2}\displaystyle\int_0^\infty \rho l_1^2 dy \qquad J_2 = \dfrac{1}{2}\displaystyle\int_0^\infty G l_1^2 dy \\ J_3 = \dfrac{1}{2}\displaystyle\int_0^\infty G \cdot \left(\dfrac{dl_1}{dy}\right)^2 dy \end{cases} \tag{23a}$$

For *Rayleigh waves*:
$$\begin{cases} I_1 = \dfrac{1}{2}\displaystyle\int_0^\infty \rho(r_1^2 + r_2^2)dy \qquad I_2 = \dfrac{1}{2}\displaystyle\int_0^\infty [(\lambda + 2G)r_1^2 + Gr_2^2]dy \\ I_3 = \displaystyle\int_0^\infty \left(\lambda r_1 \dfrac{dr_2}{dy} - Gr_2 \dfrac{dr_1}{dy}\right)dy \\ I_4 = \dfrac{1}{2}\displaystyle\int_0^\infty \left[(\lambda + 2G)\cdot\left(\dfrac{dr_2}{dy}\right)^2 + G\left(\dfrac{dr_1}{dy}\right)^2\right]dy \end{cases} \tag{23b}$$

Equations (22a) and (22b) show that the quantity ($\omega^2 J_1$-$k^2 J_2$-$J_3$) for Love waves and ($\omega^2 I_1$-$k^2 I_2$-$kI_3$-$I_4$) for Rayleigh waves is *stationary* for arbitrary perturbation of $l_1(y,k,\omega)$ and $r_1(y,k,\omega)$, $r_2(y,k,\omega)$ around an eigenfunction. Since the latter is the only quantity to be perturbed it is possible to write:

For *Love waves*: $$\delta(\mathscr{I}) = \delta\int_0^\infty \langle\mathscr{L}\rangle dy = (\omega^2 \delta J_1 - k^2 \delta J_2 - \delta J_3) = 0 \quad (24a)$$

For *Rayleigh waves*: $$\delta(\mathscr{I}) = \delta\int_0^\infty \langle\mathscr{L}\rangle dy = (\omega^2 \delta I_1 - k^2 \delta I_2 - k\delta I_3 - \delta I_4) = 0 \quad (24b)$$

Finally, from the equations of motion (13a), (13b) and the associated boundary conditions (14a) and (14b) it can be proved (Aki and Richards, 2002) that at the stationary point where $\delta(\mathscr{I}) = 0$ holds the following result:

For *Love waves*: $$\int_0^\infty \langle\mathscr{L}\rangle dy = (\omega^2 J_1 - k^2 J_2 - J_3) = 0 \quad (25a)$$

For *Rayleigh waves*: $$\int_0^\infty \langle\mathscr{L}\rangle dy = (\omega^2 I_1 - k^2 I_2 - kI_3 - I_4) = 0 \quad (25b)$$

Equations (25a) and (25b) written in the form ($\omega^2 J_1$=$k^2 J_2$+$J_3$) and ($\omega^2 I_1$=$k^2 I_2$+$kI_3$+$I_4$) can be interpreted as a statement of *conservation of energy*, i.e., the average kinetic energy associated with a given surface wave mode of propagation equals the average elastic strain energy.

The combined results given by Eqs.(24a), (24b) and (25a), (25b) will be referenced hereafter as the *variational principle of Love and Rayleigh waves* respectively. This should not be confused with *Love* and *Rayleigh principles* which assert that first-order perturbations in either a Love or a Rayleigh eigenvalue (namely the wavenumber *k*) will only result in second-order perturbations of the corresponding eigenfunctions (Ben-Menahem and Singh, 2000) and may, in fact, be derived from Eqs.(24) and (25) if desired.

**Analytical Jacobian of Rayleigh phase velocity.** An important application of the variational principle of Love and Rayleigh waves is the derivation of closed-form expressions for the *Jacobian* of phase velocity of surface waves with respect to medium parameters α and β which will be useful for the solution of the Love and Rayleigh inverse problem (see Section 4). Due to lack of room the demonstration of this result will be limited to the case of Rayleigh waves which is also the most significant for the applications and to a certain extent also the more involved.

Let the triplet {λ(y), G(y), ρ(y)} represent the material properties of a linear elastic, vertically inhomogeneous medium $\mathscr{M}$. The set $\{k_m, r_n(\lambda, G, \rho, y, k_m, \omega)\}$, m=1,$M^4$; n=1,4 represents the solution of the Rayleigh eigenvalue problem associated with this medium. In the functional dependence of the eigenfunctions $r_n$, the material parameters λ(y), G(y), and ρ(y) have been considered as the independent variables. Now let {λ(y)+δ[λ(y)], G(y)+ δ[ G(y)], ρ(y)} denote the material properties of a medium $\tilde{\mathscr{M}}$ whose Lamè's parameters differ slightly from those of the medium $\mathscr{M}$. At a given frequency ω, the set $\{k_m + \delta k_m, r_n(\lambda+\delta\lambda, G+\delta G, \rho, y, k_m+\delta k_m, \omega)\}$ represents the solution of the Rayleigh eigenproblem associated with this new medium.

In the sequel it will be shown that the problem of determining the partial derivatives of the modal Rayleigh phase velocity $[V_R(\omega)]_m$ with respect to the medium parameters $\alpha$ and $\beta$ is essentially reduced to that of computing $\delta k_m$. The latter task is accomplished by using the variational principle of Rayleigh waves, namely Eqs.(24b) and (25b).

For ease of notation in the sequel the *modal* parameters $k_m$ and $[V_R(\omega)]_m$ will be denoted without the subscript *m*. Furthermore, the medium properties $\alpha$ and $\beta$ will also be indicated with the symbols $V_P$ and $V_S$ respectively.

Application of Eq.(25b) to the medium $\tilde{\mathscr{M}}$ yields:

$$\begin{aligned}
&\frac{1}{2}\omega^2\int_0^\infty \rho\left[(r_1+\delta r_1)^2+(r_2+\delta r_2)^2\right]dy= \\
&\frac{1}{2}(k+\delta k)^2\int_0^\infty\left\{\left[(\lambda+\delta\lambda)+2(G+\delta G)\right](r_1+\delta r_1)^2+(G+\delta G)(r_2+\delta r_2)^2\right\}dy+ \\
&+(k+\delta k)\int_0^\infty\left\{(\lambda+\delta\lambda)(r_1+\delta r_1)\frac{d}{dy}(r_2+\delta r_2)-(G+\delta G)(r_2+\delta r_2)\frac{d}{dy}(r_1+\delta r_1)\right\}dy+ \\
&+\frac{1}{2}\int_0^\infty\left\{\left[(\lambda+\delta\lambda)+2(G+\delta G)\right]\cdot\left[\frac{d}{dy}(r_2+\delta r_2)\right]^2+(G+\delta G)\cdot\left[\frac{d}{dy}(r_1+\delta r_1)\right]^2\right\}dy
\end{aligned} \tag{26}$$

Expanding Eq.(26) to include only first order terms and taking into account Eqs.(24b) and (25b) leads to:

$$\begin{aligned}
&\delta k\int_0^\infty\left\{\lambda r_1\frac{dr_2}{dy}-Gr_2\frac{dr_1}{dy}+k\left[(\lambda+2G)r_1^2+Gr_2^2\right]\right\}dy+ \\
&+\int_0^\infty\left\{\left[\frac{1}{2}\left(\frac{dr_2}{dy}\right)^2+kr_1\frac{dr_2}{dy}+\frac{1}{2}k^2r_1^2\right]\delta(\lambda+2G)\right\}dy+ \\
&+\int_0^\infty\left\{\left[\frac{1}{2}\left(\frac{dr_1}{dy}\right)^2-k\left(r_2\frac{dr_1}{dy}+2r_1\frac{dr_2}{dy}\right)+\frac{1}{2}k^2r_2^2\right]\delta G\right\}dy=0
\end{aligned} \tag{27}$$

Now from Eq.(25b):

$$\begin{aligned}
&\delta\left(\omega^2 I_1-k^2I_2-kI_3-I_4\right)= \\
&=2\omega I_1\delta\omega+\omega^2\delta I_1-2I_2k\delta k-k^2\delta I_2-I_3\delta k-k\delta I_3-\delta I_4=0
\end{aligned} \tag{28}$$

which considering Eq.(24b) yields:

$$U = \frac{\delta\omega}{\delta k} = \frac{2kI_2 + I_3}{2\omega I_1} \tag{29}$$

Equation (29) is an important result because it provides an alternative procedure to compute the (modal) *group velocity* of Rayleigh waves without using numerical differentiation which is from a numerical point of view an unstable operation. In light of Eq.(29), Eq.(27) can be re-written as:

$$\delta k = -\frac{1}{4\omega U I_1}\left\{\int_0^\infty \left[\left(kr_2 - \frac{dr_1}{dy}\right)^2 - 4kr_1\frac{dr_2}{dy}\right]\delta G dy + \int_0^\infty \left(kr_1 + \frac{dr_2}{dy}\right)^2 \delta(\lambda + 2G) dy\right\} \tag{30}$$

Now since $V_R=\omega/k$, it follows that $\delta k = -(k^2/\omega)\cdot\delta V_R$, and hence from Eq.(30) one obtains:

$$\delta V_R = \frac{1}{4k^2 U I_1}\left\{\int_0^\infty \left[\left(kr_2 - \frac{dr_1}{dy}\right)^2 - 4kr_1\frac{dr_2}{dy}\right]\delta G dy + \int_0^\infty \left(kr_1 + \frac{dr_2}{dy}\right)^2 \delta(\lambda + 2G) dy\right\} \tag{31}$$

Equation (31) allows to compute the change in Rayleigh phase velocity $\delta V_R$ resulting from a small perturbation of the Lamè's parameters $\lambda(y)$ and $G(y)$. This equation may also be expressed as:

$$\delta V_R = \left\{\int_0^\infty \left[\frac{\partial V_R}{\partial G}\right]_{\omega,M} \delta G dy + \int_0^\infty \left[\frac{\partial V_R}{\partial M}\right]_{\omega,G} \delta M dy\right\} \tag{32}$$

where $M = (\lambda + 2G)$ is the constrained modulus and the terms in brackets are defined as:

$$\begin{cases} \left[\dfrac{\partial V_R}{\partial G}\right]_{\omega,M} = \dfrac{1}{4k^2 U I_1}\cdot\left[\left(kr_2 - \dfrac{dr_1}{dy}\right)^2 - 4kr_1\dfrac{dr_2}{dy}\right] \\ \left[\dfrac{\partial V_R}{\partial M}\right]_{\omega,G} = \dfrac{1}{4k^2 U I_1}\cdot\left(kr_1 + \dfrac{dr_2}{dy}\right)^2 \end{cases} \tag{33}$$

The subscripts outside the brackets are used to indicate the parameters that are held constant. Equations (33) are known in seismology as the partial derivatives of the modal Rayleigh phase velocity $V_R(\omega)$ with respect to the medium parameters G and M. However, the term partial derivative should be used with caution because $V_R = V_R\{M(y), G(y), \rho(y), k, \omega\}$ is a *functional* rather than a function of the parameters M(y) and G(y). Accordingly, the partial derivatives in Eqs. (33) are implicitly understood as referred to a particular depth $y^*$.

From Eqs.(33) it is easy to compute the partial derivatives of $V_R$ with respect to the medium parameters $V_P$ and $V_S$. Since $G = \rho V_S^2$ and $M = \rho V_P^2$, using the chain-rule it is straightforward to obtain:

$$
\begin{aligned}
\left[\frac{\partial V_R}{\partial V_S}\right]_{\omega,V_P} &= \frac{\partial V_R}{\partial V_S}(y,k,\omega)=\left[\frac{\partial V_R}{\partial G}\right]_{\omega,M}\cdot\left[\frac{\partial G}{\partial V_S}\right]_{\rho} = \frac{\rho V_S}{2k^2UI_1}\cdot\left[\left(kr_2-\frac{dr_1}{dy}\right)^2-4kr_1\frac{dr_2}{dy}\right] \\
\left[\frac{\partial V_R}{\partial V_P}\right]_{\omega,V_S} &= \frac{\partial V_R}{\partial V_P}(y,k,\omega)=\left[\frac{\partial V_R}{\partial M}\right]_{\omega,G}\cdot\left[\frac{\partial M}{\partial V_P}\right]_{\rho} = \frac{\rho V_P}{2k^2UI_1}\cdot\left(kr_1+\frac{dr_2}{dy}\right)^2
\end{aligned}
\tag{34}
$$

Equations (34) provide a *measure* of the sensitivity of modal Rayleigh phase velocity $V_R(\omega)$ to small changes of the medium parameters $V_P$ and $V_S$ at a specific depth $y^*$. The remarkable feature of these equations which makes them so important in the solution of the Rayleigh inverse problem, is that these partial derivatives can be computed using Rayleigh wave parameters referred to the original and not the perturbed $V_P$ and $V_S$ profiles.

Conversely, it would be very expensive to compute the above partial derivatives numerically with, say, a four-point central finite difference scheme (Spang, 1995): a single computation of $\partial V_R/\partial V_S$ would require the solution of *four* Rayleigh eigenproblems instead of just one eigenproblem using the variational approach.

A careful examination of Eqs.(34) shows that the phase velocity of Rayleigh waves is relatively insensitive to changes in the parameter $V_P$ (Lee and Solomon, 1979; Ben-Menahem and Singh, 2000), and thus the partial derivative $\partial V_R/\partial V_P$ is relatively small compared to $\partial V_R/\partial V_S$. This is not surprising because a similar conclusion has already been obtained in homogeneous media where it was shown (see Figure 9) that the variation of $V_R/\beta$ with Poisson ratio is rather the weak as opposed to the ratio $\alpha/\beta$. A confirmation of this result is given by Figure 10 which shows a plot of the partial derivatives given by Eq.(34) for the case of a homogeneous medium at a frequency of 40 Hz (Lai, 1998).

It is apparent from Figure 10 that the transversal phase velocity $V_S$ governs the value of Rayleigh phase velocity $V_R$. The largest values obtained for $\partial V_R/\partial V_S$ and $\partial V_R/\partial V_P$ are 0.418 and 0.105 respectively. Both maxima occur at the free-boundary ($y = 0$).

In a stratified medium composed by a finite number of homogeneous layers overlaying a homogeneous half-space it may be of interest to evaluate how Rayleigh phase velocity $V_R$ is affected by a small change in the medium parameters $V_P$ and $V_S$ of a given layer. This quantity can be computed by integrating Eq. (34) over the layer thickness. Thus, for a layered medium composed by $N_L$ layers $L_j$, $j=1,N_L$ the results is:

$$
\begin{aligned}
\left[\frac{\partial V_R}{\partial V_S}\right]_{L_j} &= \frac{\rho V_S}{2k^2UI_1}\cdot\int_{y_{j-1}}^{y_j}\left[\left(kr_2-\frac{dr_1}{dy}\right)^2-4kr_1\frac{dr_2}{dy}\right]dy \\
\left[\frac{\partial V_R}{\partial V_P}\right]_{L_j} &= \frac{\rho V_P}{2k^2UI_1}\cdot\int_{y_{j-1}}^{y_j}\left(kr_1+\frac{dr_2}{dy}\right)^2 dy
\end{aligned}
\tag{35}
$$

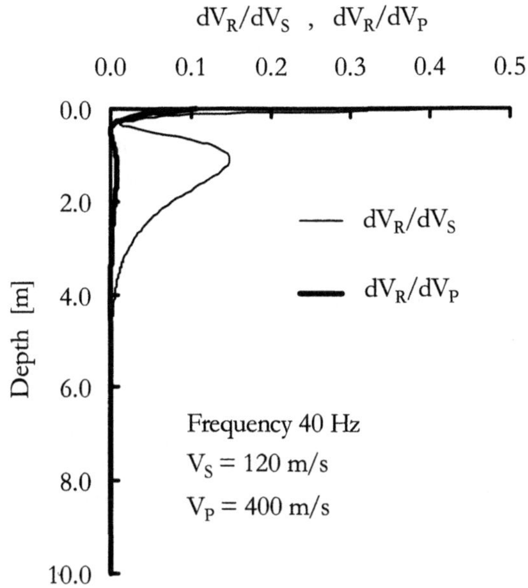


**Figure 10.** Sensitivity of partial derivatives of Rayleigh phase velocity with respect to material parameters $V_P$ and $V_S$ for an homogeneous medium. Results obtained for a frequency of 40 Hz (Lai, 1998).

### 2.4 Inhomogeneous Boundary Value Problems of Elastodynamics

**Introduction.** The linear eigenvalue problems (13a), (13b) and their associated boundary conditions (14a), (14b) correspond formally to *homogeneous* boundary value problems whose solutions are known as *free* Love and Rayleigh waves (Ewing et al., 1957). Under appropriate generating conditions however, surface waves can be generated by point or line sources applied at a free boundary of an elastic half-space. The corresponding *inhomogeneous* boundary value problems form the guts of one classical themes of elastodynamics[9] which is commonly referred to as the "*Lamb's problem*"[10].

Given the vastness of the subject it is impossible in these notes to treat in detail the solution of Lamb's problem in all its ramifications. The interested reader is referred to the extensive literature existing for this topic in addition to the already cited references[11]. In this section the attention is restricted to showing some features of the solution of Lamb's problem for a normal point source in relation to the Rayleigh surface wavefield followed by a brief introduction to the concept of *Green's function* which is of fundamental importance in elastodynamics. The last topic to be

[9] In reality the *wavefield* generated by shallow or buried sources at the free-boundary of an elastic half-space includes also other types of waves other than surface waves such as P, S and head waves.

[10] Although nowadays for *Lamb's problem* is intended a somewhat more general meaning (Achenbach, 1984; Ewing et al., 1957; Aki and Richards, 2002).

[11] See for instance Lamb (1904); Pekeris (1955); Eringen and Suhubi (1975); Hudson (1980); Kennett (1983); Bleistein (1984); Båth and Berkhout (1984); Malischewsky (1988); Philippacopoulos (1988); Sato and Fehler (1997).

discussed and that closes this section on surface wave propagation in linear elastic, vertically inhomogeneous media is that of *apparent* phase velocity, a concept arising in connection with mode superposition of surface Rayleigh waves generated by harmonic sources.

**Lamb problem for time-harmonic point source.** The problem of determining the displacement field induced by a vertical, time-harmonic point load applied at the free boundary of an homogeneous, isotropic, linear elastic half-space, was first solved by Lamb (1904) in his classical paper entitled "*On the Propagation of Tremors over the Surface of an Elastic Solid*". Lamb used complex variable theory and contour integration to find the solution of what can be considered as the dynamic analogue of another classical problem of linear elasticity: the *Boussinesq's* problem.

In reality Lamb in his work has worked out several other boundary value problems including that of the line loads applied normally to the free-surface or inside the half-space. For both the vertical point and line loads he considered harmonic and impulsive time variations. Figure 11 shows Lamb's results for the horizontal and vertical components of the displacement generated at the free-boundary (in the far-field approximation) by an impulse vertical line loading function. This plot has been considered by many as the first *synthetic seismogram* ever computed.

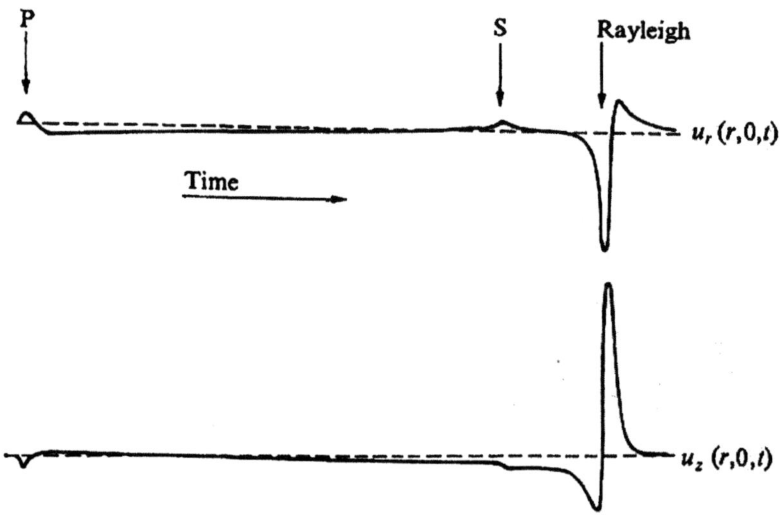


**Figure 11.** Horizontal (top) and vertical (bottom) components of surface displacement obtained in the far-field by Lamb (1904) for an impulse-type vertical line loading function. In the top figure are indicated the time of arrival at a specific location of longitudinal (P), transversal (S) and Rayleigh waves.

The wavefield shown in the figure has been computed by Lamb at a particular spatial location, far from the source. The first disturbance is due to the arrival of longitudinal (P) wave followed by the transversal (S) wave. The greatest disturbance corresponds to the arrival of the Rayleigh wave which is referred in the original Lamb's paper as the "*major tremor*" whereas body P and S waves are named "*minor tremors*".

Returning to the case of time-harmonic, vertical point source applied at the free-boundary of a linear elastic, homogeneous half-space, the solution given by Lamb for the Rayleigh vertical displacement field $u_y(r,\omega)$ in the far-field approximation can be written as follows:

$$u_y(r,\omega) = \frac{Fe^{i\omega t}}{2iG} \cdot k_R \cdot \Phi(k_R) \cdot H_0^{(2)}(k_R r) \tag{36}$$

where $r$ is the distance from a vertically oscillating harmonic force $Fe^{i\omega t}$, $G$ is the tangential modulus of the elastic medium, $k_R$ is the Rayleigh wavenumber, and $H_0^{(2)}(\cdot)$ denotes the *Hankel* function of the second kind of zero order[12]. The function $\Phi(k_R)$ is defined as follows:

$$\Phi(k_R) = -\frac{k_S^2 \sqrt{k_R^2 - k_P^2}}{R'(k_R)} \tag{37}$$

where $k_P$ and $k_S$ are the wavenumbers of P and S waves respectively. Finally, the function $R(k_R)$ is given by the following expression:

$$R(k_R) = \left(2k_R^2 - k_S^2\right)^2 - 4k_R^2 \sqrt{\left(k_R^2 - k_P^2\right)\left(k_R^2 - k_S^2\right)} \tag{38}$$

where $k_R = \omega / V_R$ and $V_R$ is the frequency-independent Rayleigh phase velocity to be determined from the solution of the *Rayleigh dispersion equation* (18).

Lamb's solution was computed for the medium properties illustrated in Table 2 and for a vertical force of unit magnitude. Figure 12 shows the synthetic seismograms obtained from the inverse Fourier transform of Eq.(36) calculated at five distances from the source: 100, 200, 300, 400 and 500 m. The vertical component of the displacement field $u_y(r,\omega)$ was computed for a frequency range from 0.1 to 20 Hz. The seismograms shown in Figure 12 illustrate the response of the half-space to a vertical point source with time-variation given by a *Dirac* $\delta$-function.

**Table 2** Medium properties used to compute Lamb's solution.

| Half-Space | $\alpha$ (m/s) | $\beta$ (m/s) | $\rho$ (t/m$^3$) |
|---|---|---|---|
| Homogeneous | 500 | 250 | 1.8 |

It is important to emphasize that waveforms computed from Eq.(36) represent *only* to the contribution of the Rayleigh surface wavefield. Other types of contributions (mainly the body wave field) are not accounted for in Eq.(36). The figure also shows an approximate calculation of the Rayleigh phase velocity from the slope of a straight line joining points of the seismograms of

[12] The *Hankel* function $H_0^{(2)}(\cdot)$ is defined as $H_0^{(2)}(z) = J_0(z) - iY_0(z)$ where $J_0(z)$ and $Y_0(z)$ are the *Bessel* functions of the first kind and second kind, respectively, of zero order.

equal phase. The amplitudes in each seismogram have been normalized with respect to the maximum of the absolute value of vertical displacement.

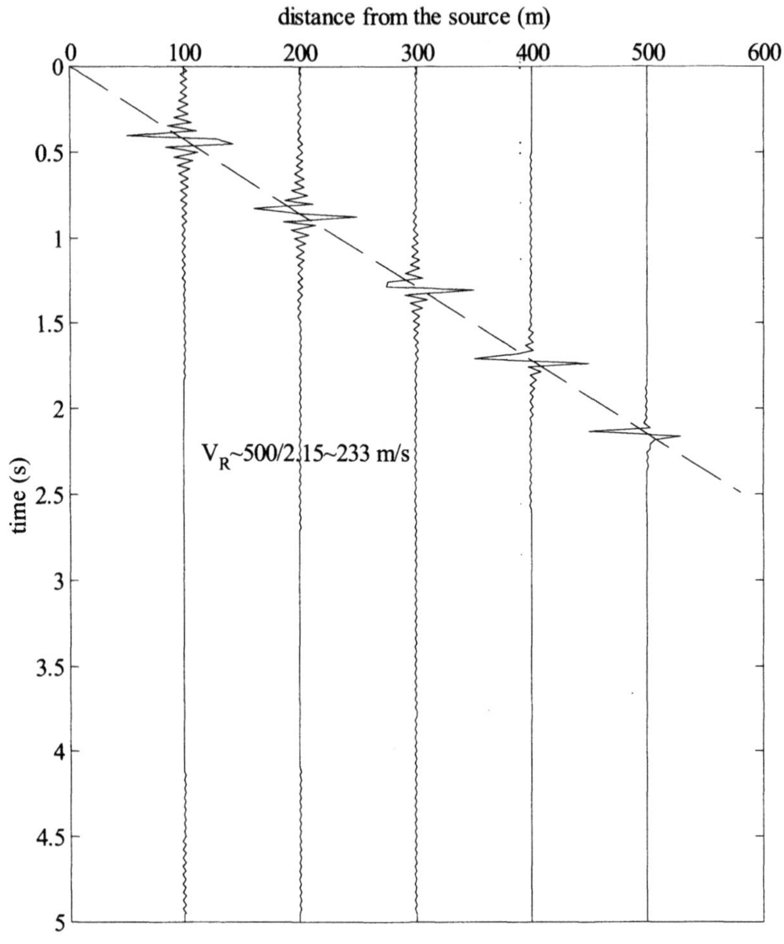


**Figure 12.** Synthetic seismograms obtained from the inverse Fourier transform of *Lamb*'s solution computed at distances of 100, 200, 300, 400 and 500 m from the source and for f = 0.1 –20 Hz. The signals represent the medium response to a vertical point load varying with time as a *Dirac* δ-function.

Lamb's solution represented by Eq.(36) has also been plotted as it stands (i.e. in the frequency domain). Figure 13 shows the variation of displacement amplitude and phase with distance from the source at a frequency of 10 Hz. As expected, the displacement *amplitude* decreases with distance from the source with a rate of decay that decreases as the distance increases. The displacement *phase* on the other hand exhibits jump discontinuities at regular intervals corresponding to the wavelength of the propagating Rayleigh wave ($\lambda_R = V_R/f \sim 233/10 \sim 23$ m).

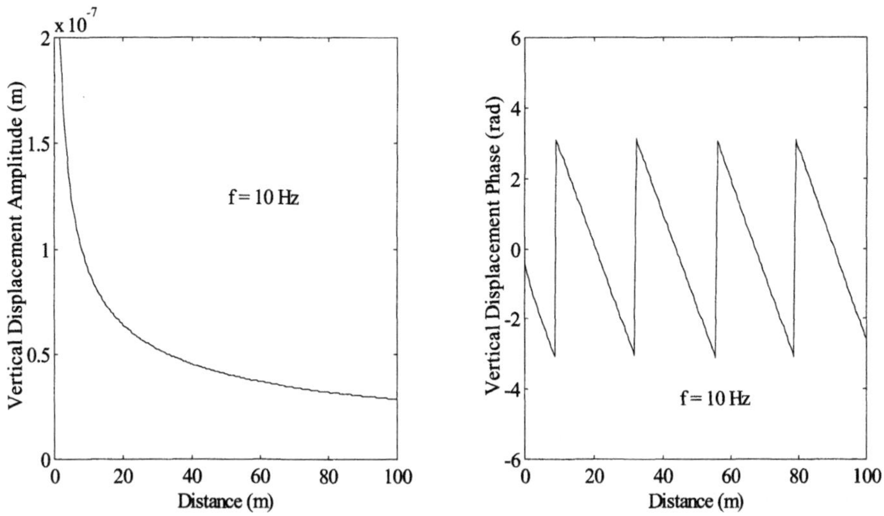


**Figure 13.** Vertical component of displacement amplitude and phase as functions of distance from the source as computed from *Lamb*'s solution. The calculation has been carried out at a frequency of 10 Hz.

The rate of decay of displacement amplitude with distance from the source may be better appreciated by considering the *asymptotic expansion* of the *Hankel* function of the second kind of order $\nu$ $H_\nu^{(2)}(z)$ (Lebedev, 1972):

$$H_\upsilon^{(2)}(z) = \left(\frac{2}{\pi \cdot z}\right)^{0.5} \cdot e^{-i\cdot(z-0.5\cdot\upsilon\cdot\pi-0.25\cdot\pi)} \cdot \left[\sum_{k=0}^{n}(\upsilon,k)(2iz)^{-k} + O\left(\left|z\right|^{-n-1}\right)\right] \quad \text{for } \left|\arg z\right| \le \pi - \delta \tag{39}$$

where $\delta$ is an arbitrarily small positive number.

$$(\upsilon,k) = \frac{(-1)^k}{k!}\left(\tfrac{1}{2}-\upsilon\right)_k\left(\tfrac{1}{2}+\upsilon\right)_k = \frac{\left(4\upsilon^2-1\right)\left(4\upsilon^2-3^2\right)\cdots\left(4\upsilon^2-(2k-1)^2\right)}{2^{2k}k!}, \quad (\upsilon,0)=1. \tag{40}$$

In Lamb's solution represented by Eq.(36) the argument $z = k_R \cdot r$ and $\nu = 0$. For sufficiently large values of $|z|$[13] the Hankel function $H_0^{(2)}(z)$ can be satisfactory approximated by the product of a complex exponential (i.e. the phase factor) by a term proportional to $z^{-0.5}$. Thus in the far-field approximation, Rayleigh waves generated by a vertical, time-harmonic point load applied at the free-boundary of a linear elastic, homogeneous half-space attenuate geometrically with the increase of the distance *r* from the source as $r^{-0.5}$ and hence they propagate along *cylindrical* wave fronts (see Figure 1).

By the same argument of asymptotic expansions, Lamb showed that for the same setting P and S body waves suffer at the free-surface a geometrical decay proportional to $r^{-2}$, therefore body waves propagate along *hemispherical* wave fronts. On the other hand it is easy to show that

[13] On the order of one half of the wavelength $\lambda_R$.

Rayleigh waves generated by a surface time-harmonic vertical, line source do not undergo geometrical attenuation whereas cylindrical body waves decay with increasing distance from the source as $r^{-1.5}$ (Achenbach, 1984).

It should be remarked however that these laws of attenuation are not applicable, in general, to *transient* wave-forms because for the latter, the spatial attenuation of the wave results from a combination of both geometrical spreading and spreading of the signal with time (Keilis-Borok, 1989).

Before ending this section a brief remark is made on a peculiar property of surface waves radiating from a point source locate at the free-boundary of a half-space. This property concerns with an intrinsic feature of radiation in two dimensions which has no counterpart in three-dimensional wave propagation. In a three-dimensional Euclidean space, a space-time localized impulse is reproduced at all other positions throughout the space after a finite time $t^*$ which is a function of the distance from the source and of the speed of propagation of disturbances of that space. The original signal of finite duration emitted from the source will be reproduced identical at later times at different points of the space.

The only difference will be the amplitude decay of the signal due to geometrical and possibly material (i.e. dissipative) spreading. This phenomenon is known as the *Huyghens's principle.* In two dimension wave propagation Huyghens's principle does not hold and an external disturbance of finite duration generates a signal that does not ceases to persist even after the external disturbance has ended its action. Geometrical (and possibly material) attenuation will still yields its effects, however in a two-dimensional space a single impulse is perceived at a particular position as a *blurred* signal that persists, although decaying in time, beyond the duration of the impulse emitted at the source[14].

As mentioned several times in these notes, surface waves propagating along the free-boundary of a half-space are essentially two-dimensional phenomenon and as such they share all the features described above about radiation in two dimensions. Lamb's solution, although it is not immediately apparent from Eq.(36), reflects these features as well, and if one were to compute the inverse, analytical Fourier transform of Eq.(36) and than convolve it with a source function of finite time duration he/she would find out that the response at a particular position of the half-space will have all the described attributes of a *blurred* signal.

Figure 14 shows the half-space response at 100 m distance from a vertical point source whose time variation is a *smooth*[15] square wave function of 3 s duration (see left of Figure 14). The medium response has been computed from the inverse Fourier transform of Lamb's solution (36) convolved with the smooth square wave source function. For the half-space were used the material properties of Table 2.

It is readily apparent from the figure that the time window of 3 seconds of the source function is enlarged in the response to more than 5 seconds. This effect is also visible in Figure 12 where

[14] If humans were to live like *cartoons* in a two-dimensional world they could hardly communicate as their conversations would overlie in time thereby producing unclear, indistinct dialogues. This feature of wave propagation in two dimensions can be proved to hold in *any* Euclidean space of *even* dimensions. Likewise the validity of Huyghens's principle in a three-dimensional world can be shown to hold in *any* Euclidean space of *odd* dimensions (Courant and Hilbert, 1994).

[15] A *smooth* square function is a square function where the discontinuities of the first derivative at the four corners of the function have been eliminated by smooth transitions to avoid the *Gibbs's effect.*

the seismograms of finite time duration represent the half-space response to a source function varying as a Dirac δ-function, i.e. an impulse of infinitesimally short time window.

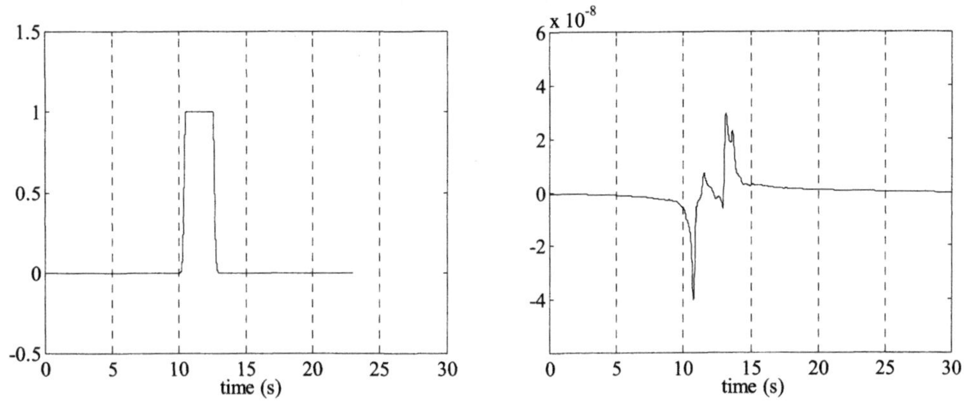


**Figure 14.** (Left). Source function representing a *smooth* square wave of time duration of 3 s. (Right). Half-space response at 100 m distance from the source obtained from the inverse Fourier transform of Lamb's solution (36) convolved with the source function. Medium properties are shown in Table 2.

**Green's functions.** The term Green's function in elastodynamics denotes a scalar or a vector field, say for instance the displacement field **u**(x,t) at time t and spatial position **x** representing the response of an elastic medium caused by an impulsive unit point source varying in space and time as a Dirac δ-function (Eringen and Suhubi, 1975; Achenbach, 1984). In this definition the domain maybe bounded or unbounded. If it is bounded the Green's function should satisfy the boundary conditions. Furthermore. the unit point source may be a force, a couple but also a dislocation[2]. The usefulness of this concept stems from the fact that in a *linear* system knowledge of the Green's function is all what is needed to construct via a convolution integral, the response of the system to an arbitrary distribution of sources in space and time.

A field where the notion of Green's function has been exploited to a great extent is seismology. Seismic sources are represented by complicated fault systems of finite spatial extension and of various types of motion which make almost natural the recourse to a Green's function for the computation of synthetic seismograms (Aki and Richards, 2002). Figure 15 shows an example (Spudich and Hartzell, 1985) for such computation where a seismic source has been represented by a finite fault plane that has been further subdivided in an array of small fault elements. At each of these elements (say for instance A) one may prescribe a relative, instantaneous, unit slip or dislocation[16]. The ground motion generated at the observation point by the activation of the fault segment A is the elastodynamic Green's function of the problem. If this function is known the motion at the observer induced by any specified slip-time function $\psi(t)$ at point A can be easily

[16] More rigorously the Green's function is the response yielded by a seismic slip whose time variation is that of a *Dirac* δ-function, and thus interpretable as the derivative of the assumed *Heaviside* time variation. This difference however does not change the essence of the definition of Green's function.

obtained through a time-convolution of ψ(t) with the corresponding elastodynamic Green's function.

To evaluate the effect of the entire fault rupture the motion at the observer generated by a single fault element is superimposed for the all array of fault segments. The Green's functions at the different points A, B, C will be in general different because different will be the distances between the source and the observer, as well as the geological characteristics of the medium traversed by the seismic waves, despite the fact that the slip functions are assumed, by hypothesis, identical at all points of the fault.

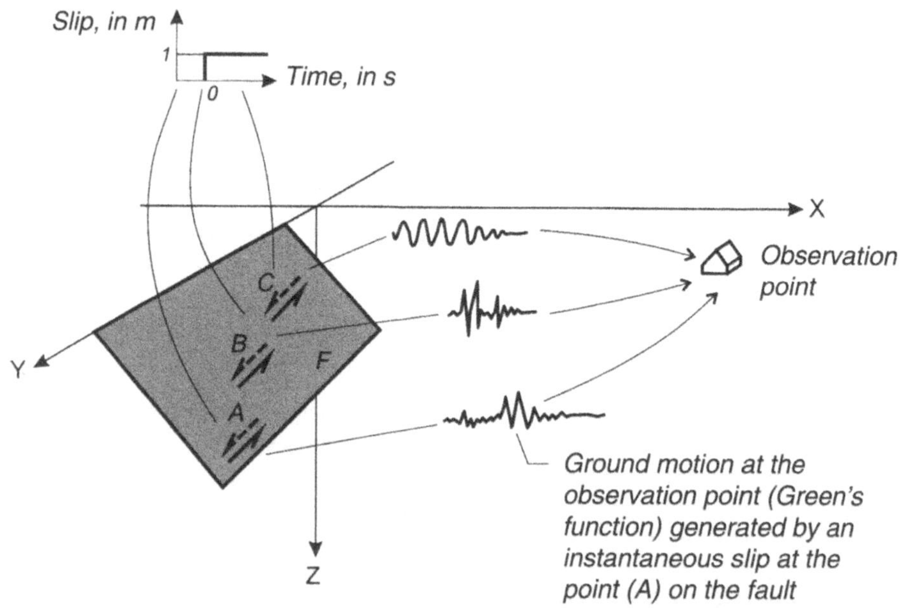


**Figure 15.** Simplified representation of the *Green's function* associated to a seismic fault rupture. The individual points A, B, C located on the fault plane F, undergo an instantaneous slip of unit amplitude, like that shown in the plot placed in the upper left side of the figure (from Spudich and Hartzell, 1985).

Now a question arises about what relation exists, if any, between surface waves and the concept of Green's function. To answer this question it is first necessary to recall that as the Green's function represents the fundamental solution to resolve virtually any type of initial-boundary value problems, solving the latter for homogenous boundary conditions (i.e. solving an eigenvalue problem) constitutes a fundamental step towards the computation of the Green's function. This because solving an elastodynamic eigenproblem corresponds to finding the totality of solutions that are admissible (i.e. solutions satisfying homogenous boundary conditions) in a given elastic body.

These solutions are the eigenfunctions of that particular eigenproblem. A Green's function is nothing but a particular solution of an elastodynamic problem satisfying appropriate initial and boundary conditions. This particular solution can be determined as a *weighted* sum of eigenfunctions because in linear elastodynamics even the most complicated source problem may in general be resolved by a series or integral expansion of solutions pertaining to the homogeneous problem.[17] Each term of the expansion is a measure of the contribution to the wavefield given by a particular mode of propagation.

In the end, solution of the eigenproblem represents almost always the most challenging task in elastodynamics because any other problem, including the computation of a Green's function, may be derived from that solution. In this sense, surface waves follow the same rule and the solution of a source problem (or inhomogeneous boundary value problem) involving surface waves may be obtained from the results of solving the corresponding eigenvalue problem (i.e. free surface waves) which has been thoroughly discussed in Section 2.2.

After this long premises, the section proceeds in the definition of the Green's function associated with surface Rayleigh waves. As shown by Figure 15 although the correct definition of Green's function requires that the source function be defined as an impulsive unit point source varying in space and time as a Dirac δ-function, sometimes with regards to the time variation it is more convenient to use other definitions. Hereinafter the term *Green's function* is use to denote the response of a linear elastic half-space to a *harmonic* unit vertical point source.

Of particular interest in this section is the *displacement Green's function*, which is defined as the displacement $\hat{\mathbf{u}}(r, y, \omega)$ induced in a linear elastic (possibly vertically inhomogeneous) half-space by a harmonic unit vertical point load $1_y \cdot e^{i\omega t}$ placed at the position $\{r = 0,\ y = y_S\}$. The subscript *y* in $1_y$ denotes the direction of action of the unit vertical point load.

The particle displacement field $\hat{\mathbf{u}}(r, y, \omega)$ represented by the Green's function can be separated into two components $\hat{\mathbf{u}}(r, y, \omega) = \hat{\mathbf{u}}_B(r, y, \omega) + \hat{\mathbf{u}}_S(r, y, \omega)$ [18]. The first component $\hat{\mathbf{u}}_B(r, y, \omega)$ represents the body wave field and is composed by a superposition of *P* and *S* waves. The second component $\hat{\mathbf{u}}_S(r, y, \omega)$ is the surface wave field and, in general, is composed by a superposition of Love and Rayleigh waves. As discussed in the previous section the body wave field attenuates with distance from the source at a much higher rate than the surface wave field. Thus it is reasonable to assume that for harmonic oscillations at large distances from the source, the surface wave field dominates the overall particle motion and $\hat{\mathbf{u}}(r, y, \omega) \approx \hat{\mathbf{u}}_S(r, y, \omega)$.

In surface waves studies, the distance from the source where the body wave field is not negligible is usually called the *near-field.* Numerical studies by Holzlohner (1980), Vrettos (1991), Tokimatsu (1995), Foti (2000) and Strobbia (2003) of surface wave propagation in vertically inhomogeneous, layered half-spaces have shown that in *normally* dispersive media the near-field effects are important up to a distance from the source equal to $\lambda_R/2$ (where $\lambda_R = \lambda_R(\omega)$ is the wavelength of Rayleigh waves). However, in *inversely*[19] dispersive media the near-field is larger and may extend up to $2\lambda_R$. In the continuation it will be assumed that near-field effects to be

[17] The technique based on using the concept of Green's function represents just one way of solving an elastodynamic problem. Other methods are also possible and an initial boundary-value problem of elastodynamics could very well be solved for instance by attacking directly the system of governing field equations and boundary-initial conditions (Achenbach, 1984).

[18] The additive decomposition of the wavefield in body and surface waves contributions is possible only in the far-field.

[19] By *normally* dispersive media are denoted half-spaces where the *mechanical impedance* ρ·β is either constant or increases smoothly with depth. Conversely *inversely* dispersive media are vertically inhomogeneous half-spaces where the mechanical impedance ρ·β varies irregularly with depth.

negligible. Furthermore, since the excitation is caused by a harmonic, vertical point source Love waves do not contribute to the wavefield.

With these assumptions the Rayleigh displacement Green's function $\hat{\mathbf{u}}(r, y, \omega)$ can be computed from the solution of the Rayleigh eigenproblem using the concept of mode superposition. The procedure is outlined in the sequel.

In an isotropic, linear elastic, vertically inhomogeneous half-space, the wavefield at large distances generated by a harmonic point source $F_y \cdot e^{i\omega t}$ acting in a direction perpendicular to the free boundary and located at a position $\{r = 0, y = y_S\}$ can be expanded, in the radial direction, in a series of $\nu^{th}$ order Hankel functions ($p$ is an integer)[20].

For large values of $r$ the $\nu^{th}$ order Hankel functions can be approximated by their asymptotic expansions involving complex exponential functions as shown by Eqs.(39) and (40). As a result, the particle displacement $\mathbf{u}(r,y,\omega) = [u_r(r,y,\omega)\cdot\mathbf{e}_r + u_y(r,y,\omega)\cdot\mathbf{e}_y]$ resulting from the superposition of $M$ distinct Rayleigh modes, can be written in cylindrical coordinates $\{r,y,\theta\}$, as follows (Ben-Menahem and Singh, 2000; Aki and Richards, 2002):

$$u_\beta(r, y, \omega) = \sum_{j=1}^{M} \left[A_\beta(r, y, \omega)\right]_j \cdot e^{i(\omega \cdot t - k_j \cdot r + \varphi_\beta)} \tag{41}$$

where $\beta = r,y$ denotes the radial and the vertical directions, respectively, and $[A_\beta(r,y,\omega)]_j$ , $k_j(\omega)$ are the Rayleigh displacement amplitudes and wavenumber, respectively, associated with the $j^{th}$ mode of propagation. Finally, $\phi_\beta = -\pi/4$ for $\beta = r$ and $\phi_\beta = \pi/4$ for $\beta = y$. Equation (41) shows, as expected for cylindrical symmetry of the problem, that $u_\beta(r,y,\omega)$ is independent of the azimuthal angle $\theta$.

The actual particle displacement is obtained by taking either the real or imaginary part of Eq.(41). By choosing the latter this equation may be re-written as follows (Lai, 1998):

$$\Im\left[u_\beta(r, y, \omega)\right] = \Im\left\{\sum_{j=1}^{M} \left[A_\beta(r, y, \omega)\right]_j e^{i(\omega \cdot t - k_j \cdot r + \varphi_\beta)}\right\} = \sum_{j=1}^{M}\left[\left(C_\beta\right)_j \sin(\omega t) - \left(D_\beta\right)_j \cos(\omega t)\right] \tag{42}$$

where the symbol $\Im(\cdot)$ denotes the imaginary part of the argument, and $(C_\beta)_j = (A_\beta)_j \cdot\cos(k_j \cdot r + \phi_\beta)$ and $(D_\beta)_j = (A_\beta)_j \cdot\sin(k_j \cdot r + \phi_\beta)$. Now using simple trigonometric identities Eq.(42) becomes:

$$\Im\left[u_\beta(r, y, \omega)\right] = \mathscr{U}_\beta(r, y, \omega) \cdot \sin\left[\omega t - \psi_\beta(r, y, \omega)\right] \tag{43}$$

where:

[20] This expansion is in a sense *natural* given the solution of the Lamb's problem in a homogeneous half-space represented by Eq.(36).

$$\begin{cases} \mathscr{U}_\beta(r,y,\omega) = \left\{ \sum_{i=1}^{M} \sum_{j=1}^{M} [A_\beta(r,y,\omega)]_i \cdot [A_\beta(r,y,\omega)]_j \cdot \cos[r \cdot (k_i - k_j)] \right\}^{0.5} \\ \psi_\beta(r,y,\omega) = \tan^{-1} \left\{ \dfrac{\sum_{i=1}^{M} [A_\beta(r,y,\omega)]_i \cdot \sin(k_i \cdot r + \varphi_\beta)}{\sum_{j=1}^{M} [A_\beta(r,y,\omega)]_j \cdot \cos(k_j \cdot r + \varphi_\beta)} \right\} \end{cases} \tag{44}$$

Taking the real part of Eq.(41) and repeating the procedure that lead to Eq.(43) one obtains after combining the results:

$$u_\beta(r,y,\omega) = \mathscr{U}_\beta(r,y,\omega) \cdot e^{i[\omega t - \psi_\beta(r,y,\omega)]} \tag{45}$$

The expressions for $\mathscr{U}_\beta(r,y,\omega)$ and $\psi_\beta(r,y,\omega)$ defined in Eq.(44) are functions of the modal amplitudes $[A_\beta(r,y,\omega)]_j$ of Rayleigh waves. For an harmonic vertical point source $F_y \cdot e^{i\omega t}$ located at $\{r = 0, y = y_S\}$, the Rayleigh displacement amplitudes $[A_\beta(r,y,\omega)]_j$ of the individual modes of propagation are related to the displacement eigenfunctions $r_1(y,k,\omega)$ and $r_2(y,k,\omega)$ (see Section 2.2) and to other modal parameters by the following expression (Aki and Richards, 2002):

$$[A_\beta(r,y,\omega)]_j = \begin{bmatrix} A_r(r,y,\omega) \\ A_y(r,y,\omega) \end{bmatrix}_j = \frac{F_y \cdot r_2(y_S, k_j, \omega)}{4V_j \cdot U_j \cdot I_j \cdot \sqrt{2\pi r \cdot k_j}} \cdot \begin{bmatrix} r_1(y,k_j,\omega) \\ r_2(y,k_j,\omega) \end{bmatrix} \tag{46}$$

where $V_j$, $U_j$, and $k_j$ are the phase, group velocity and wavenumber of the Rayleigh $j^{th}$ mode of propagation ($j = 1, M$), respectively. The term $I_j(y, k_j, \omega) = I_{1,j}$ is the *first* Rayleigh energy integral associated with the $j^{th}$ mode of propagation and it is defined by Eq.(23b). By substituting Eq.(46) into Eq.(44) after setting $F_y = 1$, one obtains after straightforward algebra (Lai, 1998):

$$\begin{cases} \hat{\mathscr{U}}_r(r,y,\omega) = \dfrac{1}{4\sqrt{2\pi \cdot r}} \left\{ \sum_{i=1}^{M} \sum_{j=1}^{M} \dfrac{r_1(k_i,y) r_1(k_j,y) r_2(k_i,y_S) r_2(k_j,y_S) \cos[r(k_i - k_j)]}{\sqrt{k_i k_j}(V_i U_i I_i)(V_j U_j I_j)} \right\}^{0.5} \\ \hat{\mathscr{U}}_y(r,y,\omega) = \dfrac{1}{4\sqrt{2\pi \cdot r}} \left\{ \sum_{i=1}^{M} \sum_{j=1}^{M} \dfrac{r_2(k_i,y) r_2(k_j,y) r_2(k_i,y_S) r_2(k_j,y_S) \cos[r(k_i - k_j)]}{\sqrt{k_i k_j}(V_i U_i I_i)(V_j U_j I_j)} \right\}^{0.5} \end{cases} \tag{47}$$

and

$$\begin{cases} \hat{\psi}_r(r,y,\omega) = \tan^{-1}\left[\dfrac{\displaystyle\sum_{i=1}^{M}\frac{r_1(k_i,y,\omega)r_2(k_i,y_S,\omega)}{\sqrt{k_i}\cdot(V_iU_iI_i)}\cdot\sin\left(k_i\cdot r-\frac{\pi}{4}\right)}{\displaystyle\sum_{j=1}^{M}\frac{r_1(k_j,y,\omega)r_2(k_j,y_S,\omega)}{\sqrt{k_j}\cdot(V_jU_jI_j)}\cdot\cos\left(k_j\cdot r-\frac{\pi}{4}\right)}\right] \\ \\ \hat{\psi}_y(r,y,\omega) = \tan^{-1}\left[\dfrac{\displaystyle\sum_{i=1}^{M}\frac{r_2(k_i,y,\omega)r_2(k_i,y_S,\omega)}{\sqrt{k_i}\cdot(V_iU_iI_i)}\cdot\sin\left(k_i\cdot r+\frac{\pi}{4}\right)}{\displaystyle\sum_{j=1}^{M}\frac{r_2(k_j,y,\omega)r_2(k_j,y_S,\omega)}{\sqrt{k_j}\cdot(V_jU_jI_j)}\cdot\cos\left(k_j\cdot r+\frac{\pi}{4}\right)}\right] \end{cases} \tag{48}$$

where the "*hat*" above the symbols $\mathscr{U}_\beta(r,y,\omega)$ and $\psi_\beta(r,y,\omega)$ has been used to emphasize that these quantities have been computed from Eq.(46) using $F_y = 1$. The expression given by:

$$\hat{u}_\beta(r,y,\omega) = \hat{\mathscr{U}}_\beta(r,y,\omega)\cdot e^{i[\omega t-\psi_\beta(r,y,\omega)]} \tag{49}$$

completely defines the components $\beta$ = r,y of the *Rayleigh displacement Green's function* $\hat{\mathbf{u}}(r,y,\omega)$. Equation (49) is instructive because it shows that a multiplicative decomposition of the displacement Green's function of the type $\hat{u}_\beta = \hat{\mathscr{U}}_\beta\cdot e^{i\cdot\arg(\hat{u}_\beta)}$ is possible even in vertically inhomogeneous half-spaces where surface wave propagation is multi-modal. However, because the wavenumber $\hat{k}_\beta(r,y,\omega)$ is no longer a constant as it happens in homogeneous media, the spatial variation of the displacement field is no longer harmonic even though the temporal variation of the source is harmonic.

Equations (47), (48) and (49) also show the remarkable result that the three main factors in the expression of $\hat{\mathbf{u}}(r,y,\omega)$, namely the source depth $\{y_S\}$, the receiver depth $\{y\}$, and the distance from the source $\{r\}$, are uncoupled in the sense that their contribution is independent from each other.

With the definition of the Rayleigh displacement Green's function given above, computation of the displacement $\mathbf{u}(r,y,\omega)$ induced by a harmonic vertical point source $F_y\cdot e^{i\omega t}$ located at $\{r = 0, y = y_S\}$ becomes trivial. In fact from Eq.(49):

$$\mathbf{u}(r,y,\omega) = F_y\cdot\hat{\mathbf{u}}(r,y,\omega) \qquad \text{or} \qquad u_\beta(r,y,\omega) = F_y\cdot\mathscr{G}_\beta(r,y,\omega)\cdot e^{i[\omega t-\psi_\beta(r,y,\omega)]} \tag{50}$$

The function $\mathscr{G}_\beta(r,y,\omega) = \hat{\mathscr{U}}_\beta(r,y,\omega)$ with $\beta$ = r,y can be named *Rayleigh geometrical attenuation function*. This function has the important physical interpretation of modeling the geometric attenuation of Rayleigh surface waves in vertically inhomogeneous half-spaces.

As mentioned in the previous section in a homogeneous half-space, Rayleigh waves generated by a vertical, harmonic point load placed at the free-boundary attenuate with increasing distance from the source by a factor proportional to $r^{-0.5}$ as a result of geometrical spreading. This simple attenuation law, which follows directly from the principle of conservation of energy, does not hold in inhomogeneous media. The reason is that in the latter the Rayleigh wavefield originates from the superposition of several modes of propagation whose existence is caused by geometric dispersion. As a result of this the geometric spreading law of Rayleigh waves in homogeneous media $r^{-0.5}$ is altered and it is replaced by the new function $\mathscr{G}_\beta(r,y,\omega)$.

A graphical representation of $\mathscr{G}_\beta(r,y,\omega)$ at the free-boundary ($y = 0$) is shown in Figure 16 in a series of numerical simulations that have been conducted for different types of half-spaces: a homogeneous medium, a three-layer normally dispersive system, and a three-layer inversely dispersive system[19]. The simulations were carried out at frequencies of 5, 20, 50 and 100 Hz and using the material properties illustrated in Table 3a, 3b and 3c.

**Table 3a.** Medium properties used to construct the Rayleigh geometrical attenuation function $\mathscr{G}_y(r,0,\omega)$ plotted in Figure 16. Case A – *Normally dispersive* half-space.

| Stratum | Thickness (m) | $\alpha$ (m/s) | $\beta$ (m/s) | $\rho$ (t/m$^3$) |
|---|---|---|---|---|
| 1 | 10 | 400 | 200 | 1.8 |
| 2 | 10 | 600 | 300 | 1.8 |
| 3 | 10 | 800 | 400 | 1.8 |
| half-space | $\infty$ | 1000 | 500 | 1.8 |

**Table 3b.** Medium properties used to construct the Rayleigh geometrical attenuation function $\mathscr{G}_y(r,0,\omega)$ plotted in Figure 16. Case B – *Inversely dispersive* half-space.

| Stratum | Thickness (m) | $\alpha$ (m/s) | $\beta$ (m/s) | $\rho$ (t/m$^3$) |
|---|---|---|---|---|
| 1 | 10 | 800 | 400 | 1.8 |
| 2 | 10 | 400 | 200 | 1.8 |
| 3 | 10 | 600 | 300 | 1.8 |
| half-space | $\infty$ | 1000 | 500 | 1.8 |

**Table 3c.** Medium properties used to construct the Rayleigh geometrical attenuation function $\mathscr{G}_y(r,0,\omega)$ plotted in Figure 16. Case C – *Homogeneous* (non dispersive) half-space.

| Half-Space | $\alpha$ (m/s) | $\beta$ (m/s) | $\rho$ (t/m$^3$) |
|---|---|---|---|
| Homogeneous | 600 | 300 | 1.8 |

Figure 16 clearly shows that the geometric attenuation law $r^{-0.5}$ is mostly altered in inversely dispersive media and as the frequency increases. This because by increasing the frequency it increases the number of propagation modes caused by constructive interference. These findings have also been observed in other numerical studies (Gucunski and Woods, 1991; Tokimatsu et al., 1992; Strobbia, 2003).

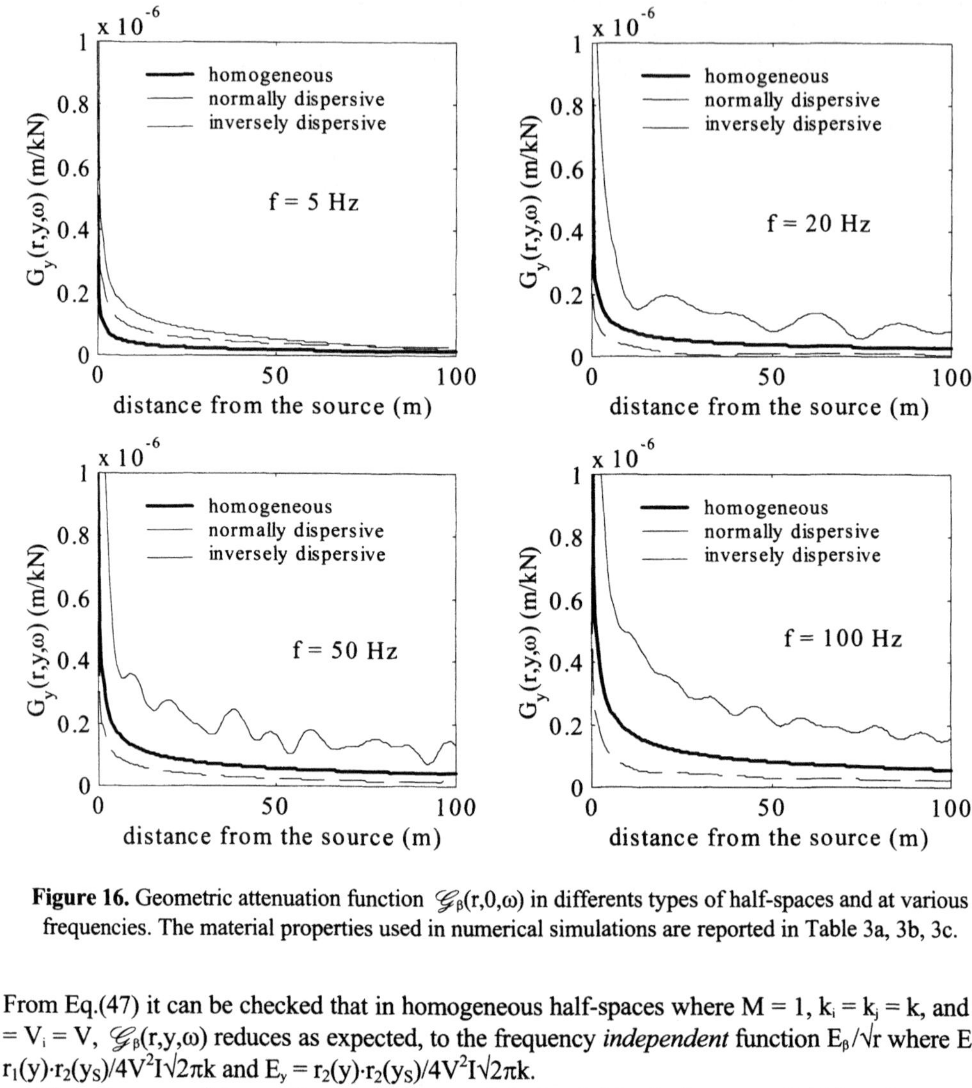


**Figure 16.** Geometric attenuation function $\mathcal{G}_\beta(r,0,\omega)$ in differents types of half-spaces and at various frequencies. The material properties used in numerical simulations are reported in Table 3a, 3b, 3c.

From Eq.(47) it can be checked that in homogeneous half-spaces where $M = 1$, $k_i = k_j = k$, and $U_i = V_i = V$, $\mathcal{G}_\beta(r,y,\omega)$ reduces as expected, to the frequency *independent* function $E_\beta/\sqrt{r}$ where $E_r = r_1(y)\cdot r_2(y_S)/4V^2I\sqrt{2\pi k}$ and $E_y = r_2(y)\cdot r_2(y_S)/4V^2I\sqrt{2\pi k}$.

The relevance of the explicit factorization of the Rayleigh displacement field into the product of source magnitude, geometric spreading function, and phase factor put in evidence by Eq.(50) will become apparent in Section 3 when discussing the attenuation of Rayleigh waves in weakly dissipative media.

**Modal and apparent phase velocity.** This section will demonstrates that in a vertically inhomogeneous half-space surface Rayleigh waves generated by harmonic sources propagate at a speed different from the modal phase velocity that is called *apparent* or *effective* Rayleigh phase velocity (Tokimatsu, 1995; Lai, 1998; Foti, 2000; Strobbia, 2003). This kinematical quantity describes the speed of propagation of a waveform composed by the superposition of various Rayleigh modes of propagation and it will shown to be a *local* quantity for it varies depending on the spatial position where it is measured.

To derive an explicit, closed-form expression for the apparent Rayleigh phase velocity, let Eq.(43) above denote the Rayleigh particle displacement $u_\beta(r,y,\omega)$ with $\beta = r,y$ denoting the radial and the vertical directions respectively, resulting from the superposition of *M* distinct Rayleigh modes with M=M(ω). In Eq. (43), which is reported here for convenience:

$$\Im\left[u_\beta(r,y,\omega)\right] = \mathscr{U}_\beta(r,y,\omega)\cdot\sin\left[\omega t - \psi_\beta(r,y,\omega)\right] \tag{43bis}$$

the expression:

$$\left[\omega t - \psi_\beta(r,y,\omega)\right] = \text{constant} \tag{51}$$

represents the equation of a *wave front*, since it is the locus of points having *constant phase*. Assuming the function $\psi_\beta(r,y,\omega)$ to be sufficiently smooth, Eq.(51) can be differentiated with respect to time, to obtain:

$$\omega - \frac{\partial\psi_\beta}{\partial r}(r,y,\omega)\cdot\frac{dr}{dt} = 0 \tag{52}$$

which yields:

$$\hat{V}_\beta(r,y,\omega) = \frac{\omega}{[\psi_\beta(r,y,\omega)]_{,r}} \tag{53}$$

where the symbol $\hat{V}_\beta(r,y,\omega)$ denotes the *apparent* or *effective* Rayleigh phase velocity and the notation $[\cdot]_{,r}$ stands for differentiation of the argument with respect to *r*. Equation (53) shows that the apparent Rayleigh phase velocity is a *local quantity*, that is its magnitude is a function of the spatial position where it is calculated.

At a fixed position $y = y^*$, the function $\hat{V}_\beta(r,y,\omega)$ describes a two dimensional surface showing the variation of the apparent Rayleigh phase velocity with frequency and distance from the source. Equation (53) also shows that since the apparent Rayleigh phase velocity is a vector quantity ($\beta = r,y$), different components of $\hat{V}_\beta(r,y,\omega)$ will, in general, travel at different apparent phase velocities. Furthermore, since the quantity:

$$\frac{\partial\hat{V}_\beta}{\partial t} = -\frac{\omega\hat{V}_\beta\cdot\left(\psi_\beta\right)_{,rr}}{\left(\psi_\beta\right)_{,r}\cdot\left(\psi_\beta\right)_{,r}} \tag{54}$$

is not, in general, equal to zero, the waveform accelerates as it propagates along the free-boundary of the half-space.

In Eq.(53) the function $(\psi_\beta)_{,r}$ can be interpreted as an effective Rayleigh wavenumber and denoted by $\hat{k}_\beta(r,y,\omega)$. However, a decomposition of the argument of Eq.(43) in the form $(\omega t - \hat{k}_\beta \cdot r)$ which is standard for monochromatic harmonic waves, is no longer possible because the *apparent wavenumber* $(\psi_\beta)_{,r}$ being a *local* quantity, must be integrated over *r* to yield the phase $\psi_\beta(r,y,\omega)$.

From Eq.(53) it is now possible to obtain an explicit definition of the apparent Rayleigh phase velocity expressed in terms of quantities derived from the solution of the Rayleigh eigenproblem (i.e. modal phase and group velocities, eigenfunctions, etc.). Substitution of Eq.(44b) into Eq.(53) yields:

$$\hat{V}_\beta(r,y,\omega) = 2\omega \cdot \frac{\sum_{i=1}^{M}\sum_{j=1}^{M}\left\{(A_\beta)_i (A_\beta)_j \cdot \cos\left[r\left(k_i - k_j\right)\right]\right\}}{\sum_{r=1}^{M}\sum_{s=1}^{M}\left\{(A_\beta)_r (A_\beta)_s (k_r + k_s)\cdot \cos\left[r\left(k_r - k_s\right)\right]\right\}} \tag{55}$$

Finally a replacement of Eq.(46) into Eq.(55) yields the final result:

$$\begin{cases} \hat{V}_r(r,y,\omega) = \dfrac{2\omega \sum_{i=1}^{M}\sum_{j=1}^{M}\left\{\dfrac{r_1(y,k_i)r_1(y,k_j)r_2(y_S,k_i)r_2(y_S,k_j)\cos\left[r(k_i-k_j)\right]}{(V_iU_iI_i)(V_jU_jI_j)\sqrt{k_ik_j}}\right\}}{\sum_{r=1}^{M}\sum_{s=1}^{M}\left\{\dfrac{r_1(y,k_r)r_1(y,k_s)r_2(y_S,k_r)r_2(y_S,k_s)(k_r+k_s)\cos\left[r(k_r-k_s)\right]}{(V_rU_rI_r)(V_sU_sI_s)\sqrt{k_rk_s}}\right\}} \\[2em] \hat{V}_y(r,y,\omega) = \dfrac{2\omega \sum_{i=1}^{M}\sum_{j=1}^{M}\left\{\dfrac{r_2(y,k_i)r_2(y,k_j)r_2(y_S,k_i)r_2(y_S,k_j)\cos\left[r(k_i-k_j)\right]}{(V_iU_iI_i)(V_jU_jI_j)\sqrt{k_ik_j}}\right\}}{\sum_{r=1}^{M}\sum_{s=1}^{M}\left\{\dfrac{r_2(y,k_r)r_2(y,k_s)r_2(y_S,k_r)r_2(y_S,k_s)(k_r+k_s)\cos\left[r(k_r-k_s)\right]}{(V_rU_rI_r)(V_sU_sI_s)\sqrt{k_rk_s}}\right\}} \end{cases} \tag{56}$$

where $\hat{V}_r(r,y,\omega)$ and $\hat{V}_y(r,y,\omega)$ denote the components of the apparent Rayleigh phase velocity along the directions *r* and *y* respectively. To reduce the length of the above expressions, the frequency dependence of the eigenfunctions $r_1(y,k,\omega)$ and $r_2(y,k,\omega)$ has been left unspecified.

As shown by Eq.(56), the apparent Rayleigh phase velocity is completely determined from the solution of the Rayleigh eigenvalue problem. Indeed all the modal quantities appearing in Eq.(56) including $V_j = \omega/k_j$ and $U_j = d\omega/dk_j$ , j = 1, M can be computed from the set $\{k_j, r_i(y, k_j, \omega)\}$, i = 1,2 (see Section 2.2 for more details).

An example of application of the concept of *apparent* Rayleigh phase velocity is illustrated in Figure 17 where the modal dispersion curves associated with two different stratified half-spaces, one normally dispersive, the other inversely dispersive[19], have been overlapped to the corresponding *apparent* dispersion curves. Material properties used in numerical simulations are those reported in Table 4a and 4b. These data represent *Case 1* (normally dispersive medium) and *Case 3* (inversely dispersive medium) treated by Tokimatsu et al. (1992).

**Table 4a.** Medium properties used to illustrate the difference between modal and apparent dispersion curves of Figure 17. Case 1 – *Normally dispersive* half-space (data from Tokimatsu et al., 1992).

| Stratum | Thickness (m) | $\alpha$ (m/s) | $\beta$ (m/s) | $\rho$ (t/m$^3$) |
|---|---|---|---|---|
| 1 | 2 | 300 | 80 | 1.8 |
| 2 | 4 | 1000 | 120 | 1.8 |
| 3 | 8 | 1400 | 180 | 1.8 |
| half-space | $\infty$ | 1400 | 360 | 1.8 |

**Table 4b.** Medium properties used to illustrate the difference between modal and apparent dispersion curves of Figure 17. Case 3 – *Inversely dispersive* half-space (data from Tokimatsu et al., 1992).

| Stratum | Thickness (m) | $\alpha$ (m/s) | $\beta$ (m/s) | $\rho$ (t/m$^3$) |
|---|---|---|---|---|
| 1 | 2 | 300 | 80 | 1.8 |
| 2 | 4 | 1000 | 180 | 1.8 |
| 3 | 8 | 1400 | 120 | 1.8 |
| half-space | $\infty$ | 1400 | 360 | 1.8 |

The dependence in Figure 17, of $\hat{V}_y(r,o,\omega)$ on the distance from the source has been eliminated via an averaging procedure which also eliminates the effects of the near-field (Lai, 1998). Details on practical determination of modal and apparent Rayleigh dispersion curves are given in the contributions by Rix (2005) and Foti (2005) of this volume where the concept of apparent phase velocity is further investigated in relation to applications of surface Rayleigh waves for near-surface characterization of soil deposits.

Figure 17 (left) is instructive as it shows that in a *normally* dispersive half-space the apparent dispersion curve almost coincides with the dispersion curve associated with the fundamental mode of propagation. This result is the direct consequence of the fact that in normally dispersive media most of the surface wave energy travel with the *group velocity* of the first mode of propagation since the contribution of higher modes is small.

Although this cannot be assumed as a conclusion of general validity as there are examples of normally dispersive sites where multi-mode wave propagation is important (Strobbia, 2003), it is certainly a rather common situation (Gucunski and Woods, 1991; Tokimatsu et al., 1992; Foti, 2000). Conversely, as shown in Figure 17 (right), in *inversely* dispersive media the contribution of higher modes of propagation is important since the apparent dispersion curve separates considerably from the fundamental mode of propagation in the frequency range from about 8 to 18 Hz.

Therefore an important conclusion that can be drawn in light of these considerations is that surface wave propagation in inversely dispersive half-spaces is a multi-modal phenomenon and as such any attempt to successfully determine the medium parameters (e.g. the $\beta$-profile) from the inversion of experimentally-determined Rayleigh dispersion curves should take into account modes of propagation higher than the first. See again Rix (2005) and Foti (2005) for a thorough discussion of this issue in the contest of geotechnical-geophysical applications.

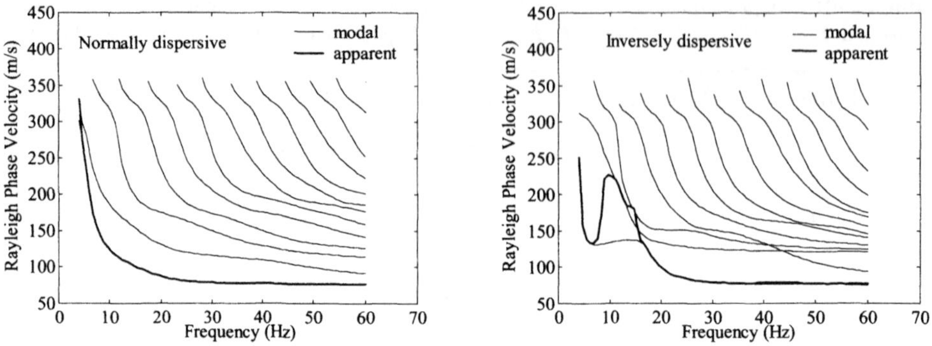


**Figure 17.** Comparison between *modal* and *apparent* Rayleigh dispersion curves in a *normally* dispersive (left) and *inversely* dispersive (right) half-space. The materials properties are reported in Table 4a and 4b.

## 3 Linear Surface Waves in Vertically Inhomogeneous Inelastic Continua

### 3.1 Introduction

This section is aimed to illustrate the main characteristics of Rayleigh wave propagation in vertically inhomogeneous, linear viscoelastic half-spaces. Linear viscoelasticity is the simplest formal theory that can be used to describe the mechanical response of solid, dissipative materials to low-amplitude dynamic excitations. It only requires the validity of a) the small strain assumption, b) the time translation invariance hypothesis and finally the current value of the Cauchy stress tensor is assumed to depend solely on the current value of the strain tensor and on the past strain history. Despite its simplicity the theory of linear viscoelasticity has proved to be quite effective in describing phenomena of wave propagation in dissipative materials like soils and rocks at low-strain levels (Pipkin, 1986; Ishihara, 1996; Ben-Menahem and Singh, 2000). Experimental evidence shows in fact that geomaterials subjected to dynamic excitations exhibit both the ability to store strain energy and to dissipate strain energy over a finite period of time even at very small strain levels, below the so-called *linear cyclic threshold shear strain*[21]. Both these phenomena can quite accurately be described by the theory of linear viscoelasticity.

In order to fulfill the fundamental *principle of causality*, phase velocity and attenuation of a mechanical disturbance propagating in a linear viscoelastic medium cannot be assigned independently. They must satisfy the *Kramers-Krönig* dispersion relation which in essence is a statement that dissipative materials are *dispersive* as the speed of propagation of a viscoelastic pulse is frequency dependent. Illustration of these and other features concerned with the propagation of viscoelastic bulk and surface Rayleigh waves in vertically inhomogeneous continua is preceded by an introductory section dedicated to a review of viscoelastic constitutive modelling with definition of the associated model parameters.

---

[21] A value of shear strain below which particulate materials like soils are not dilatant. *Dilatancy* (or its reverse *contractancy*) is the natural tendency of materials formed by a discrete assemblages of particles to exhibit volume changes under states of pure shear stress.

The mathematics of field and constitutive equations of linear viscoelasticity is fairly involved if described in the time-domain. However if these equations are either Fourier or Laplace transformed they become much simpler as they resemble those of linear elasticity with the field variables extended to the complex domain. The formalism can then take advantage of the powerful and elegant methods of complex analysis and in particular of the theory of *holomorphic* functions.

As a natural prelude before entering the subject of linear *viscoelastodynamics* a brief review is first presented on some of the most significant results of the theory of linear viscoelasticity. Will then follows a discussion on certain interesting properties of surface wave propagation in *weakly* attenuating media which are derived as *byproduct* of the variational theorems illustrated in the study of surface waves in linear elastic media. The final topic to be discussed in this section is concerned with surface Rayleigh waves in strongly dissipative, vertically inhomogeneous continua.

### 3.2 Constitutive Modeling of Linear Dissipative Materials

Experimental evidence shows that under a dynamic excitation most geomaterials like soils and rocks exhibit a mechanical response that is strongly dependent upon the magnitude of the deviatoric strain tensor[22]. Figure 18 illustrates this statement using the concept of *threshold strain boundary surfaces* in the principal strain space. The linear and volumetric threshold strain boundary surfaces $\varepsilon^t_l$ and $\varepsilon^v_l$ represent a generalization of the uniaxial notions of threshold shear strain[21] (Vucetic, 1994) to the case of multiaxial loading.

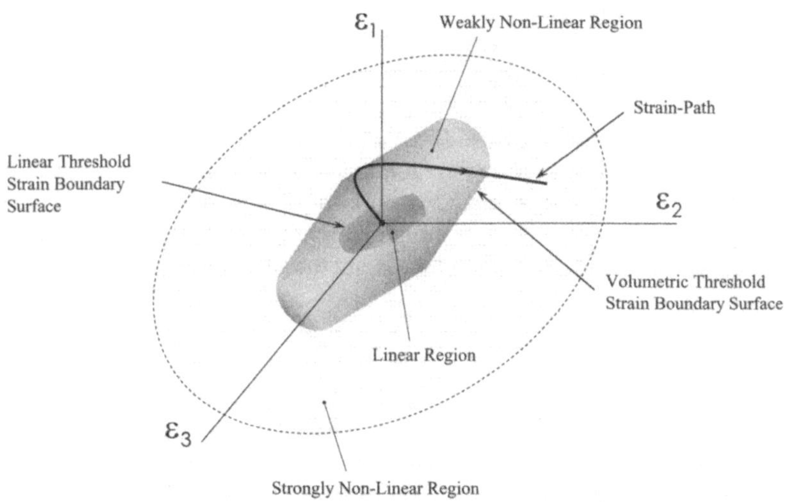


**Figure 18.** Qualitative illustration of the mechanical response of geomaterials to a dynamic excitation. Dependence of response from the magnitude of the deviatoric strain tensor illustrated through the concepts of *linear* and *volumetric threshold strain boundary surfaces* $\varepsilon^t_l$ and $\varepsilon^v_l$ in the principal strain space.

---

[22] For more details on experimental behaviour of geomaterials see the contribution by Lancellotta (2005) of this volume.

For strain states within the linear threshold strain boundary surface, geomaterials exhibit *linear* behaviour under both static-monotonic and dynamic loading when the phenomena of energy dissipation cannot be neglected. As mentioned in the introduction of Section 3, the simplest, linear constitutive model that can be used to describe the mechanical response of solid, dissipative, materials is linear viscoelasticity. Formulation of the theory of linear viscoelasticity requires essentially three assumptions:

- validity of small strain theory, namely $sup\,|\, u_{i,j}(\mathbf{x}, \tau)| \le \varepsilon^t_l(\varepsilon_1, \varepsilon_2, \varepsilon_3, \tau)$ with $\tau \in\, ]-\infty, t]$;
- current Cauchy stress tensor $\sigma_{ij}(t)$ is postulated to solely depend upon the current value of the strain tensor $\varepsilon_{ij}$ and on its past strain history, namely $\sigma_{ij}(t) \leftarrow \Phi\!: \{\, \varepsilon_{kl}(\tau)\, \}^t_{-\infty}$;
- validity of the time translation invariance hypothesis by which the material response is assumed to be independent of any shift along the time axis.

Continuity of strain history is desirable but not strictly required as discontinuous strain histories can be handled by means of integral operators intended in the *Stieltjes* sense (Fung, 1965). Under these assumptions the *Riesz* representation theorem of functional analysis guarantee the existence of a unique relationship between the Cauchy stress tensor $\sigma_{ij}(t)$ and the strain history $\{\varepsilon_{kl}(\tau)\}^t_{-\infty}$ via the following linear functional (Christensen, 1971):

$$\sigma_{ij}(t) = \int_{-\infty}^{t} G_{ijkl}(t-\tau)\frac{d\varepsilon_{kl}(\tau)}{d\tau}d\tau \tag{57}$$

where the summation convention is implied on repeated indices, $\varepsilon_{kl}$ is the infinitesimal strain tensor and $G_{ijkl}(t)$ is a fourth order tensor-valued function called the *relaxation tensor function* of the material. In deriving Eq.(57) it was assumed that the strain history is continuous, however as mentioned above discontinuities in the strain history may be represented as well if the integral appearing in Eq.(57) is intended in the Stieltjes sense.

The constitutive relationship described by Eq.(57) is sometimes called *Boltzmann's equation* since it can also be derived by applying the *Boltzmann's superposition principle*. The relaxation tensor function $G_{ijkl}(t)$ has 81 components; however, only 21 are independent due to the symmetry of the stress and strain tensors in a general anisotropic material. Boltzmann's equation can be inverted to express the strain tensor $\varepsilon_{ij}$ as a function of the stress history as follows:

$$\varepsilon_{ij}(t) = \int_{-\infty}^{t} J_{ijkl}(t-\tau)\frac{d\sigma_{kl}(\tau)}{d\tau}d\tau \tag{58}$$

where $J_{ijkl}(t)$ is a fourth order tensor-valued function called the *creep tensor function* of the material. For an isotropic, linear, viscoelastic material the creep and relaxation tensor functions have only *two* independent components and they are sufficient to completely describe the mechanical response of the material. In this case constitutive equation (57) can be rewritten as (Lai, 1998):

$$\begin{cases} s_{ij}(t) = \int_{-\infty}^{t} 2G_S(t-\tau)\dfrac{de_{ij}(\tau)}{d\tau}d\tau \\ \sigma_{kk}(t) = \int_{-\infty}^{t} 3G_B(t-\tau)\dfrac{d\varepsilon_{kk}(\tau)}{d\tau}d\tau \end{cases} \tag{59}$$

where $s_{ij} = \sigma_{ij} - \frac{1}{3}\delta_{ij}\sigma_{kk}$ and $e_{ij} = \varepsilon_{ij} - \frac{1}{3}\delta_{ij}\varepsilon_{kk}$ are the components of the deviatoric stress and strain tensors, respectively, and $\delta_{ij}$ is the Kronecker symbol. The scalar functions $G_S(t)$ and $G_B(t)$ are the shear and bulk relaxation functions, respectively. As expected, shear and volume deformations of viscoelastic isotropic materials are uncoupled, mimicking a well-known fact of linear isotropic elasticity. From Eq.(58) it is possible to obtain a pair of relationships analogous to Eq.(59) with the relaxation functions $G_S(t)$ and $G_B(t)$ replaced by the creep functions $J_S(t)$ and $J_B(t)$.

Relaxation and creep functions are material response functions. They are the analogous respectively of the shear and bulk moduli and compliances of linear elasticity with the important difference that creep and relaxation functions are no longer constants but they are time-dependent. The relaxation function $G_S(t)$ can be viewed as the shear stress response of a material subjected to a shear strain history specified as an *Heaviside function*. The creep function $J_S(t)$ on the other hand can be viewed as the shear strain response of a material subjected to a shear stress history specified as an Heaviside function. Analogous interpretations holds for the bulk relaxation and creep functions $G_B(t)$ and $J_B(t)$.

Although often to viscoelastic constitutive relationships are given a physical interpretation in terms of rheological models made up by various combinations of (linear) springs and viscous dashpots, this is not strictly necessary and far more general constitutive relationships for viscoelastic materials may be constructed which are not linked to networks of springs and dashpots (Christensen, 1971).

For an isotropic material, a complete description of a linear viscoelastic constitutive model requires specification of two material functions. These may be either the shear and bulk relaxation functions $G_S(t)$ and $G_B(t)$ or the shear and bulk creep functions $J_S(t)$ and $J_B(t)$. Sometimes the creep and relaxation functions are replaced by the shear and bulk *relaxation spectra* $H_S(\tau)$ and $H_B(\tau)$ which provide valuable information about the dissipation mechanisms associated to material relaxation times[23]. Once any two shear and bulk material functions have been specified, Eq.(59) may be used to compute stress response of the material to a prescribed strain history.

If the strain or stress history is specified as a harmonic function of time, the viscoelastic constitutive relationships assume a very simple form. Let for instance assume that the strain history in Eq.(57) is specified by the function $\varepsilon_{kl}(t) = \varepsilon_{0kl} \cdot e^{i\omega t}$ where $\varepsilon_{0kl}$ is the amplitude of the strain component. Then, the integral equation degenerates into the following set of algebraic equations:

$$\sigma_{ij}(t) = G^{*}_{ijkl}(\omega) \cdot \varepsilon_{0kl} e^{i\omega t} \tag{60}$$

[23] The inverse response functions of the *relaxation spectra* $H_S(\tau)$ and $H_B(\tau)$ are the shear and bulk *retardation spectra* and they are usually denoted by $L_S(\tau)$ and $L_B(\tau)$.

where $G^*_{ijkl}(\omega)$ is the *complex tensor modulus*. Its reciprocal is the *complex tensor compliance* and it is usually denoted by the symbol $J^*_{ijkl}(\omega)$. The components of $G^*_{ijkl}(\omega)$ are related to the Fourier sine and cosine transforms of the relaxation tensor function via the following relations:

$$\begin{cases} G_{(1)ijkl}(\omega) = G_{(e)ijkl} + \omega \cdot \int_0^\infty G_{ijkl}(\tau) \cdot \sin \omega\tau \cdot d\tau \\ \\ G_{(2)ijkl}(\omega) = \omega \cdot \int_0^\infty G_{ijkl}(\tau) \cdot \cos \omega\tau \cdot d\tau \end{cases} \tag{61}$$

where $G_{(e)ijkl} = G_{ijkl}(t \to \infty)$ is the *equilibrium response* and $G_{(1)ijkl}(\omega)$, $G_{(2)ijkl}(\omega)$ are respectively the real and the imaginary parts of the complex tensor modulus, namely:

$$G^*_{ijkl}(\omega) = [G_{(1)ijkl}(\omega) + i \cdot G_{(2)ijkl}(\omega)] \tag{62}$$

In Eqs.(61) it is implied that the relaxation tensor function $G_{ijkl}(t)$ subtracted from the equilibrium response $G_{(e)ijkl}(\omega)$ is bounded as $t \to \infty$ in such a way that the improper integral converges.

Equation (61) clearly shows that the real and the imaginary components of the complex tensor modulus are not independent. Their relationship can be easily found to be (Christensen, 1971):

$$G_{(1)ijkl}(\omega) = G_{(e)ijkl} + \frac{2}{\pi} \cdot \int_0^\infty \frac{G_{(2)ijkl}(\tau) \cdot \omega^2}{\tau \cdot (\omega^2 - \tau^2)} \cdot d\tau \tag{63}$$

This equation is known in the literature as one form of the *Kramers-Krönig relations*. They are important because in essence they states that viscoelastic materials are *inherently dispersive* that is the speed of propagation of a mechanical disturbance is frequency-dependent.

It can be shown (Tschoegl, 1989) that Eq.(63) is nothing but a statement that the real and the imaginary component of the complex tensor modulus are the *Hilbert transforms* of each other. It can also be demonstrated (Bracewell, 1965) that this result constitutes the necessary and sufficient condition for the material function $G^*_{ijkl}(\omega)$ to satisfy the *fundamental principle of causality* which states that in a physical system the *reaction* can never precede the *action*[24].

If the material is isotropic constitutive relationship (60) simplifies as follows:

$$\begin{cases} s_{ij}(t) = 2G^*_S(\omega) \cdot e_{0ij} \cdot e^{i\omega t} \\ \\ \sigma_{kk}(t) = 3G^*_B(\omega) \cdot \varepsilon_{0kk} \cdot e^{i\omega t} \end{cases} \tag{64}$$

[24] Applied to wave propagation the *principle of causality* postulates that a disturbance originated at a point in a medium (*the source*), is not allowed to arrive at a different point of the same medium (*the observer*) before the time *d/c* has elapsed, where *d* is the distance between the source and the observer and *c* is the speed of propagation of the disturbance in the medium.

where $e_{0ij}$ and $\varepsilon_{0kk}$ are respectively the amplitudes of the deviatoric and volumetric strains, and $G^*_S(\omega)$ and $G^*_B(\omega)$ are the *complex shear* and *bulk moduli*. A relevant feature hidden in Eq.(64) is that the stress components of viscoelastic materials undergoing steady state harmonic oscillations are in general *out of phase* with the corresponding strain component. The amount by which the stress lags behind the strain is measured by the argument of the complex modulus, which is also a measure of the amount of energy dissipated by the viscoelastic material during harmonic oscillations. To demonstrate this, let Eq.(64) be re-written for the simple case of shear, *uniaxial* oscillation:

$$\sigma_S(\omega) = G^*_S(\omega) \cdot \varepsilon_S(\omega) \tag{65}$$

where $\varepsilon_S(\omega) = \varepsilon_{S0} \cdot e^{i\omega t}$, with $\varepsilon_{S0} \in \mathbf{R}$. By now considering the real part of $\varepsilon_S(\omega)$, Eq.(65) can be rewritten as:

$$\sigma_S(\omega) = \left|G^*_S(\omega)\right| \cdot \varepsilon_{S0} \cdot \cos\left[\omega t - \varphi_S(\omega)\right] \tag{66}$$

where

$$\tan\left[\varphi_S(\omega)\right] = \arg\left[G^*_S(\omega)\right] = \frac{G_{(2)S}}{G_{(1)S}} \tag{67}$$

is the *loss shear angle* or the *loss shear tangent*. The real and imaginary parts $G_{(1)S}$ and $G_{(2)S}$ of the complex modulus $G^*_S(\omega)$ are often referred to as the *storage* and *loss* shear moduli respectively (Pipkin, 1986). Using trigonometric identities Eq.(66) can be rewritten as:

$$\sigma_S(\omega) - \left|G^*_S(\omega)\right| \cdot \varepsilon_{S0}(\omega) \cdot \cos\left[\varphi_S(\omega)\right] = \left|G^*_S(\omega)\right| \cdot \varepsilon_{S0} \cdot \sin\left[\varphi_S(\omega)\right] \cdot \sin(\omega t) \tag{68}$$

Equation (68) combined with the relation $\varepsilon_S(\omega) = \varepsilon_{S0} \cdot \cos(\omega t)$ yields:

$$\left(\frac{\varepsilon_S}{\varepsilon_{S0}}\right)^2 + \left(\frac{\sigma_S - G_{(1)S} \cdot \varepsilon_S}{G_{(2)S} \cdot \varepsilon_{S0}}\right)^2 = 1 \tag{69}$$

This is the equation of an *ellipse* rotated by an angle $\psi_S(\omega)$ with respect to the strain axis assumed to coincide with the x-axis. This ellipse represents the stress-strain hysteretic loop exhibited by a general linear viscoelastic material subjected to harmonic oscillations.

It can be shown that the inclination of the principal axes of the ellipse $\psi_1 = \psi_S(\omega)$ and $\psi_2 = (\pi/2 + \psi_1)$ is given by the following relation (Lai, 1998):

$$\tan\psi_{1,2}(\omega) = \frac{\left[|G_S^*|^2 - 1\right] \mp \sqrt{\left[|G_S^*|^2 + 1\right]^2 - 4G_{(2)S}^2}}{2 \cdot G_{(1)S}} \tag{70}$$

The energy dissipated by the material (per unit volume) during a cycle of harmonic oscillation can be computed from the area enclosed by the elliptic hysteretic loop. In a stress-controlled test this area denoted by $\Delta W_S^{dissip}$ is defined by the following relation:

$$\Delta W_S^{dissip}(\omega) = \oint_\ell dW_S = \oint_\ell \Re(\sigma_S) \cdot \Re(d\varepsilon_S) \tag{71}$$

where the symbol $\Re(\cdot)$ denotes the real part of a complex quantity, and $\ell$ is length of the hysteretic loop. In Eq.(71) the term $dW_S$ represents the work done by the stress (per unit volume of material) for an infinitesimal variation of strain. Considering Eq.(66) and the fact that $\Re(d\varepsilon_S) = \omega \cdot \varepsilon_{S0} \cdot \cos(\omega t + \pi/2)\, dt$, Eq.(71) can be rewritten as follows:

$$\Delta W_S^{dissip}(\omega) = \int_0^{2\pi/\omega} \omega \cdot \left|G_S^*\right| \cdot \varepsilon_{S0}^2 \cdot \cos(\omega t - \varphi_S) \cdot \cos(\omega t + \pi/2) \cdot dt \tag{72}$$

Making use of trigonometric identities the integral above can be easily solve to give:

$$\Delta W_S^{dissip}(\omega) = \pi \cdot G_{(2)S} \cdot \left|\varepsilon_S\right|^2 \tag{73}$$

Therefore the quantity $\Delta W_S^{dissip}$ is directly proportional to the loss modulus $G_{(2)S}(\omega)$. From a thermodynamical point of view $\Delta W_S^{dissip}$ is equal to the amount of entropy[25] produced in one cycle of harmonic loading and due to unrecoverable mechanical work. Although the above results have been obtained for the simple case of shear uniaxial oscillation, they can be easily generalized also for the case of *multiaxial* loading.

Figure 19 shows an experimental stress-strain hysteretic loop measured during a cyclic torsional shear test performed on a cylindrical clay specimen. Despite the maximum shear strain amplitude is well below $\gamma_l^t$, the *cyclic threshold shear* strain[26] (see also Figure 18), the stress-strain loop encloses a non-vanishing area which is a clear evidence that in geomaterials energy dissipation takes place even at very low amplitude dynamic excitations[27]. In this particular experiment it was measured a value of damping ratio of about 2 %.

[25] The amount of *unrecoverable* energy produced by an inelastic material during a thermodynamical process of deformation.

[26] For normally consolidated clays with plasticity index of 50, $\gamma_l^t$ is on the order of $10^{-4}$ (Lo Presti, 1987).

[27] At the *microscopic* level different mechanisms have been proposed to explain the process of energy dissipation at very small strain levels occurring in geomaterials subjected to dynamic excitations (Biot, 1956; Stoll, 1974; Johnston et al., 1979; White, 1983; Leurer, 1997). These studies indicate that an interactive combination of several individual mechanisms is responsible for most of the phenomena *macroscopically* called *energy dissipation*. For *coarse-grained soils* the two mechanisms that have been postulated to account for the internal entropy production are frictional losses between soil particles and

Another important feature shown by Figure 19 is that the elliptical shape of cyclic stress-strain loops predicted by the theory of linear viscoelasticity is matched fairly well by geomaterials in experimental measurements performed at very small strains as demonstrated also by other researchers (Dobry, 1970).

Based on Eq.(73) several definitions have been proposed in the literature as measures of energy dissipation for geomaterials (O'Connel and Budiansky, 1978; Ishihara, 1996; Aki and Richards, 2002). All of them are consistent with each other only when applied to *weakly* dissipative viscoelastic materials[28]. Seismologists and geophysicists have adopted definitions of energy losses borrowed from disciplines such as electrical engineering (Cole and Cole, 1941). Geotechnical engineers and earth scientists have invented their own measures. However despite the wide range of followed approaches most of these definitions involve dimensionless parameters proportional to the ratio between the energy dissipated during a cycle of harmonic oscillation and some measures of the stored energy per unit volume.

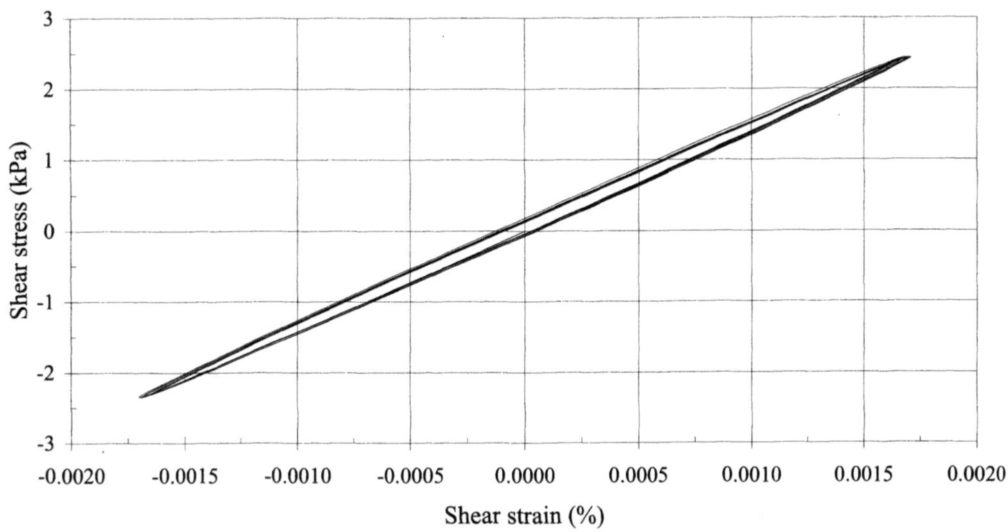


**Figure 19.** Experimental stress-strain hysteretic loop measured during a cyclic torsional shear test on a cylindrical clay specimen. Maximum shear strain amplitude $1.7 \cdot 10^{-5}$ at $f = 0.01$ Hz (from Lai et al., 1999).

The measure of energy dissipation adopted in this chapter is given by the following relation:

fluid flow losses due to the relative movement between the solid and fluid phases. *Fine-grained soils* exhibit more complex phenomena, which are controlled by electromagnetic interactions between water molecules and microscopic solid particles.

[28] A weakly dissipative (or *low-loss* or *loss-less*) material will be defined in the next section as a material for which *sup*(D) $\leq 0.05$ where *sup*($\cdot$) denotes the least upper bound of the argument and $D$ is the *material damping ratio*, also defined in the next section.

$$D_\chi(\omega) = \frac{\Delta W_\chi^{dissip}(\omega)}{8\pi \cdot W_\chi^{ave}(\omega)} \qquad \text{with } \chi = P, S \tag{74}$$

where $D_\chi(\omega)$ is a dimensionless parameter called *material damping ratio*[29], $\Delta W_\chi^{dissip}$ is the energy dissipated (per unit volume) during a cycle of harmonic oscillation, and $W_\chi^{ave}$ is the average stored energy per one cycle of harmonic oscillation. It can be shown that in a linear viscoelastic material the average stored energy $W_\chi^{ave}$ in the denominator of Eq.(74) is equal to (Tschoegl, 1989):

$$W_\chi^{ave}(\omega) = \frac{G_{(1)\chi} \cdot \left|\varepsilon_\chi\right|^2}{4} \qquad \text{with } \chi = P, S \tag{75}$$

where the complex-valued *constrained modulus*, $G^*_P = G^*_B + 4/3 \cdot G^*_S = [G_{(1)P} + i \cdot G_{(2)P}]$ has been introduced to simplify the notation. Substitution of Eqs.(73) and (75) into Eq.(74) yields:

$$D_\chi(\omega) = \frac{G_{(2)\chi}}{2 \cdot G_{(1)\chi}} \tag{76}$$

Thus, the definition of energy dissipation given by Eq.(74) has the desirable feature of linking directly to the constitutive parameters of linear viscoelasticity. Furthermore Eq.(76) is independent from the magnitude of the energy losses.

In seismology and geophysics it is customary to adopt an energy dissipation parameter analogous to material damping ratio defined by Eq.(67) and that is called *quality factor* (O'Connel and Budiansky, 1978) and denoted by $Q_\chi(\omega)$. Material damping ratio and quality factor are simply related by the expression:

$$Q_\chi(\omega) = \frac{1}{2 \cdot D_\chi(\omega)} \tag{77}$$

From Eq.(64) it is readily apparent that *complex moduli* or *complex compliances* are the fundamental material parameters (in isotropic materials) that need to be specified for the solution of any linear viscoelastic boundary value problem where all field variables are harmonic functions of time. However, because of the relationship existing between complex moduli and relaxation functions via Eq. (61) and its inverse, the complex moduli may also be considered as alternative definitions of material viscoelastic parameters.

[29] The subscript $\chi$ = P, S attached to the symbol *D* is used to denote *material damping ratio* associated with *irrotational* (or longitudinal) and *equivoluminal* (or transversal) wave motion respectively.

Explicit definition of material functions G(t) or $G^*(\omega)$ or their inverses[30] J(t) and $J^*(\omega)$, is generally accomplished through the help of experimental measurements. However in contrast to linear elasticity where the elastic constants are simply determined from the slopes of experimental stress-strain curves[31], in viscoelastic materials the process of defining a material function is considerably more subtle for there are infinitely many ways to practically construct it. Each model will have its own features, which may or may not be desirable for modeling a specific material behaviour. If for instance from a series of experimental tests a given material exhibits *hysteretic*[32] behaviour, a potential viscoelastic model should be able to reproduce this important feature (Lai, 1998).

Usually the construction of a viscoelastic model proceeds by way of what might be called a *forward* approach: one chooses a particular constitutive model (e.g. Kelvin-Voigt, Maxwell, standard linear solid, etc.) and checks *a posteriori* how the model fits the experimental data and what are the most relevant flaws. An alternative strategy would be to reverse the methodology and follow an *inverse* approach where the *a priori* starting point are the experimental data. These will follow a certain pattern that will try to be captured by a certain response function (say for instance the complex modulus or the relaxation function) dependent upon a set of material parameters (which would be the equivalent of the elastic constants of linear elasticity). The problem of defining a particular viscoelastic model has then been reduced to a *curve fitting* operation.

The *a priori* approach has the great advantage of generality, which make it suitable to accommodate a large variety of experimental results while remaining consistent with the fundamental postulates of the theory of linear viscoelasticity. Examples of application of this approach to geomaterials include the work of Liu et al. (1976) who assumed a hyperbolic distribution of the relaxation spectrum, and that of Kjartansson (1979) who adopted a power law time dependence for the creep response function. Both models are able to predict quite accurately several features of geomaterial response including the seismic bandwidth hysteretic behaviour and material dispersion effects.

### 3.3 Viscoelastic Waves in Unbounded Homogeneous Continua

A major result of the previous section is that the constitutive relation of a linear, isotropic viscoelastic solid is completely defined by two *material functions* which may be given either in the *time* or in the *frequency* domain. In the time domain they are the *relaxation functions* $G_S(t)$ and $G_B(t)$ or the *creep functions* $J_S(t)$ and $J_B(t)$. In the frequency domain they are the *complex moduli* $G^*_S(\omega)$ and $G^*_B(\omega)$ or the *complex compliances* $J^*_S(\omega)$ and $J^*_B(\omega)$. Alternatively but less frequently, material properties may also be specified by the *relaxation spectra* $H_S(\tau)$ and $H_B(\tau)$ or by the *retardation spectra* $L_S(\tau)$ and $L_B(\tau)$.

Mimicking the situation in linear elasticity, knowledge of *a pair* of material functions in one domain enables one to determine (at least in principle) the corresponding material functions in

30 From the viscoelastic constitutive equations illustrated in this section it is apparent that whereas $J^*(\omega) = 1/G^*(\omega)$ the same simple relation doesn't hold for creep and relaxation functions J(t) and G(t).

31 In linear elasticity the difference in mechanical response of two materials is simply that one material is stiffer than the other, either in bulk or in shear or in both modes of deformation.

32 Frequency or rate-independent material response. The term *hysteretic* is often used in physics and other sciences to denote memory effects processes that are scale independent (Visintin, 1994).

any other domain. However since $G^*_S(\omega)$ and $G^*_B(\omega)$ are complex-valued, the real and imaginary parts of the complex moduli cannot be *independent* from each other as stated by Eq.(63). This point will be further discussed in the sequel of this section.

Albeit the three types of material functions are in a sense equivalent for they carry the same kind of information, it is often advantageous to specify them in the frequency domain via the complex moduli $G^*_S(\omega)$ and $G^*_B(\omega)$.

In the frequency domain, the constitutive relations of linear viscoelasticity become simple and compact algebraic equations, which resemble those of linear elasticity. But this resemblance goes farther for it can be shown that the Fourier or Laplace-transformed field equations of linear viscoelasticity are formally identical to those associated to linear elasticity except that the elastic shear and bulk moduli G and B are replaced by the complex moduli $G^*_S(\omega)$ and $G^*_B(\omega)$ in case of Fourier transform, or by $s\cdot G^*_S(s)$ and $s\cdot G^*_B(s)$ in case the field equations are transformed using the Laplace transform[33] (Christensen, 1971; Pipkin, 1986).

This analogy between the field equations of linear elasticity and viscoelasticity forms the essence of what is known as the *elastic-viscoelastic correspondence principle* (Read, 1950; Fung, 1965; Christensen, 1971; Ben-Menahem and Singh, 2000). According to this principle, elastic solutions to steady state harmonic boundary-value problems can be easily converted into viscoelastic solutions for identical boundary conditions by simply replacing the elastic shear and bulk moduli G and B with the corresponding complex-valued and frequency dependent parameters $G^*_S(\omega)$ and $G^*_B(\omega)$[34]. The validity of the correspondence principle however, is restricted to problems where the boundary conditions (e.g. prescribed displacements and tractions) are *time-invariant* (Christensen, 1971).

Application, in a homogeneous medium, of the elastic-viscoelastic correspondence principle to the *Navier's* equations of motion of linear elastodynamics (Achenbach, 1984) with no body forces yields:

$$\left(G^*_B + \tfrac{4}{3}G^*_S\right)\text{grad div } \hat{\mathbf{u}} - G^*_S \text{curl curl } \hat{\mathbf{u}} = -\rho\omega^2\,\hat{\mathbf{u}} \tag{78}$$

where the vector $\hat{\mathbf{u}} = \hat{\mathbf{u}}(\mathbf{x},\omega)$ is the Fourier-transformed displacement vector ($\mathbf{x}$ is the position vector), and $\rho$ is the mass density of the homogeneous medium assumed to be time-independent. Applying the *divergence* and *curl* operators to Eq.(78) yields[35]:

$$\begin{cases} \left(V^*_P\right)^2 \nabla^2\left(\text{div } \hat{\mathbf{u}}\right) = -\omega^2 \cdot \text{div } \hat{\mathbf{u}} \\ \\ \left(V^*_S\right)^2 \nabla^2\left(\text{curl } \hat{\mathbf{u}}\right) = -\omega^2 \cdot \text{curl } \hat{\mathbf{u}} \end{cases} \tag{79}$$

---

[33] The variable *s* is used here to denote the complex-valued frequency characteristic of Laplace transform.

[34] An analogous result holds for the solution of viscoelastic *initial*-boundary value problems using the Laplace transform which is suitable when the prescribed boundary conditions are arbitrary functions of time. In this case G and B must be replaced respectively by the products $s\cdot G^*_S(s)$ and $s\cdot G^*_B(s)$.

[35] An identical procedure would be that of applying to Eq.(78) the *Helmholtz's* decomposition theorem.

where $\nabla^2$ denotes the *Laplacian* operator and $V_P^*(\omega)$, $V_S^*(\omega)$ are respectively the complex-valued longitudinal (or irrotational) and transversal (or equivoluminal) speeds of propagation of P and S waves. They define *phase velocity* and *spatial attenuation* of monochromatic, bulk waves propagating in a linear, homogeneous, viscoelastic unbounded medium, and are expressed by the following relations:

$$\begin{cases} V_P^*(\omega) = \sqrt{\dfrac{G_B^*(\omega) + \frac{4}{3} \cdot G_S^*}{\rho}} \\ \\ V_S^*(\omega) = \sqrt{\dfrac{G_S^*(\omega)}{\rho}} \end{cases} \tag{80}$$

Equations (79) shows that *distortional* and *volume* deformations in linear, isotropic viscoelastic materials are *uncoupled* from each other as occurring in linear elasticity. A general solution of Eqs.(79) may be written as follows:

$$\mathbf{u}(\mathbf{x}, t) = \mathbf{A}_1 \exp\left[i\left(\omega t - \mathbf{k}_\chi^* \cdot \mathbf{x}\right)\right] + \mathbf{A}_2 \exp\left[i\left(\omega t + \mathbf{k}_\chi^* \cdot \mathbf{x}\right)\right] \tag{81}$$

where $\mathbf{A}_1$ and $\mathbf{A}_2$ are two arbitrary constant vectors or *bivectors* to be determined from boundary conditions, $\chi$ = P, S is a subscript denoting as usual, longitudinal and transversal wave motion respectively, and finally $\mathbf{k}^*_\chi = \mathbf{k}_\chi - i \cdot \boldsymbol{\alpha}_\chi$ is the *bivector* wavenumber and defines for the $\chi$-wave both the *direction of propagation* through the vector $\mathbf{k}_\chi$,and the *direction of attenuation* through the vector $\boldsymbol{\alpha}_\chi$.

It can be shown that the vector $\mathbf{k}_\chi$ is normal to planes of constant phase which are defined by the equation $\mathbf{k}_\chi \cdot \mathbf{x}$ = *constant*. Conversely the vector $\boldsymbol{\alpha}_\chi$ is normal to planes of constant amplitude which are defined by the equation $\boldsymbol{\alpha}_\chi \cdot \mathbf{x}$ = *constant* (Ben-Menahem and Singh, 2000). The *phase velocity* of the monochromatic $\chi$-wave is equal to $\omega/|\mathbf{k}_\chi|$.

The two vectors $\mathbf{k}_\chi$ and $\boldsymbol{\alpha}_\chi$ do not need to be parallel (Aki and Richards, 2002). When the vector $\mathbf{k}_\chi$ is parallel to the vector $\boldsymbol{\alpha}_\chi$ the corresponding $\chi$-wave is called *simple* or *homogeneous* (Lockett, 1962). In a simple $\chi$-wave the direction of propagation is always coincident with the direction of maximum attenuation as it happens for instance in one-dimensional wave propagation.

*Non-simple* waves may arise as a result of boundary effects (e.g. reflection and refraction of monochromatic waves at a plane interface) combined with special types of viscoelastic materials (Christensen, 1971). All viscoelastic waves considered in this chapter are assumed to be *simple*.

If Eq.(81) is particularized for *one-dimensional* wave propagation all vector and bivector terms degenerate into scalars yielding the following simplified relation:

$$u(x, t) = A_1 \exp\left[i\left(\omega t - k_\chi^* \cdot x\right)\right] + A_2 \exp\left[i\left(\omega t + k_\chi^* \cdot x\right)\right] \tag{82}$$

where the complex-valued, scalar, wavenumber $k^*_\chi$ associated with the propagation of the $\chi$-wave is defined by:

$$k^*_\chi = \frac{\omega}{V^*_\chi} = \left(\frac{\omega}{V_\chi} - i\alpha_\chi\right) \tag{83}$$

where $V_\chi$ and $\alpha_\chi$ are respectively the real-valued, physical *phase velocity* and *attenuation coefficient* of the $\chi$-wave. This is easily verified by substituting Eq.(83) into Eq.(82). The attenuation coefficient is a measure of the spatial amplitude decay of the $\chi$-wave as it propagates through a viscoelastic though dissipative medium. From Eqs.(80) and (83) the phase velocities and attenuation factors of bulk P and S waves are given by the following expressions (Fung, 1965, Lai, 1998):

$$\begin{cases} V_\chi(\omega) = \left[\Re\left(\sqrt{\dfrac{\rho}{G^*_\chi(\omega)}}\right)\right]^{-1} \\ \\ \alpha_\chi(\omega) = \omega \cdot \Im\left(\sqrt{\dfrac{\rho}{G^*_\chi(\omega)}}\right) \end{cases} \tag{84}$$

where again $\chi$ = P, S, $G^*_P = (G^*_B + 4/3 \cdot G^*_S)$ is the complex-valued *constrained modulus* and the symbols $\Re(\cdot)$ and $\Im(\cdot)$ denote respectively the real and the imaginary part of the argument.
Thus in linear viscoelastic, unbounded media the mechanics of wave propagation is completely defined in the frequency domain either by the complex-valued phase velocities $V^*_P$, $V^*_S$, or by the real-valued phase velocities $V_P$, $V_S$ and attenuation coefficients $\alpha_P$, $\alpha_S$. Whereas $V_P$ and $V_S$ give a measure of the speed at which irrotational and equivoluminal disturbances propagate in a viscoelastic medium, $\alpha_P$ and $\alpha_S$ describe the spatial attenuation of these waves as they propagate through a dissipative material and hence they are directly related to the physical mechanisms responsible for the energy losses.

However, since two material functions are required to specify the constitutive parameters of a linear, isotropic, viscoelastic material, $V_\chi$ and $\alpha_\chi$ may not be prescribed arbitrarily, just like $\Re(G^*_\chi)$ and $\Im(G^*_\chi)$. They are linked by the *Kramers-Krönig* relationship through Eq.(63) which "*translated*" into wave propagation parameters states that the real and the imaginary parts of the complex wavenumber $k^*_\chi$ need to be a *Hilbert transform* pair. This in order to fulfill causality. Note as additional remark to Eq.(84), that because of the frequency dependence of $G^*_\chi$, $V_\chi$ and $\alpha_\chi$ are also frequency-dependent thereby causing a pulse to change its shape as it propagates through a viscoelastic medium. This the phenomenon of *material dispersion*.

It is instructive now to express $V_\chi$ and $\alpha_\chi$ in a form that makes their dependence on the dissipative properties of the viscoelastic medium more explicit. At this purpose let Eq.(84) be rewritten as follows:

$$\begin{cases} V_\chi(\omega) = \sqrt{\dfrac{G_{(1)\chi}}{\rho}} \cdot \sqrt{\dfrac{2\left(1 + \dfrac{G^2_{(2)\chi}}{G^2_{(1)\chi}}\right)}{\left[1 + \sqrt{1 + \dfrac{G^2_{(2)\chi}}{G^2_{(1)\chi}}}\right]}} \\ \\ \alpha_\chi(\omega) = \dfrac{\omega}{V_\chi(\omega)} \cdot \dfrac{\sqrt{1 + \dfrac{G^2_{(2)\chi}}{G^2_{(1)\chi}}} - 1}{\dfrac{G_{(2)\chi}}{G_{(1)\chi}}} \end{cases} \tag{85}$$

Substitution of Eq.(76) into Eq.(85) yields:

$$\begin{cases} V_\chi(\omega) = V_\chi^e \cdot \sqrt{\dfrac{2[1 + 4D_\chi^2(\omega)]}{[1 + \sqrt{1 + 4D_\chi^2(\omega)}]}} \\ \\ \alpha_\chi(\omega) = \dfrac{\omega}{V_\chi(\omega)} \cdot \dfrac{\sqrt{[1 + 4D_\chi^2(\omega)]} - 1}{2D_\chi(\omega)} \end{cases} \tag{86}$$

where $V^e_\chi = (G_{(1)\chi}/\rho)^{0.5}$ are the phase velocities of $\chi$ = P, S waves in the corresponding elastic medium. Equation (86) clearly shows that phase velocity $V_\chi(\omega)$ and material damping ratio $D_\chi(\omega)$ are not independent quantities. Substitution of Eq.(86) into Eq.(83) yields a relation for the complex-valued phase velocities of bulk waves:

$$V_\chi^*(\omega) = \frac{V_\chi(\omega)}{\sqrt{[1 + 4D_\chi^2(\omega)]}} \cdot \left[\frac{1 + \sqrt{[1 + 4D_\chi^2(\omega)]}}{2} + i \cdot D_\chi\right] \tag{87}$$

The analogous of Eq.(87) expressed in terms of *quality factor* $Q_\chi(\omega)$ defined by Eq.(77) is:

$$V_\chi^*(\omega) = \frac{V_\chi(\omega)}{2 \cdot \sqrt{1 + Q_\chi^{-2}(\omega)}} \cdot \left[1 + \sqrt{1 + Q_\chi^{-2}(\omega)} + i \cdot Q_\chi^{-1}(\omega)\right] \tag{88}$$

Equations (87) and (88) are *exact* and hence they are valid for arbitrary values of material damping ratio $D_\chi(\omega)$ and quality factor $Q_\chi(\omega)$. However, for strain levels *within* the linear cyclic threshold strain boundary surface (see Figure 19), experimental measurements in geomaterials show for $D_\chi(\omega)$ that (Vucetic and Dobry, 1991; Ishihara, 1996):

$$\sup[D_\chi(\omega)] \le 0.05 \tag{89}$$

where *sup*(·) denotes the least upper bound of the argument. Equation (89) is considered in this chapter a definition of a *low-loss*, *weakly* dissipative medium (Lai and Rix, 2002).

After expanding Eqs. (86) in *Mac Laurin* series about $D_\chi$ and substituting into Eq.(83) by retaining only terms up to *second order* one obtains (Lai, 1998):

$$V_\chi^*(\omega) = \frac{V_\chi(\omega)\cdot[1 + i\cdot D_\chi(\omega)]}{[1 + D_\chi^2(\omega)]} \tag{90}$$

By retaining only *first-order* terms, the result is:

$$V_\chi^*(\omega) = V_\chi(\omega)\cdot[1 + i\cdot D_\chi(\omega)] \tag{91}$$

In Figure 20 are shown *magnitude* and *phase* of the three relations for the complex-valued phase velocity normalized with respect to the value of phase velocity in an elastic medium (i.e. for $D_\chi$ = 0). With regards to magnitude, Figure 20 (left) shows that the expression that includes terms up to second-order is valid for $D_\chi \le 20\%$, whereas the relation that includes only first-order terms differ considerably from the other two expressions for $D_\chi \ge 5\%$. Concerning the phase, Figure 20 (right) shows that both approximate expressions yield identical values that however deviate from the exact solution for $D_\chi \ge 15\%$.

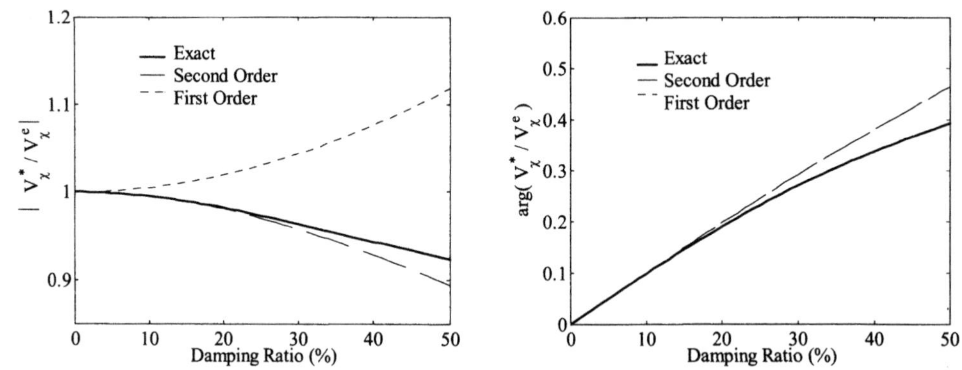


**Figure 20.** Magnitude (*left*) and phase (*right*) of the normalized complex-valued phase velocity $V^*_\chi$.

The results obtained by comparing different approximations of Eq.(87) help to link a correct formal estimate of the complex-valued phase velocity, with the definition of weakly dissipative media given by Eq.(89) on a purely empirical basis. In particular Figure 20 provides the rationale for the adoption in *low-loss* materials (i.e. for $D_\chi \le 5\%$) of Eq.(91) as a formal definition of material functions. This result will be profitably used in the next section devoted to surface wave propagation in weakly dissipative media.

As implied by the *Kramers-Krönig* relation represented by Eq.(63), the real and imaginary parts of the complex modulus, $G_{(1)\chi}(\omega)$ and $G_{(2)\chi}(\omega)$, are *not independent*, and hence from Eq.(76) material damping ratio $D_\chi(\omega)$ cannot be specified arbitrarily. If for instance $D_\chi(\omega)$ is assumed to be *constant* for the entire frequency range $\omega \in ]-\infty, +\infty[$ then Eq.(86a) would predict a frequency independent speed of propagation $V_\chi$. This conclusion is unacceptable not only because unsupported by experimental data, but also because it violates the principle of causality, since no Hilbert transform pair may satisfy the Kramers-Krönig relation with a constant material damping ratio $D_\chi$ or quality factor $Q_\chi$ (Aki and Richards, 2002).

By substituting Eqs.(76) and (86a) into Eq.(63) written for an isotropic viscoelastic material it is possible to write the Kramers-Krönig relation in terms of the physical parameters $V_\chi(\omega)$ and $D_\chi(\omega)$ to obtain (Meza, 2005):

$$V_\chi^2(\omega)\cdot\frac{1+\sqrt{1+4D_\chi^2(\omega)}}{2\cdot[1+4D_\chi^2(\omega)]}+\frac{4}{\pi}\omega^2\cdot\int_0^\infty\left[\frac{\frac{1+\sqrt{1+4D_\chi^2(\tau)}}{2\cdot[1+4D_\chi^2(\tau)]}\cdot D_\chi(\tau)}{\tau\cdot(\tau^2-\omega^2)}\right]\cdot V_\chi^2(\tau)d\tau = G_{\chi(e)} \tag{92}$$

where $\chi = P, S$ and again $G_{\chi(e)} = G_\chi(t\to\infty)$ is the *equilibrium response* of the relaxation function. Equation (92) is a *singular, Fredholm integral equation of 2nd kind with Cauchy kernel* for the unknown function $V_\chi(\omega)$ and defines the mathematical constraint imposed by causality on the two material functions $V_\chi(\omega)$ and $D_\chi(\omega)$[36]. If solved for $V_\chi(\omega)$ this equation would provide an explicit, non-integral form of the material dispersion relation in arbitrary dissipative, linear viscoelastic materials.

An explicit solution of Eq.(92) is known under the assumption that material damping ratio $D_\chi(\omega)$ is approximately frequency-independent over the seismic bandwidth (i.e. 0.001 – 10 Hz). The result is given by the following expression (Ben-Menahem and Singh, 2000; Aki and Richards, 2002):

$$V_\chi(\omega)=\frac{V_\chi(\omega_{ref})}{\left[1+\frac{2D_\chi}{\pi}\ln\left(\frac{\omega_{ref}}{\omega}\right)\right]} \tag{93}$$

where $\omega_{ref}$ denotes a reference angular frequency usually assumed in seismology equal to $2\pi$. The above *dispersion relation* predicts values of phase velocity $V_\chi(\omega)$ that increases monotonically with material damping ratio for a prescribed frequency. For a particular value of damping ratio Eq.(93) predicts an asymptotic increase of $V_\chi(\omega)$ with frequency. This form of the dispersion relation is often invoked in seismology in light of the fact that a large number of experimental data on geomaterials seem to support the frequency-independence or hysteretic assumption on

[36] Note that the integral equation (92) is *linear* in $V^2_\chi(\omega)$.

material damping ratio over the seismic band (Shibuya et al., 1995; Lo Presti and Pallara, 1997; Aki and Richards, 2002).

A remarkable feature of the Kramers-Krönig relation even in the integral form of Eq.(92) is that it establishes a functional dependence between two basic material functions allowing to compute one, say $V_\chi(\omega)$, as a function of the other, say $D_\chi(\omega)$. This can be very useful in the applications since both material functions $V_\chi(\omega)$ and $D_\chi(\omega)$ are important and in the usual practice they are determined independently. The reader interested on this subject is referred to a promising methodology recently proposed in the work by Meng (2003). Other interesting aspects of wave propagation in linear viscoelastic media have been described in detail by Borcherdt (1971) and (1973a) and Buchen (1971).

### 3.4 Surface Rayleigh Waves in Weakly Dissipative Media

Section 2.3 has shown the usefulness of variational methods in deriving closed-form, analytical relations for the *Jacobian* of (modal) Rayleigh phase velocity with respect to the medium parameters (e.g. $\alpha$ and $\beta$) which is important in the solution of the surface wave inverse problem (see Section 4). Another useful result obtained using variational methods was the derivation of an alternative relation for computing (modal) *group* velocity without the need to adopt unstable numerical differentiation.

Variational techniques can also be used to determine important results in problems of surface wave propagation in *weakly* dissipative media as it will be shown in this section. At this purpose Eq.(32) of Section 2.3, which is reported here for convenience:

$$\delta V_R = \left\{ \int_0^\infty \left[ \frac{\partial V_R}{\partial G} \right]_{\omega,M} \delta G dy + \int_0^\infty \left[ \frac{\partial V_R}{\partial M} \right]_{\omega,G} \delta M dy \right\} \tag{32bis}$$

can be re-written replacing the independent variables *G* and *M* by the closely-related parameters $V_S$ and $V_P$ ($V_P = \alpha$ and $V_S = \beta$) in light of the relations[6] $G = \rho \cdot V_S^2$ and $M = \rho \cdot V_P^2$. The result is:

$$\delta V_R = \left\{ \int_0^\infty \left[ \frac{\partial V_R}{\partial V_S} \right]_{\omega,V_P} \delta V_S dy + \int_0^\infty \left[ \frac{\partial V_R}{\partial V_P} \right]_{\omega,V_S} \delta V_P dy \right\} \tag{94}$$

According to Eq.(89) in weakly dissipative media material damping ratio $D_\chi$ ($\chi = P, S$) is a small quantity and Figure 20 legitimize the adoption of Eq.(91) as a definition of the material function $V^*_\chi$. Thus, if $V^e_\chi$ denotes the phase velocity of a $\chi$-wave of a linear elastic medium, the existence of material damping may be thought (Anderson and Archambeau, 1964; Anderson et al., 1965) to introduce a small change in $V^e_\chi$ given by:

$$\delta V^*_\chi = \left[ \left( V_\chi - V^e_\chi \right) - i \cdot V_\chi D_\chi \right] \qquad \text{with } \chi = P, S \tag{95}$$

The quantity $\delta V_\chi$ has been denoted with an asterisk to emphasize that it is complex-valued. Then, substitution of Eq.(95) into Eq.(94) for $\delta V_S$ and $\delta V_P$ yields:

$$\delta V_R^* = \int_0^\infty \left[ \frac{\partial V_R}{\partial V_S} \right]_{\omega, V_P} \left[ (V_S - V_S^e) - iV_S D_S \right] dy + \int_0^\infty \left[ \frac{\partial V_R}{\partial V_P} \right]_{\omega, V_S} \left[ (V_P - V_P^e) - iV_P D_P \right] dy \tag{96}$$

where

$$\delta V_R^* = \left[ (V_R - V_R^e) - iV_R D_R \right] \tag{97}$$

Taking the real and the imaginary components of Eq.(96) yields:

$$\begin{cases} V_R(\omega) = V_R^e + \left\{ \int_0^\infty V_S \left[ \frac{\partial V_R}{\partial V_S} \right]_{\omega, V_P} \left[ 1 - \frac{V_S^e}{V_S} \right] dy + \int_0^\infty V_P \left[ \frac{\partial V_R}{\partial V_P} \right]_{\omega, V_S} \left[ 1 - \frac{V_P^e}{V_P} \right] dy \right\} \\ D_R(\omega) = \frac{1}{V_R(\omega)} \cdot \left\{ \int_0^\infty V_S D_S \left[ \frac{\partial V_R}{\partial V_S} \right]_{\omega, V_P} dy + \int_0^\infty V_P D_P \left[ \frac{\partial V_R}{\partial V_P} \right]_{\omega, V_S} dy \right\} \end{cases} \tag{98}$$

In Eq.(98) the expression:

$$\left[ 1 - \frac{V_\chi^e}{V_\chi(\omega)} \right] \quad \text{with } \chi = P, S \tag{99}$$

specifies the frequency dependence law of the speed of propagation of longitudinal P and transversal S waves; in other words Eq.(99) defines a (material) *dispersion relation*. As illustrated in the previous section, for a waveform propagating in a dissipative medium to be causal, this relation cannot be assigned arbitrarily but it must satisfy the Kramers-Krönig constraint.

For small values of material damping ratio, i.e. for values $D_\chi$ satisfying Eq.(89), an approximated expression for the attenuation coefficient $\alpha_\chi(\omega)$ can be obtained by expanding Eq.(86b) in a Mac Laurin series about $D_\chi$. By retaining only terms up to first order the result is[37]:

$$\alpha_\chi(\omega) = \frac{\omega}{V_\chi(\omega)} \cdot D_\chi \qquad \text{with } \chi = P, S \tag{100}$$

If now is postulated that Eq.(100) holds also for surface Rayleigh waves, Eq.(98) can finally be re-written as follows:

[37] The approximated relation (100) for $\alpha_\chi(\omega)$ could alternatively be obtained from Eqs.(83) and (91) after neglecting terms higher than first order in $D_\chi$.

$$\begin{cases} V_R(\omega) = V_R^e + \left\{ \int_0^\infty V_S \left[ \dfrac{\partial V_R}{\partial V_S} \right]_{\omega, V_P} \left[ 1 - \dfrac{V_S^e}{V_S} \right] dy + \int_0^\infty V_P \left[ \dfrac{\partial V_R}{\partial V_P} \right]_{\omega, V_S} \left[ 1 - \dfrac{V_P^e}{V_P} \right] dy \right\} \\ \alpha_R(\omega) = \dfrac{\omega}{\left[ V_R(\omega) \right]^2} \cdot \left\{ \int_0^\infty V_S D_S \left[ \dfrac{\partial V_R}{\partial V_S} \right]_{\omega, V_P} dy + \int_0^\infty V_P D_P \left[ \dfrac{\partial V_R}{\partial V_P} \right]_{\omega, V_S} dy \right\} \end{cases} \quad (101)$$

It is noted as a remark that in the above equations, the material dispersion law for both P and S waves has been deliberately left *unspecified.* Equation (101)[38] is an important result because it shows that in vertically heterogeneous, weakly dissipative media Rayleigh phase velocity $V_R(\omega)$ and the associated attenuation coefficient $\alpha_R(\omega)$ can be computed from the solution of the elastic Rayleigh eigenvalue problem including the partial derivates of $V_R(\omega)$ with respect to medium parameters $V_P$ and $V_S$ (see Section 2.2 and 2.3). Indeed, Eq. (101) forms the basis of some of the most common procedures used by seismologists and geophysicists to solve surface wave propagation problems in the inelastic Earth and they are all based on the assumption of weak dissipation (Lee and Solomon, 1979; Keilis-Borok, 1989; Ben-Menahem and Singh, 2000; Aki and Richards, 2002; Herrmann, 2002).

These procedures can be used to solve both the *forward* surface wave problem and the *inverse* problem associated with surface wave motion. In the forward or *direct* problem the objective is to determine Love or Rayleigh dispersion and attenuation curves $V_R(\omega)$ and $\alpha_R(\omega)$ given a model (generally one-dimensional) of medium parameters specified by the set $\{\rho(y), \alpha(y), \beta(y)\}$. Equation (101) is written in a form already suitable for the solution of the forward problem. Conversely, the inverse or *backward* problem associated with Eq.(101) is a problem where the goal is to determine which set or subset of model parameters $\{\rho(y), \alpha(y), \beta(y)\}$ correspond to a given data ensemble $V_R(\omega)$ and $\alpha_R(\omega)$. As will be illustrated in Section 4, generally solution of the Rayleigh inverse problem is mathematically more involved than the solution of the corresponding forward problem.

With regards to the latter, Figure 21 shows the Rayleigh dispersion and attenuation curves $V_R(\omega)$ and $\alpha_R(\omega)$ obtained from the application of Eq.(101) to a flat-layered model of the continental Earth crust. The model refers to south-east continental United States and has been calculated through the computer program CRUST 5.1 (Mooney et. al., 1998) which allows to determine a global, flat-crustal model of the Earth anywhere in the world with a 5 by 5 degrees resolution. Model parameters $\{\rho(y), \alpha(y), \beta(y)\}$ and layer thickness of the Earth crust are specified by the code except the values of the quality factors $Q_\alpha$ and $Q_\beta$ which were obtained from Ben-Menahem and Singh (2000). Table 5 reports a summary of these data.

The modal Rayleigh dispersion and attenuation curves $V_R(\omega)$ and $\alpha_R(\omega)$ shown in Figure 21 were calculated for a frequency range from 0.0001 to 10 Hz which is considered significant at this scale. The dispersion curves appear numerous and regular with the expected tendency to align, as the frequency increases, to the asymptote represented by the Rayleigh phase velocity of the stratum of soft sediments having $\beta = 1{,}100$ m/s.

[38] A result perfectly analogous can be obtained also for Love waves (e.g. Ben-Menahem and Singh 2000).

**Table 5.** Earth crust properties of the 5-layer flat-crustal model used to calculate the Rayleigh dispersion and attenuation curves $V_R(\omega)$ and $\alpha_R(\omega)$ plotted in Figure 21. The crustal structure was generated using the computer program CRUST 5.1 (Mooney et. al., 1998) with reference to south-east continental United States. The values of the quality factors $Q_\alpha$ and $Q_\beta$ were obtained from Ben-Menahem and Singh (2000).

| Stratum | Thickness (m) | $\alpha$ (m/s) | $\beta$ (m/s) | $Q_\alpha$ (-) | $Q_\beta$ (-) | $\rho$ (t/m$^3$) |
|---|---|---|---|---|---|---|
| soft sediments | 500 | 2,500 | 1,100 | 225 | 100 | 2.10 |
| hard sediments | 500 | 5,300 | 3,100 | 1125 | 500 | 2.60 |
| upper crust | 19,000 | 6,100 | 3,500 | 1575 | 700 | 2.75 |
| middle crust | 12,000 | 6,600 | 3,800 | 1350 | 600 | 2.90 |
| lower crust | 7,000 | 7,200 | 4,000 | 900 | 400 | 3.10 |
| upper mantle | $\infty$ | 8,150 | 4,670 | 100 | 110 | 3.38 |

At very low frequencies instead, the dispersion curve associated with the fundamental mode tend to approximate the Rayleigh phase velocity of the upper mantle with $\beta = 4{,}670$ m/s.

As for the Rayleigh attenuation curves computed from Eq.(101), they appear more irregular though the attenuation coefficient $\alpha_R(\omega)$ exhibits a natural tendency to increase with frequency. This is expected in light of Eq.(100). Although it is not immediately apparent from Figure 21, the *cut-off* frequencies above which higher modes cease to exist are the same for both Rayleigh dispersion and attenuation curves. This result is inherent to Eq.(101) and to the procedure used to compute the Rayleigh attenuation coefficient under the assumption of weak dissipation. As will be shown in the next section in arbitrarily dissipative media cut-off frequencies for dispersion and attenuation curves are not necessarily coincident.

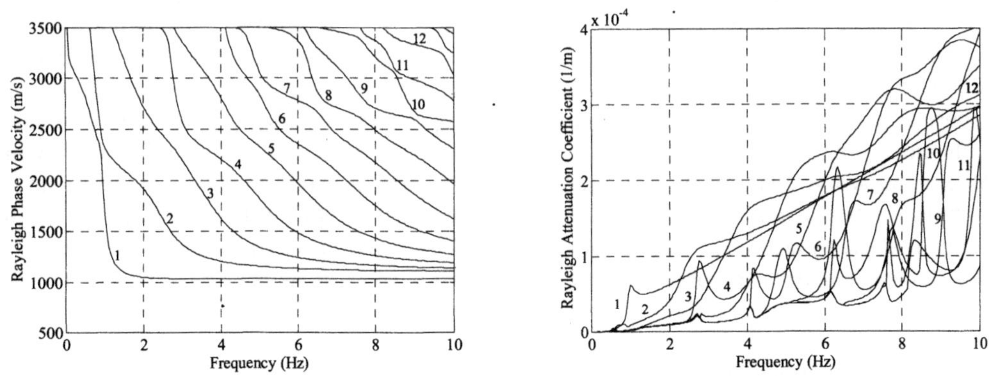


**Figure 21.** Rayleigh modal dispersion (left) and attenuation (right) curves for the continental Earth crust model of Table 5. Solution is obtained using the theory of Rayleigh waves in *weakly* dissipative media.

Concerning the Rayleigh *inverse problem* a particularly advantageous feature of Eq.(101) is that it *uncouples* the two problems of inverting dispersion and attenuation data. The procedure, whose details will be described in Section 4 in relation to *seismic prospecting* and *near-surface geophysics*, involves three major steps. The first is the experimental determination of the dispersion and attenuation curves, namely $V_R(\omega)$ and $\alpha_R(\omega)$ from surface wave measurements[39]. In the second step the experimental dispersion curve $V_R(\omega)$ is inverted to obtain the profile $V_S(y)$ of transversal phase velocity of an unknown elastic layered medium.

The third and final step involves the use of Eq.(101) as the basis of the inversion of the experimentally measured attenuation curve $\alpha_R(\omega)$ to obtain the material damping ratio profile $D_S(y)$ or quality factor $Q_S(y)$. An important remark of this procedure is that whereas the inversion of $\alpha_R(\omega)$ to obtain $D_S(y)$ is linear, the inversion of $V_R(\omega)$ to determine $V_S(y)$ is not.

Section 4 will illustrate an elegant technique for the *joint-causal* inversion of the dispersion and the attenuation curves $V_R(\omega)$ and $\alpha_R(\omega)$. The technique is quite general as it can be adopted to solve surface wave inverse problems in both *weakly* and *strongly* dissipative layered continua. With regards to the approximate solution represented by Eq.(101), *causality* in the uncoupled inversion of $V_R(\omega)$ and $\alpha_R(\omega)$ may be enforced iteratively after assigning, through Eq.(99), a material dispersion relation that satisfies the Kramers-Krönig constraint.

## 3.5 Surface Rayleigh Waves in Strongly Dissipative Media

Generalization to viscoelastic media of the procedure illustrated in Section 2.2 to define the Love and Rayleigh eigenvalue problems in elastic media can be established through the employment of the variational technique. However integral transform methods provide a more straightforward approach to the solution of the problem as it leads to the elegant and powerful *elastic-viscoelastic correspondence principle* (see Section 3.3). When this association between field equations of elasticity and viscoelasticity as implied by the principle, is applied to the homogeneous boundary value problem of surface waves, it naturally leads to a pair of complex-valued eigenproblems. Formally, Eqs.(13) and boundary conditions (14) are still valid provided that vectors $\mathbf{f}(y)$, $\mathbf{g}(y)$ and matrices $\mathbf{A}(y)$, $\mathbf{B}(y)$ are intended to be complex-valued arrays.

As a result most of the features described in Section 2.2 for the elastic eigenproblems carry over to their viscoelastic counterparts with the important difference that non-trivial solutions of the latter are now complex-valued wavenumbers and eigenfunctions[40]. Even the numerical techniques used to solve the elastic and viscoelastic eigenproblems are essentially the same, the main difference being that in the latter case use of complex arithmetic can no longer be avoided and algorithms such as root finding techniques must be properly generalized to remain applicable for complex values of the arguments[41]. The next section below will illustrate a technique for the solution of the surface wave eigenproblems in viscoelastic media which is based on the exploitation of the *holomorphic* properties of Love and Rayleigh secular functions $\mathscr{F}_L[\cdot]$ and $\mathscr{F}_R[\cdot]$.

---

39 In seismological applications this is done through observed earthquake recordings. In shallow geophysics and geotechnical engineering the dispersion and attenuation curves are obtained via non-destructive *active* or *passive* testing methods (see the chapters by Rix, 2005 and Foti, 2005 for more details).

40 It is observed however that certain properties of the real-valued eigenproblem require careful consideration before they can be generalized to the complex-valued case. One example in this sense is the definition of *orthogonality* between eigenfunctions (Lai, 1998).

41 For a detailed description of the numerical methods used to solve the Love and Rayleigh eigenproblems in viscoelastic media see the chapter by Kausel (2005) of this volume.

Surface wave propagation in linear viscoelastic media enclose a rather interesting special case. If the complex-valued *Lamè*'s parameters $\lambda^*$, $G^*$ are specified in such a way that the corresponding Poisson's ratio is a frequency-independent, real-valued constant, the roots of Love and Rayleigh dispersion equations (Christensen, 1971). Therefore in this special circumstance the solution of the complex-valued eigenproblem can be obtained using the same procedures employed for the solution of the elastic eigenproblem, albeit the ensuing wavenumbers will still be complex-valued.

In the most general situation, Poisson's ratio is a complex-valued, frequency-dependent function and the solution of the surface wave eigenproblems is not trivial, particularly with regards to the computation of the eigenvalues, namely the zeros of Love and Rayleigh secular functions $\mathcal{F}_L[\cdot]$ and $\mathcal{F}_R[\cdot]$.

**A solution based on Cauchy's theorem of complex analysis.** In general, the problem of computing the zeros of a complex-valued function of a complex variable is non trivial, especially if the function is highly non-linear and is known only numerically as for the case of Love and Rayleigh secular functions. No general method currently exists, and in most cases the adopted strategy consists in breaking up the complex nature of the function and transforming it into an equivalent system of two real-valued equations in two real variables. The zeros of such a system are then found by root-finding conventional techniques (Henrici, 1974).

Alternative techniques are available if the function $f(z)$: $\mathbf{C}\rightarrow\mathbf{C}$ satisfies the condition of *analyticity* inside an open set $\mathcal{D} \subset \mathbf{C}$[42]. In this circumstance the zeros of the function $f(z)$ may be determined using a different class of algorithms which take full advantage of the numerous properties enjoyed by *holomorphic* functions (Kravanja and Barel, 2000). So now the question is to verify whether Love and Rayleigh secular functions $\mathcal{F}_L[V^*_L]$: $\mathbf{C}\rightarrow\mathbf{C}$, and $\mathcal{F}_R[V^*_R]$: $\mathbf{C}\rightarrow\mathbf{C}$ are holomorphic in $\mathcal{D}$ with respect to complex phase velocities $V^*_L$ and $V^*_R$. A rigorous analysis of this problem would require examination of the *Cauchy-Riemann* equations for $\mathcal{F}_L[V^*_L]$ and $\mathcal{F}_R[V^*_R]$ a task that, given the complexity of these functions and the way they are built, would be extremely involved even for a two-layer medium. However by inspection of the terms composing $\mathcal{F}_L[V^*_L]$ and $\mathcal{F}_R[V^*_R]$ it is possible to infer that as they stand these functions cannot be holomorphic because a) they are not *single-valued* of their arguments $V^*_L$ and $V^*_R$ and b) they may posses poles and/or other types of singularities (Lai and Rix, 2002).

Yet the situation is not that severe as single-valuedness can be easily restored, within a set $\mathcal{D}$: $V^*_L, V^*_R \in \mathcal{D} \subset \mathbf{C}$, during construction of $\mathcal{F}_L[V^*_L]$ and $\mathcal{F}_R[V^*_R]$ by choosing the proper branches of these functions on the corresponding *Riemann* surfaces (Båth and Berkhout, 1984) after first identifying branch-cuts and branch-points of $\mathcal{F}_L[V^*_L]$ and $\mathcal{F}_R[V^*_R]$ in $\mathcal{D}$. Finally, the presence of poles and/or other kind of singularities does not impede a complex-valued function to be holomorphic in absolute terms. It may simply pose restrictions on the domain where this important property holds.

In light of these considerations it is assumed that both $\mathcal{F}_L[V^*_L]$ and $\mathcal{F}_R[V^*_R]$ are holomorphic functions in the domain of interest, with the exception at most of a finite number of isolated sin-

[42] For a precise definition of the necessary and sufficient conditions required for a function $f(z)$: $\mathbf{C}\rightarrow\mathbf{C}$ to be *analytic* or *holomorphic* in $\mathcal{D} \subset \mathbf{C}$ see any textbook on complex analysis (e.g. Remmert, 1998).

gularities of pole-type[43]. With this setting it is now recalled *Cauchy's residue theorem* of complex analysis, which can be written in the following form:

$$\frac{1}{2\pi i}\cdot\oint f(z)dz = \sum_{j=1}^{M} \text{Res}(z_j) \tag{102}$$

where the integral sign denotes integration along a positively oriented closed contour Γ, *f(z)* is an *holomorphic* function inside and on Γ except at the points $z_j$, $j = 1, M$ where it may have isolated singularities; $z \in \mathbf{C}$ is a point of the complex plane $z = (x + iy)$, and the symbol $Res(z_j)$ denotes the *residue* of the function *f(z)* at the point $z_j$[44]. Finally, *M* is the number of isolated singularities of *f(z)* inside Γ. Equation (102) forms the basis of the algorithm proposed by Abd-Elall et al. (1970) for computing the roots of *f(z)*. In fact Cauchy's residue theorem may also be written in the form:

$$G_N = \frac{1}{2\pi i}\cdot\oint \frac{z^N}{f(z)}dz = \sum_{j=1}^{m} \rho_j z_j^N \tag{103}$$

where $\rho_j$ are the *residues* of *1/f(z)* at the points $z_j$ that are zeros of *f(z)*, and *m* is the number of *zeros* of *f(z)* located inside Γ. For different values of *N*, for $N = 0, (2m-1)$, Eq.(103) allows to compute a sequence of complex numbers $G_N$ via operations of contour integration. From the knowledge of the $G_N$ 's one can determine the coefficients of the complex polynomial $P_m(z)$:

$$P_m(z) = c_0 + c_1 z + c_2 z^2 + \dots + c_{m-1} z^{m-1} + z^m \tag{104}$$

by solving the linear system of equations that can be constructed from the modified Newton identities (Abd-Elall et al., 1970):

$$\sum_{j=0}^{m-1} G_{r+j}\cdot c_j + G_{r+m} = 0 \qquad \text{with} \quad r = 0,\ m-1 \tag{105}$$

The zeros of $P_m(z)$ coincide with the *zeros* of *f(z)* inside Γ. Note that solution of the linear system of equations (105) does not require an *a priori* knowledge of the *residues* $\rho_j$.

By now identifying *f(z)* with $\mathcal{F}_L[V^*_L]$ and $\mathcal{F}_R[V^*_R]$ and the contour of integration Γ represents the boundaries of a region $\mathcal{D}$ where the zeros of Love and Rayleigh secular functions are located, then the roots of $P_m(z)$ are nothing but the complex-valued Love and Rayleigh (modal) phase velocities which ultimately give the eigenvalues $\omega/V^*_L$ and $\omega/V^*_R$ of the viscoelastic eigenproblems associated with surface waves (Lai and Rix, 2002).

A crucial step required for the implementation of the algorithm is the computation of the complex numbers $G_N$ for $N = 0, (2m-1)$ of Eq.(103) through numerical integration. To perform this task it is necessary to define the region $\mathcal{D}$ delimited by the boundary Γ where the zeros of $\mathcal{F}_L$

[43] Such type of functions are usually named *meromorphic* which is then the most appropriate designation for the functions $\mathcal{F}_L[V^*_L]$ and $\mathcal{F}_R[V^*_R]$.

[44] For ease of notation the use of an *asterisk* to denote a complex quantity will hereinafter be dropped.

$\mathcal{F}_L[V^*_L]$ and $\mathcal{F}_R[V^*_R]$ are located. For this purpose it is useful to determine the lower and upper bounds of the real and imaginary parts of $V^*_L$ and $V^*_R$. In the plane $w_R$: $\{V_R, D_R\}$ lower and upper bounds for $V_R$ may be easily established from the Rayleigh phase velocities in homogeneous half-spaces using $min[V_S(y)]$ and $max[V_S(y)]$, respectively[45]. Lower and upper bounds for $D_R$ are obtained from the observing that $0 \leq D_R \leq max(D_R) < max[(D_S(y)]$.

However in defining the region $\mathcal{D}$, it must be considered that the *domain of analyticity* of the Rayleigh secular function is not the plane $w_R$: $\{V_R, D_R\}$ but rather the plane $z_R$: $\{x_R, y_R\}$, whose relations with $w_R$: $\{V_R, D_R\}$ are given by Eq.(87), namely:

$$z_R : \begin{cases} x_R = x_R(V_R, D_R) = \dfrac{V_R\left(1+\sqrt{1+4D_R^2}\right)}{2\sqrt{1+4D_R^2}} \\ y_R = y_R(V_R, D_R) = \dfrac{V_R \cdot D_R}{\sqrt{1+4D_R^2}} \end{cases} \tag{106}$$

with $V^*_R = z_R = (x_R + iy_R)$. In Eq.(103) the numerical evaluation of the integral involving the Rayleigh secular function $\mathcal{F}_R[V^*_R]$ along the contour $\Gamma$ must be performed in the $z_R$-plane. Equations (106) provide the relationship of the *mapping* between a point $w_R$: $\{V_R, D_R\}$ of the $w_R$-plane and the corresponding point of the $z_R$-plane. Thus, from Eq.(106) it is possible to map a region $\mathcal{C}$ of the $w_R$-plane into another region $\mathcal{D}$ of the $z_R$-plane. Figure 22 shows the two regions $\mathcal{C}$ and $\mathcal{D}$ enclosing the zeros of $\mathcal{F}_R[V^*_R]$. Note from the figure the opposite orientation of the boundaries $\gamma$ and $\Gamma$ of the two regions $\mathcal{C}$ and $\mathcal{D}$ of the complex plane.

Numerical evaluation of the integral in Eq.(103) may be carried out by introducing an *admissible* parametrization of the contour $\Gamma$. Once the numbers $G_N$ have been computed for $N = 0, (2m-1)$ the remaining steps for the complete the implementation of the algorithm are:

- solution of the linear systems of equations (105) for the coefficients $c_j$ (j = 0, m-1);
- computation of the zeros of the complex polynomial in Eq.(104) and thus the complex-valued Rayleigh phase velocities $V^*_R(\omega)$.

A detailed description of the technical aspects related to the practical implementation of the algorithm are not discussed here. Likewise it is not treated the important problem of determining the number of zeros $m$ of $\mathcal{F}_R[V^*_R]$ at each frequency. The interested reader is referred to the article by Lai and Rix (2002).

[45] To simplify the formalism, the description of the algorithm hereinafter will be restricted *only* to the Rayleigh secular function even though most statements are equally valid also for $\mathcal{F}_L[V^*_L]$.

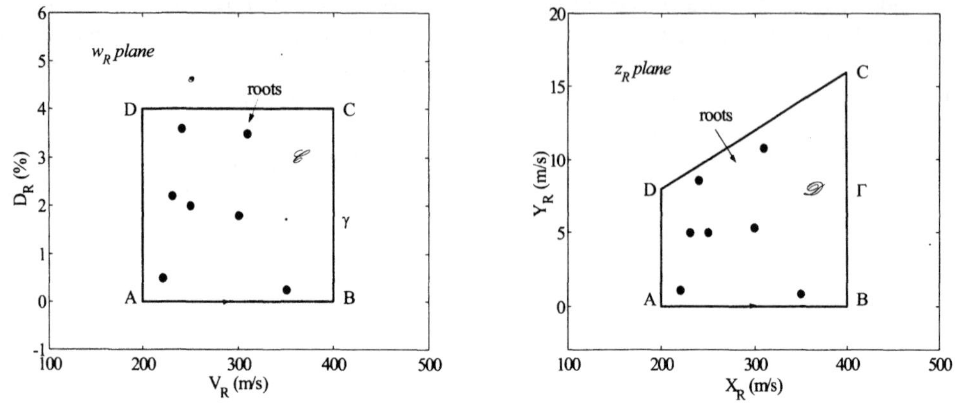


**Figure 22.** Complex-valued roots of the Rayleigh dispersion equation $\mathcal{F}_R[V^*_R]$ in (left) region $\mathcal{C}$ of the $w_R$-plane and (right) region $\mathcal{D}$ of the $z_R$-plane (from Lai and Rix, 2002).

After determining the roots $V^*_R(\omega) = (x_R + iy_R)$ of $\mathcal{F}_R[V^*_R]$ the Rayleigh phase velocity and attenuation coefficients can finally be determined by inverting Eq.(106) thereby obtaining:

$$w_R : \begin{cases} V_R = V_R(x_R, y_R) = \dfrac{(x_R^2 + y_R^2)}{x_R} \\ D_R = D_R(x_R, y_R) = \dfrac{x_R \cdot y_R}{(x_R^2 - y_R^2)} \end{cases} \tag{107}$$

Application of Eq.(107) for a discrete set of frequencies allows to construct the Rayleigh *dispersion* and *attenuation* curves. The ultimate task of computing the Rayleigh *eigenfunctions* $r^*_n(y, k_j, \omega)$, $n = 1, 4$ is straightforward and can be performed using the same type of algorithms used to solve the elastic eigenproblem. The only difference being that now *all* the operations must be performed using *complex arithmetic*[46].

However, the fact that the Rayleigh eigenfunctions in viscoelastic media are complex-valued implies that the phase difference between the horizontal and the vertical components of the displacement field is no longer equal to $\pi/2$ as in the elastic case. As a result, the principal axes of the *ellipse* describing the *orbit* of the Rayleigh particle motion are now rotated *forward* or *backward* with respect to the free boundary of the half-space depending on the properties of the medium (Båth and Berkhout, 1984).

Figure 23 shows the trajectories of Rayleigh particle motion obtained from numerical simulations in a strongly dissipative, linear viscoelastic, layered half-space with material properties listed in Table 6.

[46] In a layered system these operations consist essentially in matrix multiplications.

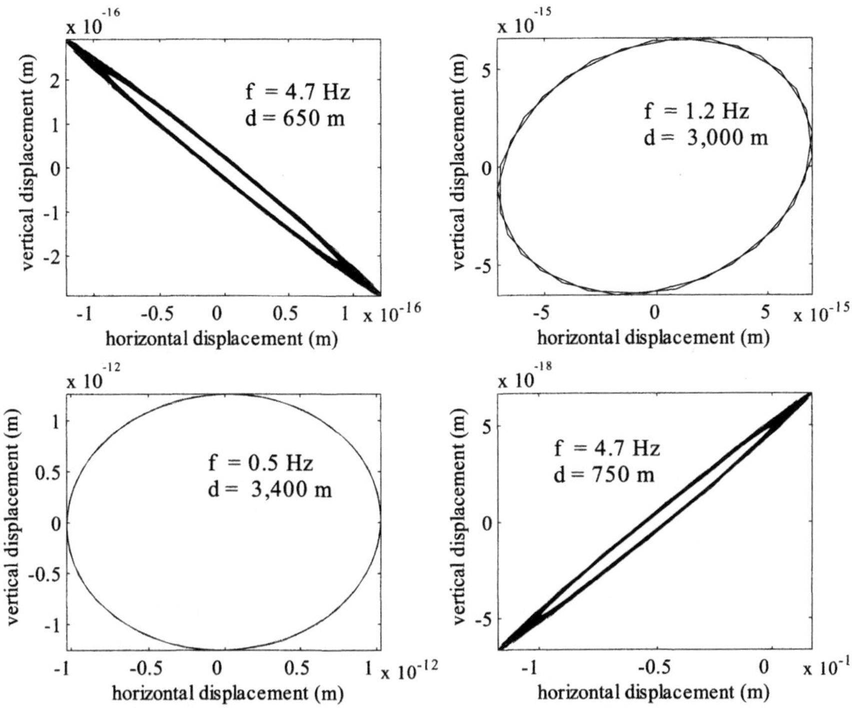


**Figure 23.** Rayleigh particle *orbits* at the free-boundary of a *strongly dissipative*, linear viscoelastic, layered half-space at variable frequencies (*f*) and distances (*d*) from a time-harmonic vertical point source. Material properties used in numerical simulations are listed in Table 6.

The numerical simulations have been conducted using a time-harmonic, vertical point source applied at the free-boundary of the half-space. Horizontal and vertical Rayleigh displacement were computed from the superposition of different modes of propagation of surface waves. Figure 23 shows the results obtained at various frequencies and distances from the source. They clearly show the inclination of the principal axes of the *ellipse*[47] with respect to the free-boundary of the half-space and also a variable semi-axes ratio.

The degree of the inclination and the semi-axes ratio varies *independently* with frequency and distance from the source. The figure on top left corner refers to the same frequency of the figure shown on bottom right corner however is rotated by $\pi$ due to a different distance. As for the influence in sign of these two variables it is noted as expected that the amplitude of particle motion decreases with the increase of the excitation frequency[48] and distance from the source.

[47] At the free-boundary of a vertically inhomogeneous half-space, the particle orbit of a Rayleigh wave is not necessarily *retrograde* even in a perfectly elastic medium (Haskell, 1953).

[48] As the frequency increases, the *wavelength* of the disturbance decreases, and thus it increases the number of *cycles* traveled by the signal to cover the same distance. Since the medium is inelastic this results

**Table 6.** Material properties used for numerical simulations of Rayleigh waves in a *strongly dissipative*, linear viscoelastic, layered half-space. Medium parameters correspond to Table 1 of Lai and Rix (2002).

| Stratum | Thickness (m) | $V_P$ (m/s) | $V_S$ (m/s) | $D_P$ (-) | $D_S$ (-) | $\rho$ (t/m$^3$) |
|---|---|---|---|---|---|---|
| 1 | 5.0 | 300 | 150 | 0.15 | 0.30 | 1.7 |
| 2 | 5.0 | 400 | 200 | 0.10 | 0.20 | 1.7 |
| 3 | 5.0 | 500 | 250 | 0.15 | 0.25 | 1.8 |
| 4 | 7.5 | 600 | 300 | 0.10 | 0.20 | 1.8 |
| 5 | 7.5 | 700 | 350 | 0.10 | 0.15 | 1.9 |
| half-space | $\infty$ | 800 | 400 | 0.15 | 0.20 | 1.9 |

Even though the results shown in Figure 23 were obtained from numerical simulations conducted in a medium having dissipation parameters *one-order of magnitude* greater than those exhibited by geomaterials at small strain levels[49], the *rotation* at the free boundary of the elliptical orbit of Rayleigh particle motion predicted by the theory is a fact that has been observed experimentally even for $\varepsilon_{ij} \leq \varepsilon^t_l$ (Strobbia, 2003).

Returning to the algorithm for solving the Rayleigh viscoelastic eigenproblem, determination of the *wavenumbers* $k^*_j = \omega/V^*_{R(j)}$ , j = 1, M and the associated *eigenfunctions* $r^*_n(y, k_j, \omega)$, n = 1, 4 completely solves the homogeneous boundary value problem of *free* Rayleigh waves in linear viscoelastic, vertically inhomogeneous half-spaces[50]. Figure 24 shows an application of the algorithm described above and compare it with the procedure illustrated in Section 3.4 based on the assumption of *weak dissipation*. The *exact* and *approximate* solutions of the viscoelastic eigenproblem were used to compute *synthetic seismograms* for the strongly dissipative, layered half-space of Table 6. The vertical point source at the free-boundary was considered to have a time-variation of a *Dirac* δ-function. The horizontal and vertical components of Rayleigh particle displacement were computed at various distances from the source. The differences between the exact and approximate solutions are fairly small with the approximate theory that have a tendency to slightly overestimate the amplitude of the response. Figure 25 shows the *orbit* and *polarization* of the Rayleigh particle motion observed (with the exact solution) at an offset of 2,000 m from the source.

in a greater energy dissipation. The analogy in *time-domain* would be to consider a *fixed point* in space and observe the attenuation in the same time window of two *low* and *high* period signals.

49 That is strain levels $\varepsilon_{ij}$ whose magnitude is in the interior of the linear cyclic threshold strain boundary surface $\varepsilon^t_l$ (see Figure 18).

50 To properly account for *material dispersion* and satisfy *causality*, the Rayleigh secular function $\mathcal{F}_R[V^*_R]$ must be constructed using *bulk wave velocities* $\alpha = \alpha(y,\omega)$ and $\beta = \beta(y,\omega)$ and *material damping ratios* $D_P = D_P(y,\omega)$ and $D_S = D_S(y,\omega)$ satisfying an appropriate (causal) frequency-dependence law.

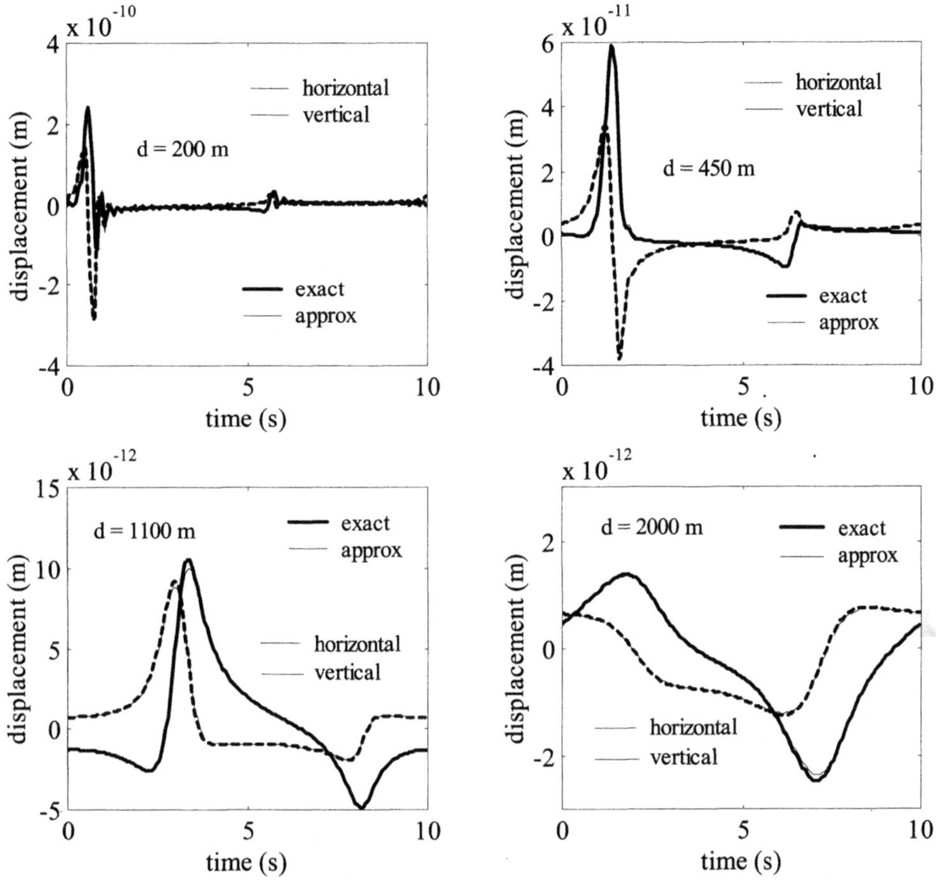


**Figure 24.** Rayleigh particle *displacement* time-histories generated at the free-boundary of a *strongly dissipative*, linear viscoelastic, layered half-space by a vertical, *Dirac*-type impulse. Comparison between *exact* and *approximate* solutions at various distances. Material properties are listed in Table 6.

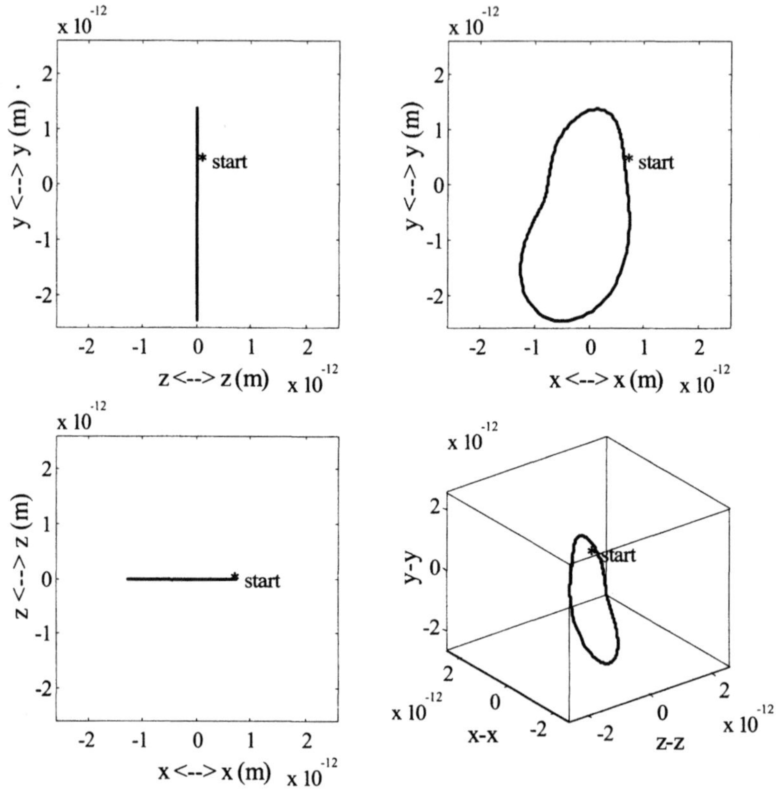


**Figure 25.** Rayleigh particle *orbit* at the free-boundary of a *strongly dissipative*, linear viscoelastic, layered half-space at *2,000 m* from a vertical *Dirac*-type impulse. Material properties are listed in Table 6.

The reader interested in examining in detail other features of Rayleigh wave propagation in linear viscoelastic half-spaces is referred to Borcherdt (1973b) and Borcherdt and Wennerberg (1985).

## 4 Inverse Problem Associated with Surface Wave Motion

### 4.1 Introduction

Thus far all of the sections of this chapter have been devoted to presenting solutions to the so-called "*direct*" or "*forward*" problem. That is for certain geometric and constitutive models of continua (e.g. homogeneous, elastic, dissipative media) were computed solutions to homogenous (i.e. the *eigenproblem*) or inhomogeneous (i.e. the *source problem*) boundary values problems associated with the propagation of surface waves.

The calculated solutions could be viewed as specific types of "*response functions*" to specific types of "*excitations*". For the latter were included small perturbations of the initial *at-rest* conditions, and sources with certain geometric and time-variation properties (e.g. vertical time-harmonic point or line forces, *Dirac*-type impulse, etc.). With regards to response functions were considered surface wave phase and group velocities, apparent phase velocities, displacement spectra, seismograms. If the "*forward problem*" is interpreted as a deterministic process, the *excitation* can be viewed as a "*cause*" whereas the response function as the "*effect*".

An "*inverse or backward problem*" is a problem where the roles of causes and effects are *inverted* and the goal is to determine what *causes* are responsible for the observed *effects* (Engl, 1993; Groetsch, 1993). Applied to surface waves the inverse problem consists in computing a set of material parameters defining a specific model of a half-space from the knowledge of response functions associated with surface wave motion, for instance *dispersion* and *attenuation* curves.

If the *deterministic process* is viewed as a formal mapping K: K·x = y, the *forward problem* corresponds to the problem of computing $y$ from knowing $K$, and $x$. In this interpretation $K$ is the model of the deterministic process, $x$ is the cause and $y$ the effect[51]. Numerical simulations are nothing but approximate solutions of direct or forward mathematical problems. As for the *inverse problem* two are the possibilities: the first is the "*causation problem*" (inverse problem of category "*a*") which corresponds to determining $x$ given $K$ and $y$, the second is "*model identification problem*" (inverse problem of category "*b*") where the objective is to compute $K$ from knowing $x$ and $y$. Figure 26 illustrates these possibilities.

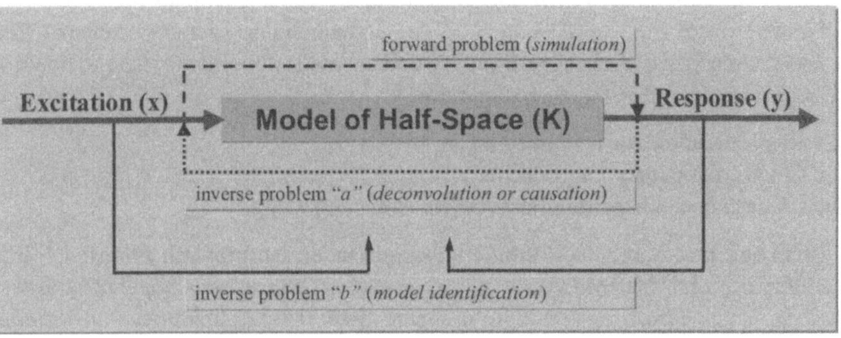


**Figure 26.** *Forward* and *inverse* modelling associated with surface wave motion. The problem is described through a *deterministic process* K formalized by K: K·x = y where x is the *cause* and y the *effect*.

The inverse problem associated with surface wave motion belongs to inverse problem of category "*b*" as it does *any* problem concerned with geophysical prospecting and seismological studies of the Earth structure where the objective is to determine some type of geological information about the earth's interior from measurements at its surface. Of category "*b*" is also the geodetic inverse problem of determining the shape of the Earth from gravimetric measurements.

Instead the problem of retrieving information on seismic source parameters from seismograms, dispersion curves and travel times data belongs to inverse problems of category "*a*" which is also called the category of "*deconvolution problems*". Determining the input signal of a seismometer (e.g. a geophone) from the knowledge of the output signal and the seismometer response is another example of a deconvolution problem.

[51] In the theory of systems, $x$ is the *input*, $y$ the *output* and $K$ represents the linear or nonlinear *system*.

Model identification is often denoted interchangingly as a *parameter-identification problem* since the model of a physical system or a process[52] is usually defined *a-priori* from the laws of physics. Consequently *model identification* actually becomes *parameter-identification* as the original objective has been converted to a problem of determining a set of "*model-parameters*".[53]

Several important inverse problems arising in modern science and engineering have spurred in recent years a renewed interest for inverse theory whose developments and ramifications, even in the restricted field of geophysical prospecting, have been spectacular. With regards to surface waves, the relevance of the associated parameter-identification problem lies at the very core of the success achieved by sophisticated techniques used in non-destructive testing and geophysical-geotechnical site characterization which are in essence inverse problem solvers.

The remaining part of these notes will be devoted to illustrate some basic features of inverse problems and contextualize them with reference to the parameter-identification problem associated with Rayleigh waves. This will also include a brief discussion on inversion strategies and a description of a solution algorithm for the joint inversion of Rayleigh dispersion and attenuation curves. The sections ends with an application of the theory to synthetic and real data.

### 4.2 Stability, Regularization and a-Priori Information

Inverse problems are known to be *ill-posed* or *unstable* which is perhaps their most relevant feature if compared with the *well-posedness* of the corresponding forward problems (Engl, 1993; Kirsch, 1996). According to Hadamard (1923) a mathematical problem is said to be *well-posed* or *stable* if it satisfies the following three conditions[54]:

- for all admissible data, a solution exists (*existence*);
- for all admissible data, the solution is unique (*uniqueness*);
- the solution depends continuously on the data (*stability*).

If one the three conditions fails to hold Hadamard called the problem *ill-posed*[55]. It should be remarked that in reality the *Hadamard's postulates* of well posedness apply to both *forward* and *backward* problems. However, it is only in recent years that the importance of *instability* in the solution of forward problems has been widely recognized in studies of non-linear dynamics[56] where small perturbations in the initial data produce unpredictable changes in the solution (Parker, 1994).

[52] Typically represented by a system of partial differential equations, integral equations, or integro-differential equations and their associated initial and boundary conditions.

[53] These are in general the constant or variable coefficients of the above differential/integral equations.

[54] To be more precise, a definition of *well-posedness* should specify the *functional space* in which the solution is supposed to exist, and the restrictions that a given set of data must satisfy to be considered admissible (Engl, 1993).

[55] At the origin of the Hadamard's definition of *well-posedness* is the idea that a mathematical model of a physical problem *must always* be well-posed. It was later recognized however that there are physical phenomena that do not necessarily satisfies this requisite (e.g. quantum physics).

[56] Typical example is the "*butterfly paradigm*" where a butterfly flapping its wings in New York generate an hurricane in San Francisco !

In inverse problems the two conditions that are most often violated are *uniqueness* and *stability*[57]. Of particular relevance in parameter-identification problems is violation of uniqueness, that is the existence for a given problem and set of data of more than *one* solution. In the causative interpretation of inverse problem given in the introduction (see Figure 26) this would corresponds to a physical process where distinct causes may yield the same effect, a situation that is not so uncommon. For the surface wave inverse problem non-uniqueness implies for instance that a given experimental dispersion or attenuation curve may correspond to more than one set of model parameters (e.g. the transversal phase velocity $V_S(y)$ or material damping ratio $D_S(y)$ profiles of a layered system).

From a mathematical point of view, non- uniqueness in the solution of an inverse problem is caused by a lack of sufficient information to constrain the solution. Another possibility is that the available information is not completely independent, at least in certain regions of the domain of definition of solution[58]. Two strategies can be used to enforce uniqueness in the solution of an inverse problem. The first is to add *a priori information* about the solution of the problem. For the parameter-identification problem of surface waves, this might be independent knowledge of model parameters like $V_S$ and $D_S$ at specific depths.

Another strategy to enforce uniqueness is to *constrain* the solution to satisfy certain requirements such as *smoothness* and *bounds*. In some cases adding a-priori information to the solution may be regarded as a constraint. An obvious example is that of requiring the model parameter to vary within a prescribed range or to be nonnegative (e.g. for material damping ratio). There are however constraints of a different nature for they enforce the solution with features of *global* behaviour.

For parameter-identification problems one example is provided by the *Occam's algorithm* which among a set of admissible solutions chooses the one having maximum *smoothness*, the latter being defined as the Euclidean norm of an appropriate *differential operator* (Constable et al., 1987; Parker, 1994). The concluding section of these notes will show an application of the Occam's algorithm to the joint inversion of Rayleigh dispersion and attenuation curves to determine the transversal phase velocity and quality factor of a viscoelastic, layered system.

Violation of the *stability* condition in the *Hadamard's definition* of well-posedness is also an important concern in the solution of inverse (but also forward) problems. Some of them like the solution of the *Fredholm integral equation of the first-kind*, are very sensitive to perturbations of the initial data and small changes of the latter, caused for instance by unavoidable measurement errors, may result in very large changes of the solution (Groetsch, 1993). It is important to emphasize that the instability of these problems is a feature inherent to their nature and have nothing to do with a particular type of algorithm used to solve them. For *linear* inverse problems with both discrete and continuous linear operators, a *stability analysis* is usually carried out through the method of *singular-value decomposition*[59] (Menke, 1989; Engl, 1993; Kirsch, 1996).

[57] This happens particularly in nonlinear problems such as the inverse problem associated with the propagation of surface waves.

[58] A quotation attributed to Lanczos (1961) states that "*A lack of information in the solution of a problem cannot be remedied by any mathematical trickery*".

[59] It turns out by this method that the *smallest singular value* controls the amplification of the perturbation errors whereas the rate of decay of the singular values arranged in order of decreasing magnitude is used as a measure, in a given inverse problem, to quantify its *degree of instability*.

Very unstable parameter-identification problems can be solved using special mathematical techniques, called *regularization methods*, that approximate the ill-posed problem with a parameter-dependent family of neighboring well-posed problems (Tikhonov and Arsenin, 1977; Engl, 1993). Because some of these regularization methods admit a variational formulation (e.g. Tikhonov regularization) where the objective is the minimization of appropriate functionals, they can successfully be applied also for the solution of *non-linear inverse problems*.

### 4.3 Inversion Strategies, LS and GS Methods

A parameter-identification problem is usually solved by converting it into a *parameter optimization* problem whose solution is then found from the stationary condition of a unconstrained or constrained functional (Parker, 1994). In general, the techniques used to solve *non-linear* optimization problems can be broadly divided into *Global-Search* (GS) and *Local-Search* (LS) methods. The functional of a nonlinear optimization problem will have in fact several stationary points in the solution space and a question arises about finding the global extremum.

Local-search procedures are iterative schemes that starting from an initial guess of the solution generate a sequence of improved approximations converging, under suitable conditions to the solution. Most LS methods are *calculus-based techniques* that linearize at each iteration a non-linear functional until a stationary point is reached.

These techniques require the functional to be sufficiently smooth so that its *Fréchet* derivatives with respect to model parameters exist and are continuous. Furthermore, even if all the smoothness requirements for the functional are satisfied, the sequence of approximations of the solution is guaranteed to converge only if the initial guess is sufficiently close to the solution. However, the most important limitation of LS procedures is that even when they succeed in finding a stationary point, there are no simple means to determine whether it is a local or a global stationary point in the solution space.

This dilemma is addressed by GS procedures, which are optimization techniques where the search for a *global* stationary point is conducted thorough an exploration of the entire *solution space*. This can be done either systematically by defining a grid, or randomly as in Monte Carlo simulations. Other GS methods include simulated annealing, genetic algorithms, fractal inversion, neural network inversion, enumerative methods, and differential semblance optimization. Usually GS techniques are computationally more expensive than LS methods, however they are more robust and reliable in finding the global extremum in the solution space. A thorough discussion of GS techniques is beyond the scope of this section and the interested reader is referred to the specialized literature of *convex analysis* and *nonlinear optimization*.

With regards to the parameter-identification problem of surface waves, ordinarily the strategies adopted for the inversion of surface wave data belong to the category of LS procedures using as response functions modal dispersion and attenuation curves. The dispersion curves are usually computed from measurements of surface phase velocity, however group velocities have also been used.

Dispersion and attenuation curves represent one possible type of response function but there in principle other choices. In the frequency domain they include displacement amplitude, displacement phase, or displacement spectra. In the time domain they might be short or long period seismograms. The ability to successfully *invert* the field variables measured at a free-boundary of an half-space to determine reliable estimates of model parameters in the interior depends to a significant degree on the response function selected to describe the medium response to a dynamic excitation. Factors governing the selection of appropriate response functions include the ability to experimentally measure the field variables, the ability to solve the corresponding inverse problem[60], and the information content associated with the selected response function.

Figure 27 shows a flow-chart illustrating the ramifications of the algorithms that can be conceived for the solution of the parameter-identification problem associated with surface wave motion. The shaded boxes denote the most commonly used procedures in geophysical geotechnical prospecting.

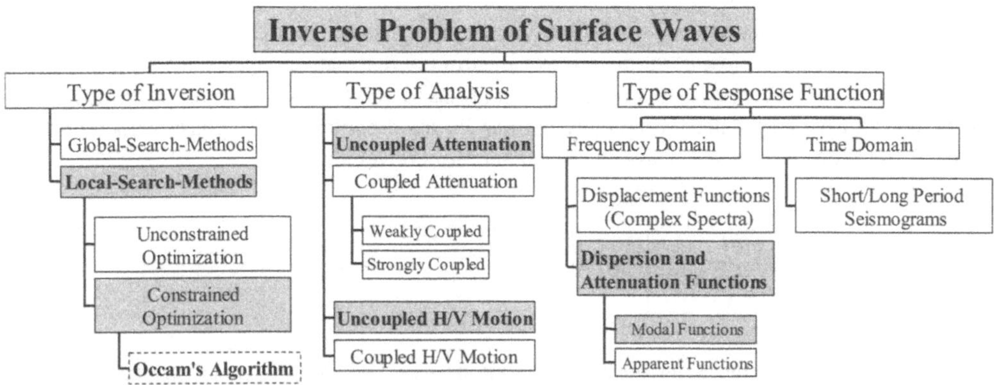


**Figure 27.** Flow-chart of *algorithms* for the solution of the parameter-identification problem associated with surface wave motion. The shaded boxes indicate the most commonly used algorithms in geophysical-geotechnical prospecting (modified from Lai, 1998).

In Figure 27 coupled and uncoupled H/V motion denotes the surface wave motion considered in the inversion analysis. Usually only the vertical component of Rayleigh particle motion is taken into account. Love waves or horizontal components of Rayleigh waves are rarely used even if it would be desirable for increasing the information content of measured surface wave data (Tokimatsu, 1995; Strobbia, 2003).

## 4.4 Joint-Causal Inversion of Rayleigh Dispersion and Attenuation Curves

Solving the parameter-identification problem associated with surface wave motion using the dispersion data as response functions is equivalent to solve an *inverse eigenvalue problem* or *inverse spectral problem* (Kirsch, 1996)[61]. In fact here the objective is to determine certain coefficients of the two sets of first-order, linear ordinary differential equations (13a) and (13b) from the knowledge of their eigenvalues $\omega/V_L$ or $\omega/V_R$.

[60] This also depends on the analytical or numerical procedures used to compute the *Jacobian* of response function with respect to model parameters.

[61] The study of *inverse eigenvalue problems* is a standard topic in inverse theory. To the interested reader is cited a classical paper on the subject by Kac (1966) entitled "*Can one hear the shape of the drum ?*"

When considering *material dissipation* (see Figure 27), the standard approach is to *uncouple* the two parameter-identification problems of inverting the experimentally measured dispersion and attenuation data. As illustrated in Section 3.4 this is particularly advantageous when the problem is solved using the assumption of weak dissipation which allows to obtain the solution of the forward problem in dissipative media as a *byproduct* of the corresponding elastic solution.

However it has been shown in Section 3.3 that in linear dissipative media phase velocity and attenuation (or material damping ratio) of viscoelastic waves are not *independent* and furthermore they are frequency-dependent as a consequence of material dispersion. Thus a correct inversion procedure of surface wave data should determine these two parameters *jointly*, at the same frequency of excitation and for arbitrary values of material damping ratio (Lai et al., 2002). A *natural* way to construct an inversion algorithm satisfying these requirements is to start from the solution of *forward problem* of surface waves in strongly dissipative, layered media illustrated in Section 3.5. In the frequency domain, a linear viscoelastic, arbitrary dissipative, isotropic material has its constitutive parameters completely defined by the complex-valued phase velocity of bulk waves P and S given by Eq.(87) and re-written here for convenience:

$$V_\chi^*(\omega) = \frac{V_\chi(\omega)}{\sqrt{[1+4D_\chi^2(\omega)]}} \cdot \left[ \frac{1+\sqrt{[1+4D_\chi^2(\omega)]}}{2} + i \cdot D_\chi \right] \tag{87-bis}$$

where $\chi$ = P, S and by the Kramers-Krönig relations Eqs. (92) or (93) specifying the functional dependence between material damping ratio $D_\chi(\omega)$, and phase velocity $V_\chi(\omega)$. An algorithm for the strongly-coupled, joint inversion of Rayleigh dispersion and attenuation data may be constructed by viewing the complex-valued Rayleigh phase velocity as an *holomorphic mapping* of the complex-valued speeds of propagation of longitudinal and transversal waves of the medium. Formally:

$$\mathbf{V}_R^* = \mathbf{V}_R^*\left(\mathbf{V}_P^*,\ \mathbf{V}_S^*\right) \tag{108}$$

where $\mathbf{V}^*_P = [(V^*_P)_1, (V^*_P)_2, \ldots\ldots (V^*_P)_{nl}]$, $\mathbf{V}^*_S = [(V^*_S)_1, (V^*_S)_2, \ldots\ldots (V^*_S)_{nl}]$, and $\mathbf{V}^*_R = [(V^*_R)_1, (V^*_R)_2, \ldots\ldots (V^*_R)_{nf}]$. The symbols *nl* and *nf* denote respectively the number of layers (including the half-space) of a stratified medium and the set of experimental frequencies at which the Rayleigh phase velocity has been measured. Equation (108) represents the formal solution to the Rayleigh *forward* problem in linear viscoelastic media. The complex valued Rayleigh phase velocity vector $\mathbf{V}^*_R$ represents jointly the *dispersion* and *attenuation* curves and it is computed from Eq.(87) after solving Eq.(86b) for $D_R$ obtaining:

$$D_R(\omega) = \left[ \frac{\dfrac{\alpha_R \cdot V_R}{\omega}}{1-\left(\dfrac{\alpha_R \cdot V_R}{\omega}\right)^2} \right] \tag{109}$$

where $\alpha_R = \alpha_R(\omega)$ is the frequency-dependent Rayleigh attenuation coefficient. For the solution of the Rayleigh *inverse* problem Eq.(108) can be expanded in a *Taylor series* about an initial guess of model parameters $\mathbf{V}^*_{P0}$ , $\mathbf{V}^*_{S0}$ thereby obtaining:

$$\begin{aligned}\mathbf{V}^*_R &= \mathbf{V}^*_{R0} + \left(\mathbf{J}^*_P\right)_{\mathbf{V}^*_{P0},\mathbf{V}^*_{S0}} \cdot \left(\mathbf{V}^*_P - \mathbf{V}^*_{P0}\right) + \left(\mathbf{J}^*_S\right)_{\mathbf{V}^*_{P0},\mathbf{V}^*_{S0}} \cdot \left(\mathbf{V}^*_S - \mathbf{V}^*_{S0}\right) + \\ &+ o\left\| \left(\mathbf{V}^*_P - \mathbf{V}^*_{P0}\right)^2 + \left(\mathbf{V}^*_S - \mathbf{V}^*_{S0}\right)^2 \right\|_2\end{aligned} \tag{110}$$

where $\|\cdot\|_2 \in \mathbf{R}$ denotes the *Euclidean* norm of a vector in $\mathbf{C}^N$, and $\mathbf{V}^*_{R0}$ is the $nf \times 1$ vector of Rayleigh phase velocities corresponding to solution of Eq.(108) with medium parameters equal to $\mathbf{V}^*_{P0}$ and $\mathbf{V}^*_{S0}$. The terms $(\mathbf{J}^*_P)$ and $(\mathbf{J}^*_S)$ are the $nf \times nl$ complex-valued *Jacobian* matrices whose elements are defined by:

$$\begin{cases} \left[\left(J^*_P\right)_{jk}\right]_{V^*_{P0},V^*_{S0}} = \left[\dfrac{\partial (V^*_R)_j}{\partial (V^*_P)_k}\right]_{V^*_{P0},V^*_{S0}} \\ \left[\left(J^*_S\right)_{jk}\right]_{V^*_{P0},V^*_{S0}} = \left[\dfrac{\partial (V^*_R)_j}{\partial (V^*_S)_k}\right]_{V^*_{P0},V^*_{S0}} \end{cases} \tag{111}$$

where $j = 1, nf$ and $k = 1, nl$. The subscripts outside the brackets of Eqs.(111) indicate the point in $\mathbf{C}^{nf \times nl}$ at which the *Jacobian* matrices are evaluated. The partial derivatives of the above *Jacobians* are calculated using Eqs.(35) extended by *analytic continuation* to complex values of field variables[62].

By neglecting in Eq.(110) terms higher than the first order and by considering $\mathbf{V}^*_R$ to be a function of only $\mathbf{V}^*_S$ this equation reduces to:

$$\left(\mathbf{J}^*_S\right)_{\mathbf{V}^*_{S0}} \cdot \mathbf{V}^*_S = \left(\mathbf{J}^*_S\right)_{\mathbf{V}^*_{S0}} \cdot \mathbf{V}^*_{S0} + \left(\mathbf{V}^*_R - \mathbf{V}^*_{R0}\right) \tag{112}$$

Equation (112) could be used as a basis of a classic *least squares* algorithm to determine the complex-valued transversal phase velocity profile of a layered medium. However, inversions performed with conventional least-squares techniques where the solution is constrained to have the minimum norm (e.g. the stochastic damped least-squares algorithm) are often inadequate, and may lead to physically unreasonable profiles of model parameters (Constable et al., 1987; Menke, 1989). The inadequacy of this class of algorithms can be often attributed to a lack of physical justification for assuming the minimum norm constraint.

A more reasonable approach for constraining the solution of an inverse *geophysical problem* is the so-called *Occam's inversion* (Constable et al., 1987; Parker, 1994). The strategy of this algorithm is to find the *smoothest* profile of model parameters subjected to the constraint that the *misfit* between observed and predicted data cannot exceed a prescribed value. The development of

---

[62] It is reminded that the closed-form relationships of the Jacobians given by Eqs.(35) have been obtained from the application of the Rayleigh variational principle in elastic, vertically inhomogeneous continua.

this algorithm was motivated by the following observations. The solution of a parameter identification problem relies on the ability to synthetically reproduce a set of experimental data by solving the *forward* problem associated to a particular mathematical model. In *discrete inverse theory* (Tarantola, 1987; Menke, 1989) the mathematical model is assumed to depend on a certain number of unknown model parameters, whose determination is the objective of the inversion algorithm.

For the Rayleigh inverse problem, the model is a linear viscoelastic, multi-layered medium. The model parameters are for instance the complex-valued bulk wave velocities of the individual layers[63]. Because the number of layers *nl* is generally assumed, the inverted profile of model parameters will depend on the *a priori* assumption about *nl*, and it may contains large discontinuities or other features that are not essential for matching the experimental data. By enforcing maximum *smoothness* and regularity in the solution, one minimizes its dependence upon the assumed number of layers, and at the same time rejects solutions that are unnecessarily complicated (Constable et al., 1987).

Implementation of the smoothed least-squares inversion algorithm requires an adequate definition of *smoothness* of a profile of model parameters. In a multi-layered medium, where the variation of the model parameters is discontinuous, smoothness or its converse *roughness*, is defined in terms of *difference* rather than *differential* operators. For complex-valued model parameters represented by the vector $\mathbf{V}^*_S$ roughness may be defined by either one of the two following expressions (Constable et al., 1987; Lai, 1998):

$$\begin{cases} R_1 = \left\| \partial \mathbf{V}_S^* \right\|_2^2 = \left( \partial \mathbf{V}_S^* \right)^H \cdot \left( \partial \mathbf{V}_S^* \right) \\ R_2 = \left\| \partial \partial \mathbf{V}_S^* \right\|_2^2 = \left\| \partial^2 \mathbf{V}_S^* \right\|_2^2 = \left( \partial^2 \mathbf{V}_S^* \right)^H \cdot \left( \partial^2 \mathbf{V}_S^* \right) \end{cases} \tag{113}$$

where the symbols $(\cdot)^H$ and $\partial$ denote respectively the *Hermetian* transpose of a complex-valued matrix and an $nl \times nl$ real-valued matrix representing the *two-point central* finite difference operator. It can be easily shown that for a *continuously* varying medium and for real-valued model parameters, the two definitions of $R_1$ and $R_2$ given by Eq.(113) correspond to the integral over depth of the square of the *first* and the *second* derivative, respectively, of the model parameter function $V_S(y)$ with respect to depth.

The misfit $\varepsilon^2$ between the measured and the predicted complex-valued Rayleigh phase velocities may be written as follows:

$$\varepsilon^2 = \left[ \mathbf{W}^* \overline{\mathbf{V}}_R^* - \mathbf{W}^* \mathbf{V}_R^* \left( \mathbf{V}_S^* \right) \right]^H \cdot \left[ \mathbf{W}^* \overline{\mathbf{V}}_R^* - \mathbf{W}^* \mathbf{V}_R^* \left( \mathbf{V}_S^* \right) \right] \tag{114}$$

where $\overline{\mathbf{V}}_R^*$ is a complex-valued vector of experimental Rayleigh phase velocities, and $\mathbf{W}^*$ is a complex-valued diagonal $nl \times nl$ matrix defined by

$$\mathbf{W}^* = \text{diag}\left\{ 1/\sigma_1^*,\ 1/\sigma_2^*, \ldots\ldots,\ 1/\sigma_{nl}^* \right\} \tag{115}$$

[63] Additional or alternative sets of model parameters may be the *mass density* ρ(y) and *layer thicknesses*.

with $\sigma^*_j$ , j = 1, *nf* being the *uncertainties* associated with the experimental data $\overline{\mathbf{V}}^*_R$ . Solution of the linearized Rayleigh inverse problem represented by Eq.(112) consists in finding a vector $\mathbf{V}^*_S$ that minimizes $R_1$ or $R_2$ while constraining the residual error $\varepsilon^2$ to be equal to $\varepsilon^{2(*)}$ , an acceptable value in light of the uncertainties. The method of *Lagrange multipliers* is employed to solve this complex-valued, constrained optimization problem resulting in (Lai, 1998):

$$\mathbf{V}^*_S = \left\{ \mu(\partial^T\partial) + \left[\mathbf{W}^* \cdot (\mathbf{J}^*_S)_{\mathbf{V}^*_{S0}}\right]^H \cdot \left[\mathbf{W}^* \cdot (\mathbf{J}^*_S)_{\mathbf{V}^*_{S0}}\right] \right\}^{-1} \cdot \left[\mathbf{W}^* \cdot (\mathbf{J}^*_S)_{\mathbf{V}^*_{S0}}\right]^H \cdot \mathbf{W}^* \overline{\mathbf{d}}^*_0 \tag{116}$$

where $\mu$ is the Lagrange multiplier which may be interpreted as a *smoothing parameter* and must be determined with the additional constraint that the specified residual error $\varepsilon^{2(*)}$ is matched with a vector $\mathbf{V}^*_S$ composed only of negative imaginary parts. This condition will insure that the transversal damping ratio obtained from the inversion algorithm will be always a positive quantity. Finally:

$$\overline{\mathbf{d}}^*_0 = \mathbf{J}^*_{\mathbf{V}^*_{S0}} \cdot \mathbf{V}^*_{S0} + \left(\overline{\mathbf{V}}^*_R - \mathbf{V}^*_{R0}\right) \tag{117}$$

Equation (116) is used iteratively to refine the estimated complex-valued transversal phase velocity profile $\mathbf{V}^*_S$ until convergence. Once the unknown complex-valued vector $\mathbf{V}^*_S$ is found, the real-valued transversal phase velocities $(V_S)_k$ and damping ratios $(D_S)_k$ , k = 1, *nl* of the *nl*-layered medium can be recovered using Eqs.(107) with the subscript R replaced by S.

In the above derivation, no specific assumption was made about the frequency-dependence law of the complex-valued vector $\mathbf{V}^*_S$. A standard technique used to account for material dispersion is to conduct the inversion process at a prescribed *reference frequency* $\omega_{ref}$ (Lee and Solomon, 1979; Herrmann, 2002). The procedure described above for the strongly coupled inversion of Rayleigh dispersion and attenuation curves is then applicable also to a *causal inversion* provided that the partial derivatives appearing in Eq.(111) are computed with respect to $(V^*_{P(ref)})_k$ and $(V^*_{S(ref)})_k$ and the vector $\mathbf{V}^*_S$ is replaced by $\mathbf{V}^*_{S(ref)} = \mathbf{V}^*_S(\omega_{ref})$ (Lai, 1998).
Figures 28 and 29 show the results obtained from the application of the Occam's algorithm to a pair of *synthetic* dispersion and attenuation curves generated using the medium parameters of Table 7.

**Table 7.** Medium parameters used to generate the Rayleigh *synthetic* dispersion and attenuation curves to be *jointly* inverted by employing the constrained, damped least-squares *Occam's* algorithm.

| Stratum | Thickness (m) | $V_P$ (m/s) | $V_S$ (m/s) | $D_P$ (-) | $D_S$ (-) | $\rho$ (t/m$^3$) |
|---|---|---|---|---|---|---|
| 1 | 5.0 | 400 | 200 | 0.020 | 0.035 | 1.7 |
| 2 | 10.0 | 600 | 300 | 0.015 | 0.030 | 1.8 |
| 3 | 10.0 | 800 | 400 | 0.010 | 0.025 | 1.8 |
| half-space | $\infty$ | 1000 | 500 | 0.010 | 0.020 | 1.8 |

The synthetic response functions were computed by solving the Rayleigh inhomogeneous forward problem for a vertical, time-harmonic point source applied at the free-boundary of the layered-medium of Table 7[64]. The displacement field was calculated at 24 receiver locations, with the closest receiver located at a distance equal to one wavelengths from the source. The frequency range considered in the analysis varied from 5 to 54 Hz. The techniques used to determine dispersion and attenuation curves from the synthetic displacement field are not discussed since they are thoroughly discussed in the contributions by Rix (2005) and Foti (2005).

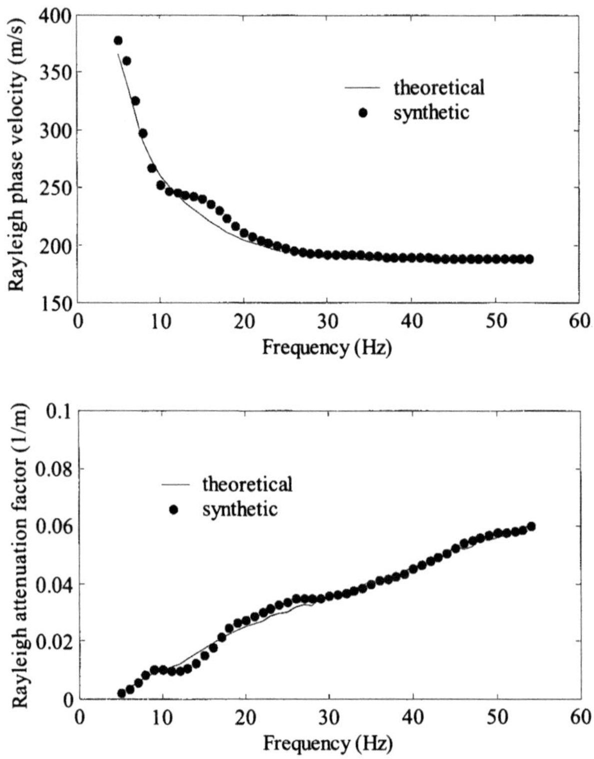


**Figure 28.** Comparison between *synthetic* Rayleigh dispersion and attenuation curves generated using the medium parameters of Table 7 and the corresponding *theoretical* curves (*fundamental* mode of propagation) obtained by employing the constrained, damped least-squares *Occam's* algorithm.

The algorithm converged after only three iterations. Figure 28 shows an excellent agreement between *synthetic* and *theoretical* Rayleigh dispersion and attenuation curves at the last iteration. The theoretical curves were computed with reference to the *fundamental* mode of propagation

[64] The synthetic dispersion and attenuation curves associated with the medium of Table 7 were computed by Prof. Glenn J. Rix of Georgia Institute of Technology who also wrote a chapter of this volume.

and the good agreement with the synthetic curves can be attributed to the fact that the layered medium of Table 7 is *normally* dispersive[19].

Figure 29 illustrates the transversal phase velocity $V_S$ and material damping ratio $D_S$ profiles resulting from the *joint* inversion of Rayleigh *synthetic* dispersion and attenuation curves. They correspond to the *theoretical* curves of Figure 28. The comparison between the values of $V_S$ and $D_S$ predicted by the inversion algorithm (bolded lines) and the $V_S$ and $D_S$ profiles of Table 7 used to generate the synthetic dispersion and attenuation curves (dashed lines) is satisfactory.

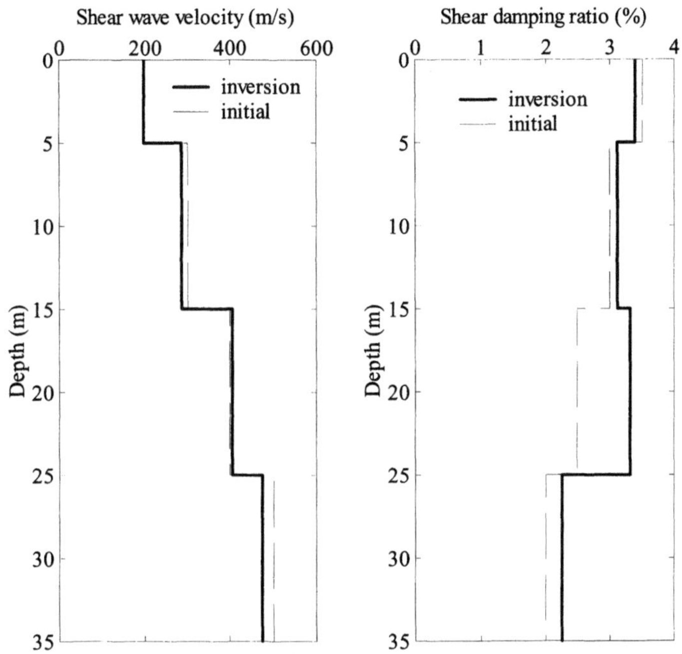


**Figure 29.** Transversal *phase velocity* and material *damping ratio* profiles obtained from the *joint* inversion of Rayleigh *synthetic* dispersion and attenuation curves of Figure 28 using the constrained, damped least-squares *Occam's* algorithm (bolded line). The dashed lines denote the $V_S$ and $D_S$ profiles of Table 7.

Earlier in this section it was stated that the joint inversion of Rayleigh dispersion and attenuation curves is superior to the corresponding uncoupled analysis because it takes implicitly into account the inherent coupling existing between phase velocities and attenuation of viscoelastic waves as a consequence of material dispersion. However there is another, more subtle reason why a coupled analysis is preferred. An uncoupled inversion of dispersion and attenuation data requires the solution of two inverse problems for a total of $2 \cdot nl$ unknown model parameters, for instance the transversal phase velocities $(V_S)_k$ and material damping ratios $(D_S)_k$ ($k = 1, nl$) of a $nl$-layer medium.

The solution of these two inverse problems is not completely independent because the $V_S$ profile obtained from the non-linear inversion of the Rayleigh dispersion curve is subsequently used as input in the linear inversion of the attenuation curve to obtain the $D_S$ profile (see Section 3.4). Therefore the amplification of the uncertainties resulting from inversion of the dispersion curve will carry over to the inversion of the attenuation curve, a process which will eventually add its own uncertainty. In other words, the uncoupled inversion suffers of a negative cross-coupling effect resulting from the solution of two inverse problems where certain input data of one problem comes from the solution of the other (Lai, 1998).

Conversely, the *joint* inversion of both dispersion and attenuation curves is *free* from this negative cross-coupling because both sets of experimental data are inverted simultaneously in a single, complex-valued, elegant inversion. Furthermore, the *coupled* analysis of surface wave data is characterized by a built-in *internal constraint* embedded in the formalism of the complex inversion and which makes it a *better-posed* mathematical problem. This internal constraint, is constituted by the *Cauchy-Riemann equations* that are satisfied by the Rayleigh phase velocity $V^*_R$ if viewed as an *holomorphic* function $V^*_R = V^*_R(V^*_S)$ of the complex-valued transversal phase velocity. In summary, application of complex variable theory for the joint inversion of surface wave data is not only an elegant procedure to account for the inherent coupling between phase velocities and dissipative properties (of arbitrary magnitude) of viscoelastic media, but it also improves the well-posedness of the associated inverse problem.

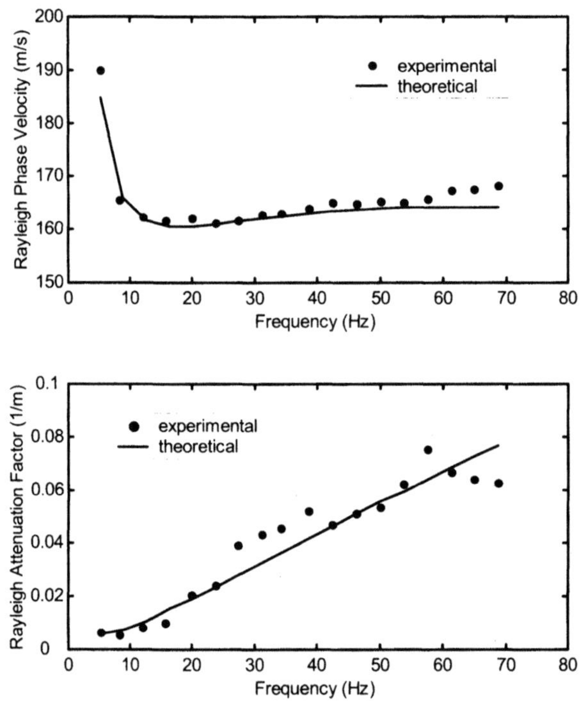


**Figure 30.** Comparison between *experimental* and *theoretical* (*fundamental* mode) dispersion and attenuation curves obtained by applying the *Occam's* algorithm at surface wave data measured at a site in USA at Memphis, Tennessee (from Lai et al.2002).

Before concluding this section are illustrated the results obtained from the application of the *Occam's* algorithm to the joint inversion of real surface wave data. Figure 30 shows a comparison between the experimental and theoretical (fundamental mode) dispersion and attenuation curves measured at a site in USA at Memphis, Tennessee at the end of the fifth and final iteration (from Lai et al.2002). Both the theoretical dispersion and attenuation curves are in satisfactory agreement with the curves measured experimentally. The resulting $V_S$ and $D_S$ profiles are shown in Figure 31 with the acronym SASW which stands for Spectral Analysis of Surface Waves. The transversal phase velocities determined from surface wave data compare well with the values obtained from independent measurements here denoted with the acronym S-CPTU which stands for Seismic Cone Penetration Test[65]. Material damping ratio varies from about 1.5 % to 4.0 %. Even though no independent measurements of $D_S$ are available for comparison, the values obtained from surface wave data are within the range of variation of this parameter in soils at small-strain levels (Ishihara, 1996).

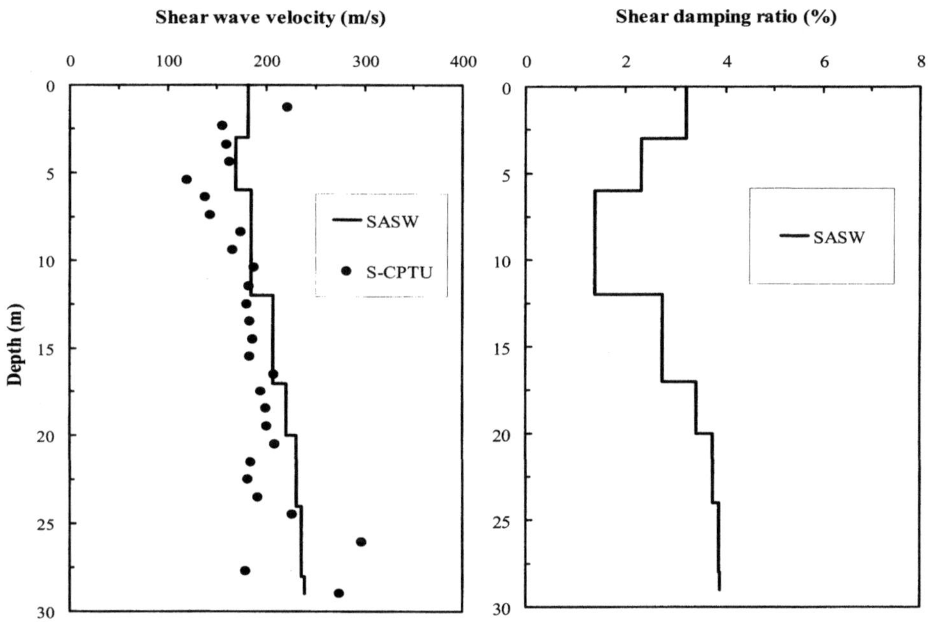


**Figure 31.** Transversal *phase velocity* and material *damping ratio* profiles obtained from the *joint* inversion of experimental dispersion and attenuation curves measured at a site in USA at Memphis, Tennessee using the *Occam's* algorithm. Comparison of $V_S$ with independent measurements (from Lai et al.2002).

[65] The "U" in the acronym S-CPTU is used to indicate that this test provides also a measure of the excess of pore water pressure U generated during the advancement of the probe.

## 5 References

Abd-Elall, L.F., Delves, L.M., Reid, J.K. (1970). A Numerical Method for Locating the Zeros and Poles of a Meromorphic Function. In *Numerical Methods for Non Linear Algebraic Equations*, by P. Rabinowitz, Ed. Gordon and Breach Science Publishers, pp. 47-59.
Abo-Zena, A.M. (1979). Dispersion Function Computations for Unlimited Frequency Values. *Geophys. J. R. Astr. Soc.*, 58, 91-105.
Achenbach, J.D. (1984). *Wave Propagation in Elastic Solids*. North-Holland, Amsterdam, Netherlands, pp. 425.
Aki, K. and Richards P.G. (2002). *Quantitative Seismology*. 2nd Edition, University Science Books, Sausalito, California, 700 pp.
Albers, B. (2005). Modelling of Surface Waves in Poroelastic Saturated Materials by means of a Two Component Continuum. In C.G. Lai and K. Wilmanski, editors, *Surface Waves in Geomechanics – Direct and Inverse Modelling for Soil and Rocks*, CISM Lectures Notes. Springer-Verlag, Wien-New York.
Al-Eqabi G.I., Herrmann R.B. (1993) Ground Roll: A potential tool for constraining shallow shear wave structure. *Geophysics*, 58, 713-719.
Anderson, D.L., and Archambeau, C.B. (1964). The Anelasticity of the Earth. *J. Geophysical Research*, 69(10), 2071-2084.
Anderson, D.L., Ben-Menahem, A., and Archambeau, C.B. (1965). Attenuation of Seismic Energy in the Upper Mantle. *J. Geophysical Research*, 70, 1441-1448.
Båth, M. and Berkhout, A.J. (1984). *Mathematical Aspects of Seismology*. Elsevier Publishing Company, Amsterdam, The Netherlands, pp.438.
Ben-Menahem, A., and Singh, S.J. (2000). *Seismic Waves and Sources*. Second Edition, Dover Publications, 1136 pp.
Biot, M.A. (1956). Theory of Propagation of Elastic Waves in a Fluid-Saturated Porous Solid. I. Lower Frequency Range; II. Higher Frequency Range, *J. Acoust. Soc. Am.*, 28, 168-178; 179-191.
Bullen, K.E. and Bolt, B.A. (1985). *An Introduction to the Theory of Seismology*. Cambridge University Press, 499 pp.
Blake R.J., Bond L.J. (1990). Rayleigh wave scattering from surface features: wedges and down steps, *Ultrasonics*, 28, 214-228.
Blake R.J., Bond L.J. (1992). Rayleigh wave scattering from surface features: up steps and toughts, *Ultrasonics*, 30, 255-265.
Bleistein, N. (1984). *Mathematical Methods for Wave Phenomena*. Academic Press, 675 pp.
Birkhoff, G. and Rota, G.C. (1978). *Ordinary Differential Equations*. Third Edition, Wiley, New York.
Boore, D.M. (1972). Finite Difference Methods for Seismic Wave Propagation in Heterogeneous Materials. *Methods of Computational Physics, Vol.11*, Edition. Bolt, B.A., Academic press, New York, pp.1-36.
Borcherdt, R. (1971). *Inhomogeneous Body and Surface Wave Plane Waves in a Generalized Viscoelastic Half Space*. PhD thesis, University of California, Berkeley.
Borcherdt, R. (1973a). Energy and plane waves in linear viscoelastic media. *Journal of Geophysical Research*, 78(14):2442–2453.
Borcherdt, R. (1973b). Rayleigh-type surface wave on a linear viscoelastic half-space. *Journal of the Acoustical Society of America*, 54(6):1651–1653.
Borcherdt, R. and Wennerberg, L. (1985). General P, type-I S, and type-II S waves in anelastic solids; inhomogeneous wave fields in low-loss solids. *Bulletin of the Seismological Society of America*, 75(6):1729–1763.
Bracewell, R. (1965). *The Fourier Transform and its Applications*. McGraw-Hill Company.
Buchen, P. (1971). Plane waves in linear viscoelastic media. *Geophysical Journal of the Royal Astronomical Society*, 23:531–542.

Bullen, K.E., Bolt, B.A. (1985). *Introduction to the Theory of Seismology*. Cambridge University Press, pp. 509.

Chen, X. (1993). A Systematic and Efficient Method of Computing Normal Modes for Multilayered Half Space. *Geophysics J. Int.*, Vol. 115, pp. 391-409.

Christensen, R.M. (1971). *Theory of Viscoelasticity - An Introduction*. Ed. Academic Press, 245 pp.

Cole, K.S., and Cole, R.H. (1941). Dispersion and Absorption in Dielectrics. I. Alternating Current Characteristics. *J.Chem.Phys.*, 9, 341-351.

Constable, S.C., Parker, R.L., and Constable, G.G. (1987). Occam's Inversion: A Practical Algorithm For Generating Smooth Models From Electromagnetic Sounding Data. *Geophysics*, 52, 289-300.

Courant, R., Hilbert, D. (1994). *Methods of Mathematical Physics*. Two Volumes, Wiley, John & Sons, Incorporated, 1390 pp.

Dobry, R. (1970). *Damping in Soils: Its Hysteretic Nature and the Linear Approximation*. Research Report R70-14, Massachusetts Institute of Technology, pp 82.

Dorman J., Ewing M., Olivier J. (1960). Study of the shear-velocity distribution in the upper mantle by mantle Rayleigh waves. *Bulletin of the Seismological Society of America*, 50, 87-115.

Engl, H.W. (1993). Regularization Methods for the Stable Solution of Inverse Problems. *Surveys on Mathematics for Industry*, Vol. 3, pp. 71-143.

Eringen, A.C. and Suhubi, E.S. (1975). *Elastodynamics*. Vol. I and Vol. II, Academic Press.

Ewing, W.M., Jardetzky, W.S., & Press, F. (1957). *Elastic Waves in Layered Media*. McGraw-Hill, 380p.

Faccioli, E., Maggio, F., Quarteroni, A., and Tagliani, A. (1996). Spectral-Domain Decomposition Methods for the Solution of Acoustic and Elastic Wave Equations. *Geophysics*, Vol.61, No.4, pp.1160-1174.

Foti S. (2000). *Multi-Station Methods for Geotechnical Characterisation Using Surface Waves*, PhD Dissertation, Politecnico di Torino, 229 pp.

Foti, S. (2005). Surface Wave Testing for Geotechnical Characterization. In C.G. Lai and K. Wilmanski, editors, *Surface Waves in Geomechanics – Direct and Inverse Modelling for Soil and Rocks*, CISM Lectures Notes. Springer-Verlag, Wien-New York.

Fung, Y.C. (1965). *Foundations of Solid Mechanics*. Prentice-Hall, New Jersey, pp.525.

Glangeaud F., Mari J.L., Lacoume J.L., Mars J., Nardin M. (1999). Dispersive seismic waves in geophysics. *Eur. J. Environ. Eng. Geophys.*, 3, 265-306.

Goldstein, H. (1980). *Classical Mechanics*. Addison-Wesley Publishing Company, 2nd Ed., pp.672.

Gucunski, N., and Woods, R.D. (1991). Use of Rayleigh Modes in Interpretation of SASW Tests. *Proceedings, 2nd International Conference on Recent Advances in Geotechnical Earthquake Engineering and Soil Dynamics*, Vol. 2, pp. 1399-1408.

Groetsch, C.W. (1993). *Inverse Problems in the Mathematical Sciences*. Vieweg Mathematics for Scientists and Engineers, pp. 152.

Hadamard, J. (1923). *Lectures on the Cauchy Problem in Linear Partial Differential Equations*. Yale University Press, New Haven.

Harvey, D. (1981). Seismogram Synthesis using Normal Mode Superposition: the Locked Mode Approximation. *Geophys. J. R. Astr. Soc.*, 66, 37-70.

Haskell, N.A. (1953). The Dispersion of Surface Waves on Multilayered Media. *Bulletin of the Seismological Society of America*, 43, 17-34.

Henrici, P. (1974). *Applied and Computational Complex Analysis*. Vol. 1, John Wiley & Sons, New York, pp. 682.

Herrmann, R.B. (2002). *Computer Programs in Seismology*. User's Manual, Ver. 3.20, St.Louis University, Missouri.

Hisada, Y. (1994). An Efficient Method for Computing Green's Functions for a Layered Half-Space with Sources and Receivers at Close Depths. *Bulletin of the Seismological Society of America*, 84(5), 1456-1472.

Hisada, Y. (1995). An Efficient Method for Computing Green's Functions for a Layered Half-Space with Sources and Receivers at Close Depths (Part 2). *Bulletin of the Seismological Society of America*, 85(4), 1080-1093.

Holzlohner, U. (1980). Vibrations of the Elastic Half-Space Due to Vertical Surface Loads. *Earthquake Engineering and Structural Dynamics*, 8, 405-414.

Hudson, J.A. (1980). *The Excitation and Propagation of Elastic Waves*. Cambridge University, 224 pp.

Ishihara, K. (1996). *Soil Behaviour in Earthquake Geotechnics*. Oxford Science Publications, Oxford, UK, pp. 350.

Jocker, J., Smeulders, D., Drijkoningen, G., Van Der Lee, C., and Kalfsbeek, A. (2004). Matrix Propagator Method for Layered Porous Media: Analytical Expressions and Stability Criteria. *Geophysics*, Vol. 69, No. 4; pp. 1071–1081.

Johnston, D.H., Toksöz, M.N., and Timur, A. (1979). Attenuation of Seismic Waves In Dry and Saturated Rocks: II. Mechanisms. *Geophysics*, 44(4), 691-711.

Jones R.B. (1962). Surface waves techniques for measuring the elastic properties and thickness of roads: theoretical development. *British Journal of Applied Physics*, 13, 21-29.

Kac, M. (1966). Can One Hear the Shape of the Drum ? *Am. Math. Mon.*, 73:1-23.

Kausel, E. (1981). *An Explicit Solution For The Green Functions For Dynamic Loads In Layered Media*. Massachusetts Institute of Technology, Research Report R81-13, 79 pp.

Kausel, E. (2005). Waves Propagation Modes: From Simple Systems to Layered Soils. In C.G. Lai and K. Wilmanski, editors, *Surface Waves in Geomechanics – Direct and Inverse Modelling for Soil and Rocks*, CISM Lectures Notes. Springer-Verlag, Wien-New York.

Kausel, E., and Roësset, J.M. (1981). Stiffness Matrices for Layered Soils. *Bulletin of the Seismological Society of America*, 71, 6, 1743-1761.

Keilis-Borok, V.I. (1989). *Seismic Surface Waves in a Laterally Inhomogeneous Earth*. Kluwer Academic Publishers, 304 pp.

Kennett, B.L.N. (1974). Reflections, Rays, and Reverberations. *Bulletin of the Seismological Society of America*, 64, 1685-1696.

Kennett, B.L.N. (1983). *Seismic Wave Propagation in Stratified Media*. Cambridge University Press, UK, pp.342.

Kennett, B.L.N., and Kerry, N.J. (1979). Seismic Waves in a Stratified Half-Space. *Geophys. J. R. Astr. Soc.*, 57, 557-583.

Kirsch, A. (1996). *An Introduction to the Mathematical Theory of Inverse Problems*. Springer-Verlag, New York. Applied Mathematical Sciences Vol. 120, pp. 282.

Kjartansson, E. (1979). Constant Q-Wave Propagation and Attenuation. *J. Geophys. Res.*, Vol.84, pp.4737-4748.

Knopoff, L. (1964). A Matrix Method for Elastic Wave Problems. *Bulletin of the Seismological Society of America*, 54, 431-438.

Komatitsch, D., and Vilotte, J.P. (1998). The Spectral Element Method: An Efficient Tool to Simulate the Seismic Response of 2D and 3D Geological Structures. *Bulletin of the Seismological Society of America*, 88(2), 368-392.

Kovach R.L. (1978). Seismic surface waves and crust and upper-mantle structure. *Rev. Geophys and Space Phys.*, 16, 1-13.

Kravanja, P., and Barel, M.V. (2000). *Computing the Zeros of Analytic Functions*. Springer-Verlag New York, 118 pp.

Lai, C.G. (1998). *Simultaneous Inversion of Rayleigh Phase Velocity and Attenuation for Near-Surface Site Characterization*. Ph.D. Dissertation, The Georgia Institute of Technology, Atlanta, Georgia, 370 pp.

Lai, C.G., Lo Presti, D.C.F., Pallara, O., Rix, G.J. (1999). Misura simultanea del modulo di taglio e dello smorzamento intrinseco dei terreni a piccole deformazioni. *Atti Convegno ANIDIS 1999* (in Italian).

Lai, C.G., Rix, G.J. (2002). Solution of the Rayleigh Eigenproblem in Viscoelastic Media. *Bulletin of Seismological Society of America*, Vol.92, No.6, pp. 2297-2309.

Lai, C.G., Rix, G.J., Foti, S., Roma, V. (2002). Simultaneous Measurement and Inversion of Surface Wave Dispersion and Attenuation Curves. *Soil Dynamics and Earthquake Engineering*, Vol. 22, No. 9-12, pp. 923-930.

Lamb, H. (1904). On the Propagation of Tremors over the Surface of an Elastic Solid. *Philosophical Transactions of the Royal Society of London* A203:1-42 pp.

Lancellotta, R. (1995). *Geotechnical Engineering*. Balkema Publisher, 436 pp.

Lancellotta, R. (2005). Experimental Soil Behavior, Its Testing by Waves, Engineering Applications. In C.G. Lai and K. Wilmanski, editors, *Surface Waves in Geomechanics – Direct and Inverse Modelling for Soil and Rocks*, CISM Lectures Notes. Springer-Verlag, Wien-New York.

Lanczos, C. (1961). *Linear Differential Operators*. Van Nostrand, New York.

Lebedev, N.N. (1972). *Special Functions & Their Applications*. Dover Publications, New York, pp. 308.

Lee, W.B., and Solomon, S.C. (1979). Simultaneous Inversion of Surface-Wave Phase Velocity and Attenuation: Rayleigh and Love Waves over Continental and Oceanic Paths. *Bulletin of the Seismological Society of America*, 69(1), 65-95.

Leurer, K.C. (1997). Attenuation in Fine-Grained Marine Sediments: Extension of the Biot-Stoll Model by the Effective Grain Model (EGM). *Geophysics*, Vol.62, No.5, 1465-1479.

Liu, H.P., Anderson, D.L., and Kanamori, H. (1976). Velocity Dispersion due to Anelasticity; Implications for Seismology and Mantle Composition. *Geophys. J.R. Astr. Soc.* 47, 41-58.

Lockett, F.J. (1962). The Reflection and Refraction of Waves at an Interface between Viscoelastic Materials. *J. Mech. Phys. Solids*, 10, 53.

Logan, J.D. (1997). *Applied Mathematics.*, John Wiley & Sons, 2nd Ed., pp.476.

Lo Presti, D.C.F. (1987). *Behavior of Ticino Sand During Resonant Column Tests*. Ph.D. Dissertation, Politecnico di Torino, Torino, Italy.

Lo Presti, D.C.F. and Pallara, O. (1997). Damping Ratio of Soils from Laboratory and In-Situ Tests. *Proceedings, 14th International Conference on Soil Mechanics and Foundation Engineering*, Hamburg, Germany, 6-12, September, 1997.

Luco, J.E. and Apsel, R.J. (1983). On the Green's Function for a Layered Half-Space. Part I. *Bulletin of the Seismological Society of America*, 73, 909-929.

Lysmer, J. and Drake, L.A. (1972). A Finite Element Method for Seismology. *Methods of Computational Physics, Vol.11*, Ed. Bolt, B.A., Academic press, New York, pp.181-215.

Lysmer, J. and Waas, G. (1972). Shear Waves in Plane Infinite Structures. *J. Eng. Mech. Div.*, ASCE, 18, 859-877.

Malagnini, L., Herrmann, R.B., Mercuri, A., Opice, S., Biella, G., and De Franco, R. (1997). Shear-Wave Velocity Structure of Sediments from the Inversion of Explosion-Induced Rayleigh Waves: Comparison with Cross-Hole Measurements. *Bulletin of the Seismological Society of America*, 87(6), 1413-1421.

Malischewsky, P., (1988). *Surface Waves and Discontinuities*. Elsevier, 230 pp.

Manolis G.D., and Beskos, D.E. (1988). *Boundary Element Methods in Elastodynamics.*, Unwin Hyman, London, pp. 282.

Maugin, G. (2005). Theory of Nonlinear Surface Waves and Solitons. In C.G. Lai and K. Wilmanski, editors, *Surface Waves in Geomechanics – Direct and Inverse Modelling for Soil and Rocks*, CISM Lectures Notes. Springer-Verlag, Wien-New York.

Meng, J. (2003). *The Influence of Loading Frequency on Dynamic Soil Properties*. Ph.D. Dissertation, The Georgia Institute of Technology, Atlanta, Georgia, 174 pp.

Menke, W. (1989). *Geophysical Data Analysis: Discrete Inverse Theory*. Revised Edition, International Geophysics Series, Vol.45, Academic Press, 289 pp.

Meza K. (2005). Personal communication. ROSE School, Pavia.

Mooney,W.D., Laske,G. and Masters, G.(1998). CRUST 5.1: A Global Crustal model at 5°x5°. *Journal of Geophysical Research*, N. 103, p. 727-747. Web site: http://quake.wr.usgs.gov/study/CrustalStructure/.
O'Connell, R.J., and Budiansky, B. (1978). Measures of Dissipation in Viscoelastic Media. *Geophys. Res. Lett.*, Vol.5, pp.5-8.
Parker, R.L. (1994). *Geophysical Inverse Theory*. Princeton University Press, New Jersey, pp.386.
Pecorari C. (2001). Scattering of a Rayleigh wave by a surface breaking crack with faces in partial contact. *Wave Motion*, 33, 259-270.
Pekeris, C.K. (1955). The Seismic Buried Pulse. *Proceedings of the National Academy of Sciences*, 41, pp. 629-639.
Philippacopoulos, A.J. (1988). Lamb's Problem for Fluid-Saturated, Porous Media. *Bulletin of Seismological Society of America*, Vol. 78, pp. 908 – 923.
Pipkin, A.C. (1986). *Lectures on Viscoelasticity Theory*. 2nd Edition, Springer-Verlag, Berlin, pp.188.
Press, W.H., Teukolsky, S.A., Vetterling, W.T., and Flannery, B.P. (1992). *Numerical Recipes in Fortran - The Art of Scientific Computing*. Cambridge University Press, 2nd Ed., pp. 963.
Rayleigh, J.W.S. (1887). On Waves Propagated along the Plane Surface of an Elastic Solid. *Proc. London Math. Soc.*, 17: 4-11.
Read, W.T. (1950). Stress Analysis for Compressible Viscoelastic Materials. *J. Appl. Phys.*, 21, 671.
Remmert, R. (1998). *Theory of Complex Functions*. New York, Springer-Verlag.
Rix, G.J. (2005). Surface Wave Testing for Near-Surface Site Characterization. In C.G. Lai and K. Wilmanski, editors, *Surface Waves in Geomechanics – Direct and Inverse Modelling for Soil and Rocks*, CISM Lectures Notes. Springer-Verlag, Wien-New York.
Sato, H., and Fehler, M.C. (1997). *Seismic Wave Propagation and Scattering in the Heterogeneous Earth*. New York, Springer-Verlag.
Shibuya, S., Mitachi, T., Fukuda, F., and Degoshi, T. (1995). Strain Rate Effects on Shear Modulus and Damping of Normally Consolidated Clay. *Geotechnical Testing Journal*, 18(3), 365-375.
Spang, A.W., Jr. (1995). *In Situ Measurements of Damping Ratio Using Surface Waves*. Ph.D. Dissertation, The Georgia Institute of Technology, pp. 347.
Spudich, P.K.P., and Hartzell, S. (1985). Prediction of Ground Motion Time Histories, in J.I. Ziony, ed., *Earthquake Hazards of the Los Angeles Region: U.S. Geological Survey Professional*. Paper 1360, pp. 249—261.
Stokoe K.H., Wright S.G., Bay J., Roesset J.M. (1994). Characterization of geotechnical sites by SASW method, in *Geophysical Characterization of Sites*. ISSMFE TC#10 by R.D. Woods Ed. Oxford & IBH Publ., pp. 15-25.
Strobbia, C. (2003). *Surface Wave Methods for Near-Surface Site Characterization*. Ph.D. Dissertation, Politecnico di Torino, Italy. Available at Dipartimento di Georisorse e Territorio.
Schwab, F., and Knopoff, L. (1970). Surface-Wave Dispersion Computations. *Bulletin of the Seismological Society of America*, 60, 321-344.
Schwab, F., and Knopoff, L. (1971). Surface Waves on Multilayered Anelastic Media. *Bulletin of the Seismological Society of America*, 61, 4, 893-912.
Stoll, R.D. (1974). *Acoustic Waves in Saturated Sediments*. Physics of Sound in Marine Sediments, Plenum Press, 19-39.
Takeuchi, H. and Saito, M. (1972). Seismic Surface Waves. *Methods of Computational Physics, Vol.11*, Ed. Bolt, B.A., Academic press, New York, pp.217-294.
Tarantola, A. (1987). *Inverse Problem Theory: Methods for Data Fitting and Model Parameter Estimation*. Elsevier Science Publisher, Amsterdam, 614 pp.
Thomson, W.T. (1950). Transmission of Elastic Waves through a Stratified Solid Medium. *J.Appl.Phys.*, 21, 89-93.
Tikhonov, A.N. and Arsenin, V.Y. (1977). *Solutions of Ill-Posed Problems*. Winston & Sons, Washington D.C., pp.258.

Tokimatsu, K. et al. (1992). Effects of Multiple Modes on Rayleigh Wave Dispersion Characteristics. *J. Geotechnical Engineering*, ASCE, 118(10), 1529-1543.
Tokimatsu, K. (1995). Geotechnical Site Characterization using Surface Waves. *Proceedings, First International Conference on Earthquake Geotechnical Engineering, IS-Tokyo '95*, Tokyo, November 14-16, Balkema, Rotterdam, 1333-1368.
Tschoegl, N.W. (1989). *The Phenomenological Theory of Linear Viscoelastic Behavior - An Introduction.* Springer-Verlag, Berlin, pp.769.
Visintin, A. (1994). *Differential Models of Hysteresis.* Springer-Verlag, Berlin, pp.407.
Vrettos, C. (1991). Time-Harmonic Boussinesq Problem For a Continuously Non-Homogeneous Soil. *Earthquake Engineering and Structural Dynamics*, 20, 961-977.
Vucetic, M., and Dobry, R. (1991). Effect of Soil Plasticity on Cyclic Response. *J. Geotechnical Engineering*, ASCE, 117(1), 89-107.
Vucetic, M. (1994). Cyclic Threshold Shear Strains in Soils. *Journal of Geotechnical Engineering*, ASCE, Vol.120, No.12, pp.2208-2228.
Viktorov, I.A. (1967). *Rayleigh and Lamb waves: physical theory and applications.* Plenum Press, NewYork, 154 pp.
White J.E. (1983). *Underground Sound. Application of Seismic Waves.* Elsevier Science Publ. 253 pp.
Wilmanski, K. (2005). Elastic Modelling of Surface Waves in Single and Multicomponent Systems. In C.G. Lai and K. Wilmanski, editors, *Surface Waves in Geomechanics – Direct and Inverse Modelling for Soil and Rocks*, CISM Lectures Notes. Springer-Verlag, Wien-New York.
Xia J., Miller R.D., Park C.B. (1999) Estimation of near-surface shear-wave velocity by inversion of Rayleigh waves. *Geophysics*, 64, 691-700.

# Waves Propagation Modes: From Simple Systems to Layered Soils

**Eduardo Kausel**
Department of Civil and Environmental Engineering
Massachusetts Institute of Technology, Cambridge, MA 02139

**Abstract.** This chapter presents a brief review of the normal modes of wave propagation in simple systems such as rods and beams, and uses these to illustrate the fundamental concepts of wave propagation, including complex wave spectra. Thereafter, it generalizes these concepts to the much more complicated case of horizontally layered media. It presents a brief review of the *Stiffness Matrix Method* (SMM) for layered media and its uses in the solution of wave propagation problems. Finally, it summarizes the fundamental elements of the discrete counterpart to the SMM, the *Thin Layer Method* (TLM) which constitutes a powerful tool to obtain the normal modes for both propagating and evanescent waves.

## 1 Introduction

As is well known, when an elastic medium is subjected to a local disturbance and then left on its own, the deformation and kinetic energies stored in the neighborhood of the disturbance give rise to waves that propagate away from this local region and carry elastic energy to other parts of the structure. Mathematically, this represents what is commonly referred to as an *initial value problem*, i.e. a wave propagation problem. In essence, an initial value problem is one that involves the free vibration of an elastic body subjected to initial conditions in velocity and displacement, and can thus be interpreted in terms of a superposition of *normal modes* of vibration. Physically, such normal modes can be interpreted as *stationary waves* of appropriate fixed wavelength such that all boundary and material continuity conditions are satisfied simultaneously at all times. Alternatively, the modes may be visualized as resulting from constructive and destructive interference of waves elicited by reflections at material discontinuities and at the external boundaries.

In the case of *finite* elastic bodies with appropriate boundary conditions, there exist in general infinitely many, albeit distinct vibration modes, each of which is characterized by a real valued, discrete natural frequency of vibration and an associated modal shape that describes the spatial variation of the motion for that mode. In such case, the free vibration for arbitrary initial conditions can be obtained by superposition over all modes of vibration, that is, by recourse to the classical *modal superposition* technique of Mechanical Vibration and Structural Dynamics. The weight assigned to each mode in this summation —the participation factor— is obtained by expressing the initial conditions in terms of the modes and then applying certain orthogonality conditions. By appropriate generalization, this method can also be used to find the response to external sources with arbitrary spatial/temporal variation, i.e. the so-called *forced vibration* problem.

By contrast, when a body is *unbounded* in some —or all— coordinate directions, say an elastic layer over an elastic half-space, it may admit a *continuous* (as opposed to discrete) spectrum of frequencies for which *guided waves* can propagate in the absence of external sources. Moreover, it is not always possible to express a source (i.e. forced vibration, or inhomogeneous wave propagation) problem solely in terms of normal modes, although very useful results can still be obtained using this method. This is so because the normal modes may not suffice to express all physical aspects of the source and response, and so additional terms (i.e. branch integrals) are needed to capture the arrival and evolution of body waves, especially at points near the source. For example, an elastic half-space has only one mode —the Rayleigh wave— which is not enough to fully describe motions elicited by a source, even if these waves dominate the response in the far field, at least near the surface.

In general, there exist two kinds of normal modes, the real-valued *propagating modes* (which may decay, but only because of geometric spreading and/or material attenuation), and the complex-valued *evanescent modes*, which even in the absence of damping decay exponentially in the direction of propagation, and thus exist only in the vicinity of the source. However, at large ranges (i.e. source-receiver or epicentral distances), a subset of the normal modes —the real-valued propagating modes— dominate the response, and they may indeed suffice to describe the wave motion.

## 2 Normal modes and dispersive properties of simple systems

To demonstrate some of the fundamental concepts underlying the propagation of waves in unbounded systems in general, and in layered soils in particular, we begin with a brief discussion of the wave types and normal modes in some simple mechanical systems. This allows us to review the concepts of wave modes, phase and group velocity, dispersion, and frequency-wavenumber spectra in the context of simple systems. For this purpose, we begin with waves in and infinite rod, and then move up gradually to other, more complicated systems.

### 2.1 Infinite rod

An infinite, homogeneous rod of constant cross-section is the simplest of all mechanical systems that can sustain waves, namely compression-dilatation (or P) waves. In the absence of external sources, and neglecting "skin effects", the rod obeys the classical wave equation

$$C_r^2 \frac{\partial^2 u}{\partial x^2} = \frac{\partial^2 u}{\partial t^2} \tag{1}$$

in which $u(x,t)$ is the axial displacement at some arbitrary point, $x$ is the axial coordinate, $C_r = \sqrt{E/\rho}$ is the *rod-wave velocity*, and $E$, $\rho$ are the Young's modulus and mass density, respectively. As is well known, this equation admits solutions of the form

$$u(x,t) = F_1(x - C_r t) + F_2(x + C_r t) \tag{2}$$

with $F_1$, $F_2$ being arbitrary functions. This expression can be interpreted as the superposition of two *wavelets* propagating in opposite directions with *celerity* (i.e. speed) $C_r$, the first of shape $F_1(x)$ moving in the positive $x$ direction, while the other with shape $F_2(x)$ propagates in the negative $x$ direction. Furthermore, if $\Delta x$ and $\Delta t$ are arbitrary space-time increments and we consider the case of a single wavelet $F_1$, then

$$\begin{aligned} u(x+\Delta x, t+\Delta t) &= F_1\left(x+\Delta x - C_r(t+\Delta t)\right) \\ &= F_1\left(x - C_r t + (\Delta x - C_r \Delta t)\right) \end{aligned} \tag{3}$$

Choosing $\Delta x - C_r \Delta t = 0$, i.e. $\Delta x / \Delta t = C_r$, we conclude that $u(x+\Delta x, t+\Delta t) = u(x,t)$, that is, the motion at $x+\Delta x$ is a delayed replica of the motion at $x$. Thus, the wavelet propagates in the rod with velocity $C_r$ and suffers no distortions, i.e. doesn't change shape as it propagates. Thus, rod waves are said to be *non-dispersive*.

Consider next the special case of harmonic waves of the form

$$u(x,t) = C \sin(\omega t - kx) \tag{4}$$

with $C$ = the amplitude, $\omega$ = frequency, $k = 2\pi / \lambda$ = wavenumber (i.e. spatial frequency), and $\lambda$ = wavelength (i.e. the spatial period) of the waves. Factoring out the frequency, this can be written as

$$u(x,t) = C \sin\left[\omega(t - xk/\omega)\right] = C \sin\left[\omega\left(t - x/V_{ph}\right)\right] \tag{5}$$

in which

$$V_{ph} = \omega / k \quad = \textit{phase velocity} \tag{6}$$

Clearly, in this case $V_{ph} = C_r$, so the phase velocity is seen to be independent of the frequency of the waves; this is another way of saying that rod waves are non-dispersive. It also follows that $V_{ph} = \omega\lambda / 2\pi$, that is,

$$V_{ph} = f\lambda \tag{7}$$

with $f$ being the frequency of the waves in cycles per second (Hz). If we now combine two harmonic waves of the same amplitude but moving in opposite directions, we obtain

$$\begin{aligned} u(x,t) &= C \sin(\omega t - kx) + C \sin(\omega t + kx) \\ &= 2C \sin \omega t \cos kx = 2C \sin \omega t \cos 2\pi x / \lambda \end{aligned} \tag{8}$$

We see that the motion is now harmonic both in space and in time. More importantly, we observe that there exist points or *nodes* $x_j = 2j\pi\lambda$, $j = 0, \pm 1, \pm 2, \ldots$ at which the motion is zero at all times, that is, these points do not move but remain stationary. Thus, the motion is confined to the regions between the nodes, and the wave exhibits no net propagation. The maximum occurs at the mid-point between nodes —the crests and troughs— at which the axial strain (i.e. the stress) is zero. Such waves are referred to as *stationary waves*, see Fig. 1.

We can now use the results of a standing wave to obtain the normal modes of a rod of *finite* length $L$ and arbitrary boundary conditions —and do so without the need to solve a boundary value problem. It suffices for us to choose a rod length $L$ equal to an even or odd multiple of a quarter wavelength $\lambda/4$ —the distance between a node (where the displacement is zero) and/or a trough (where the stress is zero).

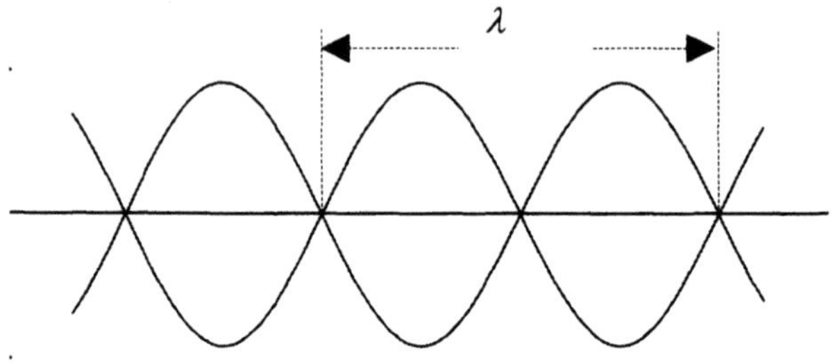


**Figure 1:** Stationary wave in rod

For a free-free, fixed-free and fixed-fixed rod, this gives

| | | | |
|---|---|---|---|
| FR-FR: | $L = 2j\dfrac{\lambda_j}{4} = j\dfrac{C_r}{2f_j}$ | $f_j = j\dfrac{C_r}{2L}$ | $j = 0,1,2,3,\cdots$ |
| FX-FR: | $L = (2j-1)\dfrac{\lambda_j}{4} = (j-\tfrac{1}{2})\dfrac{C_r}{2f_j}$ | $f_j = (j-\tfrac{1}{2})\dfrac{C_r}{2L}$ | $j = 1,2,3,\cdots$ |
| FX-FX: | $L = 2j\dfrac{\lambda_j}{4} = j\dfrac{C_r}{2f_j}$ | $f_j = j\dfrac{C_r}{2L}$ | $j = 1,2,3,\cdots$ |

In the first case, the $0^{th}$ mode ($j$=0) is the rigid-body mode, which has zero frequency, infinite period, and an infinite wavelength (i.e. constant displacement). Observe that the frequencies of the second case interlace those of the first and third. Also, while the frequencies of the first and third case appear to be identical despite the different boundary conditions, this doesn't violate the fundamental principle that states that when kinematic constraints are added to a system, then some or all of its natural frequencies must rise. Indeed, the first frequency of the free-free case (i.e. the rigid-body mode zero) rises to the first frequency of the fixed-free case, and as we add a further constraint this one rises in turn to the first frequency of the fixed-fixed case, which happens to coincide with the *second* frequency of the free-free case.

The above observation explains why the pitch of an open organ pipe (or flute) seemingly *drops* when its free end is covered with a stopper (i.e. a kinematic constraint is added): the fundamental frequency of the open pipe is zero, and can thus not be heard, so the audible tone that it emits is really the second mode, which is higher in pitch than the fundamental mode of the fixed-free pipe.

### 2.2 Infinitely deep ocean

Next, we examine the somewhat more complicated case of gravity waves on the surface of an infinitely deep ocean. From the theory of such waves, it can be shown that they satisfy

$$\lambda = 2\pi \frac{g}{\omega^2} = \frac{1}{2\pi} g T^2 \qquad = \text{wavelength} \qquad (9a)$$

$$\omega = \sqrt{kg} \qquad = \text{wave spectrum} \qquad (9b)$$

$$V_{ph} = \frac{g}{\omega} = \frac{1}{2\pi} gT \qquad = \text{phase velocity} \qquad (9c)$$

with $g$ = acceleration of gravity and $T$ = period of waves. As the sea conditions change from calm to very rough and stormy, this period increases from a few seconds, say 3s to 5s, to some 10s to 15s. Also, the height (amplitude) of open ocean waves will generally not exceed a threshold of about $\lambda/7$, because the waves would break if higher. As can be seen, the phase velocity of ocean waves is now a function of frequency, and falls in tandem with it. Hence, ocean waves are said to be *dispersive.*

### 2.3 Bending beam

As a third example, we consider an infinitely long Euler-Bernoulli bending beam, which takes only bending deformation and translational inertia into consideration. In the absence of body sources, a homogeneous beam is governed by the differential equation

$$\rho A \frac{\partial^2 u}{\partial t^2} + EI \frac{\partial^4 u}{\partial x^4} = 0 \qquad (10)$$

with $u$ = transverse displacement, $\rho$ = mass density, and $A$ = cross section, $I$ = moment of inertia. Assuming a harmonic wave of the form $u(x,t) = C \sin(\omega t - kx)$, we obtain

$$\left[-\rho A \omega^2 + EI\, k^4\right] u = 0 \qquad (11)$$

from which we deduce the phase velocity and frequency-wavenumber relationship as

$$V_{ph} = \sqrt{R\, C_r\, \omega} = kRC_r \qquad (12a)$$

$$\omega = R\, C_r\, k^2 \qquad (12b)$$

where $R = \sqrt{I/A}$ is the radius of gyration. Thus, the phase velocity is again a function of the frequency, but unlike ocean waves, the speed of flexural waves rise with frequency instead of falling. Hence, flexural waves are *dispersive.* The phase velocity can also be expressed in terms of the wavelength as

$$\frac{V_{ph}}{C_r} = 2\pi \frac{R}{\lambda} \tag{13}$$

Thus, low-frequency flexural waves whose wavelength is much greater than the beam's radius of gyration $R$ (which is of the same order of magnitude as the beam's depth) move at substantially lower speeds than the speed of rod waves $C_r$. However, for frequencies high enough to elicit waves comparable in length to the thickness of the beam, the phase speed would appear to be able to exceed the rod wave velocity without bound. However, this is spurious, because no waves in a beam can exceed the largest physical speed, which is the speed of compressional waves. We conclude that at high frequencies, the simple theory of flexural waves in an Euler-Bernoulli beam breaks down.

## 2.4 Bending beam on elastic foundation

The next level of complication arises when we consider a bending beam resting on an elastic (distributed Winkler) foundation of stiffness $K$. The dynamic equilibrium equation for this case is

$$\rho A \frac{\partial^2 u}{\partial t^2} + EI \frac{\partial^4 u}{\partial x^4} + Ku = 0 \tag{14}$$

For harmonic waves of the form $u(x,t) = C \exp i(\omega t - kx)$, we infer that

$$\left[-\rho A \omega^2 + EI\, k^4 + K\right] u = 0 \tag{15}$$

so that

$$k = \left[\frac{\rho A \omega^2 - K}{EI}\right]^{1/4} = \sqrt{\frac{\omega_0}{RC_r}} \left[\left(\frac{\omega}{\omega_0}\right)^2 - 1\right]^{1/4} \tag{16a}$$

with

$$\omega_0 = \sqrt{\frac{K}{\rho A}} \tag{16b}$$

from which it becomes clear that all waves in a beam on elastic foundation are frequency-dependent and are thus *dispersive*. Observe that the frequency $\omega_0$ is the frequency at which the beam moves up and down as a rigid body, which is analogous to the "breathing mode" of cylindrical shells.

The fourth root term has four solutions that generally show up in complex conjugate pairs. Depending on the value of the frequency compared to the frequency of the rigid-body mode, the four roots for the wavenumber $k$ are

$$k = \sqrt{\frac{\omega_0}{RC_r}}\left[\left|\left(\frac{\omega}{\omega_0}\right)^2 - 1\right|\right]^{1/4} \times \begin{Bmatrix} \pm 1, \pm \mathrm{i} & \omega \geq \omega_0 \\ \frac{1}{2}\sqrt{2}\left(\pm 1 \pm \mathrm{i}\right) & \omega < \omega_0 \end{Bmatrix} \tag{17}$$

When the frequency $\omega$ is taken as a parameter and $k$ is allowed to take on complex values, the above roots define four branches in the three-dimensional $\omega - \mathrm{Re}\,k - \mathrm{Im}\,k$ space. The combined set of branches in this space is referred to as the frequency-wavenumber spectrum, or simply the *wave spectrum*.

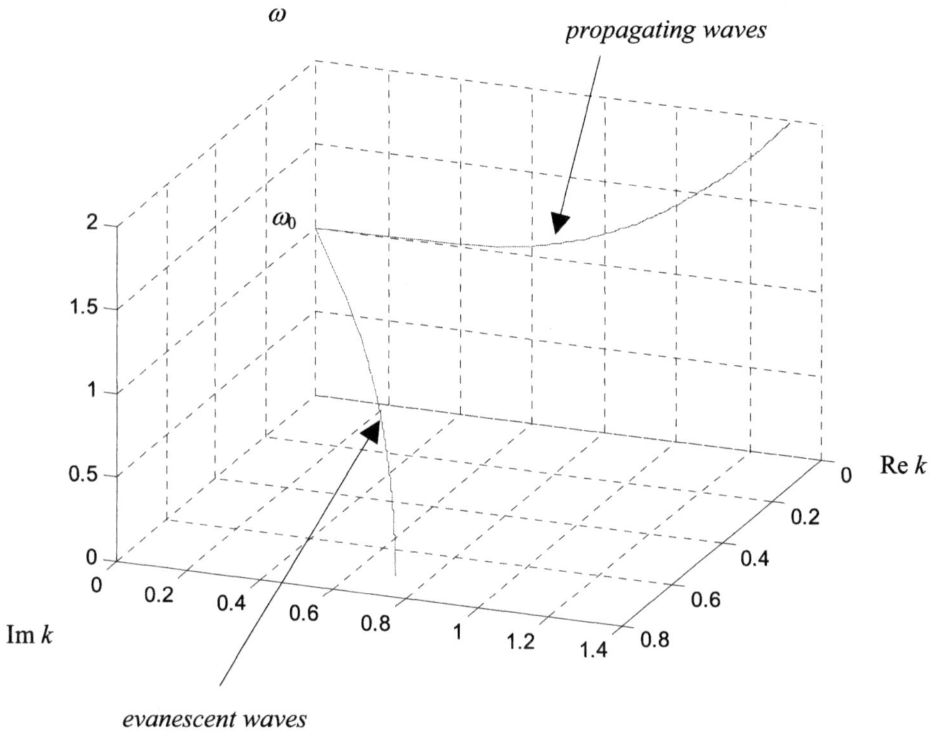


**Figure 2:** Wave spectrum for beam on elastic foundation

Fig. 2 shows one such branch, which is complex below the *cutoff frequency* $\omega_0$, and then turns purely real above this frequency. There exists also a purely imaginary branch above $\omega_0$ as well as a continuation of the complex branch along a mirror image branch that lies in the graph's third octant. In addition, branches associated with the complex conjugate of these branches exist as well, so that the wave spectrum is symmetric with respect to the vertical axis. Real branches correspond to propagating modes, purely imaginary branches are exponentially decaying (or rising modes), and complex branches are evanescent modes (i.e modes that appear to decay or rise as they propagate). Specifically, a positive real part $\mathrm{Re}\,k > 0$ indicates propagation in the positive $x$ direction, and vice-versa. Also, a negative imaginary part $\operatorname{Im} k < 0$ indicates exponential decay in the positive $x$ direction, while a positive imaginary part indicates exponential growth (i.e. decay in the negative $x$ direction).

The four normal modes of the beam on elastic foundation exhibit the following characteristics:

a) $\omega < \omega_0$

There exist four complex branches that appear in complex conjugate pairs. Two of these branches have identical *negative* imaginary parts and opposite real parts, while the other two are the complex conjugate counterparts. The first pair corresponds to waves that decay in the positive $x$ direction and appear to propagate in opposite coordinate directions, while the second pair represents waves that decay in the negative $x$ direction, and again appear to travel in opposite directions. It can be shown that in the presence of sources, each of the two branches in a pair combine with equal weights and form an evanescent standing wave. Thus, such modes do not carry any energy, and they may only exist if external sources are present in some neighborhood —otherwise, the wave would grow exponentially in one of the coordinate directions, and would violate the boundedness condition at infinity.

b) $\omega = \omega_0$

Here, the four solutions for the wavenumber $k$ degenerate to zero, which implies no spatial variation in $x$. Hence, the "wave" is simply an up and down motion of the beam executing a rigid body vibration.

c) $\omega > \omega_0$

Two branches are real, and the other two are purely imaginary. Of the two real ones, one propagates in the positive $x$ direction, the other in the negative. These are the only modes that carry energy, and the only ones that can be observed at distances from a source. By contrast, the imaginary branches correspond to exponentially decaying motions that do not carry any energy and quickly die out.

We observe that waves do not necessarily propagate at all frequencies, but that they do so only in some specific frequency bands. Thus, it is said that wave spectra exhibit so-called *starting* and *stopping bands*. In this particular case, the propagating mode exists only above the cutoff frequency. From the preceding it follows that when a low-frequency excitation is applied to a beam on elastic foundation, the beam responds only in the neighborhood of the source, and the disturbance does not propagate to any significant distance from this source. Hence, all waves are evanescent. However, as the frequency of the excitation is raised above the cutoff frequency, it begins to propagate waves which *radiate* energy away from the source.

## 2.5 Bending beam on elastic half-space

Next, we assume that the bending beam rests smoothly and without friction on an elastic half-space. This case is qualitatively similar to the preceding one and also shows a cutoff frequency associated with the rigid-body mode, but it differs from it in that the half-space —unlike the elastic foundation— now allows guided waves and body waves on its own. As a result, the wavenumber spectrum grows further in complexity, and additional propagation characteristics appear:

- At very low frequencies, all modes are evanescent, so no wave propagation takes place anywhere in either the beam or the half-space.
- At low frequencies above the cut-off frequency of the beam, slow flexural waves can be seen propagating in the beam, but in the half-space, the motion remains confined to the vicinity of the beam, and decays exponentially with depth.
- As the frequency rises further, the flexural wave speed eventually grows until it equals or exceeds the Rayleigh wave velocity of the half-space, at which point waves begin to be transmitted and radiated into the half-space. The frequency at which this transition occurs is referred to as the *coincidence frequency*. Hence, energy begins leaking into the half-space, which means that the waves in the beam *must* decay as they propagate. Flexural waves whose speed exceeds any of the physical wave velocities in the half-space are said to be *supersonic*. To a first approximation, the coincidence frequency is obtained by equating the speed of flexural waves in a free beam with the shear wave velocity of the half-space, i.e. $\sqrt{R\,C_r\,\omega} = C_S$. Clearly, this is only an approximation, because the beam and the half-space will exhibit strong coupling well before reaching this frequency.

By way of illustration, consider the problem of why we can hear somebody knocking a door: the brief (i.e. high-frequency) knocks elicit fast flexural waves in the door which are supersonic in comparison to the acoustic wave speed in air. Hence, these waves radiate strongly into the air and transmit to our ears the sound of the knocks.

## 2.6 Elastic thick plate (Mindlin plate)

As a final example, we consider a homogeneous plate of thickness $h$ subjected to plane strain. This system can sustain infinitely many wave modes —the so-called Lamb's waves— so the plate's spectrum has infinitely many branches. It can be shown that the characteristic wavenumbers for Lamb's modes can be obtained by solving the pair of equations

$$ps \tanh \tfrac{1}{2} kph - q^2 \tanh \tfrac{1}{2} ksh = 0 \qquad \text{Symmetric modes} \qquad (18a)$$

$$ps \coth \tfrac{1}{2} kph - q^2 \coth \tfrac{1}{2} ksh = 0 \qquad \text{Anti-symmetric modes} \qquad (18b)$$

in which

$$p = \sqrt{1-(k_P/k)^2} \qquad (19a)$$

$$s = \sqrt{1-(k_S/k)^2} \qquad (19b)$$

$$q = \tfrac{1}{2}\left(1+s^2\right) \qquad (19c)$$

$$k_P = \omega / C_P, \quad k_S = \omega / C_S \qquad (19d)$$

and $C_P$, $C_S$ are the compressional and shear wave velocities of the plate.

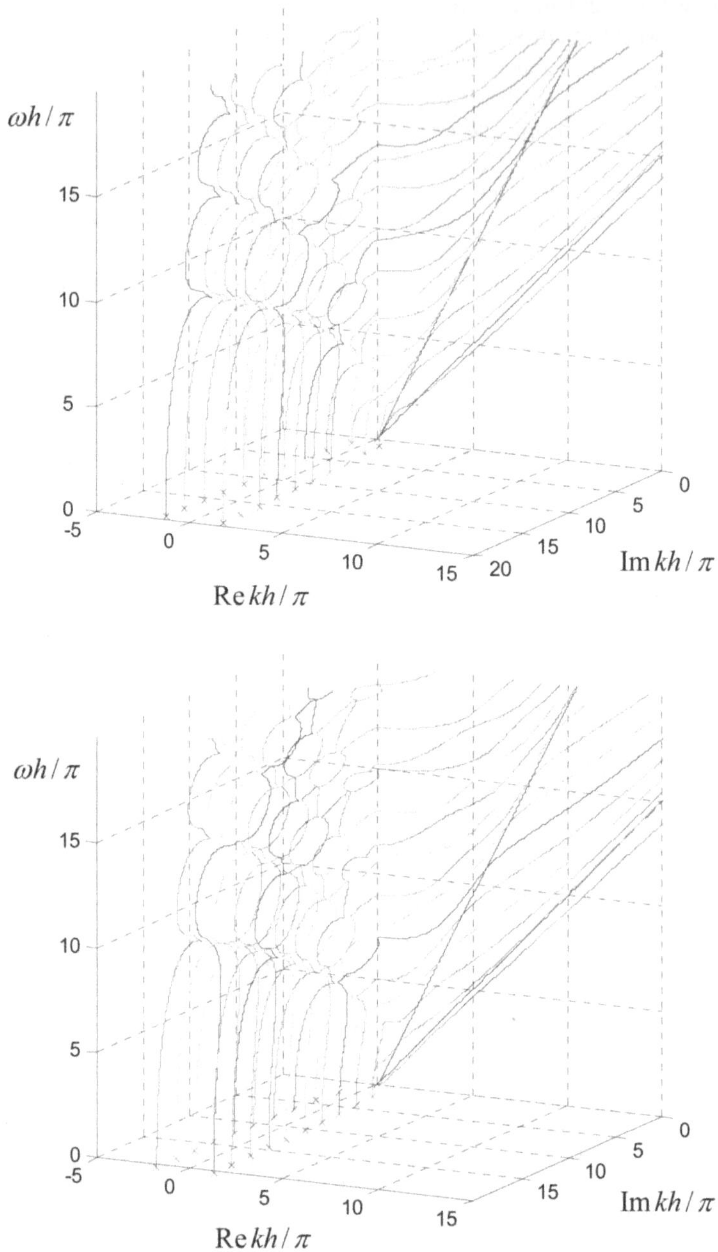


**Figure 3:** Wave spectrum for thick plate (Mindlin plate), $\nu$=0.31
Top: symmetric modes. Bottom: antisymmetric modes

This deceptively simple-looking transcendental eigenvalue problem happens to be extremely difficult to solve —at least by standard search techniques. The first investigator to fully unravel the intricacies of the complex wave spectrum of thick plates was Mindlin, so in his honor such plates are now called Mindlin plates. Today, effective solution techniques are available, some of which will be discussed briefly later on.

The wave spectrum for Lamb's waves in a thick plate is shown in Fig. 3, separated into symmetric and anti-symmetric modes. Only the branches in the first octant are shown. As can be seen, at any given frequency there exists only a *finite* number of wave modes that propagate, i.e. of real branches; their wavenumbers $k$ lie between zero and the wavenumber of shear waves, i.e. $\omega = C_S k$. As the frequency rises above each of the cutoff frequencies of the plate —which happen to coincide with the shear and dilatational resonant frequencies of the plate— one more complex branch turns real, and so an increasing number of modes participate in propagating energy.

### 2.7 Summary of concepts

- Wave spectra provide fundamental information on propagation modes in a mechanical system.
- Real branches represent propagating waves
- Imaginary branches do not propagate, but decay exponentially instead.
- Complex branches represent evanescent waves that both propagate and decay.
- Wave spectra are normally plotted in terms of frequency-wavenumber ($\omega$-$k$), or alternatively in terms of either phase velocity-frequency ($V_{ph}$-$\omega$) or phase velocity-wavelength ($V_{ph}$-$\lambda$). However, the latter two are only used for real, propagating waves.
- The *phase velocity* $V_{ph}$ equals the slope of the secant from the origin to a point on a given branch.
- The *group velocity* $V_{gr}$ (see also the next section) equals the slope of the tangent to a given branch. It can be shown that wave energy flows at the speed of the group velocity.
- If both the phase and group velocities have the same sign, the mode is normally dispersive. However, if they have opposite sign, the mode is said to exhibit *anomalous* dispersion. In general, to the right of a source in a Mindlin plate (and more generally, in a horizontally layered medium) only waves with positive group velocity can exist, because energy always radiates away from a source. Hence, anomalous modes, if any, will involve waves that while appearing to move toward the source, they still carry energy away from it.

## 3 Wave groups and group velocity

Consider once more a dispersive system, such as a bending beam or the deep ocean. Assume further that two harmonic waves of equal amplitude and nearly identical frequencies $\omega \pm \Delta\omega$ and wavenumbers $k \pm \Delta k$ propagate in such a system. Hence, the wave field is given by

$$u = \tfrac{1}{2} C \left\{ \sin\left[ (\omega + \Delta\omega)t - (k + \Delta k)x) \right] + \sin\left[ (\omega - \Delta\omega)t - (k - \Delta k)x) \right] \right\}$$
$$= C \sin(\omega t - kx) \cos(\Delta\omega t - \Delta k\, x) \qquad (20)$$

In the limit when $\Delta\omega \to 0$ and $\Delta k \to 0$, the ratio $\frac{\Delta\omega}{\Delta k} \to \frac{d\omega}{dk} \underset{def}{=} V_{gr}$. This limit is termed the *group velocity*. Hence, the combination of two waves can be written as

$$u(x,t) = C \sin \omega\left(t - x / V_{ph}\right) \cos \Delta\omega\left(t - x / V_{gr}\right) \qquad (21)$$

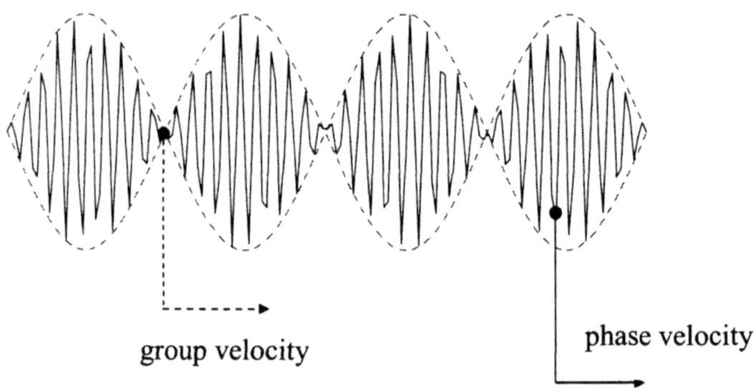


**Figure 4:** Group of two near-frequency waves

This equation can be interpreted as a *carrier wave* of low frequency $\Delta\omega$ that propagates at the group velocity $V_{gr}$ and which *modulates* the amplitude of another wave of frequency $\omega$ propagating with speed $V_{ph}$, see Fig. 4. If the phase velocity is greater than the group velocity, as in the case of ocean waves, the waves appear to emanate from a node, move forward from there and grow within the group, and then vanish at the subsequent node. The opposite is true when the phase velocity is less than the group velocity, such as in the case of flexural waves: the waves appear to move backwards relative to the group (or more precisely, be left behind by the group), so they appear to emanate from a node and vanish at the preceding node.

It is easy to show that the group and phase velocities are related as

$$\frac{V_{gr}}{V_{ph}} = \frac{1}{1 - \dfrac{\omega}{V_{ph}} \dfrac{dV_{ph}}{d\omega}} = 1 - \frac{\lambda}{V_{ph}} \frac{dV_{ph}}{d\lambda} \qquad (22)$$

It can also be shown that energy flows at the speed of the group velocity. The table below summarizes the phase and group velocities in the simple systems considered earlier.

| | Spectrum | Phase velocity $V_{ph} = \frac{\omega}{k}$ | Group Velocity $V_{gr} = \frac{d\omega}{dk}$ |
|---|---|---|---|
| **Rod** | $\omega = C_r k$ | $V_{ph} = C_r$ | $V_{gr} = V_{ph}$ |
| **Ocean** | $\omega = \sqrt{kg}$ | $V_{ph} = \frac{g}{\omega}$ | $V_{gr} = \frac{1}{2} V_{ph}$ |
| **Beam** | $\omega = R C_r k^2$ | $V_{ph} = \sqrt{R C_r \omega}$ | $V_{gr} = 2V_{ph}$ |

## 4 Guided waves in layered soils via the Stiffness Matrix Method

Having discussed the propagation of waves in simple systems, we can now elaborate on the much more complicated case of waves in unbounded, horizontally layered media. For this purpose, let us examine first the case of a full, homogeneous, isotropic, elastic space.

A full space has no boundaries, so it cannot sustain *guided waves*, nor does it exhibit any resonances or preferred directions. Hence, in the absence of sources, only *body waves* in the form of dilatational (longitudinal, primary, pressure or P) waves and shear (transverse, secondary, shear or S) waves can propagate along arbitrary directions in such a full space. Depending on the coordinate system being used, these P and S waves could take the form of plane waves, cylindrical waves, or spherical waves. A full space has no preferred propagation modes, which means that S-P waves of any frequency and any wavenumber are admissible.

By contrast, a homogeneous half-space has a free boundary that admits a non-dispersive guided wave, namely the Rayleigh wave. In addition, it may support body waves, among which we distinguish between the *SV-P waves* (vertically polarized S and P waves, which exhibit particle motions in the vertical plane defined by the normal to the surface and the direction of propagation), and *SH waves* (horizontally polarized shear waves involving particle motions in horizontal planes parallel to the free surface). In general, when a body wave hits the free surface, it both reflects and also converts partially to other waves, a phenomenon referred to as *mode conversion*. In particular, an SV wave reflects as an SV wave and also converts partially to a reflected P wave, provided the angle of incidence is not too shallow, i.e. it is greater than some critical angle.

The addition of one or more layers to the half-space greatly complicates the picture. Now a host of guided waves will exist, which are dispersive; these consist of SV-P modes (Generalized Rayleigh waves and Stoneley waves) and of SH modes (Generalized Love waves and torsional waves). In addition, complex branches for evanescent modes and leaky modes will exist as well.

A very important —indeed fundamental— observation that can be made about waves in horizontally layered media is that the wave spectra for SV-P and SH waves do *not* depend on whether plane-strain waves or cylindrical waves are being considered, even if the displacements patterns elicited by such waves are not the same. For example, guided torsional (cylindrical SH) waves in a layer over an elastic half-space exhibit *exactly* the same dispersion characteristics as plane Love waves in that *same medium*, that is, they propagate at exactly the same speed for a given frequency. Thus, it suffices for us to consider in detail only the two plane strain cases for SH and SV-P waves, which we then generalize without much ado to cylindrical waves.

### 4.1 SH waves and generalized Love waves

Consider a horizontally stratified soil consisting of layers that are laterally homogenous. Using a right-handed Cartesian coordinates with $x$ from left to right and $z$ up, the equation of motion within a layer is of the form

$$\rho \ddot{u}_y = b_y + \mu \left( \frac{\partial^2 u_y}{\partial x^2} + \frac{\partial^2 u_y}{\partial z^2} \right) \tag{23}$$

where $u_y$ is the *anti-plane* displacement caused by SH waves (i.e. along the y direction into the paper), $b_y$ is the body force, and $\rho$, $\mu$ are the mass density and shear modulus of the layer.

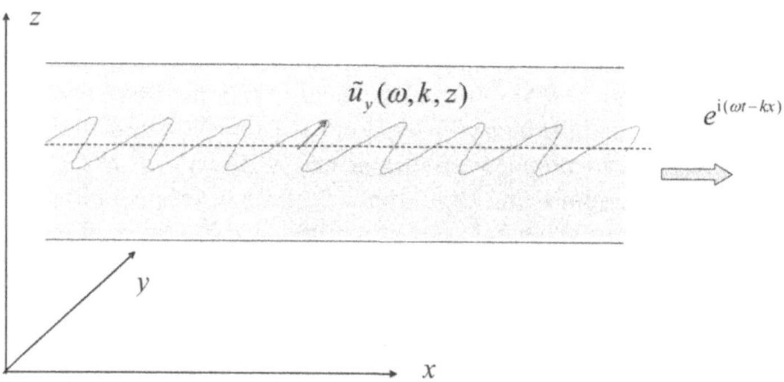


**Figure 5:** Displacements in frequency-wavenumber domain

Applying a Fourier transform in $x$ and $t$ of the form

$$\tilde{u}_y(k,\omega) = \int_{-\infty}^{+\infty} \int_{-\infty}^{+\infty} u_y(x,t) e^{-i(\omega t - kx)} dx\, dt \tag{24}$$

and a similar transform on the body load, we obtain

$$-\omega^2 \rho \tilde{u}_y = \tilde{b}_y + \mu \left( -k^2 \tilde{u}_y + \frac{\partial^2 \tilde{u}_y}{\partial z^2} \right) \tag{25}$$

This is the wave equation expressed in the frequency-wavenumber domain. A visualization of the transformed quantities is depicted in Fig. 5. In the absence of body sources, this equation can be written as a differential equation in $z$:

$$\frac{\partial^2 \tilde{u}_y}{\partial z^2} + \left(k_S^2 - k^2\right)\tilde{u}_y = 0 \tag{26}$$

with

$$k_S = \frac{\omega}{C_S} \tag{27a}$$

$$C_S = \sqrt{\frac{\mu}{\rho}} \tag{27b}$$

whose solution is

$$\tilde{u}_y = A_1\, e^{\mathrm{i}\beta z} + A_2\, e^{-\mathrm{i}\beta z} \tag{28}$$

with $A_1$, $A_2$ being arbitrary constants, and

$$\beta = \sqrt{k_S^2 - k^2} \qquad \begin{cases} \operatorname{Re}\beta \geq 0 \\ \operatorname{Im}\beta \leq 0 \end{cases} \qquad = \text{vertical wavenumber} \tag{29}$$

From the above we obtain the shearing stresses in horizontal planes as

$$\tilde{\tau}_{yz} = \mu \frac{\partial \tilde{u}_y}{\partial z} = \mathrm{i}\mu\beta\left(A_1\, e^{\mathrm{i}\beta z} - A_2\, e^{-\mathrm{i}\beta z}\right) \tag{30}$$

Consider now a single layer of thickness $h$ as a *free body*, and assign (temporarily) the subindices 1, 2 to the upper and lower surfaces, respectively. To preserve dynamic equilibrium, we apply external tractions $p_1$, $p_2$ at these surfaces so as to match the internal stresses at these locations. Setting the origin of coordinates at the lower horizon (at which $z$=0), writing the displacement and stresses in matrix format, and evaluating these at the upper and lower external surfaces, we obtain

$$\begin{Bmatrix} \tilde{u}_1 \\ \tilde{u}_2 \end{Bmatrix} = \begin{Bmatrix} e^{\mathrm{i}\beta h} & e^{-\mathrm{i}\beta h} \\ 1 & 1 \end{Bmatrix} \begin{Bmatrix} A_1 \\ A_2 \end{Bmatrix} \tag{31a}$$

and

$$\begin{Bmatrix} \tilde{p}_1 \\ \tilde{p}_2 \end{Bmatrix} = \begin{Bmatrix} \tilde{\tau}_1 \\ -\tilde{\tau}_2 \end{Bmatrix} = \mathrm{i}\mu\beta \begin{Bmatrix} e^{\mathrm{i}\beta h} & -e^{-\mathrm{i}\beta h} \\ -1 & 1 \end{Bmatrix} \begin{Bmatrix} A_1 \\ A_2 \end{Bmatrix} \tag{31b}$$

Here we reversed the sign of the internal stress component at the lower interface so as to express this stress as an external traction. Eliminating the integration constants $A_1$, $A_2$ between these two expressions, we obtain after brief algebra

$$\begin{Bmatrix} \tilde{p}_1 \\ \tilde{p}_2 \end{Bmatrix} = \frac{i\mu\beta}{e^{i\beta h} - e^{-i\beta h}} \begin{Bmatrix} e^{i\beta h} + e^{-i\beta h} & -2 \\ -2 & e^{i\beta h} + e^{-i\beta h} \end{Bmatrix} \begin{Bmatrix} \tilde{u}_1 \\ \tilde{u}_2 \end{Bmatrix} \tag{32}$$

which can be written as

$$\begin{Bmatrix} \tilde{p}_1 \\ \tilde{p}_2 \end{Bmatrix} = \frac{\mu\beta}{\sin\beta h} \begin{Bmatrix} \cos\beta h & -1 \\ -1 & \cos\beta h \end{Bmatrix} \begin{Bmatrix} \tilde{u}_1 \\ \tilde{u}_2 \end{Bmatrix} \tag{33}$$

This equation can be written in compact form as

$$\tilde{\mathbf{p}}_l = \mathbf{K}_l \tilde{\mathbf{u}}_l \tag{34a}$$

with

$$\boxed{\mathbf{K}_l = \mu\beta \begin{Bmatrix} \cot\beta h & -\sin^{-1}\beta h \\ -\sin^{-1}\beta h & \cot\beta h \end{Bmatrix}} \tag{34b}$$

being the symmetric *stiffness matrix* of the $l^{th}$ layer. This relationship is *exact*, and is valid for any frequency $\omega$, horizontal wavenumber $k$, and layer thickness $h$.

Before moving on to consider an ensemble of layers in a stratified soil, we first examine the special case of a homogeneous elastic half-space, i.e. of an infinitely deep layer. In this case $A_2=0$ (i.e. no waves moving up), in which case

$$\tilde{u}_y = A_1 e^{i\beta z} \tag{35a}$$

and

$$p_y = \tilde{\tau}_{yz} = i\mu\beta e^{i\beta z} A_1 \tag{35b}$$

which leads immediately to

$$p_y = i\mu\beta\,\tilde{u}_y \tag{36}$$

Hence

$$\boxed{K_H = i\mu\beta} \tag{37}$$

is the stiffness (impedance) of the half-space, as seen from its upper surface.

We can now generalize the above results to a stack of $N$ parallel layers that may or not be underlain by an elastic half-space. To obtain the stiffness matrix for the global system of layers, it suffices for us to overlap the stiffness matrices of the individual layers at the appropriate interface locations (analogous to the assembly of the global stiffness matrix of a structural system obtained by overlapping member stiffness matrices), see Fig. 6.

Re-numbering the layer interfaces from the top down, the final result is a matrix equation of the form

$$\tilde{\mathbf{p}} = \mathbf{K}\tilde{\mathbf{u}} \tag{38a}$$

$$\tilde{\mathbf{u}} = \begin{bmatrix} \tilde{u}_1 & \tilde{u}_2 & \cdots & \tilde{u}_{N+1} \end{bmatrix}^T = \tilde{\mathbf{u}}(\omega, k) \tag{38b}$$

$$\tilde{\mathbf{p}} = \tilde{\mathbf{p}}(\omega, k) \tag{38c}$$

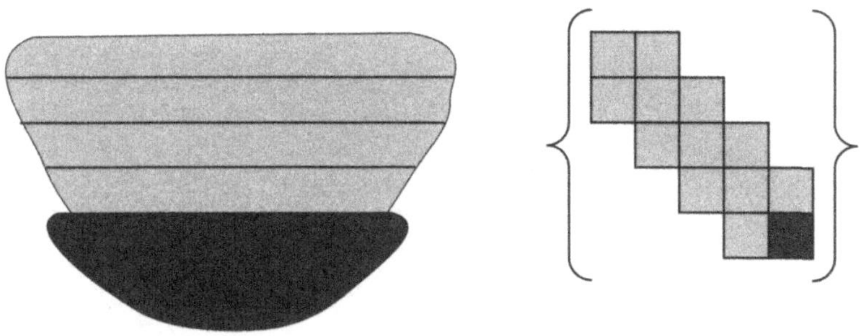

**Figure 6:** Construction of stiffness matrix by overlapping

in which **K** is the symmetric, block-tridiagonal, global stiffness matrix, $\tilde{\mathbf{p}}$ is the vector of external interface tractions (i.e. sources), and $\tilde{\mathbf{u}}$ is the vector of interface displacements. We shall discuss shortly the use of these matrices in the ensuing, but before doing so, we first provide (without proof) the expressions needed to obtain the displacements and stresses in the interior of any layer and/or half-space by means of the so-called *analytic continuation.*

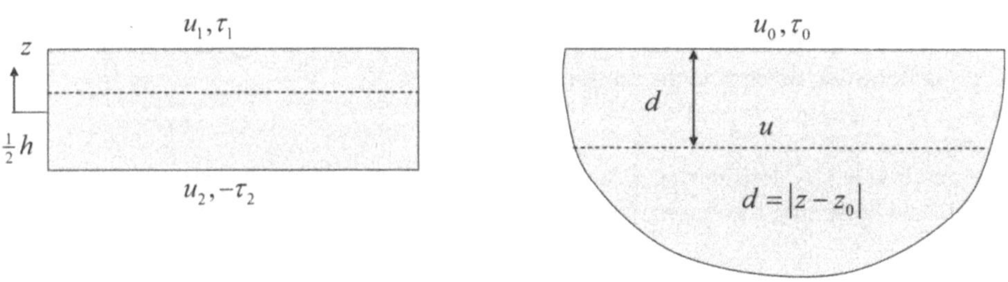


**Figure 7:** Analytic continuation in layer and half-space

With reference to Fig. 7, these are as follows:

*a) layer*

$$\tilde{u}(\zeta) = \Delta u \frac{\sin\frac{1}{2}\zeta\beta}{\sin\frac{1}{2}\beta} + u_m \frac{\cos\frac{1}{2}\zeta\beta}{\cos\frac{1}{2}\beta} \tag{39a}$$

$$\tilde{\tau}_{yz}(\zeta) = \frac{2\mu}{h}\left\{\Delta u \frac{\cos\frac{1}{2}\zeta\beta}{\sin\frac{1}{2}\beta} - u_m \frac{\sin\frac{1}{2}\zeta\beta}{\cos\frac{1}{2}\beta}\right\}\tfrac{1}{2}\beta \tag{39b}$$

in which

$$\Delta u = \frac{\tilde{u}_1 - \tilde{u}_2}{2} \tag{40a}$$

$$u_m = \frac{\tilde{u}_1 + \tilde{u}_2}{2} \tag{40b}$$

$$\zeta = \frac{2z}{h} \qquad (z\text{=0 at mid-height}) \tag{40c}$$

*b) Half-space*

$$\tilde{u} = \tilde{u}_0\, e^{-\mathrm{i}\beta d} \tag{41a}$$

$$\tilde{\tau} = \tilde{\tau}_0\, e^{-\mathrm{i}\beta d} \qquad d = \text{depth below half-space interface} \tag{41b}$$

provided that there are no sources within the half-space.

Having obtained the global equilibrium equation, we proceed to describe its uses. The following three problems can be addressed by means of the stiffness matrix formulation:

- Normal modes: Determine the generalized Love modes in a system of layers
- Source problem: Apply a dynamic source in some layer or the half-space
- Wave amplification: Given a plane wave that impinges the layers from the half-space underneath, determine the motions in the layers.

*A) Normal modes:*

These are obtained by considering free waves, that is, waves in the absence of sources. This is accomplished by setting the external tractions (source) vector $\tilde{\mathbf{p}}$ to zero, that is,

$$\mathbf{K}\tilde{\mathbf{u}} = \mathbf{0} \tag{42}$$

A non-trivial solution exists only if the determinant of **K** vanishes, that is $|\mathbf{K}| = 0$. To illustrate matters, consider the case of a single elastic layer underlain by an elastic half-space. The global stiffness matrix is

$$\mathbf{K}_{system} = \begin{Bmatrix} \mu_1\beta_1 \cot\beta_1 h_1 & -\mu_1\beta_1 \sin^{-1}\beta_1 h_1 \\ -\mu_1\beta_1 \sin^{-1}\beta_1 h_1 & \mu_1\beta_1 \cot\beta_1 h_1 + \mathrm{i}\mu_2\beta_2 \end{Bmatrix} \tag{43a}$$

$$\beta_l = h\sqrt{k_l^2 - k^2} \qquad k_l = \frac{\omega}{C_l} \qquad l = 1,2 \tag{43b}$$

in which $l$=1 refers to the layer, and $l$=2 refers to the half-space. Setting the determinant of this matrix to zero, we obtain

$$\cot\beta_1 h_1\left(\cot\beta_1 h_1 + \frac{\mathrm{i}\mu_2\beta_2}{\mu_1\beta_1}\right) - \sin^{-2}\beta_1 h_1 = 0 \tag{44}$$

which reduces to

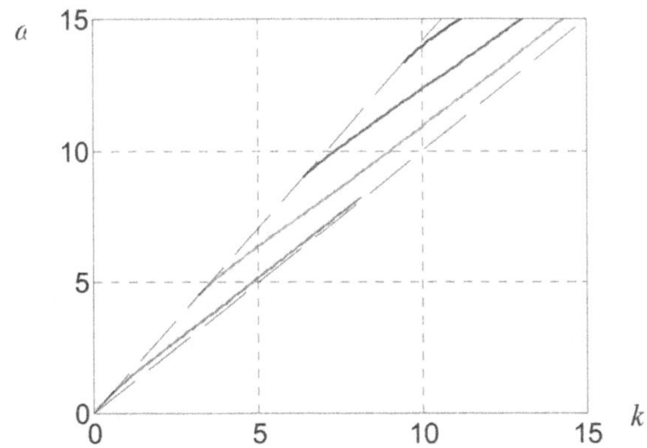


**Figure 8:** Dispersion spectrum for Love waves

$$\tan h_1\sqrt{k_1^2 - k^2} = \frac{\mu_2}{\mu_1}\frac{\sqrt{k^2 - k_2^2}}{\sqrt{k_1^2 - k^2}} \tag{45}$$

This is the classical characteristic equation for Love waves, whose solution (Fig. 8) is well known. Having obtained the characteristic eigenvalues $k$, the *modal shapes* are obtained by introducing the eigenvalues found into the characteristic equation, setting arbitrarily the first component to 1, and solving for the remaining components (in this case, only one). Hence,

$$\mathbf{f} = \begin{Bmatrix} 1 \\ \cos\beta_1 h_1 \end{Bmatrix} = \begin{Bmatrix} 1 \\ \cos h_1\sqrt{k_1^2 - k^2} \end{Bmatrix} \tag{46}$$

In general, solving the transcendental eigenvalue problem $|\mathbf{K}| = 0$ for an arbitrarily layered system is very difficult, especially so when both real and complex roots are sought. While tedious and error-prone search techniques can be used for this purpose, a much more convenient and very robust technique is the *Thin Layer Method* (TLM), which resorts to a discretization of the displacement field in the direction of layering. As a result, the eigenvalue problem changes from transcendental to algebraic, for which standard solution methods are available. As shown in Kausel[6] 1981, the stiffness matrix for a discrete layer attains the form

$$\mathbf{K}_l = \mathbf{A}_l k^2 + \mathbf{G}_l - \omega^2 \mathbf{M}_l \tag{47}$$

in which $\mathbf{A}_l$, $\mathbf{G}_l$, $\mathbf{M}_l$ are matrices that depend solely on the material properties of the layer, the layer thickness, and on the expansion order used for the TLM approximation (typically, either linear or quadratic), see the Appendix. Thus, the eigenvalue problem for a fixed frequency $\omega$ is linear in $k^2$. On the other hand, an elastic half-space that may underlie the layers can be modeled by means of a *paraxial approximation* of the form (Engquist & Majda[3] ,1977)

$$K_\mathrm{H} = \mathrm{i}\,\omega\rho C_S \left[ 1 - \frac{1}{2}\left(\frac{kC_S}{\omega}\right)^2 \right] \tag{48}$$

This approximation is necessary in the TLM to attain a half-space stiffness that is quadratic in $k$. To avoid artifacts, the paraxial approximation must be augmented by means of a *buffer layer* that has the same properties as the half-space and is subdivided into an appropriate number of thin layers. Further details on the TLM can be found in the references at the end.

*B) Source problem*

This is carried out by first casting the source in terms of interface tractions, solving for the displacements, and then carrying out an inverse Fourier transform into space-time. We illustrate the procedure in the box below by means of an example, namely a single layer underlain by an elastic half-space that contains an anti-plane line source at elevation $z_0$ (refer also to eqs. 34b used for the layer and the half-space, 37 for the half-space, and Figs. 6,7):

i) Source at depth $d = |z_0|$ in the half-space: $p_y(x, z_0, t) = \delta(x)\,\delta(t)$

ii) Fourier transform of source: $\tilde{p}_y(\omega, k, z_0) = 1$

iii) Equilibrium equation: $\mathbf{K}\tilde{\mathbf{u}} = \tilde{\mathbf{p}}$

or in full

$$\begin{Bmatrix} \beta_1\mu_1 \cot\beta_1 h & -\beta_1\mu_1/\sin\beta_1 h & 0 \\ -\beta_1\mu_1/\sin\beta_1 h & \beta_1\mu_1\cot\beta_1 h + \beta_2\mu_2\cot\beta_2 d & -\beta_2\mu_2/\sin\beta_2 d \\ 0 & -\beta_2\mu_2/\sin\beta_2 d & \beta_2\mu_2(\cot\beta_2 d + \mathrm{i}) \end{Bmatrix} \begin{Bmatrix} \tilde{u}_{y1} \\ \tilde{u}_{y2} \\ \tilde{u}_y \end{Bmatrix} = \begin{Bmatrix} 0 \\ 0 \\ 1 \end{Bmatrix}$$

iv) Solve for displacements for a dense set of $k$, $\omega$: $\tilde{\mathbf{u}} = \mathbf{K}^{-1}\tilde{\mathbf{p}}$

v) Carry out an inverse Fourier transform

$$\mathbf{u}(x,t) = \left(\tfrac{1}{2\pi}\right)^2 \int_{-\infty}^{+\infty}\int_{-\infty}^{+\infty} \tilde{\mathbf{u}}(k,\omega)\, e^{\mathrm{i}(\omega t - kx)}\, dk\, d\omega$$

Of course, the above sketch glosses over many issues, such as the numerical method used to evaluate the inverse transform, the wavenumber step or the highest wavenumber used in the computation of the improper integrals. However, inasmuch as the focus of this work is on wave propagation and the normal modes, we do not elaborate further on the source problem.

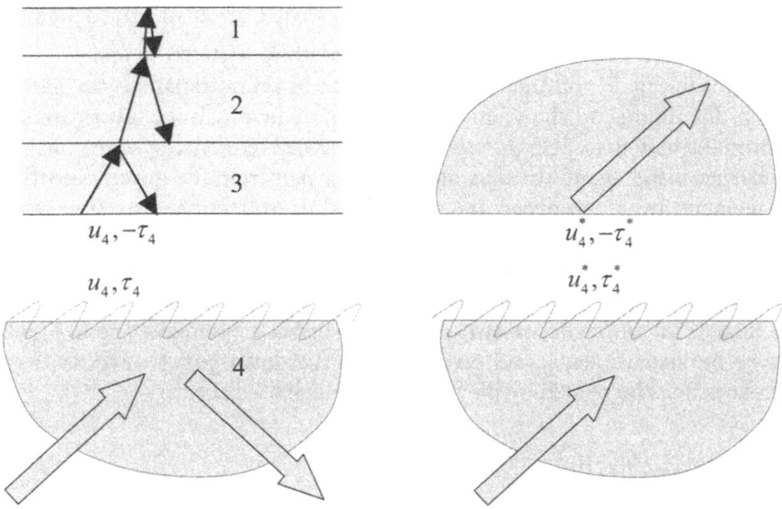


**Figure 9:** Layers over half-space vs. full space (exploded view)

*C) Wave amplification problem*

Consider a layered system underlain by an elastic half-space with shear wave velocity $C_r$ ("rock") which is subjected to plane waves of amplitude $A$ that propagate upwards in the half-space at some angle $\theta$ with respect to the vertical, as shown on the left in Fig. 9. In the absence of reflections and feedback from the layers (i.e. if the half-space were instead a full space, as shown in Fig. 9 on the right), this incident wave would elicit displacements

$$u_y^* = A e^{-\mathrm{i}(k_x x + k_z z)} \equiv A e^{-\mathrm{i} k_z z}\left(e^{-\mathrm{i} kx}\right) \tag{49a}$$

$$k_x \equiv k = \frac{\omega}{C_r}\sin\theta \tag{49b}$$

$$k_z = \frac{\omega}{C_r}\cos\theta \tag{49c}$$

We refer to these displacements as belonging to the *free field*, a fact that we emphasize by means of a superscript star. Evaluating the free field displacement at the surface of the half-space (for which we conveniently take $z$=0), and discarding the exponential term in $x$ (because it is common to all $N$ layers, so it drops out), we obtain $u_{N+1}^* = A$ . We shall now show that the solution for the layers can be obtained by applying a fictitious source at the half-space interface of the form

$$\tilde{p}_{N+1} = K_{full} u^{*}_{N+1} = K_{full} A \tag{50}$$

in which $K_{full} = K_{lower} + K_{upper} = 2\mathrm{i}\beta_r \mu_r$ is the impedance of an imagined homogeneous *full* space, as seen from the elevation of the half-space interface with the layers.

With reference to Fig. 9, consider a system of three layers underlain by an elastic half-space. On the left in this figure, we show an "exploded" view in which the layers are separated as a free body from the half-space. On the right is an imagined reference system consisting of a full, homogenous space, and again, we separate the upper part from the lower. Clearly, in the latter case, the incident wave produces no reflections or refractions, so the wave continues unperturbed to the upper part. Comparing next the lower parts on the left with that on the right, we observe that they are geometrically and materially *identical*, and that they are also subjected to *exactly* the same source, yet the displacements and stresses on the surface are still different. Clearly, it must be the *difference* in stresses on the surface (a secondary source) that is causing the *difference* in displacements, and giving rise to the down going wave on the left. Hence, these differences must be related by the stiffness of the half-space as

$$\tau_4 - \tau_4^{*} = K_{lower}\left(u_4 - u_4^{*}\right) = \mathrm{i}\beta_{full}\mu_{full}\left(u_4 - u_4^{*}\right) \tag{51}$$

On the other hand, the upper half-space on the right is free from sources, so it too must obey a proportionality condition between stresses and displacements:

$$-\tau_4^{*} = K_{upper} u_4^{*} = \mathrm{i}\beta_4 \mu_4 u_4^{*} \tag{52}$$

Combining these two expressions, we obtain

$$-\tau_4 = -K_{lower} u_4 + \left(K_{lower} + K_{upper}\right) u_4^{*} = -K_{lower} u_4 + K_{full} u_4^{*} \tag{53}$$

Introducing this result into the equilibrium equation for the upper layers and moving the term in $K_{lower} u_4$ to the left hand side, we finally obtain the global system equation

$$\begin{Bmatrix} K_{11}^1 & K_{12}^1 & 0 & 0 \\ K_{21}^1 & K_{22}^1 + K_{11}^2 & K_{12}^2 & 0 \\ 0 & K_{21}^2 & K_{22}^2 + K_{11}^3 & K_{12}^3 \\ 0 & 0 & K_{21}^3 & K_{22}^3 + K_{lower} \end{Bmatrix} \begin{Bmatrix} u_1 \\ u_2 \\ u_3 \\ u_4 \end{Bmatrix} = \begin{Bmatrix} 0 \\ 0 \\ 0 \\ K_{full} u_4^{*} \end{Bmatrix} \tag{54}$$

in which $K_{full} = 2\mathrm{i}\beta_4 \mu_4$ is the impedance of the full space, and $u_4^{*} = A$ is again the amplitude of the incident wave, and

$$\beta_l = \sqrt{k_l^2 - k^2} = \sqrt{\left(\frac{\omega}{C_l}\right)^2 - \left(\frac{\omega}{C_4}\sin\theta\right)^2} = \frac{\omega}{C_l}\sqrt{1 - \left(\frac{C_l}{C_4}\right)^2 \sin^2\theta} \qquad (55)$$

### 4.2 SV-P waves and generalized Rayleigh waves

We consider next the more complicated case of vertically polarized shear waves and P waves. In general, SV-P problems are similar to those for SH waves, but much more involved, more difficult to solve, involve double the number of degrees of freedom and double the bandwidth, and the wave spectra are far more elaborate. Still, since the developments for SV-P waves closely parallel the methods used for SH waves, we present only a sketch with the most essential parts. We begin with the in-plane dynamic equilibrium equation, which is

$$\begin{Bmatrix} b_x \\ b_z \end{Bmatrix} + \left[ \begin{Bmatrix} \lambda+2\mu & 0 \\ 0 & \mu \end{Bmatrix} \frac{\partial^2}{\partial x^2} + \begin{Bmatrix} 0 & \lambda+\mu \\ \lambda+\mu & 0 \end{Bmatrix} \frac{\partial^2}{\partial x \partial z} + \begin{Bmatrix} \mu & 0 \\ 0 & \lambda+2\mu \end{Bmatrix} \frac{\partial^2}{\partial z^2} - \begin{Bmatrix} \rho & 0 \\ 0 & \rho \end{Bmatrix} \frac{\partial^2}{\partial t^2} \right] \begin{Bmatrix} u_x \\ u_z \end{Bmatrix} = \begin{Bmatrix} 0 \\ 0 \end{Bmatrix} \qquad (56)$$

Applying a spatio-temporal Fourier transform, and —to attain *symmetric* stiffness matrices— adding an imaginary factor to the vertical components, we obtain

$$\begin{Bmatrix} \tilde{b}_x \\ \mathrm{i}\tilde{b}_z \end{Bmatrix} + \left[ \begin{Bmatrix} \mu & 0 \\ 0 & \lambda+2\mu \end{Bmatrix} \frac{\partial^2}{\partial z^2} + k \begin{Bmatrix} 0 & \lambda+\mu \\ \lambda+\mu & 0 \end{Bmatrix} \frac{\partial}{\partial z} + \rho\omega^2 \begin{Bmatrix} 1 & 0 \\ 0 & 1 \end{Bmatrix} - k^2 \begin{Bmatrix} \lambda+2\mu & 0 \\ 0 & \mu \end{Bmatrix} \right] \begin{Bmatrix} \tilde{u}_x \\ \mathrm{i}\tilde{u}_z \end{Bmatrix} = \begin{Bmatrix} 0 \\ 0 \end{Bmatrix} \qquad (57)$$

Solving the homogeneous equation (i.e. no body loads)

$$\begin{Bmatrix} \tilde{u}_x \\ \mathrm{i}\tilde{u}_z \end{Bmatrix} = \begin{Bmatrix} 1 & -s & 1 & s \\ -p & 1 & p & 1 \end{Bmatrix} \begin{Bmatrix} e^{\mathrm{i}\alpha z} & & & \\ & e^{\mathrm{i}\beta z} & & \\ & & e^{-\mathrm{i}\alpha z} & \\ & & & e^{-\mathrm{i}\beta z} \end{Bmatrix} \begin{Bmatrix} A_1 \\ A_2 \\ A_3 \\ A_4 \end{Bmatrix} \qquad (58)$$

in which

$$\alpha = -\mathrm{i}kp = \sqrt{k_P^2 - k^2} \qquad \begin{cases} \operatorname{Re}\alpha \ge 0 \\ \operatorname{Im}\alpha \le 0 \end{cases} \qquad (59a)$$

$$\beta = -\mathrm{i}ks = \sqrt{k_S^2 - k^2} \qquad \begin{cases} \operatorname{Re}\beta \ge 0 \\ \operatorname{Im}\beta \le 0 \end{cases} \qquad (59b)$$

$$p = \sqrt{1-(k_P/k)^2} \qquad \begin{cases} \operatorname{Re} kp \ge 0 \\ \operatorname{Im} kp \ge 0 \end{cases} \tag{59c}$$

$$s = \sqrt{1-(k_S/k)^2} \quad \begin{cases} \operatorname{Re} ks \ge 0 \\ \operatorname{Im} ks \ge 0 \end{cases} \tag{59d}$$

$$k_P = \omega / C_P \tag{59e}$$

$$k_S = \omega / C_S \tag{59f}$$

By differentiation, the stresses in horizontal planes are

$$\begin{Bmatrix} \tilde{\tau}_{xz} \\ \mathrm{i}\,\tilde{\tau}_{zz} \end{Bmatrix} = -k\mu \begin{Bmatrix} -2p & 1+s^2 & 2p & 1+s^2 \\ 1+s^2 & -2s & 1+s^2 & 2s \end{Bmatrix} \begin{Bmatrix} e^{\mathrm{i}\alpha z} & & & \\ & e^{\mathrm{i}\beta z} & & \\ & & e^{-\mathrm{i}\alpha z} & \\ & & & e^{-\mathrm{i}\beta z} \end{Bmatrix} \begin{Bmatrix} A_1 \\ A_2 \\ A_3 \\ A_4 \end{Bmatrix} \tag{60}$$

Applying these two results to a single free layer, evaluating both at the top and bottom surfaces of the layer, and eliminating the amplitudes, we obtain

$$\begin{Bmatrix} \tilde{p}_{x1} \\ \mathrm{i}\,\tilde{p}_{z1} \\ \tilde{p}_{x2} \\ \mathrm{i}\,\tilde{p}_{z2} \end{Bmatrix} = \begin{Bmatrix} \tilde{\tau}_{xz1} \\ \mathrm{i}\,\tilde{\tau}_{zz1} \\ -\tilde{\tau}_{xz2} \\ -\mathrm{i}\,\tilde{\tau}_{zz2} \end{Bmatrix} = \begin{Bmatrix} K_{11} & K_{12} & K_{13} & K_{14} \\ K_{21} & K_{22} & K_{23} & K_{24} \\ K_{31} & K_{32} & K_{33} & K_{34} \\ K_{41} & K_{42} & K_{43} & K_{44} \end{Bmatrix} \begin{Bmatrix} \tilde{u}_{x1} \\ \mathrm{i}\,\tilde{u}_{z1} \\ \tilde{u}_{x2} \\ \mathrm{i}\,\tilde{u}_{z2} \end{Bmatrix} \tag{61}$$

or in compact form

$$\begin{Bmatrix} \tilde{\mathbf{p}}_1 \\ \tilde{\mathbf{p}}_2 \end{Bmatrix} = \begin{Bmatrix} \mathbf{K}_{11} & \mathbf{K}_{12} \\ \mathbf{K}_{21} & \mathbf{K}_{22} \end{Bmatrix} \begin{Bmatrix} \tilde{\mathbf{u}}_1 \\ \tilde{\mathbf{u}}_2 \end{Bmatrix} \tag{62}$$

which relates the external tractions to the interface displacements through a symmetric stiffness matrix. The elements of this stiffness matrix are as follows (Kausel & Roësset[7], 1981):

$$C_p = \cosh kph \qquad S_p = \sinh kph \qquad C_s = \cosh ksh \qquad S_s = \sinh ksh$$

$$D = 2\left(1 - C_p C_s\right) + \left(\frac{1}{ps} + ps\right) S_p S_s$$

$$\mathbf{K}_{11} = 2k\mu \left[ \frac{1-s^2}{2D} \begin{Bmatrix} \frac{1}{s}\left(C_p S_s - ps\, C_s S_p\right) & -\left(1 - C_p C_s + ps\, S_p S_s\right) \\ -\left(1 - C_p C_s + ps\, S_p S_s\right) & \frac{1}{p}\left(C_s S_p - ps\, C_p S_s\right) \end{Bmatrix} - \frac{1+s^2}{2} \begin{Bmatrix} 0 & 1 \\ 1 & 0 \end{Bmatrix} \right]$$

$\mathbf{K}_{22}$ = same as $\mathbf{K}_{11}$, with off-diagonal signs reversed

$$\mathbf{K}_{12} = 2k\mu \left[ \frac{1-s^2}{2D} \begin{Bmatrix} \frac{1}{s}\left(ps\, S_p - S_s\right) & -\left(C_p - C_s\right) \\ C_p - C_s & \frac{1}{p}\left(ps\, S_s - S_p\right) \end{Bmatrix} \right] \qquad \mathbf{K}_{21} = \mathbf{K}_{12}^T$$

(63)

Also useful is the inverse of the coupling sub-matrix $\mathbf{K}_{12}$,

$$\mathbf{K}_{12}^{-1} = \frac{1}{2k\mu} \left[ \frac{2}{1-s^2} \begin{Bmatrix} \frac{1}{p}\left(ps\, S_s - S_p\right) & C_p - C_s \\ -\left(C_p - C_s\right) & \frac{1}{s}\left(ps\, S_p - S_s\right) \end{Bmatrix} \right] \tag{64}$$

The half-space stiffness matrix is obtained by discarding the terms for up-going waves. The final result is

$$\mathbf{K}_{lower} = 2k\mu \left[ \frac{1-s^2}{2(1-ps)} \begin{Bmatrix} p & 1 \\ 1 & s \end{Bmatrix} - \begin{Bmatrix} 0 & 1 \\ 1 & 0 \end{Bmatrix} \right] \quad \text{Lower half-space} \tag{65}$$

which satisfies

$$\Delta = ps - \left(\frac{1+s^2}{2}\right)^2 = \det(\mathbf{K}) = \text{Rayleigh function} \tag{66}$$

In addition

$$\mathbf{K}_{lower}^{-1} = \frac{1}{4k\mu\Delta} \begin{Bmatrix} s(1-s^2) & 1+s^2-2ps \\ 1+s^2-2ps & p(1-s^2) \end{Bmatrix} \tag{67}$$

The stiffness matrix for an upper half-space $\mathbf{K}_{upper}$ is the same as that of a lower, but with the off-diagonal signs reversed, i.e.

$$\mathbf{K}_{upper} = 2k\mu\left[\frac{1-s^2}{2(1-ps)}\begin{Bmatrix} p & -1 \\ -1 & s \end{Bmatrix} + \begin{Bmatrix} 0 & 1 \\ 1 & 0 \end{Bmatrix}\right] \tag{68}$$

Finally, the stiffness matrix (i.e. impedance) of a full space, as seen from any given horizon, is

$$\mathbf{K}_{full} = \mathbf{K}_{lower} + \mathbf{K}_{upper} = 2k\mu\frac{1-s^2}{1-ps}\begin{Bmatrix} p & 0 \\ 0 & s \end{Bmatrix} \tag{69}$$

It should be observed that the elements of the stiffness matrix are assembled with hyperbolic functions of complex argument in both the numerator and denominator, which may grow exponentially. To prevent numerical problems associated with underflow, overflow, and severe cancellation errors, it will be necessary to use the following strategy in their evaluation:

Let

$$\begin{aligned} C_p &= \cosh kph = \cosh(a+ib) = \cosh a\cos b + \mathrm{i}\,\sinh a \sin b \\ &= e^a \tfrac{1}{2}\left[\left(1+e^{-2a}\right)\cos b + \mathrm{i}\left(1-e^{-2a}\right)\sin b\right] \\ &\underset{\text{def}}{=} e^a\, C_1 \end{aligned} \tag{70a}$$

Similarly

$$C_s = \cosh ksh = \cosh\left(c+id\right) = e^c\, C_2 \tag{70b}$$

$$S_s = e^c S_2 \tag{70c}$$

$$S_p = e^a S_1 \tag{70d}$$

Hence, an individual term such as the first one in $\mathbf{K}_{11}$ would be computed as

$$\begin{aligned} \frac{C_p S_s}{D} &= \frac{e^{a+c} C_1 S_2}{2\left(1-e^{a+c}C_1C_2\right)+\left(\frac{1}{ps}+ps\right)e^{a+c}S_1S_2} \\ &= \frac{C_1 S_2}{2\left(e^{-(a+c)}-C_1C_2\right)+\left(\frac{1}{ps}+ps\right)S_1S_2} \end{aligned} \tag{71}$$

Observe that the exponential growth factor has cancelled out, and the ratio is numerically stable.

As in the anti-plane case, the global system matrices are obtained by overlapping the layer stiffness matrices, including the lower and/or upper half-spaces. This results in a symmetric, block-tridiagonal system stiffness matrix. Additional useful expressions for the element of the stiffness matrices in the limit of zero wavenumber, zero frequency, or both can be found in Kausel & Roësset[7], 1981.

After solving for the displacements, it is necessary to remove the additional imaginary factor applied to the vertical components, which was introduced to attain symmetry. We remark that without this trick and for soils with material damping (modeled via complex moduli), the stiffness matrices would have been neither symmetric nor hermitic, in which case the storage requirements and computational effort needed to solve the system equations would have been substantially greater.

*Normal modes*

As in the anti-plane case, this is achieved by setting the determinant of the global matrix to zero. We illustrate this by means of the classical problem of Rayleigh waves. For a half-space without any added layers, this implies

$$\det(\mathbf{K}_{lower}) = \Delta = ps - \left(\frac{1+s^2}{2}\right)^2 = 0 \tag{72}$$

With the substitution $C_R = \omega / k$ (= speed of Rayleigh waves) we obtain the characteristic equation for Rayleigh waves

$$\sqrt{1-\left(\frac{C_R}{C_P}\right)^2}\sqrt{1-\left(\frac{C_R}{C_S}\right)^2} - \left(1-\frac{1}{2}\left(\frac{C_R}{C_S}\right)^2\right)^2 = 0 \tag{73}$$

A very close solution to this equation valid for any Poisson's ratio is

$$C_R / C_s = 0.874 + 0.197\nu - 0.056\nu^2 - 0.027\nu^3 \tag{74}$$

Again, in the case of a layered system, finding the roots of $\det(\mathbf{K}) = 0$ is extremely difficult. A great simplification is again obtained by recourse to the Thin-Layer Method. It can be shown (Kausel, 1981) that the stiffness matrices for thin layers are of the form

$$\mathbf{K}_l = \mathbf{A}_l k^2 + \mathbf{B}_l k + \mathbf{G}_l - \omega^2 \mathbf{M}_l \tag{75}$$

where again the matrices $\mathbf{A}_l$, $\mathbf{B}_l$, $\mathbf{G}_l$, $\mathbf{M}_l$ (which are different from those for SH waves) depend solely on the material properties of the layers, the thin layer thickness, and the expansion order used for the discretization (see Appendix). Unlike the anti-plane case where the normal modes could be obtained by solving a linear eigenvalue problem in $k^2$, in the case of SV-P waves, we obtain a *quadratic* eigenvalue problem in *k*. Effective and robust algorithms for this purpose are readily available.

As in the SH wave case, when the layers are underlain by an elastic half-space, it is necessary to add a paraxial approximation at the base that is preceded by an appropriately thick buffer layer. The stiffness matrix for the paraxial approximation is (Clayton & Engquist[2], 1977)

$$\mathbf{K}_{\text{SVP}} = \omega\rho\left[ \text{i}\begin{Bmatrix} C_S & \cdot \\ \cdot & C_P \end{Bmatrix} + \begin{Bmatrix} \cdot & C_P - 2C_S \\ C_P - 2C_S & \cdot \end{Bmatrix}\left(\frac{kC_S}{\omega}\right) + \frac{\text{i}}{2}\begin{Bmatrix} C_S - 2C_P & \cdot \\ \cdot & (C_P - 2C_S)(C_P / C_s)^2 \end{Bmatrix}\left(\frac{kC_S}{\omega}\right)^2 \right] \tag{76}$$

This expression can be shown to be equal to the first three terms of the Taylor series expansion in $k$ of the exact stiffness matrix of the half-space. For values of Poisson's ratio greater than 1/3, the term in $C_p$–$2C_s$ on the diagonal of the third matrix changes sign, at which point the paraxial approximation turns unphysical (see Kausel[10], 1992). Numerical experiments suggest that this problem can be overcome by setting this term to zero.

### 4.1 Stiffness matrix method in cylindrical coordinates

It can be shown that when a horizontally layered system is described in cylindrical coordinates $r$, $\theta$, $z$, and the Fourier transform of Cartesian coordinates is replaced by a Hankel (i.e. Fourier-Bessel) transform, the resulting equilibrium equation in the frequency-wavenumber domain is *identical* to a combination of the SH and SV-P system equations, except that the vertical components do *not* carry an imaginary unit factor. Thus, the traction-displacement equation in the frequency-radial wavenumber domain is of the form

$$\begin{Bmatrix} \tilde{p}_{r1} \\ \tilde{p}_{z1} \\ \tilde{p}_{r2} \\ \tilde{p}_{z2} \\ \tilde{p}_{\theta 1} \\ \tilde{p}_{\theta 2} \end{Bmatrix} = \begin{Bmatrix} K_{11}^{SVP} & K_{12}^{SVP} & K_{13}^{SVP} & K_{14}^{SVP} & 0 & 0 \\ K_{21}^{SVP} & K_{22}^{SVP} & K_{23}^{SVP} & K_{24}^{SVP} & 0 & 0 \\ K_{31}^{SVP} & K_{32}^{SVP} & K_{33}^{SVP} & K_{34}^{SVP} & 0 & 0 \\ K_{41}^{SVP} & K_{42}^{SVP} & K_{43}^{SVP} & K_{44}^{SVP} & 0 & 0 \\ 0 & 0 & 0 & 0 & K_{11}^{SH} & K_{12}^{SH} \\ 0 & 0 & 0 & 0 & K_{21}^{SH} & K_{22}^{SH} \end{Bmatrix} \begin{Bmatrix} \tilde{u}_{r1} \\ \tilde{u}_{z1} \\ \tilde{u}_{r2} \\ \tilde{u}_{z2} \\ \tilde{u}_{\theta 1} \\ \tilde{u}_{\theta 2} \end{Bmatrix} \tag{77}$$

Observe that the "in-plane" components are uncoupled from the "anti-plane" components, so the global stiffness matrix necessarily decouples into two *separate* equations. It follows that when the normal modes in cylindrical coordinates are sought via $\det(\mathbf{K}) = 0$, one obtains exactly the same roots as $\det(\mathbf{K}_{SV-P}) = 0$ *and* $\det(\mathbf{K}_{SH}) = 0$. Hence, the wavenumber spectra for cylindrical waves, say, torsional guided waves, are *identical* to those in plane strain. Furthermore, these eigenvalue problems do not depend on the circumferential index $n$ used to accomplish the Fourier expansion in the azimuth $\theta$. In summary, finding the normal modes in cylindrical coordinates involves exactly the same eigenvalue problems as in the two plane strain cases.

For completeness, we provide also the necessary formulas and describe the computational steps for the case of *3-D* sources (say a point source) at some location in the medium. These are as follows:

*a) Express loads in ω-k via Hankel transform*

$$\tilde{\mathbf{b}}_n(k,z,\omega) = a_n \int_0^\infty r\,\mathbf{C}_n \int_0^{2\pi} \mathbf{T}_n \int_{-\infty}^{+\infty} \mathbf{b}(r,\theta,z,t)\, e^{-\mathrm{i}\omega t}\, dt\, d\theta\, dr \tag{78a}$$

$$a_n = \begin{cases} \frac{1}{2\pi} & n = 0 \\ \frac{1}{\pi} & n \neq 0 \end{cases} \tag{78b}$$

$$\mathbf{T}_n = \mathrm{diag}\left\{ \begin{pmatrix} \cos n\theta \\ \sin n\theta \end{pmatrix}, \begin{pmatrix} -\sin n\theta \\ \cos n\theta \end{pmatrix}, \begin{pmatrix} \cos n\theta \\ \sin n\theta \end{pmatrix} \right\} \tag{78c}$$

$$\mathbf{C}_n = \left\{ \begin{matrix} J_n' & \frac{n}{kr} J_n & 0 \\ \frac{n}{kr} J_n & J_n' & 0 \\ 0 & 0 & -J_n \end{matrix} \right\} \tag{78d}$$

$$J_n = J_n(kr) \quad = \text{Bessel function} \tag{78e}$$

$$J_n' = \frac{dJ_n}{d(kr)} \tag{78f}$$

*b) Solve for the displacements*

$$\tilde{\mathbf{u}}_n = \mathbf{K}^{-1} \tilde{\mathbf{p}}_n \qquad \text{(note that } \mathbf{K} \text{ is } \textit{not} \text{ a function of n!)} \tag{79}$$

*c) Obtain displacements in the spatial domain r, θ by inverse Hankel transform*

$$\mathbf{u}(r,\theta,z,t) = \frac{1}{2\pi} \int_{-\infty}^{+\infty} e^{\mathrm{i}\omega t} \sum_{n=0}^{\infty} \mathbf{T}_n \int_0^\infty k\,\mathbf{C}_n \tilde{\mathbf{u}}_n(k,z,\omega)\, dk\, d\omega \tag{80}$$

*Example:*
To illustrate matters, consider the classical problem of a point load applied tangentially at the surface of an elastic half-space (Chao's problem[1]). In cylindrical coordinates, the load can be expressed as

$$\mathbf{p}_1 = \begin{Bmatrix} p_r \\ p_\theta \\ p_z \end{Bmatrix}_{(1)} = \frac{1}{2\pi} \begin{Bmatrix} \cos\theta \\ -\sin\theta \\ 0 \end{Bmatrix} \frac{\delta(r)}{r} = \begin{Bmatrix} \cos\theta & 0 & 0 \\ 0 & -\sin\theta & 0 \\ 0 & 0 & \cos\theta \end{Bmatrix} \begin{Bmatrix} 1 \\ 1 \\ 0 \end{Bmatrix} \frac{\delta(r)}{2\pi r} = \mathbf{T}_1 \begin{Bmatrix} 1 \\ 1 \\ 0 \end{Bmatrix} \frac{\delta(r)}{2\pi r} \tag{81}$$

We begin by transforming this into the wavenumber domain via a Hankel transform:

$$\tilde{\mathbf{p}}_1 = \begin{Bmatrix} \tilde{p}_r \\ \tilde{p}_\theta \\ \tilde{p}_z \end{Bmatrix}_{(1)} = \int_0^\infty \begin{Bmatrix} J_1' & \frac{1}{kr}J_1 & 0 \\ \frac{1}{kr}J_1 & J_1' & 0 \\ 0 & 0 & -J_1 \end{Bmatrix} \begin{Bmatrix} 1 \\ 1 \\ 0 \end{Bmatrix} \frac{\delta(r)}{2\pi} dr = \int_0^\infty \begin{Bmatrix} J_0 \\ J_0 \\ 0 \end{Bmatrix} \frac{\delta(r)}{2\pi} dr = \frac{1}{2\pi} \begin{Bmatrix} 1 \\ 1 \\ 0 \end{Bmatrix} \tag{82}$$

Observe that because the load involves only the $n$=1 azimuthal component, it follows that the displacements will exist solely for that component.

Next, as in the plane-strain case, we solve separately for the SV-P and SH components of the displacements. This involves multiplication of the inverse of the half-space impedances (stiffnesses) by the loads in the frequency-wavenumber domain just obtained:

$$\begin{Bmatrix} \tilde{u}_r \\ \tilde{u}_z \end{Bmatrix} = \frac{1}{4k\mu\Delta} \begin{Bmatrix} s(1-s^2) & 1+s^2-2ps \\ 1+s^2-2ps & p(1-s^2) \end{Bmatrix} \frac{1}{2\pi} \begin{Bmatrix} 1 \\ 0 \end{Bmatrix} \tag{83a}$$

$$\tilde{u}_\theta = \frac{1}{2\mu ks}\left(\frac{1}{2\pi}\right) \tag{83b}$$

Thereafter, we apply an inverse Hankel transform

$$\mathbf{u} = \begin{Bmatrix} u_r \\ u_\theta \\ u_z \end{Bmatrix} = \mathbf{T}_1 \int_0^\infty \begin{Bmatrix} J_1' & \frac{1}{kr}J_1 & 0 \\ \frac{1}{kr}J_1 & J_1' & 0 \\ 0 & 0 & -J_1 \end{Bmatrix} \begin{Bmatrix} \tilde{u}_r \\ \tilde{u}_\theta \\ \tilde{u}_z \end{Bmatrix}_{(1)} k\,dk \tag{84}$$

Substituting the above results into the various components into the integral, we obtain ultimately

$$u_r = (\cos\theta)\left[\frac{1}{4\pi\mu}\int_0^\infty \frac{s\frac{1}{2}(1-s^2)}{\Delta}\left(J_0(kr) - \frac{J_1(kr)}{kr}\right)dk + \frac{1}{2\pi\mu}\int_0^\infty \frac{1}{s}\frac{J_1(kr)}{kr}dk\right]$$

$$u_\theta = (-\sin\theta)\left[\frac{1}{4\pi\mu}\int_0^\infty \frac{s\frac{1}{2}(1-s^2)}{\Delta}\frac{J_1(kr)}{kr}dk + \frac{1}{2\pi\mu}\int_0^\infty \frac{1}{s}\left(J_0(kr) - \frac{J_1(kr)}{kr}\right)dk\right]$$

$$u_z = (\cos\theta)\left[\frac{1}{4\pi\mu}\int_0^\infty \frac{ps - \frac{1}{2}(1+s^2)}{\Delta}J_1(kr)\,dk\right] \tag{85a-c}$$

These expressions can be integrated analytically via contour integration (as Chao[1] did), or numerically (which requires great care and computational effort).

# 5 Appendix: Brief summary of the Thin-Layer Method[2-4,6,7,9,10,12-16]

We restrict attention here to linear elements only. However, we observe that a system assembled with quadratic elements outperforms a similarly composed model with double the number of linear elements, that is, a model with the same number of degrees of freedom and a comparable numerical effort. Thus, quadratic elements are preferable, and these can be found in the references, for example, Kausel[11], 1994. Full details on the TLM are given in Kausel[6] (1981) and Kausel & Peek[8] (1982a), which document the theory underlying our (generally available) program PUNCH[17] for the computation of the Green's functions in layered media.

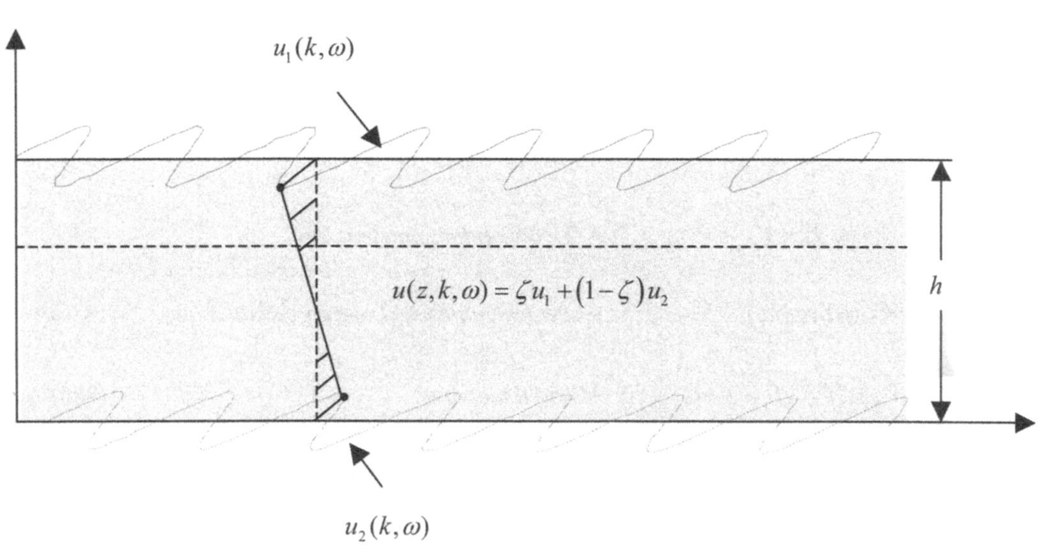


**Figure 10:** Linear discretization of $n^{\text{th}}$ thin layer

*Symbols:*

| | |
|---|---|
| $\mu$ = shear modulus | $\lambda$ = Lamé constant |
| $\rho$ = Mass density | $\xi$ = fraction of material damping |
| $h$ = sub-layer thickness | $l$ = layer index (1,2, ...$N$) |

To account for material damping, in the ensuing formulas the elastic moduli $\lambda$, $\mu$ should be replaced by their complex counterparts:

$$\lambda_c = \lambda\left(1+2\mathrm{i}\xi\,\mathrm{sgn}\,\omega\right) \tag{86a}$$

$$\mu_c = \mu\left(1+2\mathrm{i}\xi\,\mathrm{sgn}\,\omega\right) \tag{86b}$$

### 5.1 Anti-plane case (SH or generalized Love waves)

$$\mathbf{A}_l = \frac{\mu h}{6}\begin{Bmatrix} 2 & 1 \\ 1 & 2 \end{Bmatrix} \tag{87a}$$

$$\mathbf{G}_l = \frac{\mu}{h}\begin{Bmatrix} 1 & -1 \\ -1 & 1 \end{Bmatrix} \tag{87b}$$

$$\mathbf{M}_l = \frac{\rho h}{6}\begin{Bmatrix} 2 & 1 \\ 1 & 2 \end{Bmatrix} \tag{87c}$$

Overlapping the thin-layer matrices, obtain the system matrices $\mathbf{A}_y, \mathbf{G}_y, \mathbf{M}_y$

$\mathbf{C}_y = \mathbf{G}_y - \omega^2 \mathbf{M}_y$ = *Auxiliary matrix* (88a)

$\mathbf{K} = \mathbf{A}_y k^2 + \mathbf{C}_y$ = *Global stiffness matrix* (88b)

$\mathbf{K}_L = \operatorname{diag}\{k_j\}$ = *spectral matrix of Love eigenvalues* (88c)

$\mathbf{F}_y = \{\boldsymbol{f}_1 \;\; \boldsymbol{f}_2 \;\; \cdots\}$ = *modal matrix* (88d)

$k_j^2 \mathbf{A}_y \boldsymbol{f}_j = -\mathbf{C}_y \boldsymbol{f}_j$ = *Eigenvalue problem for Love waves* (88e)

$\mathbf{A}_y \mathbf{F}_y \mathbf{K}_L^2 = -\mathbf{C}_y \mathbf{F}_y$ = *Eigenvalue problem in matrix format* (88f)

$\mathbf{F}_y^T \mathbf{A}_y \mathbf{F}_y = \mathbf{I}$
$\mathbf{F}_y^T \mathbf{C}_y \mathbf{F}_y = -\mathbf{K}_L^2$ = *Orthogonality conditions* (88g)

The eigenvalues satisfy the following properties (See also Waas[18], 1972 for further details):

- If the medium has no damping ($\lambda$, $\mu$ are real) then **A** and **M** are positive definite, while **G** is positive semi-definite. Hence, all eigenvectors are real, and so are also all $k_j^2$. Thus, $k_j$ may be either purely real or purely imaginary.
- If $k_j$ is a solution, then so is also $-\square k_j$, both having the same eigenvector. They correspond to waves traveling or vanishing in opposite directions.
- If the medium has damping, then all wavenumbers are complex. Of the two solutions $\pm k_j$, one must then choose the wavenumber $k_j$ whose imaginary part is negative (which decays in $+x$).
- In the undamped case, there exist special frequencies, the so-called cutoff frequencies, at which a wavenumber can be zero. These represent vertically propagating shear waves (i.e. moving up and down).

*Perturbation of eigenvalues with respect to parameter*

Assume that the layer stiffness matrices depend on some physical parameter $\psi$. For example, the shear modulus (or Poisson's ratio) in some layer could be given by $\mu = \psi\mu_0$ It can then be shown that (this assumes normalized modes):

$$\frac{\partial k_j}{\partial \psi} = -\frac{1}{2k_j}\left[ \boldsymbol{f}_j^T \left( \frac{\partial \mathbf{A}}{\partial \psi} \right) \boldsymbol{f}_j k_j^2 + \boldsymbol{f}_j^T \frac{\partial \left( \mathbf{G} - \omega^2 \mathbf{M} \right)}{\partial \psi} \boldsymbol{f}_j \right] \tag{89}$$

This is very useful for wave inversion (NDT) problems. In particular, if $\psi = \omega$, then

$$\frac{\partial k_j}{\partial \omega} = \frac{\omega}{k_j}\left[ \boldsymbol{f}_j^T \mathbf{M} \boldsymbol{f}_j \right] \qquad = (\textit{group velocity})^{-1} \tag{90}$$

*Computational algorithm for SH eigenvalue problem*

The global stiffness matrix is tri-diagonal, symmetric, but complex. An extremely effective algorithm consists in carrying out a *determinant search* of the eigenvalues with deflation by the eigenvalues already found, using the so-called *Trace Algorithm* of Lancaster[13]. Thereafter, the eigenvectors are found by one or two steps of inverse iteration with shift by the known eigenvalues. Convergence can be accelerated by projecting the eigenvalues from one frequency to the next by means of the perturbation technique.

### 5.2 In-plane case (SV-P or generalized Rayleigh waves)

Note: The following assumes that the vertical components have an added factor $\mathrm{i} = \sqrt{-1}$ (as in the preceding).

$$\mathbf{A}_l = \frac{h}{6}\begin{Bmatrix} 2(\lambda+2\mu) & \cdot & \lambda+2\mu & \cdot \\ \cdot & 2\mu & \cdot & \mu \\ \lambda+2\mu & \cdot & 2(\lambda+2\mu) & \cdot \\ \cdot & \mu & \cdot & 2\mu \end{Bmatrix} \tag{91a}$$

$$\mathbf{B}_l = \frac{1}{2}\begin{Bmatrix} \cdot & \lambda-\mu & \cdot & -(\lambda+\mu) \\ \lambda-\mu & \cdot & \lambda+\mu & \cdot \\ \cdot & \lambda+\mu & \cdot & -(\lambda-\mu) \\ -(\lambda+\mu) & \cdot & -(\lambda-\mu) & \cdot \end{Bmatrix} \tag{91b}$$

$$\mathbf{G}_l = \frac{1}{h}\begin{Bmatrix} \mu & \cdot & -\mu & \cdot \\ \cdot & \lambda+2\mu & \cdot & -(\lambda+2\mu) \\ -\mu & \cdot & \mu & \cdot \\ \cdot & -(\lambda+2\mu) & \cdot & \lambda+2\mu \end{Bmatrix} \tag{91c}$$

$$\mathbf{M}_l = \frac{\rho h}{6}\begin{Bmatrix} 2 & \cdot & 1 & \cdot \\ \cdot & 2 & \cdot & 1 \\ 1 & \cdot & 2 & \cdot \\ \cdot & 1 & \cdot & 2 \end{Bmatrix} \tag{91d}$$

Overlapping the thin-layer matrices, obtain the system matrices $\mathbf{A},\mathbf{B},\mathbf{G},\mathbf{M}$, which will be block-tridiagonal matrices. In general, they will be stored by interface, each of which has two degrees of freedom. However, for *conceptual* purposes, we shall imagine that they have been stored instead by degrees of freedom, first all horizontal, then all vertical. In this case, the quadratic eigenvalue problem for generalized Rayleigh waves will have the form (**O, I** are the null and identity matrices)

$$\mathbf{AF\,K}_R^2 + \mathbf{BF\,K}_R + \mathbf{CF} = \mathbf{O} \tag{92a}$$

$$\mathbf{C} = \mathbf{G} - \omega^2\mathbf{M} \tag{92b}$$

$$\begin{Bmatrix} \mathbf{A}_x & \mathbf{O} \\ \mathbf{O} & \mathbf{A}_z \end{Bmatrix}\begin{Bmatrix} \mathbf{F}_x \\ \mathbf{F}_z \end{Bmatrix}\mathbf{K}^2 + \begin{Bmatrix} \mathbf{O} & \mathbf{B}_{xz} \\ \mathbf{B}_{xz}^T & \mathbf{O} \end{Bmatrix}\begin{Bmatrix} \mathbf{F}_x \\ \mathbf{F}_z \end{Bmatrix}\mathbf{K} + \begin{Bmatrix} \mathbf{C}_x & \mathbf{O} \\ \mathbf{O} & \mathbf{C}_z \end{Bmatrix}\begin{Bmatrix} \mathbf{F}_x \\ \mathbf{F}_z \end{Bmatrix} = \begin{Bmatrix} \mathbf{O} \\ \mathbf{O} \end{Bmatrix} \tag{92c}$$

$$\mathbf{C}_x = \mathbf{G}_x - \omega^2\mathbf{M}_x \tag{92d}$$

$$\mathbf{C}_z = \mathbf{G}_z - \omega^2\mathbf{M}_z \tag{92e}$$

$$\mathbf{F} = \{f_j\} = \begin{Bmatrix} \mathbf{F}_x \\ \mathbf{F}_z \end{Bmatrix} \tag{92f}$$

$\mathbf{\Phi}$ is the square modal matrix (i.e. $\mathbf{F}_x,\mathbf{F}_z$ are both rectangular), and $\mathbf{K}_R = \operatorname{diag}\{k_j\}$ is the spectral matrix (i.e. the diagonal matrix of generalized Rayleigh eigenvalues). All sub-matrices are tridiagonal.

The eigenvectors satisfy the following relationships and *orthogonality conditions*:

$$\mathbf{K}_R\mathbf{F}^T\mathbf{AF\,K}_R - \mathbf{F}^T\mathbf{CF} = \mathbf{K}_R^2 \tag{93a}$$

$$\mathbf{A}_x^{-1} = \mathbf{F}_x\mathbf{F}_x^T \tag{93b}$$

$$\mathbf{A}_z^{-1} = \mathbf{F}_z\mathbf{F}_z^T \tag{93c}$$

$$\mathbf{A}_z^{-1}\mathbf{B}_{xz}^T\mathbf{A}_x^{-1} = -\ \mathbf{F}_z\mathbf{K}\mathbf{F}_x^T \tag{93d}$$

$$\mathbf{O} = \mathbf{F}_x \mathbf{K}_R^{-1} \mathbf{F}_z^T \tag{93e}$$

$$\mathbf{C}_x^{-1} = -\ \mathbf{F}_x \mathbf{K}_R^{-2} \mathbf{F}_x^T \tag{93f}$$

$$\mathbf{C}_x = \mathbf{G}_x - \omega^2 \mathbf{M}_x \tag{93g}$$

$$\mathbf{C}_z^{-1} = -\ \mathbf{F}_z \mathbf{K}_R^{-2} \mathbf{F}_z^T \tag{93h}$$

$$\mathbf{C}_z = \mathbf{G}_z - \omega^2 \mathbf{M}_z \tag{93i}$$

$$\mathbf{C}_z^{-1} \mathbf{B}_{xz}^T \mathbf{C}_x^{-1} = \mathbf{F}_z \mathbf{K}_R^{-3} \mathbf{F}_x^T \tag{93j}$$

$$\mathbf{O} = \mathbf{F}_z \mathbf{K}_R^{-1} \mathbf{F}_x^T \tag{93k}$$

The quadratic eigenvalue problem satisfies the following properties:

- If $k_j$ is a solution, then so is also $-k_j$. These two solutions have *adjoined* eigenvectors whose polarization is opposite (i.e. the signs of the vertical components are reversed). They correspond to waves traveling and/or vanishing in opposite directions. Also, in the undamped case, the eigenvalues and eigenvectors occur in complex conjugate pairs.
- If $k_j$ is *real*, then the eigenvector is also real. $k_j > 0$ corresponds to waves with positive phase velocity. However, the group velocity need not necessarily be positive, i.e. energy propagation and wave propagation need not be in the same direction! In the TLM, one chooses the mode whose group velocity is positive, which can be shown to happen when horizontal and vertical components have the same sign. Since the vertical components include an implied factor $\mathrm{i} = \sqrt{-1}$, this means that the actual vertical components are -90° out of phase with the horizontal, which in turn implies retrograde elliptic orbits.
- If $k_j$ is *purely imaginary*, the mode does not propagate, but vanishes to the right or left depending on the sign of $k_j$.
- If $k_j$ is *complex*, the eigenvectors are complex, the motions are not in phase, and the wave is evanescent. The Rayleigh mode propagates to either the right or left (i.e. in $\pm x$) depending on the sign of Re $k_j$. Also, it decays (or vanishes) to the right or left depending on the sign of Im $k_j$. Observe that the wave may propagate and decay in different directions!

*Perturbation of eigenvalues with respect to parameter*

Assume that the layer stiffness matrices depend on some physical parameter $\psi$, say a material constant, or even the frequency. For example, the shear modulus (or Poisson's ratio) in some layer could be given by $\mu = \psi\, \mu_0$. It can then be shown that (this assumes normalized modes):

$$\frac{\partial k_j}{\partial \psi} = -\frac{1}{2k_j} \boldsymbol{f}_j^T \left[ \frac{\partial \mathbf{A}}{\partial \psi} k_j^2 + \frac{\partial \mathbf{B}}{\partial \psi} k_j + \frac{\partial \left( \mathbf{G} - \omega^2 \mathbf{M} \right)}{\partial \psi} \right] \boldsymbol{f}_j \tag{94}$$

In particular, if $\psi = \omega$, then we have once more

$$\frac{\partial k_j}{\partial \omega} = \frac{\omega}{k_j} \left[ \boldsymbol{f}_j^T \mathbf{M} \boldsymbol{f}_j \right] \qquad = 1/\textit{group velocity} \tag{95}$$

*Computational algorithm for SV-P eigenvalue problem*
The global stiffness matrix is block-tridiagonal, symmetric and complex. The most robust algorithm has proven to be *inverse iteration with shift by Rayleigh quotient.* Since the eigenvalue problem must be repeated for each frequency of interest, a very effective strategy consists in projecting the current eigenvalues to the next frequency by means of the perturbation technique just described, and using the current eigenvectors as starting vectors for the iteration. This saves more than half of the typical 6 to 10 iterations per mode. Also, it is slightly faster and the iterations are more stable when one descends instead of ascends in frequencies, especially for free plates containing rigid-body modes (such modes, if any, are obtained by inspection).

## 6 List of symbols

| | |
|---|---|
| $A$ | cross-section (area) of beam or rod |
| $C_r$ | rod wave velocity, or shear wave velocity of rock |
| $C_p$, $C_s$ | hyperbolic cosines, defined by eqs. 63 |
| $C_P$, $C_S$ | shear and pressure wave velocity |
| $d$ | depth of source (a positive quantity) |
| $E$ | Young's modulus |
| $f$ | frequency, in Hz |
| $g$ | acceleration of gravity |
| $h$ | layer thickness |
| i | imaginary unit = $\sqrt{(-1)}$ |
| $I$ | moment of inertia of beam |
| $j$ | modal index |
| $k$ | wavenumber |
| $k_P$, $k_S$ | wavenumber for P and S waves , defined by eqs. 19d or 59e-f |
| $K$ | (may include subscripts) stiffness or impedance |
| $\mathbf{K}$ | stiffness (impedance) matrix |
| $l$ | layer index |
| $L$ | length of beam or rod |
| $n$ | azimuthal index |
| $\mathbf{O}$ | null matrix |
| $p$ | number defined by eq. 19a or 59c |
| $p_y$ | (or other subscript) traction applied at layer interface |
| $\mathbf{p}$ | vector of external tractions (usually subscripted) |
| $q$ | number defined by eq. 19c |
| $r$ | radial distance (range) |
| $s$ | number defined by eq. 19b or 59d |
| $t$ | time |
| $u$ | displacement |
| $\mathbf{u}$ | displacement vector (a tilde denotes frequency-wavenumber domain) |
| $V_{gr}$ | group velocity (eq. 22) |
| $V_{ph}$ | phase velocity (eqs. 6, 7) |
| $x,y,z$ | spatial coordinates |
| $\alpha$, $\beta$ | numbers defined by eqs. 59a, 59b (or 29) |
| $\delta$ | Dirac-delta singularity function |
| $\Delta$ | Rayleigh function (eq. 66) |
| $\zeta$ | dimensionless vertical coordinate |
| $\theta$ | angle of incidence of waves, or horizontal azimuth |
| $\lambda$ | wavelength. Alternatively, Lamé constant |
| $\mu$ | shear modulus |
| $\rho$ | mass density |
| $\tau$ | stress (usually subscripted, say $\tau_{xz}$, $\tau_{zz}$) |
| $\omega$ | frequency (rad/s) |

## 7 References and bibliography

1. Chao, C.C. (1960). Dynamical response of an elastic half-space to tangential surface loading, *Journal of Applied Mechanics*, **27**, 559-567 (April)
2. Clayton, R. and Engquist, B. (1977). Absorbing boundary conditions for acoustic and elastic wave equations, *Bulletin of the Seismological Society of America*, **67**, 6, 1529-1540.
3. Engquist, B. and Majda, A. (1977). Absorbing boundary conditions for the numerical simulation of waves, *Math. Comp.*, **31** (139), 629-651.
4. Kausel, E. (1974). Forced vibrations of circular foundations on layered media, *MIT Research Report R74-11*, Department of Civil Engineering, MIT, Cambridge, MA, 02139.
5. Kausel, E. and J.M. Roësset (1977). Semi-analytical hyperelement for layered strata, *Journal of Engineering Mechanics*, ASCE, **103**, 4, 569-588, August.
6. Kausel, E. (1981). An explicit solution for the Green's functions for dynamic loads in layered media, *MIT Research Report R81-13*, Department of Civil Engineering, MIT, Cambridge, MA, 02139.
7. Kausel, E. and J.M. Roësset (1981). Stiffness matrices for layered soils, *Bulletin of the Seismological Society of America*, **71**, 6, 1743-1761, December.
8. Kausel, E. and R. Peek (1982a). Dynamic loads in the interior of a layered stratum: An explicit solution, *Bulletin of the Seismological Society of America*, **72**, 5, 1459-1481, October.
9. Kausel, E. and R. Peek (1982b). Boundary integral method for stratified soils, *MIT Research Report R82-50*, Department of Civil Engineering, MIT, Cambridge, MA, 02139.
10. Kausel, E (1992): Physical interpretation and stability of paraxial boundary conditions, *Bulletin of the Seismological Society of America*, **82**, 2, 899-913, April.
11. Kausel, E. (1994). Thin-layer method: Formulation in the time domain, *International Journal for Numerical Methods in Engineering*, **37**, 927-941.
12. Kausel, E. (1999). Dynamic point sources in laminated media via the thin-layer method, *Internal Journal of Solids and Structures*, **36**,4725-4742
13. Lancaster, P. (1966). Lambda matrices and vibrating systems, Pergamon Press, 83
14. Lysmer, J. (1970). Lumped mass method for Rayleigh waves, *Bulletin of the Seismological Society of America*, **60**, 1, 89-104, February.
15. Lysmer, J. and G. Waas (1972). Shear waves in plane infinite structures, *Journal of Engineering Mechanics*, ASCE, **98**, 85-105.
16. Park, J. (2002). Wave motion in finite and infinite media using the Thin-Layer Method, *Ph.D. Thesis*, Department of Civil and Environmental Engineering, MIT, February 2002.
17. PUNCH: A computer program for the determination of the Green's functions in layered media, by E. Kausel. A version based on linear elements is available as part of this publication.
18. Waas, G. (1972). Linear two-dimensional analysis of soil dynamic problems in semi-infinite layer media, *Ph.D. Thesis*, University of California, Berkeley.

# Elastic modelling of surface waves in single and multicomponent systems

Krzysztof Wilmański

Weierstrass Institute for Applied Analysis and Stochastics
Mohrenstr. 39, D-10117 Berlin, Germany

**Abstract** The main aim of this article is to present a review of most important acoustic surface waves which are described by linear one- and two-component models. Among the waves in one-component linear elastic media we present the classical Rayleigh waves on a plane boundary, Rayleigh waves on a cylindrical surface, Love waves, Stoneley waves (solid/solid and fluid/solid interface). In the second part of the article we discuss two two-component models of porous materials (Biot's model and a simple mixture model). We indicate basic differences of the models and demonstrate qualitative similarities. We introduce as well some fundamental notions yielding the description of surface waves in two-component systems (saturated porous materials) and review certain (porous materials with impermeable boundaries) asymptotic results for such waves. However, the full discussion of this subject including numerous results of computer calculations can be found in the article of B. Albers (2005) also included in this volume.

## Contents

**1 Introduction** **204**

**2 Water waves and classical Rayleigh waves** **206**
2.1 Water waves in an ideal incompressible fluid model . . . . . . . . . . . . . 206
2.2 Remark on the choice of the independent variable in dispersion relations . 212
2.3 Rayleigh waves on plane boundaries of linear elastic homogeneous materials 215
2.4 Rayleigh waves on cylindrical boundaries . . . . . . . . . . . . . . . . . . . 223

**3 Waves in a layer of an ideal fluid and Love waves on plane boundaries** **226**
3.1 Layer of an ideal compressible fluid on a semiinfinite rigid body . . . . . . 226
3.2 Love waves on plane boundaries . . . . . . . . . . . . . . . . . . . . . . . . 227
3.3 Arrival times for packages of waves – example . . . . . . . . . . . . . . . . 231
3.4 Rayleigh waves in a layer of elastic material . . . . . . . . . . . . . . . . . 235

**4 Stoneley waves** **239**
4.1 Interface of two semiinfinite elastic solids . . . . . . . . . . . . . . . . . . . . 239
4.2 Interface of a semiinfinite elastic solid and a semiinfinite ideal fluid . . . . 242
4.3 Interface of a semiinfinite elastic solid and a layer of an ideal fluid . . . . 244
4.4 A few remarks on leaky waves . . . . . . . . . . . . . . . . . . . . . . . . 245

**5 Elastic two-component media** **246**
5.1 Biot's model and a simple mixture model of two-component poroelastic materials . . . . . . . . . . . . . . . . . . . . . . . . . . . . . . . . 246
5.2 Bulk monochromatic waves in two-component poroelastic materials . . . . 255
5.3 Some remarks on the simple mixture model . . . . . . . . . . . . . . . . . 262

**6 Surface waves in two-component media** **262**
6.1 Preliminaries . . . . . . . . . . . . . . . . . . . . . . . . . . . . . . . . 262
6.2 Compatibility conditions and dispersion relation . . . . . . . . . . . . . . 263
6.3 Boundary value problems for surface waves . . . . . . . . . . . . . . . . . 267
6.4 High frequency approximation . . . . . . . . . . . . . . . . . . . . . . . . 268
6.5 Low frequency approximation . . . . . . . . . . . . . . . . . . . . . . . . 269
6.6 Remarks on modelling surface waves by Biot's model and the simple mixture model . . . . . . . . . . . . . . . . . . . . . . . . . . . . . . . . 270

**Bibliography** **272**

# 1 Introduction

Surface waves exist in an extremely wide range of frequencies over some 10 orders of magnitude. This is schematically shown in the diagram below (Hess (2002)).

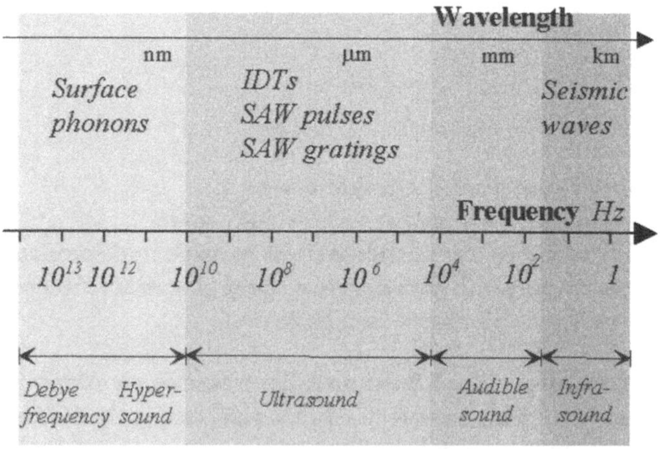


**Figure 1.** Ranges of frequencies for surface waves.

Acoustic surface waves (SAWs) were discovered in 1885 by Rayleigh (1887) and they were the main subject of studies for some decades due to their appearance in the form of seismic waves. The development of interdigital transducers (IDTs) and focused short laser pulses around 1960 extended their study and applications from the low frequency range to the ultrasound. Recent research work goes beyond this frequency range even to the hypersound region corresponding to quantized lattice vibrations (surface phonons).

BOX 1: RAYLEIGH, *John William Strutt, 3rd Baron*

*Born:* Terling, Essex, November 12, 1842,
*Died:* Witham, Essex, June 30, 1919

"Strutt at the age of thirty-one inherited his father's title, so that he is almost invariably referred to as Lord Rayleigh" (I. Asimov; *Asimov's Biographical Encyclopedia of Science and Technology*, Doubleday & Co., Inc., Garden City, N.Y., 1964). In 1873 he was elected to the Royal Society, in 1879 suceeded Maxwell as director of the Cavendish Laboratory at Cambridge. All his life he was interested in wave motion of all varieties: electromagnetic waves, black-body radiation, sound waves, water waves. Important discoveries in chemistry but the Nobel Prize in the year 1904 in physics. Since 1905 he was the president of the Royal Society, and in 1908 he became chancellor of Cambridge University.

This article is devoted to the presentation of models of basic types of surface waves in homogeneous elastic materials and in two-component poroelastic materials. As the main field of application of these models is assumed to be geomechanics (see: Lancellotta (2005)), we concentrate on models appropriate for the frequency regime of ultra-, audible- and infrasound. The reader interested in applications of surface waves to testing electronic materials should consult the review article of R. M. White (1970). Modelling, numerical evaluation and experimental verification of surface waves in heterogeneous materials, both elastic and viscoelastic, do not enter at all these notes in spite of the fact that they are particularly important in nondestructive testing of soils (e.g. Lai (1998), Rix et al. (2001)). However, they are the subject of detailed presentations in other articles of this volume ( Foti (2005), Kausel (2005), Lai (2005), Rix (2005)). One should mention an area of application of surface waves in geotechnics which shall not be quoted further in these notes either in spite of its rash development in the last few years. It is related to the detection of buried objects, particularly land mines (e.g. Schroeder and Scott (2000), Zeng and Liu (2001b), Zeng and Liu (2001a)). Modelling in this area is similar to this presented in the present book but practical aspects, in particular – configurations of experimental sites, are different from those characteristic for the investigation of morphology of soils.

Theoretical and numerical details concerning the propagation of surface waves in two-component porous materials are also presented separately (see: Albers (2005)). For this reason, we concentrate here solely on the explanation of some guidelines for these subjects without going much into details.

We leave out entirely the problem of nonlinear surface waves. This subject develops in the recent years very vehemently (e.g. compare Parker and Maugin (1988)) but it requires different theoretical methods than these applied for linear waves. However, some hints on this subject can be found in one of the articles of this volume (Maugin (2005)).

In spite of the vast literature on surface waves, there exists no comprehensive presentation of this subject for scientists working on engineering and materials sciences and applying the technique of wave propagation in, for example, nondestructive testing. This article aims at the theoretical description of the most popular surface waves, in particular Rayleigh, Love, Stoneley, in one-component elastic media as well as the appropriate extensions to the saturated porous materials. We introduce as well some terminology used in this theory which is peculiar for seismology, geotechnics, and nondestructive testing in materials science. Due to the lack of space, we leave out the important problem of sources of waves. However, an extensive presentation of this subject can be found elsewhere ( Aki and Richards (2002), Ben-Menahem and Singh (2000), Sato and Fehler (1998)).

Each surface wave is presented in a self-contained manner and the corresponding Section can be read independently of other Sections of the article. It means that the derivation of the fundamental equation – dispersion relation, follows from the field equations and boundary conditions for each case anew. In the literature, this is sometimes unified by the ray method and the so-called transition and reflection conditions, following from boundary conditions, which altogether yield the method of the **constructive interference** (e.g. Aki and Richards (2002), Brekhovskikh and Goncharov (1994), Udias (1999)). We do not apply this method in this article in order to have an opportunity for the discussion of the physical inside of both field equations and of boundary conditions for such different systems as the contact of a linear elastic solid with an ideal fluid vs., say, the contact of porous materials with a permeable boundary with an ideal fluid.

## 2 Water waves and classical Rayleigh waves

### 2.1 Water waves in an ideal incompressible fluid model

There exists a vast literature on this subject (e.g. Ben-Menahem and Singh (2000), Elmore and Heald (1985), Coulson and Jeffrey (1977), Whitham (1974)). In contrast to all other waves considered in this article which are produced by an unspecified source far away from the space-time point $(\mathbf{x}, \mathbf{t})$ of analysis, water waves result from the action of gravity. We present them here because they are well known to all who observed the motion of water on the beach (e.g. Trefil (1984)), and, simultaneously, they possess all properties characteristic for surface waves.

We consider an ideal (inviscid) incompressible fluid which means that its mass density $\rho$ is constant. We assume that the motion is irrotational, i.e. the velocity $\mathbf{v}(\mathbf{x}, t)$ possesses a potential $\phi$. Then the mass conservation yields

$$\nabla^2\phi = 0, \quad \mathbf{v} = \text{grad}\phi, \quad \rho = const. \tag{2.1}$$

Simultaneously, the momentum balance equation has the form

$$\rho\left(\frac{\partial\mathbf{v}}{\partial t} + \mathbf{v}\cdot\text{grad}\mathbf{v}\right) = -\text{grad}p + \rho\mathbf{b}, \tag{2.2}$$

where $p$ – pressure, $\mathbf{b}$ – mass force. Integration of this equation with respect to spacial variable $\mathbf{x}$ yields the Bernoulli equation

$$\frac{\partial\phi}{\partial t} + \frac{1}{2}v^2 + \chi + \frac{p}{\rho} = C(t), \quad \mathbf{b} = -\text{grad}\chi, \tag{2.3}$$

where $C(t)$ is an arbitrary function of time and we have used the identity $\mathbf{v}\cdot\text{grad}\mathbf{v} = \frac{1}{2}\text{grad}v^2$, $v^2 \equiv \mathbf{v}\cdot\mathbf{v}$, which holds for irrotational velocity fields.

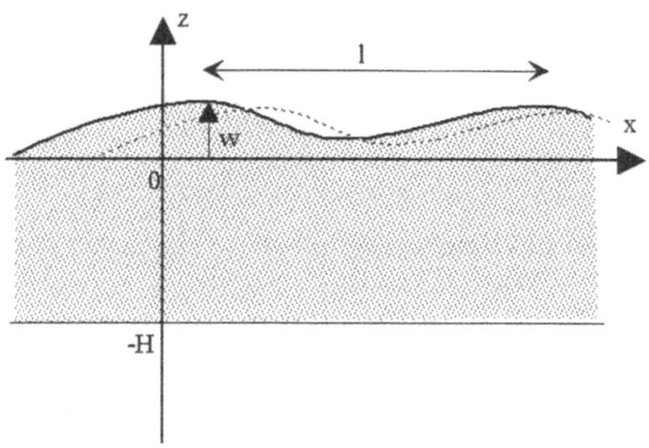


**Figure 2.** Water profile at the instant of time $t$.

We investigate the motion of the fluid shown schematically in Fig. 2. For the two-dimensional motion assumed in this problem the upper boundary surface is described by the relations

$$\begin{aligned} f_s(\mathbf{x},t) &= z - w(x,t) = 0 \quad \Rightarrow \\ &\Rightarrow \quad \mathbf{n} = \frac{\text{grad}f_s}{|\text{grad}f_s|} \approx \mathbf{e}_z, \quad \mathbf{n}\cdot\mathbf{v}_s = -\frac{\frac{\partial f_s}{\partial t}}{|\text{grad}f_s|} \approx \frac{\partial w}{\partial t}\mathbf{e}_z, \end{aligned} \tag{2.4}$$

where $\mathbf{n}$ is the unit outward normal vector of the surface, $\mathbf{v}_s$ is the velocity of this surface, and $\mathbf{e}_z$ the unit vector in the direction of $z$-axis. The approximation in the above relations means that we linearize the problem with respect to the elevation function $w$ and its derivatives.

Let $p_a$ denote the atmospheric pressure acting on the fluid. Then, under the assumption that we can neglect the surface tension, the continuity of the mass flux on the surface (the surface is **material** with respect to the fluid) and the Bernoulli equation (2.3) yield

$$\begin{aligned} (\mathbf{v}-\mathbf{v}_s)\cdot\mathbf{n} &= 0, \\ p_a &= -\rho g w - \rho\left(\frac{\partial\phi}{\partial t}\right)_{z=w} + \rho C(t), \end{aligned} \tag{2.5}$$

where the nonlinear contribution $\frac{1}{2}v^2$ has been neglected[1].

Instead of the potential $\phi$, we use the potential $\phi' = \phi + \frac{p_a}{\rho}t - \int C(t)\,dt$ because $\mathbf{v} = \text{grad}\phi \equiv \text{grad}\phi'$. Then, omitting the prime,

$$w = -\frac{1}{g}\left.\frac{\partial\phi}{\partial t}\right|_{z=0}, \tag{2.6}$$

where we place the boundary at $z=0$ instead of $z=w$ which yields the error of the same order of magnitude as in other approximations.

Simultaneously, the first condition (2.5) and the relation $(2.4)_3$ yield

$$\frac{\partial w}{\partial t} = \left.\frac{\partial\phi}{\partial z}\right|_{z=0}. \tag{2.7}$$

Combination of (2.6) and (2.7) yields the following kinematic boundary condition for $\phi$

$$\frac{\partial^2\phi}{\partial t^2} + g\frac{\partial\phi}{\partial z} = 0 \quad \text{at} \quad z=0. \tag{2.8}$$

The second boundary condition which must be fulfilled by solutions of equation (2.1) is formulated at the bottom

$$\mathbf{v}\cdot\mathbf{e}_z|_{z=-H} = \left.\frac{\partial\phi}{\partial z}\right|_{z=-H} = 0. \tag{2.9}$$

[1]Approximations which we make in this derivation are based on the comparison with the wavelength $l$ ($=\frac{2\pi}{k}$, $k$ – wave number), wave period $T$ ($=\frac{2\pi}{\omega}$, $\omega$ – frequency), and the amplitude $a$. Namely

$$|\mathbf{v}| \sim \frac{a}{T}, \quad \left|\frac{\partial\mathbf{v}}{\partial t}\right| \sim \frac{a}{T^2}, \quad |\mathbf{v}\cdot\text{grad}\mathbf{v}| \sim \frac{a^2}{T^2 l} \quad \Rightarrow$$

$v^2$ can be neglected if $a \ll l$. Simultaneously

$$|w| \sim a, \quad \left|\frac{\partial w}{\partial x}\right| \sim \frac{a}{l} \ll 1.$$

BOX 2: *Water waves*

**Conditions satisfied by the velocity potential $\phi$**

$$\begin{aligned}
\nabla^2\phi &= 0 \quad \text{in the fluid,}\\
\frac{\partial\phi}{\partial z} &= 0 \quad \text{on the bottom } z=-H,\\
\frac{\partial^2\phi}{\partial t^2}+g\frac{\partial\phi}{\partial z} &= 0 \quad \text{on the free surface } z=0,
\end{aligned}$$

$$\begin{aligned}
w &= -\frac{1}{g}\frac{\partial\phi}{\partial t} \quad \text{on the free surface } z=0,\\
\frac{\partial w}{\partial t} &= \frac{\partial\phi}{\partial z} \quad \text{on the free surface } z=0.
\end{aligned}$$

Among the last three conditions two are independent.

**Dispersion relation**

$$\omega^2 = gk\tanh kH.$$

Now we are in the position to make an ansatz for solutions of the problem. We seek it in the form of the wave progressive in $x$-direction

$$\phi(x,z,t)=\left(Ae^{kz}+Be^{-kz}\right)\cos(kx-\omega t). \tag{2.10}$$

Then, according to (2.6), the elevation $w$ satisfies the one-dimensional wave equation.

Boundary conditions (2.8) and (2.9) yield

$$\begin{aligned}
-\omega^2(A+B)+gk(A-B) &= 0,\\
Ae^{-kH}-Be^{kH} &= 0.
\end{aligned} \tag{2.11}$$

Consequently, from the determinant of this homogeneous set we obtain the following **dispersion relation**

$$\omega^2 = gk\tanh(kH). \tag{2.12}$$

Simultaneously, the potential can be written in the form

$$\phi=\phi_{\max}\frac{\cosh k(H+z)}{\cosh kH}\cos(kx-\omega t), \tag{2.13}$$

where $\phi_{\max}$ is a constant of integration.

According to (2.6) we obtain for the elevation

$$w = -w_{\max} \sin\left[k\left(x - c_{ph}t\right)\right], \quad c_{ph} := \frac{\omega}{k}, \quad w_{\max} := \frac{\omega}{g}\phi_{\max}. \tag{2.14}$$

Hence the elevation changes in the $x$-direction as it were the wave moving with the **phase velocity**

$$c_{ph} = \frac{\omega}{k} = \sqrt{\frac{gl}{2\pi}\tanh\left(2\pi\frac{H}{l}\right)}, \quad l := \frac{2\pi}{k}, \tag{2.15}$$

where $l$ is the wavelength. The phase velocity of this wave depends on the frequency $\omega$ (or on the wave number $k$) and, therefore, the wave is called **dispersive**. This property is characteristic for all surface waves which propagate in systems with a characteristic **length scale** (e.g. the depth of the layer, the characteristic length of heterogeneous materials whose properties depend on the location in space, etc.).

In order to find orbits of material points we use the following relation for displacements

$$\begin{aligned} \frac{\partial u_x}{\partial t} &= v_x = \frac{\partial \phi}{\partial x} \quad \Rightarrow \quad u_x = -\frac{k}{\omega}\phi_{\max}\frac{\cosh k\left(H+z\right)}{\cosh kH}\cos\left(kx - \omega t\right), \\ \frac{\partial u_z}{\partial t} &= v_z = \frac{\partial \phi}{\partial z} \quad \Rightarrow \quad u_z = -\frac{k}{\omega}\phi_{\max}\frac{\sinh k\left(H+z\right)}{\cosh kH}\sin\left(kx - \omega t\right). \end{aligned} \tag{2.16}$$

Elimination of time yields

$$\frac{u_x^2}{\alpha_x^2} + \frac{u_z^2}{\alpha_z^2} = 1, \tag{2.17}$$

where

$$\alpha_x := \frac{k}{\omega}\phi_{\max}\frac{\cosh k\left(H+z\right)}{\cosh kH}, \quad \alpha_z := \frac{k}{\omega}\phi_{\max}\frac{\sinh k\left(H+z\right)}{\cosh kH}. \tag{2.18}$$

Consequently, the orbit of each particle is an ellipse with semiaxes $\alpha_x, \alpha_z$. The largest ellipse appears at $z = 0$, and at the bottom $z = -H$, it degenerates into a straight line. The orbits are schematically shown in Fig. 3.

In the short-wave limit (the deep water!): $kH \to \infty$, we have $\tanh kH \approx 1$, i.e. $\omega^2 \approx gk$. Consequently, the phase velocity is given by the relation

$$c_{ph} = \frac{\omega}{k} = \sqrt{\frac{g}{k}} = \frac{g}{\omega}. \tag{2.19}$$

Simultaneously, the velocity potential becomes

$$\phi \approx \phi_{\max} e^{kz} \cos\left(kx - \omega t\right). \tag{2.20}$$

Hence the motion of the fluid is negligible at the depth of about a wavelength $l = \frac{2\pi}{k}$. For this reason, these waves are called **surface waves**.

The above dispersive wave gives rise to a structure of propagation which has a very important bearing. The arrival of such waves to receivers is observed in form of wave packages rather than in the form of single monochromatic waves or wave fronts. In order to illustrate this property on our simple example of deep water waves we consider the

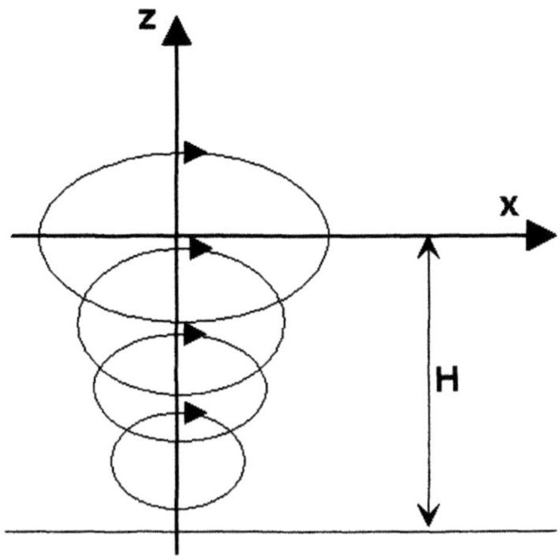


**Figure 3.** Orbits of particles given by (2.17). Particles travel in the clockwise (prograde) direction.

wave consisting of a narrow band of frequencies near the middle frequency $\omega_0$ rather than a single frequency considered above. The solution (2.20) must be now replaced by the Fourier integral which accounts for all frequencies entering the band

$$\begin{aligned}\phi(x,t) &= \frac{1}{2\pi}\int_{-\infty}^{\infty}\phi_{\max}(\omega)\,e^{kz}\cos(kx-\omega t)\,d\omega \approx \\ &\approx \frac{1}{2\pi}\phi_{\max}(\omega_0)\,e^{k_0 z}\int_{-\infty}^{\infty}\cos(kx-\omega t)\,d\omega,\end{aligned} \tag{2.21}$$

where

$$k_0 = \frac{\omega_0^2}{g}. \tag{2.22}$$

Integration yields for small $\left|\frac{\Delta\omega}{\omega_0}\right|$

$$\phi = \frac{\Delta\omega}{2\pi}\phi_{\max}(\omega_0)\,e^{k_0 z}M(\omega_0,\Delta\omega)\cos\left[k_0(x-c_{ph}t)\right], \tag{2.23}$$

where

$$M(\omega_0,\Delta\omega) := \frac{\sin\left[\frac{\Delta\omega}{c_g}(x-c_g t)\right]}{\frac{\Delta\omega}{c_g}(x-c_g t)}, \tag{2.24}$$

and we have the relations

$$\begin{aligned}k-k_0 &= \frac{1}{g}\left(\omega^2-\omega_0^2\right) \approx \frac{1}{c_g}(\omega-\omega_0), \\ c_{ph} &= \frac{g}{\omega_0}, \quad c_g := \left.\frac{d\omega}{dk}\right|_{\omega=\omega_0} = \frac{1}{2}c_{ph} = \frac{g}{2\omega_0}.\end{aligned} \tag{2.25}$$

The quantity M is called the **modulator**, and as shown on an example in Fig. 4, it has an extremum at $x - c_g t = 0$. The modulator is an envelope of the band of waves and propagates with the **group velocity** $c_g = \frac{d\omega}{dk}$. The **carrier** which in our example is described by the cosine function in (2.23) describes the motion at the frequency $\omega_0$ with the phase velocity $c_{ph}$. In Figure 4 we show the wave in two instances of time. Clearly, due to different velocities, the shape of the full wave plotted as the solid line moved differently from the envelope of the modulator indicated by dotted curves.

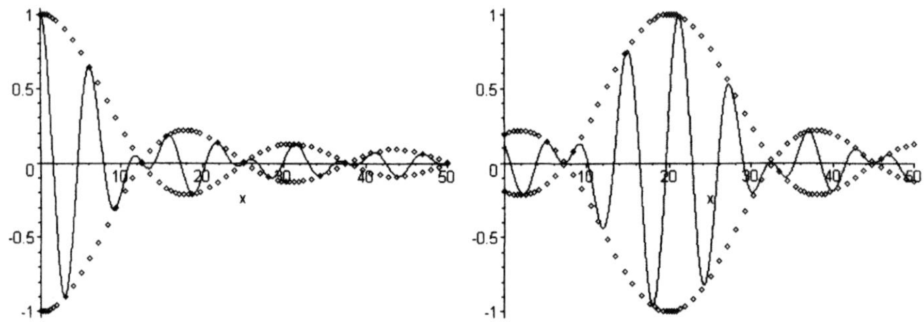


**Figure 4.** Narrow band wave: configurations at two different instances of time (arbitrary units).

Due to its practical importance, this form of propagation shall be discussed further in this article in some details.

### 2.2 Remark on the choice of the independent variable in dispersion relations

It is easy to observe in the above example that in a general dispersion relation $f(\omega, k) = 0$ we may choose the frequency $\omega$ as an independent variable and calculate the wave number $k$ as a function of $\omega$, or, *vice versa*, we can choose the wave number $k$ (or, equivalently, the wave length $l = \frac{2\pi}{k}$) as the independent variable and calculate the frequency $\omega$ as a function of $k$. A transformation from one choice to the other yields the same results provided both $\omega$ and $k$ are real and the relation between these variables is monotonous. In the seismological literature these properties are taken for granted and, on the same page, one can find sometimes $\omega$ and sometimes $k$ chosen as independent. Obviously, in the above presented example this is legitimate because the dispersion relation (2.12) is invertible.

However, problems with complex solutions of the dispersion relation cannot be treated so carelessly. Such solutions appear, for instance, in problems with damping. We present in this remark two very simple examples in order to illustrate the problem. In practical applications to surface waves this problem appears, for instance, in heterogeneous systems and in two-component systems with a relative motion of components (diffusion).

Let us first consider the following one-dimensional partial differential equation for the unknown function $u$

$$\frac{\partial^2 u}{\partial t^2} + \eta \frac{\partial u}{\partial t} - c^2 \frac{\partial^2 u}{\partial x^2} = 0, \tag{2.26}$$

where $\eta$ is a positive constant (viscosity) and $c$ is a real constant. Obviously, in the particular case $\eta = 0$ this equation becomes the classical linear wave equation.

We seek the solution of the above equation in the form of a monochromatic wave

$$u = Ue^{i(kx-\omega t)}, \tag{2.27}$$

where $U, k, \omega$ are constant. Substitution of (2.27) in (2.26) yields the dispersion relation

$$\omega^2 + i\eta\omega + c^2 k^2 = 0. \tag{2.28}$$

We consider two cases: the given real frequency $\omega$, and the given real wave number $k$.

1) The given real frequency $\omega$. Then the solution of the dispersion relation has the form

$$\begin{aligned} \mathrm{Re}k &= \frac{\omega}{c}\frac{1}{\sqrt{2}}\sqrt{1+\sqrt{1+\left(\frac{\eta}{\omega}\right)^2}}, \\ \mathrm{Im}k &= \frac{\frac{\eta}{c}}{\sqrt{2}\sqrt{1+\sqrt{1+\left(\frac{\eta}{\omega}\right)^2}}}. \end{aligned} \tag{2.29}$$

It is clear that Re$k$ is always different from zero and inspection of (2.27) shows that $\frac{\omega}{\mathrm{Re}k}$ defines the phase velocity while Im$k$ is the **attenuation** of the wave, i.e. the rate of decay of the amplitude in space due to the damping $\eta$.

2) The given real wave number $k$. Then the solution of the dispersion relation is as follows

$$\begin{aligned} \mathrm{Re}\omega &= \begin{cases} ck & \text{for} \quad k > \frac{\eta}{2c}, \\ 0 & \text{for} \quad k \leq \frac{\eta}{2c}, \end{cases}, \\ \mathrm{Im}\omega &= \begin{cases} -\frac{\eta}{2} \quad \text{for} \quad k > \frac{\eta}{2c}, \\ -\frac{\eta}{2} \mp ck\sqrt{\frac{\eta^2}{4c^2k^2} - 1} \quad \text{for} \quad k \leq \frac{\eta}{2c}. \end{cases} \end{aligned} \tag{2.30}$$

Hence, the wave propagates with the phase velocity $\frac{\mathrm{Re}\omega}{k}$ solely for sufficiently large wave numbers (i.e. for sufficiently short waves $l < \frac{\pi c}{\eta}$). The wave number $k_c = \frac{\eta}{2c}$ is **critical**. Below this number the equation (2.26) describes a pure damping and waves cannot propagate at all.

Secondly let us consider the following one-dimensional partial differential equation

$$\frac{\partial^2 u}{\partial t^2} + \gamma \frac{\partial u}{\partial x} - c^2 \frac{\partial^2 u}{\partial x^2} = 0, \tag{2.31}$$

where $\gamma$ is a positive constant (spacial damping) and $c$ is a real constant.

Again we consider a solution in the form (2.27). We obtain immediately the following dispersion relation

$$\omega^2 - i\gamma k - c^2 k^2 = 0. \qquad (2.32)$$

As before we consider two cases: the given real frequency $\omega$, and the given real wave number $k$.

1) The given real frequency $\omega$. We obtain the following relations

$$\begin{aligned} \mathrm{Re}k &= \begin{cases} \pm\frac{\omega}{c}\sqrt{1-\frac{\gamma^2}{4c^2\omega^2}} & \text{for} \quad \omega > \frac{\gamma}{2c}, \\ 0 \quad \text{for} \quad \omega \le \frac{\gamma}{2c}, \end{cases}, \\ \mathrm{Im}k &= \begin{cases} -\frac{\gamma}{2c^2} \quad \text{for} \quad \omega > \frac{\gamma}{2c}, \\ -\frac{\gamma}{2c^2} \pm \frac{\omega}{c}\sqrt{\frac{\gamma^2}{4c^2\omega^2}-1} \quad \text{for} \quad \omega > \frac{\gamma}{2c}. \end{cases} \end{aligned} \qquad (2.33)$$

Hence, in contrast to the previous case, the given real frequency yields a critical damping $\omega_c = \frac{\gamma}{2c}$. For higher frequencies, we have the propagation of a wave with the phase velocity $\frac{\omega}{\mathrm{Re}k}$ and the decay of the amplitude in space while in the range of lower frequencies the wave cannot propagate and we have a pure damping.

2) The given real wave number $k$. Then it follows

$$\begin{aligned} \mathrm{Re}\omega &= \frac{ck}{\sqrt{2}}\sqrt{1+\sqrt{1+\left(\frac{\gamma}{c^2k}\right)^2}}, \\ \mathrm{Im}\omega &= -\frac{1}{\sqrt{2}}\frac{\frac{\gamma}{c}}{\sqrt{1+\sqrt{1+\left(\frac{\gamma}{c^2k}\right)^2}}}. \end{aligned} \qquad (2.34)$$

Here the wave propagates for any choice of $k$ with the phase velocity $\frac{\mathrm{Re}\omega}{k}$.

The above simple examples show that these problems in which we have a kind of a viscosity contribution corresponding to a lower time derivative in the governing equations the choice of the wave number $k$ as the independent variable in the dispersion relation leads to critical phenomena specified by the critical wave number $k_c$. In their range the relation $k \to \omega$ is not invertible. This is not the case when we choose the frequency $\omega$ as the independent variable.

On the other hand, problems in which we deal with spacial heterogeneities yielding the presence of lower spacial derivatives in the governing equations lead to critical phenomena specified by the critical frequency $\omega_c$. In their range the relation $\omega \to k$ is not invertible. It is not the case when we choose the wave number $k$ as the independent variable.

In this article we deal primarily with homogeneous problems in which we may have the damping caused by diffusion (viscosity). Consequently, it is natural to choose the frequency $\omega$ as the independent variable. We do not pay attention to this choice solely in cases when the relation between $\omega$ and $k$ is unconditionally invertible.

Certainly, equations which describe simultaneously viscosity and heterogeneity possess critical damping related to the frequency as well as to the wave number.

### 2.3 Rayleigh waves on plane boundaries of linear elastic homogeneous materials

The surface wave described in Section 2.1 is not typical for solids. As we see in the rest of this article, models of surface waves appear primarily as a combination of bulk waves which, in turn, follow as solutions of **hyperbolic** field equations.

We illustrate this statement by the classical example of the wave discovered by Rayleigh (1887) (for example, see as well: Kolsky (1953), Viktorov (1967), Achenbach (1973)).

Let us consider the linear elastic material described by the following equations for the unknown fields of velocity $\mathbf{v}(\mathbf{x},t)$ and deformation $\mathbf{e}(\mathbf{x},t)$

◊ momentum balance

$$\rho\frac{\partial \mathbf{v}}{\partial t} = \operatorname{div}\mathbf{T}, \tag{2.35}$$

◊ Hooke's law (constitutive relation for the Cauchy stress tensor $\mathbf{T}$)

$$\mathbf{T} = \lambda \operatorname{tr}\mathbf{e}\mathbf{1} + 2\mu\mathbf{e}, \tag{2.36}$$

where $\mathbf{e}$ is the Almansi-Hamel tensor of small deformations (i.e. $\|\mathbf{e}\| \ll 1$[2]), $\lambda$, $\mu$ are Lamé's moduli,

◊ kinematic compatibility condition

$$\frac{\partial \mathbf{e}}{\partial t} = \operatorname{symgrad}\mathbf{v}. \tag{2.37}$$

Let us first seek bulk waves, described by the above equations in the infinite medium. It is sufficient to exploit the case of the monochromatic wave

$$\begin{aligned}\mathbf{v} &= \mathbf{V}e^{i(\mathbf{k}\cdot\mathbf{x}-\omega t)} \equiv \mathbf{V}\exp\left[ik\left(\frac{1}{k}\mathbf{k}\cdot\mathbf{x} - c_{ph}t\right)\right], \quad k := \sqrt{\mathbf{k}\cdot\mathbf{k}}, \\ \mathbf{e} &= \mathbf{E}e^{i(\mathbf{k}\cdot\mathbf{x}-\omega t)} \equiv \mathbf{E}\exp\left[ik\left(\frac{1}{k}\mathbf{k}\cdot\mathbf{x} - c_{ph}t\right)\right], \quad c_{ph} := \frac{\omega}{k},\end{aligned} \tag{2.38}$$

where $\omega$ is the given frequency, $\mathbf{k}$ the **wave vector**, $k = \sqrt{\mathbf{k}\cdot\mathbf{k}}$ is the **wave number**, and $\mathbf{V}, \mathbf{E}$ are constant **amplitudes**.

Substitution in (2.37) yields the following relation between the amplitudes

$$-\omega\mathbf{E} = \frac{1}{2}\left(\mathbf{k}\otimes\mathbf{V} + \mathbf{V}\otimes\mathbf{k}\right). \tag{2.39}$$

Hence the momentum balance (2.35) leads to the following equation

$$-\omega\rho\mathbf{V} = \lambda\operatorname{tr}\mathbf{E}\mathbf{k} + 2\mu\mathbf{E}\mathbf{k} = -\lambda\frac{1}{\omega}\mathbf{V}\cdot\mathbf{k}\mathbf{k} - \mu\frac{1}{\omega}\left(\mathbf{V}\cdot\mathbf{k}\mathbf{k} + k^2\mathbf{V}\right). \tag{2.40}$$

[2] The norm of the tensor $\mathbf{e}$ is defined by its eigenvalues which are identical with the principal stretches $\lambda_{(i)}$

$$\det\left(\mathbf{e} - \lambda_{(i)}\mathbf{1}\right) = 0, \quad i = 1,2,3,$$
$$\|\mathbf{e}\| := \max\left\{\left|\lambda_{(1)}\right|, \left|\lambda_{(2)}\right|, \left|\lambda_{(3)}\right|\right\}.$$

We split this equation into the component parallel to $\mathbf{k}$ and perpendicular to $\mathbf{k}$

$$\left(\rho\omega^2 - (\lambda + 2\mu)\, k^2\right) \mathbf{V}\cdot\mathbf{k} = 0, \tag{2.41}$$

$$\left(\rho\omega^2 - \mu k^2\right) \left(\mathbf{V} - \frac{1}{k^2}\mathbf{V}\cdot\mathbf{k}\mathbf{k}\right) = 0. \tag{2.42}$$

Obviously, we obtain two solutions

$$1)\ \mathbf{V} = \frac{1}{k^2}\mathbf{V}\cdot\mathbf{k}\mathbf{k} \quad \Rightarrow \quad \left(\frac{\omega}{k}\right)^2 = \frac{\lambda + 2\mu}{\rho}, \tag{2.43}$$

$$2)\ \mathbf{V}\cdot\mathbf{k} = 0 \quad \Rightarrow \quad \left(\frac{\omega}{k}\right)^2 = \frac{\mu}{\rho}. \tag{2.44}$$

The first solution describes **longitudinal waves** or P **waves** (P for primary; the amplitude is parallel to the direction of propagation: $\mathbf{V}\|\mathbf{k}$) whose phase velocity is

$$c_L = \sqrt{\frac{\lambda + 2\mu}{\rho}}, \tag{2.45}$$

and this is not dependent on the frequency. Consequently, the wave is **non-dispersive** – each monochromatic wave propagates with the same velocity.

The second solution describes **transversal waves** or S **waves** (S for secondary or shear; the amplitude is perpendicular to the direction of propagation: $\mathbf{V}\cdot\mathbf{k} = \mathbf{0}$) whose phase velocity is

$$c_T = \sqrt{\frac{\mu}{\rho}}. \tag{2.46}$$

Consequently, this wave is also **non-dispersive.**

The system of equations of linear elasticity is **hyperbolic** provided $\mu > 0, \lambda + 2\mu > 0$. Then both velocities of the above described bulk waves are **real**.

Let us mention in passing that the separation of the eigenvalue problem presented above can be done in many other ways. Depending on a particular problem, they appear in the literature on the classical elasticity as well as in other problems of mechanics. Let us mention two of them.

If we introduce the field of **displacement** $\mathbf{u}(\mathbf{x}, t)$ the problem is described by equations (2.35), (2.36) due to the following relations

$$\mathbf{v} = \frac{\partial \mathbf{u}}{\partial t}, \quad \mathbf{e} = \frac{1}{2}\left(\mathrm{grad}\mathbf{u} + (\mathrm{grad}\mathbf{u})^T\right), \tag{2.47}$$

and the relation (2.37) is identically satisfied. Simultaneously, the vector $\mathbf{u}$ can be represented as a sum of a **potential** part $\mathbf{u}_L$ and a **solenoidal** part $\mathbf{u}_T$ which, as can be shown by a straightforward calculation, satisfy the following relations

$$\begin{aligned} \mathbf{u} &= \mathbf{u}_L + \mathbf{u}_T, \quad \mathrm{curl}\mathbf{u}_L = 0, \quad \mathrm{div}\mathbf{u}_T = 0, \\ \frac{\partial^2 \mathbf{u}_L}{\partial t^2} &= c_L^2 \nabla^2 \mathbf{u}_L, \quad \frac{\partial^2 \mathbf{u}_T}{\partial t^2} = c_T^2 \nabla^2 \mathbf{u}_T. \end{aligned} \tag{2.48}$$

Hence we obtain two wave equations whose solutions have the form of two waves discussed before.

One can use as well the following identity satisfied by any differentiable vector field (the so-called Helmholtz decomposition theorem)

$$\mathbf{u} = \mathrm{grad}\varphi + \mathrm{curl}\boldsymbol{\psi}, \tag{2.49}$$

where $\varphi, \boldsymbol{\psi}$ are the so-called **scalar and vector potentials**. Again one can easily show that they satisfy the following equations

$$\frac{\partial^2 \varphi}{\partial t^2} = c_L^2 \nabla^2 \varphi, \quad \frac{\partial^2 \boldsymbol{\psi}}{\partial t^2} = c_T^2 \nabla^2 \boldsymbol{\psi}. \tag{2.50}$$

We obtain again the same result.

We proceed to construct a solution for a semiinfinite linear elastic medium, i.e. a medium with a **boundary**. The presence of a boundary leads to important wave effects. First of all, there are bulk wave reflection and transmission phenomena. We shall not discuss them in any details in this article. An interested reader should consult, for instance, Aki and Richards (2002), Udias (1999).

In order to analyze the problem we need boundary conditions on the plane boundary. We choose the Cartesian coordinates with the $z$-axis perpendicular to the boundary. The boundary is defined by $z = 0$. We consider the case of the boundary free of loading. Hence

$$\mathbf{T}\mathbf{n}\big|_{z=0} = 0, \quad \mathbf{n} = -\mathbf{e}_z, \quad \mathbf{u}\big|_{z\to\infty} = 0, \tag{2.51}$$

where $\mathbf{e}_z$ is the unit basis vector of the $z$-axis. It means that the $z$-axis is oriented into the medium.

We seek the solution by splitting the displacement $\mathbf{u}$ into the potential and solenoidal parts, $\mathbf{u}_L, \mathbf{u}_T$. Then we make the following ansatz

$$\begin{aligned} \mathbf{u}_L &= A_L e^{-\gamma z} e^{i(kx-\omega t)} \mathbf{e}_x + B_L e^{-\gamma z} e^{i(kx-\omega t)} \mathbf{e}_z, \\ \mathbf{u}_T &= A_T e^{-\beta z} e^{i(kx-\omega t)} \mathbf{e}_x + B_T e^{-\beta z} e^{i(kx-\omega t)} \mathbf{e}_z, \end{aligned} \tag{2.52}$$

where $A_L, A_T, B_L, B_T$ are constant amplitudes. It means that we anticipate a progressive wave solution in the $x$-direction and the decay of the solution in the $z$-direction provided both $\gamma, \beta$ are positive. If this should not be the case a solution in the form of the surface wave would not exist. Obviously, the form of the solution (2.52) indicates that particles move in the $xz$-plane. This is the characteristic feature of Rayleigh waves.

Substitution of the above solution in the wave equations $(2.48)_{4,5}$ yields the compatibility conditions

$$\frac{\gamma^2}{k^2} = 1 - \frac{c_R^2}{c_L^2}, \quad \frac{\beta^2}{k^2} = 1 - \frac{c_R^2}{c_T^2}, \quad c_R := \frac{\omega}{k}. \tag{2.53}$$

Now conditions $(2.48)_{2,3}$ characterizing the potential and solenoidal part of displacement lead to the following form of the solution

$$\begin{aligned} B_L &= i\frac{\gamma}{k} A_L \quad \Rightarrow \quad \mathbf{u}_L = \left(\mathbf{e}_x + i\frac{\gamma}{k}\mathbf{e}_z\right) A_L e^{-\gamma z} e^{i(kx-\omega t)}, \\ B_T &= i\frac{k}{\beta} A_T \quad \Rightarrow \quad \mathbf{u}_T = \left(\mathbf{e}_x + i\frac{k}{\beta}\mathbf{e}_z\right) A_T e^{-\beta z} e^{i(kx-\omega t)}. \end{aligned} \tag{2.54}$$

In order to find the amplitudes we apply the boundary conditions (2.51). Obviously, the last condition, the so-called **Sommerfeld condition**, is satisfied identically provided both $\gamma$ and $\beta$ are positive. The remaining relations yield

$$\begin{aligned} \left(c_L^2 - 2c_T^2\right) \frac{\partial \mathbf{u}\cdot\mathbf{e}_x}{\partial x} + c_L^2 \frac{\partial \mathbf{u}\cdot\mathbf{e}_z}{\partial z}\bigg|_{z=0} &= 0, \\ \frac{\partial \mathbf{u}\cdot\mathbf{e}_x}{\partial z} + \frac{\partial \mathbf{u}\cdot\mathbf{e}_z}{\partial x}\bigg|_{z=0} &= 0. \end{aligned} \tag{2.55}$$

Substitution of (2.54) gives rise to the following homogeneous set for the unknown constants

$$\begin{aligned} \left(2 - \frac{c_R^2}{c_T^2}\right) A_L + 2A_T &= 0, \\ 2\frac{\gamma\beta}{k^2} A_L + \left(2 - \frac{c_R^2}{c_T^2}\right) A_T &= 0. \end{aligned} \tag{2.56}$$

Hence, if we choose the positive signs for $\beta, \gamma$ in roots following from the compatibility relations (2.53), the determinant leads to the following **Rayleigh dispersion relation**

$$\mathcal{P}_R := \left(2 - \frac{c_R^2}{c_T^2}\right)^2 - 4\sqrt{1 - \frac{c_R^2}{c_T^2}}\sqrt{1 - \frac{c_R^2}{c_L^2}} = 0. \tag{2.57}$$

Clearly, solutions $c_R$ of this equation are independent of the frequency $\omega$. In other words, Rayleigh waves in a semiinfinite medium are **nondispersive**.

We show (e.g. see Ben-Menahem and Singh (2000)) that this equation possesses one real solution $c_R < c_T$. Namely, it can be easily written in the form

$$f(y,n) := y^3 - 8y^2 + 8(3-2n)y - 16(1-n) = 0, \quad y := \frac{c_R^2}{c_T^2}, \quad n := \frac{c_T^2}{c_L^2} \tag{2.58}$$

This function is concave in the interval $(0,1)$ for $n \in (0, 0.5)$, and $f(0,n) = -16(1-n) < 0$, $f(1,n) = 1$. Consequently, there exists one root in the interval determined by the condition $c_R < c_T$. However, it may possess two other real roots bigger than 1 for $c_T/c_L \equiv \sqrt{\mu/(\lambda+2\mu)}$ bigger than app. 0.57. It is instructive to calculate this coefficient in terms of the Poisson's ratio $\nu$

$$\nu = \frac{\lambda}{2(\lambda+\mu)} \equiv \frac{1}{2}\frac{1 - 2\frac{\mu}{\lambda+2\mu}}{1 - \frac{\mu}{\lambda+2\mu}} \quad \Rightarrow \quad n \equiv \frac{\mu}{\lambda+2\mu} = \frac{1}{2}\frac{1-2\nu}{1-\nu}, \tag{2.59}$$

i.e.

$$0 < \nu \leq \frac{1}{2} \quad \Rightarrow \quad \frac{1}{2} > \frac{\mu}{\lambda+2\mu} \geq 0. \tag{2.60}$$

Consequently, the value $\sqrt{\mu/(\lambda+2\mu)} = \sqrt{0.5} \approx 0.707$ corresponds to the minimum value of $\nu$ equal to zero. For growing $\nu$ the fraction $\mu/(\lambda+2\mu)$ decays. Hence the range $0.570 < \sqrt{\mu/(\lambda+2\mu)} < 0.707$ is physically not empty. The function $f(y;\nu)$ is plotted in Fig. 5 for Poisson's ratio $\nu \in (0.1, 0.45)$.

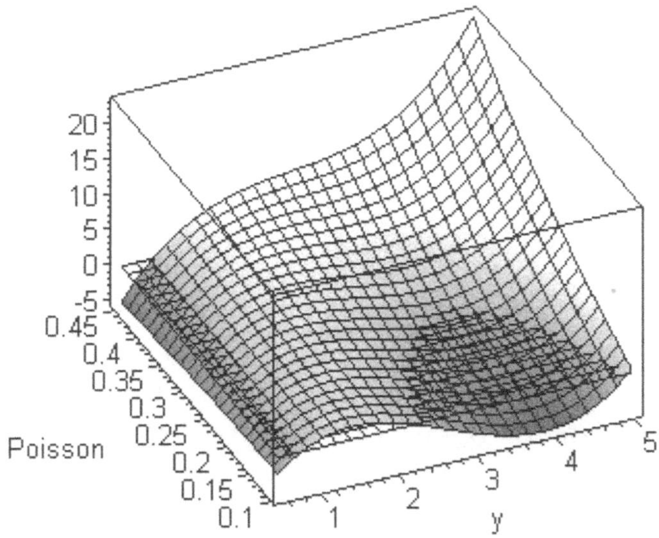


**Figure 5.** The function $f(y, \nu)$ with the indication of its zero values.

Clearly, there exist real solutions smaller than one for all values of $\nu$, and two real solutions bigger than one (i.e. the corresponding velocity is bigger than the shear velocity) for $\nu \gtrapprox 0.26309$. It is easy to check that these real solutions of (2.58) are not solutions of the dispersion relation (2.57) and they follow due to taking the square of this equation. However, they do become solutions of the dispersion relation if we substitute for $\beta$ and $\gamma$ values following from (2.53) with opposite signs: $\text{sign}\beta = -\text{sign}\gamma$. Such solutions of dispersion relation have been rejected as violating physical postulates (e.g. Achenbach (1973): "... these roots are extraneous; they arise from the rationalization process of squaring." See also: Kausel (2004)). In Section 4.4, we show that these solutions correspond to the so-called **leaky** waves intermediate between surface waves and bulk waves which loose their energy to both Rayleigh wave and two usual body waves (e.g. Nesvijski (2000)). They were observed experimentally and explain the existence of certain critical valuces of the angle of incident (Schroeder and Scott (2001)). Some details of properties of leaky waves can be also found in the books of Brekhovskikh and Godin (1998), Viktorov (1981), and possible practical applications in nondestructive testing in Ash and Paige (1985).[3]

[3]There is an elegant analytical method to check how many solutions a dispersion relation may possess and it is based on the *argument principle* of the complex analysis. Namely, an arbitrary functions $f(z)$ possesses the root $z_0$ if $f(z_0) = 0$. An analytic function possesses the pole $z_0$ of order $n$ if the function $(z - z_0)^n f(z)$ is holomorphic at $z_0$ (i.e. this function possesses the derivative in a neighborhood of $z_0$). According to the argument principle, if

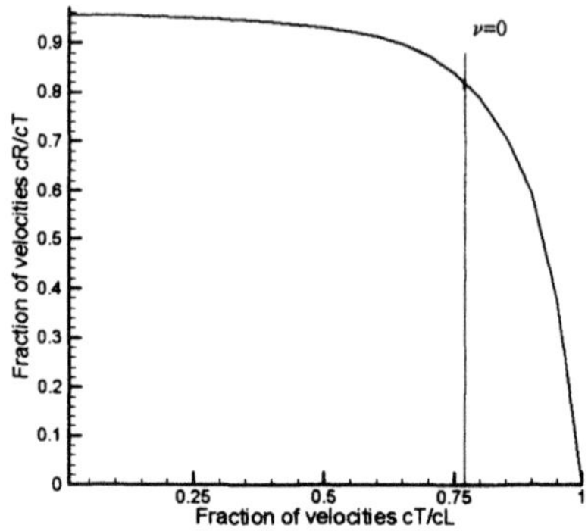


**Figure 6.** Dimensionless velocity of Rayleigh waves $\frac{c_R}{c_T}$ as a function of the fraction of shear and longitudinal velocities $\frac{c_T}{c_L}$

The solution slower than the shear velocity is shown in Fig. 6, where the fraction $c_R/c_T$ is plotted as a function of the fraction of bulk velocities $c_T/c_L$. Clearly, the value of the Rayleigh velocity lies very near the velocity of shear waves for most values of the material parameters.

Let us consider the orbits of particles of Rayleigh waves. According to $(2.48)_1$, and

---

$f(z)$ is meromorphic in a region $R$ enclosed by a contour $\Gamma$, then

$$\frac{1}{2\pi i}\int_{\Upsilon}\frac{dw}{w} = N - P, \quad w := f(z), \quad \Upsilon := f(\Gamma),$$

and $N$ is the number of complex roots of $f(z)$, and $P$ the number of poles in $\Gamma$. A meromorphic function $f(z)$ is defined as a single valued function analytic in all but possibly a discrete subset of its domain, and at those singularities it must go to infinity like a polynomial.

Achenbach (1973) defines the function

$$R(s) := \left(2s^2 - s_T^2\right)^2 + \left(s_L^2 - s^2\right)^{\frac{1}{2}}\left(s_T^2 - s^2\right)^{\frac{1}{2}} = 0,$$
$$s_L = \frac{1}{c_L}, \quad s_T = \frac{1}{c_T}, \quad s = \frac{1}{c},$$

where $s$ is the so-called slowness, and proves that the relation

$$\frac{1}{2\pi i}\int_C \frac{dv}{v} = N - P, \quad v = R(s),$$

where $C$ is a sufficiently large contour around the origin, yields a single solution for $c^2$ and it is real and positive.

The above method is particularly useful for complicated dispersion relations and its results considerably simplify a numerical effort.

(2.54) the displacement is given by the relations

$$u_x = \left(A_L e^{-\gamma z} + A_T e^{-\beta z}\right) e^{i(kx-\omega t)}, \tag{2.61}$$
$$u_z = i\left(\frac{\gamma}{k} A_L e^{-\gamma z} + \frac{k}{\beta} A_T e^{-\beta z}\right) e^{i(kx-\omega t)}.$$

We choose the motion in which the $x$-component propagates as $\cos(kx-\omega t)$. Then taking real parts of the above relations and eliminating time we obtain

$$\frac{(\mathrm{Re}u_x)^2}{\alpha_x^2} + \frac{(\mathrm{Re}u_z)^2}{\alpha_z^2} = 1, \tag{2.62}$$

where

$$\alpha_x := A_L e^{-\gamma z} + A_T e^{-\beta z}, \tag{2.63}$$
$$\alpha_z := \frac{\gamma}{k} A_L e^{-\gamma z} + \frac{k}{\beta} A_T e^{-\beta z},$$

and

$$\mathrm{Re}u_x = \alpha_x \cos(kx-\omega t), \quad \mathrm{Re}u_z = -\alpha_z \sin(kx-\omega t). \tag{2.64}$$

Hence, as in the case of water waves (comp. (2.16)), the orbit of each particle is an ellipse with semiaxes $|\alpha_x|, |\alpha_z|$. However, the motion is anticlockwise (retrograde). The semiaxes decay exponentially with the depth (see: Fig. 7).

Obviously, the vector of the amplitude of the Rayleigh wave lies on the $xz$-plane. This plane is defined by the direction of propagation $x$ (called in seismology the P direction, for primary), and the direction of decay $z$ (called in seismology SV-direction: for shear vertical). It is called in seismology the **plane of incidence** or the **sagittal plane** in materials sciences. In the general case, the progressive wave solution in the $x$-direction may also possess an amplitude in the direction perpendicular to the plane of incidence – the so-called SH-direction (for shear horizontal). This part of the solution would have the form

$$u_y(x,z,t) = Ce^{-\delta z} e^{i(kx-\omega t)}. \tag{2.65}$$

Such waves are called Love waves. It is easy to check that the boundary condition (2.51) yields immediately $C=0$. Consequently, Love waves do not exist in homogeneous semiinfinite media. As we see in the next Section they do exist in layers on semiinfinite media.

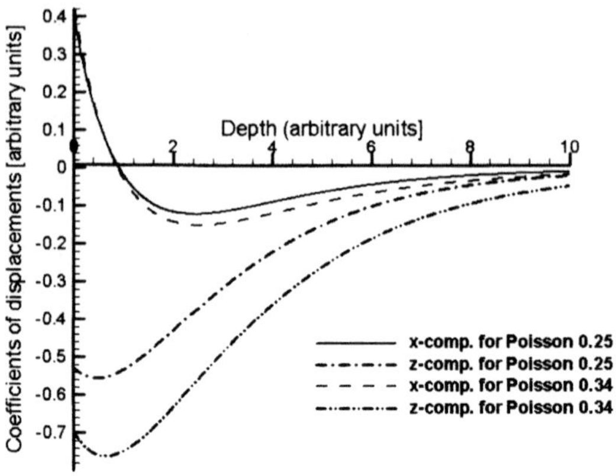


**Figure 7.** Coefficients of components of the displacement $\alpha_x, \alpha_z$ as functions of the depth $z$ for two values of the Poisson's ratio: $\nu = 0.25, 0.34$ (i.e. $c_T/c_L = 0.5773, 0.4924$).

BOX 3: *Rayleigh waves*

**Dispersion relation**

$$\mathcal{P}_R := \left(2 - \frac{c_R^2}{c_T^2}\right)^2 - 4\sqrt{1 - \frac{c_R^2}{c_T^2}}\sqrt{1 - \frac{c_R^2}{c_L^2}} = 0.$$

$$\text{where } c_R := \frac{\omega}{k}.$$

**Displacement**

$$\operatorname{Re} u_x = \alpha_x \cos(kx - \omega t), \quad \operatorname{Re} u_z = -\alpha_z \sin(kx - \omega t),$$

where

$$\alpha_x := A_L e^{-\gamma z} + A_T e^{-\beta z}, \quad \alpha_z := \frac{\gamma}{k} A_L e^{-\gamma z} + \frac{k}{\beta} A_T e^{-\beta z}.$$

## 2.4 Rayleigh waves on cylindrical boundaries

The problem of surface waves on cylindrical surfaces is quite common in geotechnics and appears, for instance, in the analysis of waves in boreholes. We present a simple example of a solid cylinder in order to see the influence of curvature on propagation of surface waves.

Let us investigate the problem of propagation of surface waves in an infinite cylinder of the radius $R$, i.e. waves which satisfy the following conditions:

1) the surface of the cylinder is free of stresses,

2) waves propagate in the circumferential direction and decay in the radial direction.

This problem has been solved in 1958 by Viktorov (1958b), Viktorov (1958a) and it demonstrates the influence of the curvature of the surface on the propagation of surface waves.

It is convenient to use the potentials $\varphi$ and $\boldsymbol{\psi}$ in this case. As the problem is two-dimensional the vector potential has solely one component $\psi_z$ in the direction of the axis of cylinder. These potentials satisfy the equations (2.50) which have the following form in **polar coordinates**

$$\begin{aligned} \frac{\partial^2 \varphi}{\partial t^2} &= c_L^2 \left\{ \frac{1}{r} \frac{\partial}{\partial r} \left( r \frac{\partial \varphi}{\partial r} \right) + \frac{1}{r^2} \frac{\partial^2 \varphi}{\partial \theta^2} \right\}, \\ \frac{\partial^2 \psi_z}{\partial t^2} &= c_T^2 \left\{ \frac{1}{r} \frac{\partial}{\partial r} \left( r \frac{\partial \psi_z}{\partial r} \right) + \frac{1}{r^2} \frac{\partial^2 \psi_z}{\partial \theta^2} \right\}, \end{aligned} \tag{2.66}$$

where $r, \theta$ are the radius and the angle coordinate, respectively.

It is convenient to consider the extension of the problem to the infinite interval for the angle $\theta$: $-\infty < \theta < \infty$. Then the axis $r = 0$ is the branch cut of the infinite order. This extension allows to consider the propagation of surface waves on an infinite plane of repetitions of the circumferential surface of the cylinder. In practical applications it may correspond, for instance, to spiral surface waves whose amplitude is weakly dependent on the $z$-variable.

In polar coordinates the physical radial and circumferential components of displacement are given by relations

$$u_r = \frac{\partial \varphi}{\partial r} + \frac{1}{r} \frac{\partial \psi_z}{\partial \theta}, \quad u_\theta = \frac{1}{r} \frac{\partial \varphi}{\partial \theta} - \frac{\partial \psi_z}{\partial r}. \tag{2.67}$$

We seek solutions in the form

$$\varphi = \Phi(r) e^{i(p\theta - \omega t)}, \quad \psi_z = \Psi(r) e^{i(p\theta - \omega t)}, \tag{2.68}$$

where $p$ plays the role of the angular wave number. It is related to the wave length $l$ by the relation $p = 2\pi R / l$. The usual wave number is then given by the relation $k = 2\pi / l \equiv p/R$. Due to the extension of domain for angle $\theta$, the variable $p$ can be considered as continuous. Certainly, the limit $p \to 0$ corresponds to very long waves ($l \gg R$) and the limit $p \to \infty$ is the limit of short waves.

Substitution of the above ansatz in the equations (2.66) yields the following Bessel equations

$$\begin{aligned}\frac{d^2\Phi}{d\xi_L^2}+\frac{1}{\xi_L}\frac{d\Phi}{d\xi_L}+\left(1-\frac{p^2}{\xi_L^2}\right)\Phi &= 0, \quad \xi_L := r\frac{\omega}{c_L}\equiv\frac{r}{R}p\frac{c}{c_L},\\ \frac{d^2\Psi}{d\xi_T^2}+\frac{1}{\xi_T}\frac{d\Psi}{d\xi_T}+\left(1-\frac{p^2}{\xi_T^2}\right)\Psi &= 0, \quad \xi_T := r\frac{\omega}{c_T}\equiv\frac{r}{R}p\frac{c}{c_T}, \quad c:=\frac{\omega R}{p}.\end{aligned} \tag{2.69}$$

The solution of this set which remains finite in the middle points of the cylinder has the following form

$$\Phi = AJ_p(\xi_L), \quad \Psi = BJ_p(\xi_T), \tag{2.70}$$

where $A, B$ are constants and $J_p$ is the Bessel function of order $p$.

The constants appearing in the solution should be determined from the boundary condition on the free surface $r = R$. These boundary conditions – the boundary is free of stresses – have the following form in the cylindrical coordinates

$$\begin{aligned}&\left(c_L^2-2c_T^2\right)\left(\frac{\partial^2\varphi}{\partial r^2}+\frac{1}{r}\frac{\partial\varphi}{\partial r}+\frac{1}{r^2}\frac{\partial^2\varphi}{\partial\theta^2}\right)+\\ &+2c_T^2\left(\frac{\partial^2\varphi}{\partial r^2}+\frac{1}{r}\frac{\partial^2\psi_z}{\partial r\partial\theta}-\frac{1}{r^2}\frac{\partial\psi_z}{\partial\theta}\right)=0, \quad \text{for} \quad r=R,\\ &\frac{2}{r}\frac{\partial^2\varphi}{\partial r\partial\theta}-\frac{1}{r^2}\frac{\partial\varphi}{\partial\theta}+\frac{1}{r^2}\frac{\partial^2\psi_z}{\partial\theta^2}-\frac{\partial^2\psi_z}{\partial r^2}+\frac{1}{r}\frac{\partial\psi_z}{\partial r}=0,\end{aligned} \tag{2.71}$$

i.e. the radial stress and the shear stress are zero at the circumferential surface $r = R$.

Substitution of (2.70) in (2.68) and, subsequently, in (2.71) yields a homogeneous set of equations for the constants $A, B$

$$\begin{aligned}b_{11}A+b_{12}iB &= 0,\\ b_{21}A+b_{22}iB &= 0,\end{aligned} \tag{2.72}$$

where[4]

$$b_{11} := -2pJ_p^L-\frac{c}{c_T}\frac{c_L}{c_T}p\left(2p+1-2\frac{c_T^2}{c_L^2}\right)J_{p+1}^L+\frac{c^2}{c_T^2}p^2J_{p+2}^L,$$

$$b_{12} := 2\left[\left(p^2-1\right)J_p^T-\frac{c}{c_T}p^2J_{p+1}^T\right], \quad b_{21} := p\left(2p-1\right)J_p^L-2\frac{c}{c_T}\frac{c_T}{c_L}p^2J_{p+1}^L,$$

$$b_{22} = 2p\left(p-1\right)J_p^T-\frac{c}{c_T}p\left(2p-1\right)J_{p+1}^T+\frac{c^2}{c_T^2}J_{p+2}^T, \tag{2.73}$$

[4]We have used here the following identity for Bessel functions

$$\frac{dJ_p(x)}{dx}=\frac{p}{x}J_p(x)-J_{p+1}(x).$$

$$J_p^L := J_p\left(p\frac{c}{c_L}\right), \quad J_p^T := J_p\left(p\frac{c}{c_T}\right),$$

and similarly for the other Bessel functions.

Consequently, for the existence of nontrivial solutions the determinant of the set (2.72) must be zero and we obtain the following dispersion relation for this problem

$$\mathcal{P}_{RC}(c,p) := b_{11}b_{22} - b_{12}b_{21} = 0. \tag{2.74}$$

This relation is plotted in Fig. 8 for the data $c_T/c_L = 0.4924$ which corresponds to Poisson's ratio $\nu = 0.34$. This value has been chosen to coincide with the value chosen by Viktorov (1981) in his numerical example. Not visible in the Figure 8 are infinitely many zero points for relative velocities bigger than $c/c_T > c_L/c_T = 2.0310$.

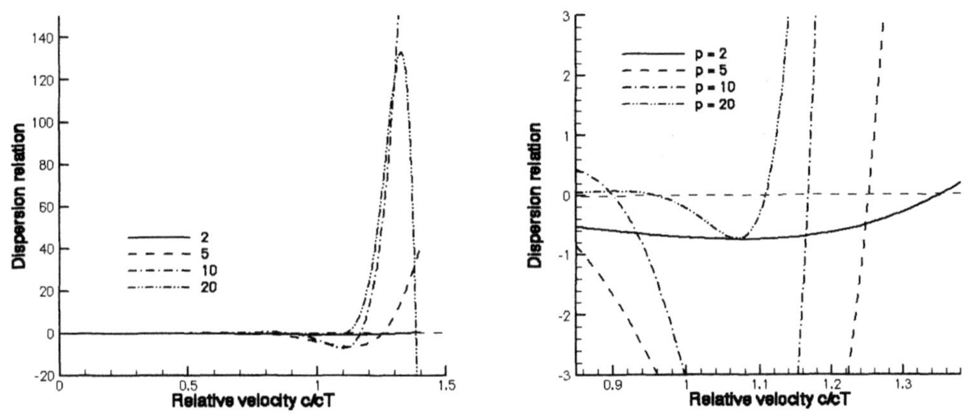


**Figure 8.** Zeros of the dispersion relation (2.74) for four different values of the wave number $p$ : 2, 5, 10, 20. The drawing on the right is the magnification of the part of the picture on the left.

Rayleigh waves on the cylinder are the waves whose velocity $c$ is bigger than $c_T$ and correspond to the first zero points above the point $\frac{c}{c_T} = 1$. This is demonstrated on the right hand side of Fig. 8 which is a magnification of the diagram on the left hand side. These solutions of the dispersion relation exist for $p > 1$ (i.e. $l < 2\pi R$) and decay to the Rayleigh velocity on the plane boundary for $p \to \infty$. This is a purely geometrical effect. The velocity of surface waves can be bigger than the velocity of shear waves because the path of the surface wave is longer than the path of the shear wave in the case of curved boundaries. For instance, for points of the boundary lying on the same diameter, the shear wave covers the distance $2R$ and the surface wave the distance $\pi R$. Consequently, the arrival time of the shear wave is shorter than the arrival time of the surface wave and, in this way, we fulfil the condition for constructive interference.

This effect was demonstrated by Viktorov who claims as well that an influence of the Poisson's ratio (i.e. relative bulk velocity $\frac{c_T}{c_L}$) is rather small.

Numerous zero points in this problem indicate that, in addition to Rayleigh waves, some additional waves may exist due to interactions of bulk waves with the surface. We shall discuss some aspects of this property of dispersion relations in Sec. 4.4.

The problem of propagation of surface waves on cylindrical surfaces has an important bearing in nondestructive testing of shafts. These circumferential waves belong then to the class of **guided** waves (e.g. Achenbach (1973), Liu and Qu (1998)). The dispersion relation for surface waves in a hollow cylinder with an inner shaft was obtained numerically by Valle et al. (1999). Valle, Qu and Jacobs compare the first five modes of propagation of these waves with the corresponding modes for the solid and hollow cylinders and with the Rayleigh wave on the plane boundary. It comes out that the behavior of waves in the layered cylinder coincides with this of the solid cylinder in the low frequency regime and it differs substantially in the range of high frequencies. In this range, the first mode tends asymptotically to the Rayleigh mode on the plane surface, while the second mode approaches in this limit a Rayleigh mode appearing on sliding interfaces. Further references can be found in the original work Valle et al. (1999).

## 3 Waves in a layer of an ideal fluid and Love waves on plane boundaries

### 3.1 Layer of an ideal compressible fluid on a semiinfinite rigid body

In order to appreciate the influence of heterogeneities on the propagation of surface waves, we investigate first a simple example of a layer of an **ideal fluid** $-\infty < x < \infty$, $0 \leq z \leq H$. The upper surface $z = H$ ($z$-axis is oriented upward in this case) is free of loading and the lower surface $z = 0$ is in contact with a **rigid body.** The problem is described by the equations of mass and momentum conservation

$$\frac{\partial \rho}{\partial t} + \rho_0 \mathrm{div}\mathbf{v} = 0, \quad \rho_0 \frac{\partial \mathbf{v}}{\partial t} = -\mathrm{grad}p, \quad p = p_0 + \kappa\left(\rho - \rho_0\right), \tag{3.1}$$

where $\rho_0, p_0$ are reference constant values of the mass density and pressure, respectively, and $\kappa$ denotes a constant compressibility coefficient of the fluid.

Simple manipulations lead to the following wave equation for the pressure $p$

$$\frac{\partial^2 p}{\partial t^2} = \kappa \nabla^2 p, \quad (x, z) \in (-\infty, \infty) \times (0, H). \tag{3.2}$$

The solution of this equation must satisfy the following boundary conditions

$$p\left(x, z = H, t\right) = 0, \quad v_z\left(x, z = 0, t\right) = 0. \tag{3.3}$$

As before, we seek the solution in the form of a monochromatic wave of the frequency $\omega$

$$p = \left(Ae^{irkz} + Be^{-irkz}\right) e^{i(kx - \omega t)}. \tag{3.4}$$

Then, due to the momentum balance, the second boundary condition can be replaced by the following one

$$\frac{\partial p}{\partial z}(x, z=0, t) = 0. \tag{3.5}$$

Substitution of (3.4) in the equation (3.2) yields the compatibility relation

$$r^2 = \frac{c_{ph}^2}{c^2} - 1, \quad c_{ph} := \frac{\omega}{k}, \quad c := \sqrt{\kappa}. \tag{3.6}$$

Simultaneously the evaluation of boundary conditions with the ansatz (3.4) yields the set of homogeneous algebraic relations for the constants $A$ and $B$

$$\begin{aligned} Ae^{irkH} + Be^{-irkH} &= 0, \\ A - B &= 0. \end{aligned} \tag{3.7}$$

Consequently, the determinant of this set must be equal to zero and we obtain the **dispersion relation**

$$\cos(rkH) = 0. \tag{3.8}$$

In order to obtain nontrivial solutions we have to require that $r$ is real. This means, however, that the dispersion relation given in term of the periodic function yields infinitely many solutions. Each solution is called a **mode of propagation**. This is the characteristic feature of **heterogeneous** systems.

Simultaneously, it follows from (3.8) that the phase velocities $c_{ph}$ are bigger than the velocity of propagation $c$ appearing in the wave equation for the pressure (3.2) ($\kappa \equiv c^2$), and that they go to infinity as the frequency approaches certain critical values. If we require that waves of the form (3.4) do exist then this seems to violate the basic property of the hyperbolic problem. This result follows from the assumption that the foundation of the fluid is a **rigid body** in which all disturbances propagate with an **infinite velocity**. As we see further a modification of the boundary condition $(3.3)_2$ for the case of contact with an elastic body which we make for the so-called Love waves eliminates this paradox.

In details, solution of the equation (3.8) yields immediately the following relation between the phase velocity and the frequency

$$c_{ph} = \frac{c}{\sqrt{1 - \frac{\omega_{cr}^2}{\omega^2}}}, \quad \omega_{cr} := \left(n + \frac{1}{2}\right)\pi\frac{c}{H}, \quad n = 1, 2, \ldots \tag{3.9}$$

This relation is illustrated in Figure 9.

### 3.2 Love waves on plane boundaries

The paradox of infinite phase velocities does not appear anymore in the case of surface waves which propagate in an elastic layer over an elastic half-space. Transversal waves in such a system have been described by Love (1911). We proceed to present briefly these results. They form the simplest illustration of the problem of surface waves in heterogeneous solids.

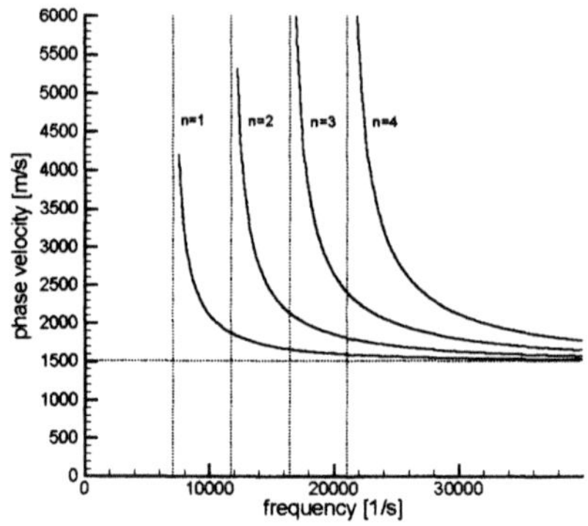


**Figure 9.** Phase velocity for a layer of an ideal fluid. Numerical data: $c = 1500\frac{m}{s}, H = 1m$. Modes: $n = 1, 2, 3, 4$ are shown in the Figure.

We consider the propagation of a wave whose amplitude has solely an $\mathbf{e}_y$-component $u_y \equiv \mathbf{u} \cdot \mathbf{e}_y$ (perpendicular to the $(x, z)$-plane; hence it corresponds to SH amplitude for waves with P-SV incident plane in seismological terminology). The body consists of a layer of thickness $H$ in the $z$-direction in which the mass density is $\rho'$ and the velocity of shear waves is $c'_T$. This layer is connected to the elastic half-space $z \leq 0$ whose mass density is $\rho$ and the velocity of shear waves $c_T$. We seek the solution of wave equations

$$\begin{aligned} \frac{\partial^2 u'_y}{\partial t^2} &= c'^2_T \nabla^2 u'_y, \quad 0 < z < H, \\ \frac{\partial^2 u_y}{\partial t^2} &= c^2_T \nabla^2 u_y, \quad z < 0, \end{aligned} \tag{3.10}$$

in the form

$$\begin{aligned} u'_y &= \left(A' e^{iks'z} + B' e^{-iks'z}\right) e^{i(kx-\omega t)} \equiv \\ &\equiv 2\left(\mathrm{Re}A' \cos ks'z - \mathrm{Im}A' \sin ks'z\right) e^{i(kx-\omega t)}, \\ u_y &= B e^{ksz} e^{i(kx-\omega t)}, \end{aligned} \tag{3.11}$$

i.e. in the form of a monochromatic wave which propagates in the direction of the $x$-axis with the frequency $\omega$, wave number $k$ in this direction, and with the phase velocity $c := \frac{\omega}{k}$. The wave should decay in the $z$-direction, i.e. $s$ must be positive. We check now if the ansatz (3.11) can fulfil equations (3.10), and the following boundary conditions

1) shear stress on the plane $z = H$ is equal to zero, i.e.

$$\frac{\partial u'_y}{\partial z}(x, z = H, t) = 0, \tag{3.12}$$

2) shear stress and the displacement must be continuous on the interface $z = 0$

$$\begin{aligned} \rho' c_T'^2 \frac{\partial u_y'}{\partial z}(x, z = 0, t) &= \rho c_T^2 \frac{\partial u_y}{\partial z}(x, z = 0, t), \\ u_y'(x, z = 0, t) &= u_y(x, z = 0, t). \end{aligned} \tag{3.13}$$

Substitution of the ansatz (3.11) in equations (3.10) yields

$$s'^2 = \frac{c^2}{c_T'^2} - 1, \quad s^2 = 1 - \frac{c^2}{c_T^2}, \quad c \equiv \frac{\omega}{k}. \tag{3.14}$$

Boundary condition (3.12) leads immediately to the following relation for the displacement in the layer

$$u_y' = 2\mathrm{Re}A' \frac{\cos(ks'(H - z))}{\cos(ks'H)} e^{i(kx - \omega t)}. \tag{3.15}$$

Then the boundary conditions (3.13) yield a homogeneous set of two algebraic relations for the constants $\mathrm{Re}A', B$. Consequently, its determinant must be zero and this condition yields the **Love dispersion relation**

$$\omega = \frac{c}{Hs'}\left[\arctan\left(\frac{\rho c_T^2 s}{\rho' c_T'^2 s'}\right) + n\pi\right], \quad n = 1, 2, 3, \ldots, \tag{3.16}$$

where both $s$, and $s'$ must be real, i.e.

$$c_T' \leq c \leq c_T. \tag{3.17}$$

This is the condition for the *existence* of Love waves. Hence the Love waves can propagate solely in layers which are **softer than the foundation**. In addition there exist infinitely many modes of propagation whose existence is limited from below by corresponding **critical frequencies**. All these modes are dispersive because the phase velocities depend on the frequency given by the inverse relation to (3.16).

Box 4: *Love waves*

**Dispersion relation**

$$\omega = \frac{c}{Hs'}\left[\arctan\left(\frac{\rho c_T^2 s}{\rho' c_T'^2 s'}\right) + n\pi\right], \quad n = 1, 2, 3, \ldots,$$

where

$$s^2 = 1 - \frac{c^2}{c_T^2}, \quad s'^2 = \frac{c^2}{c_T'^2} - 1, \quad c_T' \leq c \leq c_T.$$

**Displacement**

$$\begin{aligned} u_y' &= u_{y0}' \frac{\cos\left(ks'\left(H - z\right)\right)}{\cos\left(ks'H\right)} e^{i(kx - \omega t)}, \\ u_{y0}' &= 2\mathrm{Re}A'. \end{aligned}$$

In Fig. 10 we show an example of the solution of relation (3.16) for the following data:

$$c_T = 5\frac{km}{s}, \quad c_T' = 3\frac{km}{s}, \quad \frac{\rho'}{\rho} = 0.875, \quad H = 10km. \tag{3.18}$$

We plot the relative velocities $c_{ph}/c_T$ as functions of the frequency $\omega$ for four values of $n$.

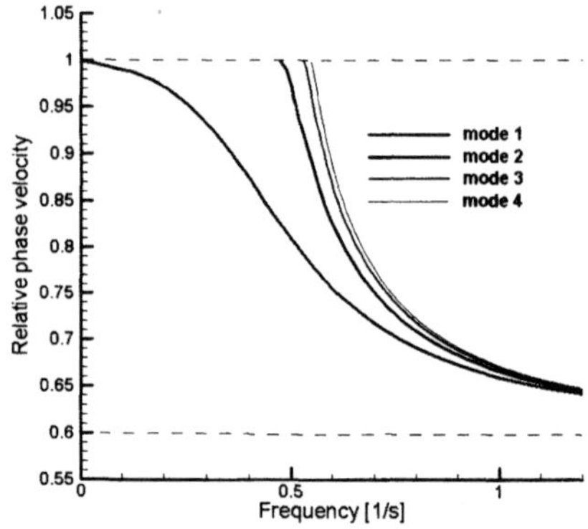


**Figure 10.** Relative phase velocities for four modes ($n = 0, 1, 2, 3$) of the Love wave given by relation (3.16).

The data were chosen to be identical with the data used in the Box 7.2. by Aki and Richards (2002). For these data the phase and group velocities of the first mode are shown in Fig. 11.

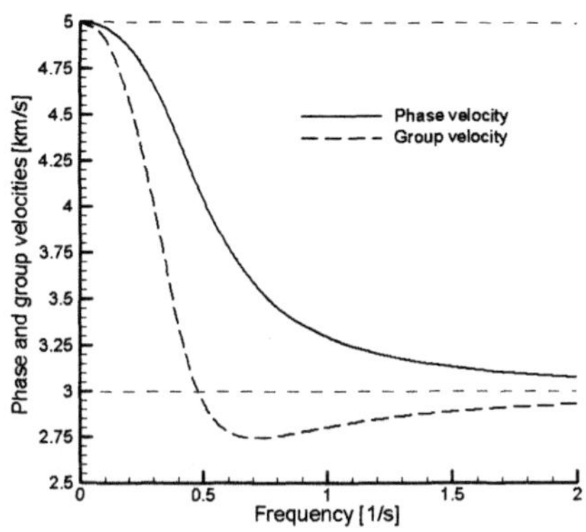


**Figure 11.** The phase and group velocities of the first mode of Love wave.

The group velocity as a function of the phase velocity was calculated by means of the following relation

$$c_g = \frac{\frac{d\omega}{dc_{ph}} c_{ph}^2}{\frac{d\omega}{dc_{ph}} c_{ph} - \omega}, \tag{3.19}$$

which follows immediately from the definitions. Then we can directly use the relation (3.16) and invert variables $c_{ph} \to \omega$ in the graphical program. In this way, we avoid a numerical differentiation.

For completeness, we present in Fig. 12 the wave number $k$ as a function of frequency for the first mode of the Love wave. Clearly, the relation is monotonous and, according to our earlier remarks, we may choose $\omega$ as well as $k$ as independent.

### 3.3 Arrival times for packages of waves – example

Bearing above results in mind, we investigate an example for the procedure of calculating the arrival times of a package of dispersive waves. This problem of propagation of packages of dispersive waves was already indicated in Sec. 2.1. Instead of assuming a narrow band structure we consider a problem in which an arbitrary displacement $u(x, z, t)$ has the following Fourier representation

$$u(x, z, t) = \frac{1}{2\pi} \int_{-\infty}^{\infty} A(\omega; z) e^{i(kx - \omega t)} d\omega, \quad k = k(\omega), \tag{3.20}$$

where a dependence on the variable $z$ is parametric. In order to evaluate this integral, we make the assumption that the amplitude $A(\omega; z)$ is a slowly varying function of $\omega$

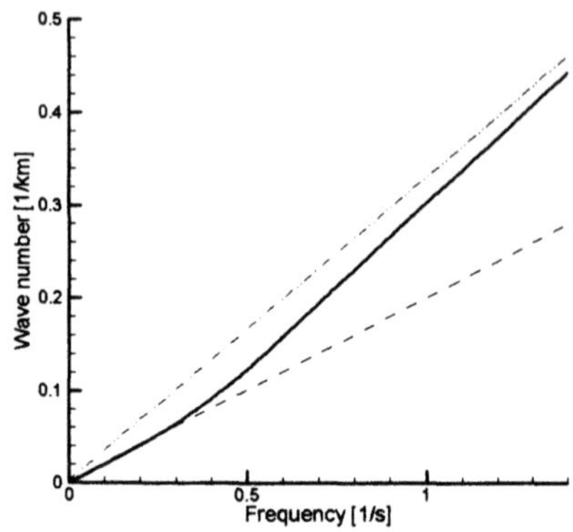


**Figure 12.** The dispersion relation $k = k(\omega)$ for the first mode of the Love wave. The dashed line corresponds to the initial velocity, i.e. $k_0 = \omega/5$, and dashdotted line to the limit velocity in infinity $k_\infty = \omega/3$.

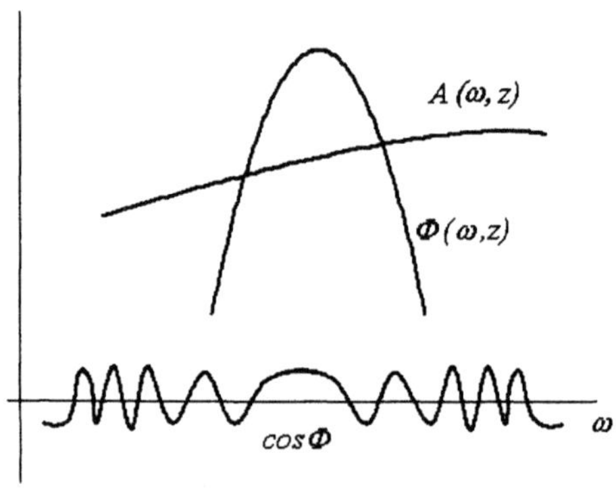


**Figure 13.** Illustration of the principle of stationary phase. Changes of the amplitude $A$, the trigonometric contribution to the integral $\cos\Phi$ (or $\sin\Phi$), and the phase $\Phi$: $\Phi := kx - \omega t$.

in comparison with variations caused by the phase $\Phi = kx - \omega t$. This assumption is schematically illustrated in Figure 13.

In such a case, similarly to the method of saddle point of integration, we can assume that the essential contribution to the integral comes from the vicinity of **stationary points** of the phase $\Phi$. For a chosen point $x$ and a chosen instant of time $t$ these points, say $\omega_s$, are defined by the relation

$$\frac{d\Phi}{d\omega}(x, t, \omega_s) = 0 \quad \Rightarrow \quad \frac{dk}{d\omega}(\omega_s)\, x - t = 0 \quad \Rightarrow \quad c_g(\omega_s) = \frac{x}{t}. \tag{3.21}$$

If there is a single such point then we can expand the phase

$$\Phi \approx k_s x - \omega_s t + \frac{1}{2} \left. \frac{d^2 k}{d\omega^2} \right|_{\omega=\omega_s} (\omega - \omega_s)^2, \tag{3.22}$$

and the integral (3.20) becomes

$$u(x,z,t) \approx \frac{1}{2\pi} A(\omega_s; z) e^{i(k_s x - \omega_s t)} \int_{-\infty}^{\infty} \exp\left[ -i \frac{1}{2c_g^2} \frac{dc_g}{d\omega} x (\omega - \omega_s)^2 \right] d\omega, \tag{3.23}$$

where $k_s = k(\omega_s)$ and $c_g$ as well as $\frac{dc_g}{d\omega}$ are evaluated at $\omega_s$. Easy integration[5] yields

$$u(x,z,t) = \frac{A(\omega_s; z) c_g}{\sqrt{2\pi \left| \frac{dc_g}{d\omega} \right| x}} e^{i\left(k_s x - \omega_s t - \varsigma \frac{\pi}{4}\right)}, \quad \varsigma := \operatorname{sign} \frac{dc_g}{d\omega}. \tag{3.24}$$

In the case of many stationary points the above formula transforms into the sum of contributions for each point.

If the derivative of the group velocity at the stationary point vanishes we have to make an expansion to the third order and it follows

$$u(x,z,t) = \frac{A(\omega_s; z)}{\left( \frac{9}{2} \frac{1}{c_g^2} \frac{d^2 c_g}{d\omega^2} x \right)^{\frac{1}{3}} \Gamma\left(\frac{2}{3}\right)} e^{i(k_s x - \omega_s t)}. \tag{3.25}$$

The phase determined in the above described manner is called the **Airy phase.** We proceed to present a numerical example of its evaluation from a given dispersion relation. We do so for the above presented Love surface waves. We choose the data (3.18) and the distance from the source $x = 500$ km. These are the same data as those chosen by Aki and Richards (2002) in their presentation (Box 7.2).

Let us begin from the plots of the phase $\Phi$ for different values of time (Fig.14). The wave number $k$ and the frequency $\omega$ are related by the implicit dispersion relation (3.16).

Obviously, the first signal arrives at $x = 500$ km after $t = 100$ seconds because the fastest disturbance corresponds to the monochromatic wave traveling with the velocity 5 km/s (see: Fig. 11). The last arrival at the instant approximately $t = 181$ seconds corresponds to the minimum group velocity $\min c_g \approx 2.74$ km/s.

[5]The integral can be written as follows

$$\int_{-\infty}^{\infty} \exp\left[ -i \frac{1}{2c_g^2} \frac{dc_g}{d\omega} x (\omega - \omega_s)^2 \right] d\omega = 2c_g \sqrt{2 \left( \left| \frac{dc_g}{d\omega} \right| x \right)^{-1}} \int_0^{\infty} \exp\left(-i\varsigma\sigma^2\right) d\sigma =$$

$$= 2c_g \sqrt{2 \left( \left| \frac{dc_g}{d\omega} \right| x \right)^{-1}} (1 - \varsigma i) \int_0^{\infty} \cos \sigma^2 d\sigma = c_g \sqrt{2\pi \left( \left| \frac{dc_g}{d\omega} \right| x \right)^{-1}} e^{-\varsigma i \frac{\pi}{4}},$$

where we have used the formula for the Fresnel integral

$$\int_0^{\infty} \cos \sigma^2 d\sigma = \frac{\sqrt{2\pi}}{4}.$$

Box 5: *Arrival times of wave packages*

**Procedure for calculation of arrivals:**

1) determination of the dispersion relation $k = k(\omega)$, phase velocity $c_{ph} = \frac{\omega}{k}$, and the group velocity $c_g = \left(\frac{dk}{d\omega}\right)^{-1}$,

2) for a chosen point $x$, determination of the phase $\Phi = k(\omega)x - \omega t$ for different instances of time $t$,

3) determination of stationary points $\omega_s$ : $\frac{d\Phi}{d\omega}(\omega_s) = 0 \quad \Rightarrow \quad \omega_s = \omega_s(t; x)$,

4) calculation of the derivative $\frac{dc_g}{d\omega}(\omega = \omega_s)$,

5) calculation of the amplitude $u(t; x)$ as a function of time.

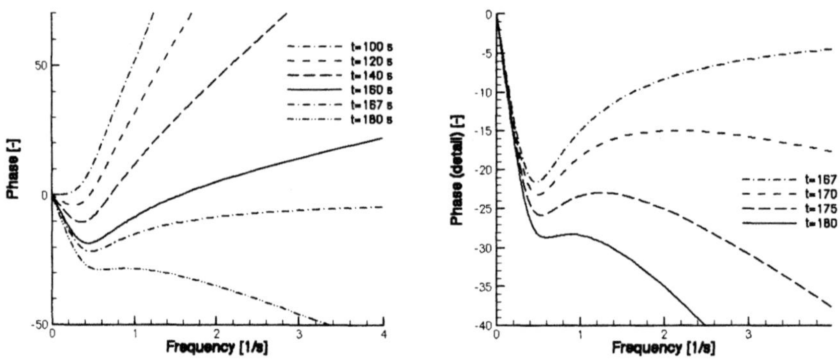


**Figure 14.** Plots of the phase $\Phi := kx - \omega t$ as a function of frequency $\omega$ for the data (3.18), $x = 500$ km, and different instances of time $t$. On the right hand side, we present some details in the important region of time (see: text).

By means of the above curves for the phase we can find the location of extrema of phases for different times. These are shown in Fig. 15. There is no minimum before the first arrival at $t = 100$ s. Then the value of frequency $\omega_s$ at which the phase possesses a minimum grows to the limit value at the arrival time with the minimum group velocity $t = 181$ s. In the time range between the arrival time with the limit phase velocity for

$\omega \to \infty$ (i.e. $500/3 = 167$ s.) and the largest arrival time $t = 181$ s the phase possesses also a maximum. Corresponding frequencies are also plotted in Fig. 15.

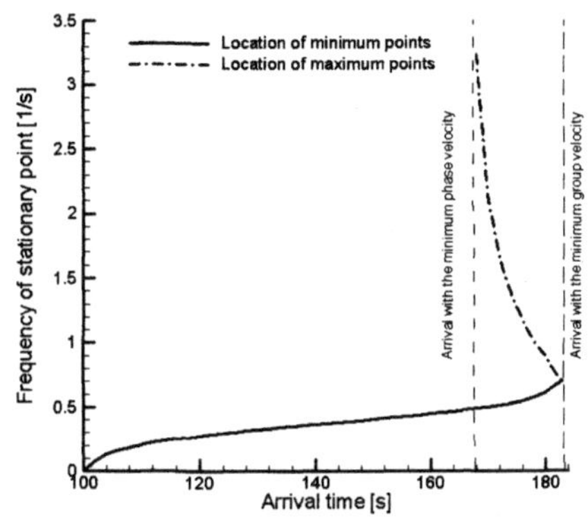


**Figure 15.** Frequencies $\omega_s$ of stationary points of the phase $\Phi$ as a function of arrival times.

The monotonous relation between the time $t$ and the frequency $\omega_s$ of the phase minimum is sometimes used by seismologists to replace on seismograms the time scale by a scale based on the frequency.

The rest of steps required in the procedure presented in Box 5 we perform numerically. We choose the time step to be 2 seconds and calculate the group velocity and its derivative at frequencies $\omega_s$ for different instances of time, and subsequently the amplitude. The results are presented in Fig. 16.

The quality of this calculation is rather poor and we can easily spot reasons for numerical problems. First of all, in order to evaluate the integral (3.20) we have used solely the approximation given by the formula (3.24). This is, of course, very bad in the vicinity of points $t = 100$ s and $t = 167$ s, where $\frac{dc_g}{d\omega}$ is equal to zero. It is visible in particular at the first instant of arrival time $t = 100$ s where the amplitude should be zero (Fig. 16 right). The procedure of calculation should be changed in such regions to this based on the formula (3.25) or to even higher order approximations. Simultaneously, in the range of the Airy phase where the variations of the amplitude are very strong the time step should be made much smaller to get a sufficient accuracy.

However, in spite of their bad quality the above results reflect the most important features of the arrival of a package of dispersive waves. An example of a real seismogram from the earthquake Oklahoma is shown in Fig. 17.

### 3.4 Rayleigh waves in a layer of elastic material

Now we investigate a problem similar to the propagation of Love waves, i.e. the propagation in a semiinfinite elastic body with a layer of thickness $H$ and different material properties but we assume the amplitude of the wave to lie in P-SV-plane rather than in

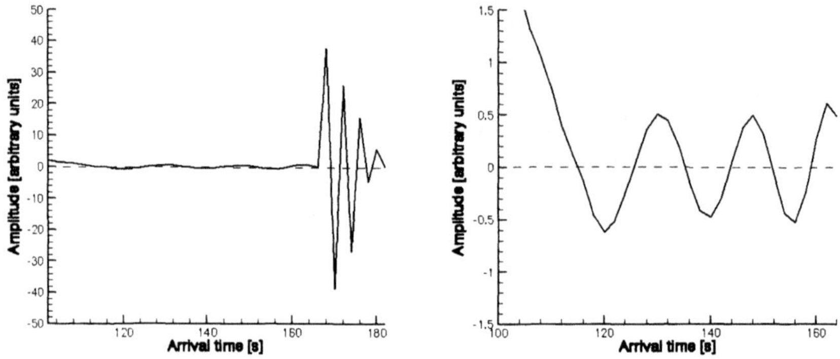


**Figure 16.** Amplitude as a function of arrival time. The left panel presents the amplitude in the whole range of existence of a nonzero amplitude while the right panel is a magnification of the amplitude in the range before the arrival with the limit group velocity $c_g = 3$ km/s ($t \approx 167$ s).

the SH-direction. This is the same assumption as in the case of Rayleigh waves on the plane boundary of the semiinfinite medium.

The material properties are as follows. The mass density of the semiinfinite medium is $\rho$, the velocities of the longitudinal and transversal waves $c_L, c_T$, and in the layer they are $\rho', c'_L, c'_T$. We use the potentials for the description of displacements, i.e.

$$u_x = \frac{\partial \varphi}{\partial x} + \frac{\partial \psi_y}{\partial z}, \quad u_z = \frac{\partial \varphi}{\partial z} - \frac{\partial \psi_y}{\partial x} \quad \text{for} \quad z \leq 0,$$

$$u'_x = \frac{\partial \varphi'}{\partial x} + \frac{\partial \psi'_y}{\partial z}, \quad u'_z = \frac{\partial \varphi'}{\partial z} - \frac{\partial \psi'_y}{\partial x} \quad \text{for} \quad 0 \leq z \leq H. \tag{3.26}$$

Anticipating the existence of a surface wave in the semiinfinite medium, we make the following ansatz for solution

$$\varphi = A e^{krz} e^{i(kx - \omega t)}, \quad \psi_y = B e^{ksz} e^{i(kx - \omega t)},$$

$$\begin{aligned} \varphi' &= \left( A'_1 \sin(kr'z) + A'_2 \cos(kr'z) \right) e^{i(kx - \omega t)}, \\ \psi'_y &= \left( B'_1 \sin(ks'z) + B'_2 \cos(ks'z) \right) e^{i(kx - \omega t)}. \end{aligned} \tag{3.27}$$

These potentials fulfil wave equations of the form (2.50) with the appropriate material parameters. This requirement yields

$$r^2 = 1 - \frac{c^2}{c_L^2}, \quad s^2 = 1 - \frac{c^2}{c_T^2}, \quad c := \frac{\omega}{k},$$

$$r'^2 = \frac{c^2}{c_L'^2} - 1, \quad s'^2 = \frac{c^2}{c_T'^2} - 1. \tag{3.28}$$

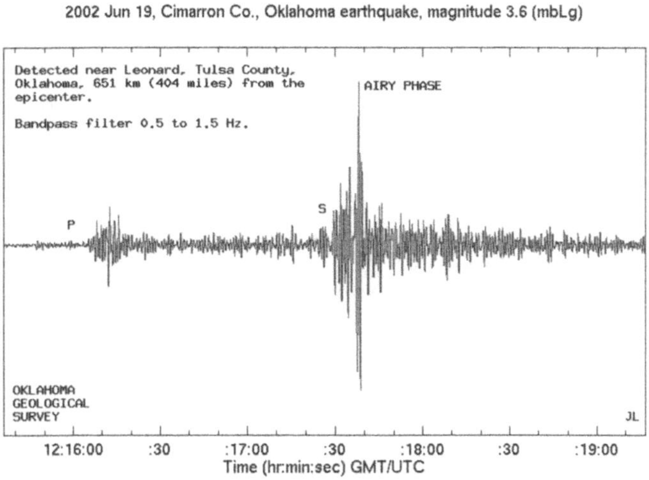


**Figure 17.** Arrivals recorded by Oklahoma (Tulsa County) earthquake in 2002 illustrating the notion of the Airy phase (Oklahoma Geological Survey, A State Agency for Research and Public Services).

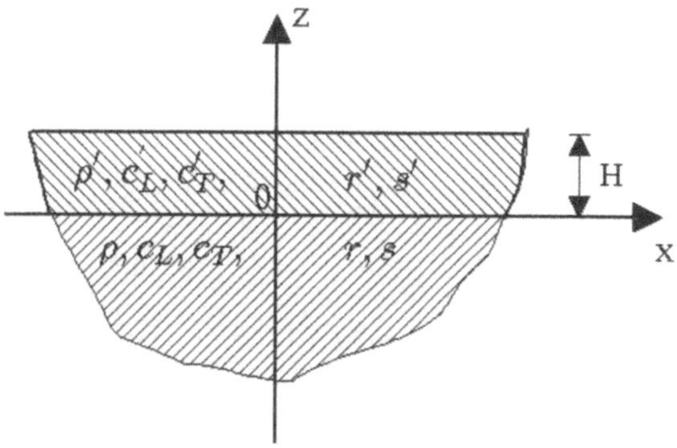


**Figure 18.** The geometry considered in the Rayleigh wave problem in the medium with the layer.

Solutions of the problem should fulfil the following boundary conditions
1) The stress vector on the free surface $z = H$ must be zero

$$\begin{aligned} \rho' c_L'^2 \left( \frac{\partial^2 \varphi'}{\partial x^2} + \frac{\partial^2 \varphi'}{\partial z^2} \right) - \rho' c_T'^2 \left( \frac{\partial^2 \varphi'}{\partial x^2} + \frac{\partial^2 \psi_y'}{\partial x \partial z} \right) &= 0, \\ \rho' c_T'^2 \left( 2\frac{\partial^2 \varphi'}{\partial x \partial z} + \frac{\partial^2 \psi_y'}{\partial z^2} - \frac{\partial^2 \psi_y'}{\partial x^2} \right) &= 0, \end{aligned} \tag{3.29}$$

2) The displacement and the stress vector must be continuous on the interface $z = 0$

$$\frac{\partial \varphi}{\partial x} + \frac{\partial \psi_y}{\partial z} = \frac{\partial \varphi'}{\partial x} + \frac{\partial \psi'_y}{\partial z}, \quad \frac{\partial \varphi}{\partial z} - \frac{\partial \psi_y}{\partial x} = \frac{\partial \varphi'}{\partial z} - \frac{\partial \psi'_y}{\partial x}, \tag{3.30}$$

$$\rho c_L^2 \left( \frac{\partial^2 \varphi}{\partial x^2} + \frac{\partial^2 \varphi}{\partial z^2} \right) - \rho c_T^2 \left( \frac{\partial^2 \varphi}{\partial x^2} + \frac{\partial^2 \psi_y}{\partial x \partial z} \right) =$$

$$= \rho' c_L'^2 \left( \frac{\partial^2 \varphi'}{\partial x^2} + \frac{\partial^2 \varphi'}{\partial z^2} \right) - \rho' c_T'^2 \left( \frac{\partial^2 \varphi'}{\partial x^2} + \frac{\partial^2 \psi'_y}{\partial x \partial z} \right), \tag{3.31}$$

$$\rho c_T^2 \left( 2 \frac{\partial^2 \varphi}{\partial x \partial z} + \frac{\partial^2 \psi_y}{\partial z^2} - \frac{\partial^2 \psi_y}{\partial x^2} \right) = \rho' c_T'^2 \left( 2 \frac{\partial^2 \varphi'}{\partial x \partial z} + \frac{\partial^2 \psi'_y}{\partial z^2} - \frac{\partial^2 \psi'_y}{\partial x^2} \right), \tag{3.32}$$

3) The solution vanishes for $z \to -\infty$.

These conditions form a homogeneous set of algebraic equations for constants $iA, B, iA'_1, B'_1, iA'_2, B'_2$. The matrix of coefficients has the following form

$$\mathbf{D}_L := \tag{3.33}$$

$$= \begin{bmatrix} 0 & 1 & s' & 0 & -1 & -s \\ r' & 0 & 0 & -1 & -r & 1 \\ 0 & \frac{\rho'}{\rho}\left( \frac{c^2}{c_T^2} - 2\frac{c_T'^2}{c_T^2} \right) & -2\frac{\rho'}{\rho}\frac{c_T'^2}{c_T^2} s' & 0 & 2 - \frac{c^2}{c_T^2} & 2s \\ 2\frac{\rho'}{\rho}\frac{c_T'^2}{c_T^2} r' & 0 & 0 & 2 - \frac{c^2}{c_T'^2} & -2r & -\left( 2 - \frac{c^2}{c_T^2} \right) \\ \left( 2 - \frac{c^2}{c_T'^2} \right) S_{r'} & \left( 2 - \frac{c^2}{c_T'^2} \right) C_{r'} & s' C_{s'} & -s' S_{s'} & 0 & 0 \\ 2r' C_{r'} & -2r' S_{r'} & \left( 2 - \frac{c^2}{c_T'^2} \right) S_{s'} & \left( 2 - \frac{c^2}{c_T'^2} \right) C_{s'} & 0 & 0 \end{bmatrix},$$

where for typographical reasons we have introduced the notation

$$S_{r'} = \sin kr'H, \quad C_{r'} = \cos kr'H, \quad S_{s'} = \sin ks'H, \quad C_{s'} = \cos ks'H. \tag{3.34}$$

Consequently, the determinant of this matrix yields the dispersion relation

$$\mathcal{P}_{RL} := \det \mathbf{D}_L = 0. \tag{3.35}$$

Solutions of this kind of equations are discussed in other contributions to this volume and, for this reason, we skip here a complicated analysis. However, two properties of this problem are immediately visible in the above relations. First of all, if there exists a solution of the above dispersion relation then, similarly to Love waves, it is **dispersive** and there exist infinitely many modes due to the contributions of trigonometric functions to the equation (3.35). Secondly, the relations (3.28) and the boundary condition in infinity yield the following necessary condition for the existence of solutions in the form of surface waves

$$c_L \geq c_T \geq c. \tag{3.36}$$

Further we discuss in some details the problem of existence of solutions solely for the much simpler problem of Stoneley waves.

# 4 Stoneley waves

## 4.1 Interface of two semiinfinite elastic solids

Properties of waves on an interface were considered for the first time by Stoneley (1924). The interface investigated in this work is a classical ideal plane boundary between two linearly elastic isotropic solids and boundary conditions follow from the so-called jump conditions on the ideal singular surface. This means that all surface effects like surface tension, boundary layers, intermediate layers of a different material, etc. are neglected. In contrast to Love waves, amplitudes of the wave should decay with the distance from the interface in both directions: for the coordinate $z$ going to $+\infty$ and to $-\infty$ (see: Fig. 19). Formally, one can construct the solution of the problem by taking the infinite limit of depth $H$ of the layer in the solution for Rayleigh waves in a semiinfinite medium with the layer. However, following the work of Stoneley, we formulate the problem anew.

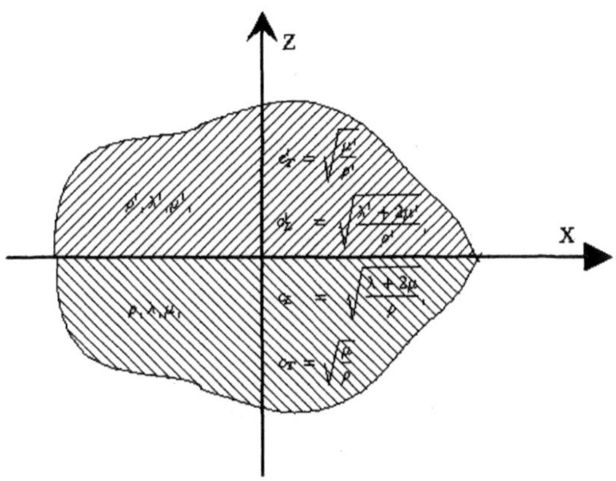


**Figure 19.** Interface of two elastic solids – Stoneley wave.

We consider two semiinfinite elastic media whose properties are described, as before, by the mass density $\rho'$, the velocities of propagation of bulk waves: $c'_L, c'_T$ in the upper part ($z > 0$), and, respectively, $\rho, c_L, c_T$ in the lower part ($z < 0$). The bulk waves satisfy the equations (2.50) for potentials with appropriate constants in each part of the system. For technical reasons, it is convenient to choose the potentials $\varphi, \psi \equiv \psi_y, \varphi', \psi' \equiv \psi'_y$ for the components $u_x, u_z, u'_x, u'_z$ of the displacement and the components of displacement $u_y, u'_y$ as unknown. For the two-dimensional problem, the latter would be given solely by the difference of derivatives of the two components of the vector potentials: $\psi_x(x,z), \psi_z(x,z)$ and, correspondingly, by $\psi'_x(x,z), \psi'_z(x,z)$. Certainly, this can be replaced by combined functions $u_y, u'_y$.

Consequently, we make the following ansatz for the solution

$$\begin{aligned}
\varphi &= Ae^{krz}e^{ik(x-ct)} \quad \text{for} \quad z<0, \quad \varphi' = A'e^{-kr'z}e^{ik(x-ct)} \quad \text{for} \quad z>0, \\
\psi &= Be^{ksz}e^{ik(x-ct)} \quad \text{for} \quad z<0, \quad \psi' = B'e^{-ks'z}e^{ik(x-ct)} \quad \text{for} \quad z>0, \\
u_y &= Ce^{kqz}e^{ik(x-ct)} \quad \text{for} \quad z<0, \quad u'_y = C'e^{-kqs'z}e^{ik(x-ct)} \quad \text{for} \quad z>0.
\end{aligned} \tag{4.1}$$

Hence, the wave is supposed to propagate in the $x$-direction, and the problem is assumed to be two-dimensional.

Substitution in equations (2.50) yields the following compatibility conditions for the exponents of potentials

$$\begin{aligned}
r^2 &= 1-\frac{c^2}{c_L^2}, \quad s^2 = 1-\frac{c^2}{c_T^2}, \\
r'^2 &= 1-\frac{c^2}{c_L'^2}, \quad s'^2 = 1-\frac{c^2}{c_T'^2},
\end{aligned} \tag{4.2}$$

As we see in a moment, the relations for $q$ and $q'$ are immaterial.

Boundary conditions follow from the continuity of displacements on the interface

$$[[\mathbf{u}]] \equiv \mathbf{u}' - \mathbf{u} = 0 \quad \text{for} \quad z=0, \tag{4.3}$$

and from the jump condition for the stress vector on the plane boundary $z=0$ as well as from the Sommerfeld conditions for $|z| \to \infty$

$$\begin{aligned}
[[\mathbf{T}]]\,\mathbf{n} &\equiv (\mathbf{T}'-\mathbf{T})\,\mathbf{n} = 0, \quad \text{for} \quad z=0, \\
\lim_{z\to\infty} \mathbf{u}' &= 0, \quad \lim_{z\to-\infty} \mathbf{u} = 0,
\end{aligned} \tag{4.4}$$

where $\mathbf{n} = \mathbf{e}_z$ is the unit vector normal to the interface.

Hence, for real $k$, Stoneley waves may exist solely if $r, s, q, r', s', q'$ are nonnegative.

The boundary condition for shear stresses in $y$-direction yields immediately

$$C = C' = 0, \tag{4.5}$$

i.e. a Stoneley wave with the amplitude in SH-direction (horizontal polarization) does not exist. The amplitude must lie in the vertical P-SV incident plane.

Bearing the relations (2.49) in mind, we obtain for $z=0$

$$\begin{aligned}
(iA' - s'B') - (iA + sB) &= 0, \\
(-r'iA' + B') - (riA + B) &= 0,
\end{aligned} \tag{4.6}$$

$$\left[-\rho' c_L'^2 \left(1-r'^2\right) + 2\rho' c_T'^2\right](iA') - 2\rho' c_T'^2 s' B' = \tag{4.7}$$

$$= \left[-\rho c_L^2\left(1-r^2\right)+2\rho c_T^2\right](iA')+2\rho c_T^2 sB,$$

$$\rho' c_T'^2\left[-2iA'r'+B'\left(1+s'^2\right)\right]=\rho c_T^2\left[2iAr+B\left(1+s^2\right)\right].$$

As usual, this set of homogeneous algebraic equations yields the dispersion relation. It can be written in the following compact form

$$\mathbf{D}_{St}\mathbf{X}=0, \quad \mathbf{X}:=\left[iA',B',iA,B\right]^T, \tag{4.8}$$

where the matrix $\mathbf{D}_{St}$ is defined as follows

$$\mathbf{D}_{St}:= \tag{4.9}$$

$$= \begin{bmatrix} -1 & \sqrt{1-\frac{c^2}{c_T'^2}} & 1 & \sqrt{1-\frac{c^2}{c_T^2}} \\ \sqrt{1-\frac{c^2}{c_L'^2}} & -1 & \sqrt{1-\frac{c^2}{c_L^2}} & 1 \\ -\frac{\rho'}{\rho}\frac{c_T'^2}{c_T^2}\left(2-\frac{c^2}{c_T^2}\right) & 2\frac{\rho'}{\rho}\frac{c_T'^2}{c_T^2}\sqrt{1-\frac{c^2}{c_T'^2}} & 2-\frac{c^2}{c_T^2} & 2\sqrt{1-\frac{c^2}{c_T^2}} \\ 2\frac{\rho'}{\rho}\frac{c_T'^2}{c_T^2}\sqrt{1-\frac{c^2}{c_L'^2}} & -\frac{\rho'}{\rho}\frac{c_T'^2}{c_T^2}\left[2-\frac{c^2}{c_T'^2}\right] & 2\sqrt{1-\frac{c^2}{c_L^2}} & 2-\frac{c^2}{c_L^2} \end{bmatrix}.$$

The determinant of this matrix gives rise to the dispersion relation for Stoneley waves

$$\mathcal{P}_{St}:=\det\mathbf{D}_{St}=0. \tag{4.10}$$

One important property of Stoneley waves follows immediately from the inspection of matrix $\mathbf{D}_{St}$. Namely, if there exists a real solution $\frac{c}{c_T}$ of the equation (4.10) then it is independent of the frequency. Hence, like Rayleigh waves, Stoneley waves are **nondispersive.**

The question of existence of Stoneley waves is far from being trivial. Already in the original work Stoneley (1924) has shown that for certain combinations of material parameters these waves may not exist (see also: Achenbach (1973), Cagniard (1962)).

In Figure 20, we quote some existence results of J. G. Scholte (1947b) for two limit cases of material parameters. Solutions for the Stoneley wave exist solely in the range between the two curves $A$ and $B$.

An extensive study of existence of this wave was also carried through by Ginzbarg and Strick (1958) whose paper contains many graphs of ranges of existence.

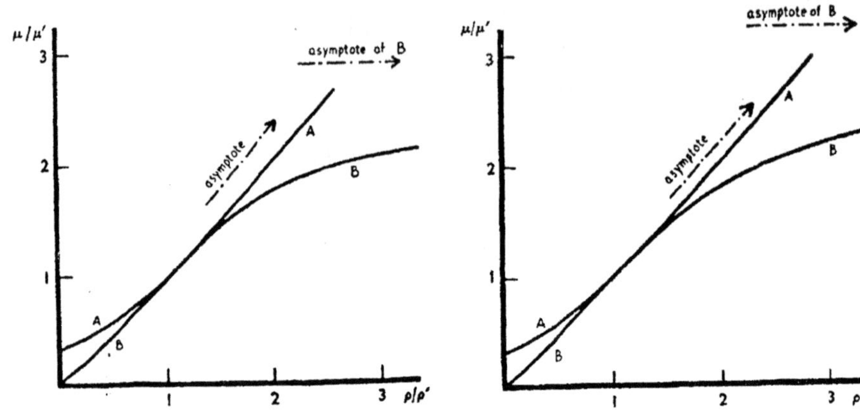


**Figure 20.** Existence results of Scholte (1947b) for two limit values of material parameters: the left panel corresponds to the case $\frac{\lambda}{\mu} = \frac{\lambda'}{\mu'} = 1$,, the right panel corresponds to the case $\lambda = \lambda' \to \infty$.

### 4.2 Interface of a semiinfinite elastic solid and a semiinfinite ideal fluid

The interface between the linear elastic solid and the ideal fluid yields the problem of surface waves which can be obtained from the above problem of Stoneley waves as a particular case. Equations for this problem were formulated by Scholte (1947a) and, for this reason, these waves are sometimes called Scholte waves.

We derive the dispersion relation again from the governing equations indicating some points which are common with modelling of surface waves for porous materials with permeable boundaries.

We choose the same coordinates as before and assume that the fluid in the range $z > 0$ is ideal, i.e. we have the following linear equations for fields $\rho', \mathbf{v}'$ in this region

$$\frac{\partial \rho'}{\partial t} + \rho_0' \mathrm{div}\mathbf{v}' = 0, \quad \rho_0' \frac{\partial \mathbf{v}'}{\partial t} = -\mathrm{grad}p', \quad p' = p_0' + \kappa' \left(\rho' - \rho_0'\right), \tag{4.11}$$

where $\kappa'$ is the compressibility coefficient. Consequently

$$\frac{\partial^2 \mathbf{v}'}{\partial t^2} = c_L'^2 \mathrm{graddiv}\mathbf{v}', \quad c_L'^2 = \kappa'. \tag{4.12}$$

It is customary to introduce for technical reasons the displacement vector $\mathbf{u}'$ also for the fluid component

$$\mathbf{v}' = \frac{\partial \mathbf{u}'}{\partial t} \quad \Rightarrow \quad \frac{\partial^2 \mathbf{u}'}{\partial t^2} = c_L'^2 \mathrm{graddiv}\mathbf{u}' + \mathbf{C}\left(\mathbf{x}\right), \tag{4.13}$$

where $\mathbf{C}(\mathbf{x})$ is an arbitrary function of the spacial variable. We can redefine the displacement in the following way

$$\mathbf{u}' \to \mathbf{u}' + \mathbf{U}\left(\mathbf{x}\right), \quad \mathrm{graddiv}\mathbf{U} = -\mathbf{C}, \tag{4.14}$$

and eliminate this arbitrary function entirely because we are interested solely in the time derivative of $\mathbf{u}'$.

It is easy to see that only the potential part of $\mathbf{u}'$ is of interest. Hence

$$\mathbf{u}' = \mathrm{grad}\varphi' \quad \Rightarrow \quad \frac{\partial^2 \varphi'}{\partial t^2} = c_L'^2 \nabla^2 \varphi', \tag{4.15}$$

where an arbitrary function of time was incorporated in the potential.

Hence, the solution for monochromatic waves in this case can be written in the form (4.1), (4.2) in which $B' = 0$.

We have to modify also the boundary conditions. Instead of the continuity of motion on the interface between two solids we have now solely the condition of continuity of the normal velocity

$$\mathbf{v}' \cdot \mathbf{n} = \frac{\partial \mathbf{u}'}{\partial t} \cdot \mathbf{n} \quad \text{for} \quad z = 0. \tag{4.16}$$

For the ideal fluid the tangential component of the velocity can be arbitrary and, consequently, we have a scalar condition instead of the vector condition (4.3).

Condition for stresses must be modified as well because now we have the shear stress in the fluid identically zero and the normal component of stresses is reduced to the pressure. The latter has to be written in terms of the displacement potential $\varphi'$. Namely, it follows by the integration of the mass balance $(4.11)_1$

$$\rho' = \rho_0' \left(1 - \mathrm{div}\mathbf{u}'\right) = \rho_0' \left(1 - \nabla^2 \varphi'\right) \quad \Rightarrow \quad p' = p_0' - \rho_0' c_L'^2 \nabla^2 \varphi'. \tag{4.17}$$

Boundary conditions for stresses at the boundary $z = 0$ have now the form

$$\rho_0' c_L'^2 \nabla^2 \varphi' = \rho c_L^2 \left( \frac{\partial^2 \varphi}{\partial x^2} + \frac{\partial^2 \varphi}{\partial z^2} \right) - 2\rho c_T^2 \left( \frac{\partial^2 \varphi}{\partial x^2} + \frac{\partial^2 \psi}{\partial x \partial z} \right), \tag{4.18}$$

$$2\frac{\partial^2 \varphi}{\partial x \partial z} + \frac{\partial^2 \psi}{\partial z^2} - \frac{\partial^2 \psi}{\partial x^2} = 0.$$

Substitution of relations (4.1) yields again the homogeneous set of equations

$$\mathbf{D}_{Sch}\mathbf{X} = 0, \quad \mathbf{X} := \left[iA', iA, B\right]^T, \tag{4.19}$$

where

$$\mathbf{D}_{Sch} := \begin{bmatrix} \sqrt{1 - \frac{c^2}{c_L'^2}} & \sqrt{1 - \frac{c^2}{c_L^2}} & 1 \\ \frac{\rho_0'}{\rho} \frac{c^2}{c_T^2} & 2 - \frac{c^2}{c_T^2} & 2\sqrt{1 - \frac{c^2}{c_T^2}} \\ 0 & 2\sqrt{1 - \frac{c^2}{c_L^2}} & 2 - \frac{c^2}{c_T^2} \end{bmatrix}. \tag{4.20}$$

The determinant of this matrix yields immediately the following

Box 6: *Stoneley (Scholte) dispersion relation*

$$\sqrt{1-\frac{c^2}{c_L'^2}}\mathcal{P}_R+\frac{\rho_0'}{\rho}\frac{c^4}{c_T^4}\sqrt{1-\frac{c^2}{c_L^2}}=0, \tag{4.21}$$

where

$$\mathcal{P}_R:=\left(2-\frac{c^2}{c_L^2}\right)^2-4\sqrt{1-\frac{c^2}{c_L^2}}\sqrt{1-\frac{c^2}{c_T^2}}. \tag{4.22}$$

The last relation is identical with the dispersion function (2.57) defining Rayleigh waves.

It is obvious that velocity $c$ is independent of the frequency $\omega$ and, consequently, also this surface wave is **nondispersive**.

The problem of existence of Stoneley-Scholte waves has been rather intensively investigated (e.g. Strick (1959), Strick and Ginzbarg (1956)). It can be shown that a single real solution exists if $c_L' < c_R$, where $c_R$ is Rayleigh velocity and it is smaller than $c_L'$, i.e. it is the smallest velocity of propagation appearing in the system.

One of the important properties of the Stoneley-Scholte wave is that the energy is carried by this wave primarily in the fluid. This can be proven by the analysis of amplitudes which we do not present here.

Apart from this real velocity, there exist complex solutions of the dispersion relation of Box 6. These were discussed in details by Ansell (1972). Some properties of complex roots yielding leaky waves (see: Phinney (1961)) shall be briefly presented in Section 4.4.

Propagation of surface waves in the system described above but either with a weak anisotropy of the solid or the fluid under a hydrostatic pressure has been investigated by Norris and Sinha (1995). They have used a perturbation method and found explicit corrections to the velocity of Scholte waves.

## 4.3 Interface of a semiinfinite elastic solid and a layer of an ideal fluid

In this case, one can consider waves intermediate between Stoneley and Rayleigh varying the height $H$ of the layer. We quote here solely a finite result which can be derived in the same way as the above presented examples of surface waves. We obtain (compare: Viktorov (1981))

$$\sqrt{\frac{c^2}{c_L'^2}-1}\mathcal{P}_R+\frac{\rho'}{\rho}\frac{c^4}{c_T^4}\sqrt{1-\frac{c^2}{c_L^2}}\tan\left(\frac{\omega H}{c}\sqrt{\frac{c^2}{c_L'^2}-1}\right)=0. \tag{4.23}$$

where $\mathcal{P}_R$ is again the Rayleigh dispersion function.

Clearly these waves are dispersive, there exist infinitely many surface modes whose velocity is bigger than the velocity in the fluid. The above dispersion relation possesses also complex solutions which we discuss in the next Subsection.

### 4.4 A few remarks on leaky waves

We have indicated already in Section 2.3. on Rayleigh waves that the dispersion relation (2.57) possesses complex roots if the signs of exponents $\beta, \gamma$ are chosen opposite. In such a case single contributions to the displacement $\mathbf{u}$ cannot satisfy the Sommerfeld condition in infinity. However, a physically reasonable solution of the problem may still exist if the combination of the longitudinal and transversal parts of displacement combine in such a way that the whole displacement decays with the distance from the surface.

Simultaneously, the solution $k(\omega)$ becomes complex which means that the factor $e^{i(kx-\omega t)} = e^{-\mathrm{Im}kx}e^{i(\mathrm{Re}kx-\omega t)}$ of the ansatz for the solution yields a decay of the amplitude characteristic for the dissipation. However, the whole system is reversible – there are no losses of energy due to the heat transfer or some other means of dissipation. Consequently, such solutions can be admissible solely under the condition that the energy of this mode of propagation must be transferred on some other modes – bulk or surface waves. If this is indeed the case we say that the wave loosing the energy is **leaky.**

In the recent paper Schroeder and Scott (2000), these complex solutions of the Rayleigh dispersion relation have been discussed in details. It has been shown that they may belong to five classes and one of these classes contains classical real solution of the problem. Other classes contain combinations of surface-like contributions and bulk-like (shear) waves which do not decay to zero in infinity. Consequently, they can exist only in finite parts of the medium and they must transfer energy to other modes of propagation. Such (pseudosurface) waves have been observed experimentally and their existence explains some critical phenomena (e.g. the critical angle of incident) on the surface.

We present briefly an example of a leaky wave considered by Viktorov (1981). It corresponds to one of the complex solutions of equation (4.23). Viktorov found analytically the complex solution of this equation under the assumption of small Poisson's ratio $\nu$. He was able to show that this solution yields attenuation of the wave $\mathrm{Im}k \sim \nu^4$ and the velocity is of the order of the longitudinal wave $c_L$. The wave consists of the longitudinal and transversal parts. The amplitude of the longitudinal part decays exponentially with the depth (i.e. it is indeed a surface wave). The transversal part behaves like a bulk wave propagating from the boundary under the angle $\beta = \frac{\pi}{4}$ (see: Fig. 20). The energy of this wave is transferred on the transversal bulk wave.

The existence of leaky waves has an important practical bearing. For instance, Norris (1983) has presented a very neat model explaining a phenomenon of backscattering of ultrasound from a fluid-solid interface. It is shown that the energy radiated back into the fluid where the incident beam originates can be observed when the angle of incidence lies near the so-called leaky wave angle. Hence, it is proven that the backscattering is due to a leaky wave reflection zone.

As already indicated, the existence of leaky modes depends on the distribution of complex roots of the dispersion relation. Phinney (1961) has designed a fairly general

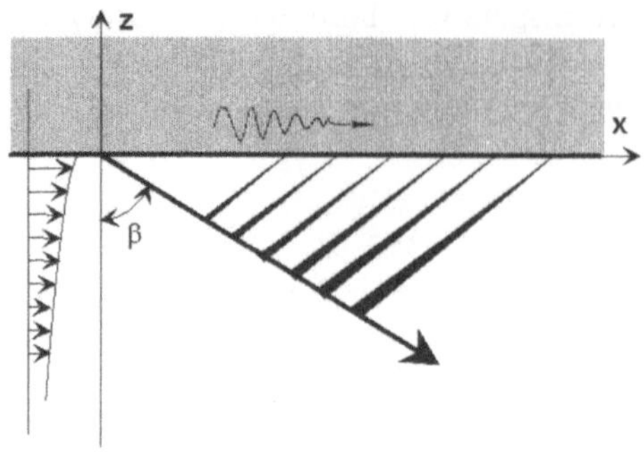


**Figure 21.** Scheme of propagation of the first leaky wave for the semiinfinite body with a layer of fluid.

method (a generalization of the so-called Rosenbaum method; compare the Footnote 3) to find these roots by the analysis of a single integral expression. The method refers to properties of such an integral when the path of integration changes the Riemann surface.

# 5 Elastic two-component media

## 5.1 Biot's model and a simple mixture model of two-component poroelastic materials

**Introduction** In contrast to single component elastic models, continuous modelling of porous materials requires the construction of a two-component model for fields describing the motion and deformation of the solid component (skeleton) and of the fluid component. In addition, it has been shown that modelling of porous media leads to the so-called **immiscible mixtures** which require an additional variable of the volume fraction of the fluid component (porosity). Porosity, in addition to the concentration of the fluid component in the mixture (i.e. the ratio of the partial mass density to the full mass density of the mixture), is the independent field in such a mixture theory.

We limit the attention to **linear poroelastic** models. Then the unknown fields are the following functions of spacial variable $\mathbf{x} \in \mathcal{B}$ and time $t \in \mathcal{T}$ ($\mathcal{B} \subset \Re^3$ is the current domain occupied simultaneously by both components, and $\mathcal{T} \subset \Re$ is the interval of time)

$$\left\{\rho^F, \mathbf{v}^F, \mathbf{v}^S, \mathbf{e}^S, n\right\}, \tag{5.1}$$

where $\rho^F$ is the current mass density of the fluid component referred to the unit macroscopic volume of the mixture, $\mathbf{v}^F, \mathbf{v}^S$ are macroscopic velocity fields of the fluid and of the skeleton, respectively, $\mathbf{e}^S$ is the Almansi-Hamel tensor of small macroscopic deformations of the skeleton, and $n$ is the current porosity. The current mass density of the skeleton $\rho^S$ does not appear in this list because in the case of small deformations it is determined by the tensor of deformations

$$\rho^S = \rho_0^S \left(1 - \operatorname{tr}\mathbf{e}^S\right), \tag{5.2}$$

where $\rho_0^S$ denotes the initial value of this mass density (in a chosen reference configuration for which $\mathbf{e}^S = 0$), and $\text{tr}\mathbf{e}^S$ reflects small volume changes of the skeleton.

In soil mechanics and geotechnics usually the set of fields does not coincide with this presented above. Instead of the mass density of the fluid, often the concepts of the volume change of the fluid $\varepsilon$ or the **increment of the fluid content** $\zeta$ are used. They are defined by the following relations

$$\varepsilon := \frac{\rho_0^F - \rho^F}{\rho_0^F}, \quad \zeta := n_0 \left( \text{tr}\mathbf{e}^S - \varepsilon \right), \tag{5.3}$$

where $\rho_0^F$ is the initial value of the fluid mass density (in the reference configuration $\mathbf{e}^S = 0$), and $n_0$ is the initial porosity in the reference configuration.

A lot of confusion arises also due to the fact that the mass densities are sometimes referred not to the common macroscopic unit volume but rather to partial volumes of components. These so-called **true mass densities** satisfy the following relations

$$\rho^{SR} = \frac{\rho^S}{1-n}, \quad \rho^{FR} = \frac{\rho^F}{n}, \tag{5.4}$$

where $n$ is the current porosity.

In our presentation we either rely on the choice (5.1) or we replace $\rho^F$ by $\varepsilon$.

Two-component models describing these fields are based on partial balance equations. In the linear model, they have the following form

1) mass balance equation of the fluid component

$$\frac{\partial \rho^F}{\partial t} + \rho_0^F \text{div}\mathbf{v}^F = 0 \quad \Rightarrow \quad \frac{\partial \varepsilon}{\partial t} = \text{div}\mathbf{v}^F, \tag{5.5}$$

2) partial momentum balance equations for both components

$$\rho_0^S \frac{\partial \mathbf{v}^S}{\partial t} = \text{div}\mathbf{T}^S + \hat{\mathbf{p}} + \rho^S \mathbf{b}^S, \quad \rho_0^F \frac{\partial \mathbf{v}^F}{\partial t} = \text{div}\mathbf{T}^F - \hat{\mathbf{p}} + \rho^F \mathbf{b}^F, \tag{5.6}$$

where $\mathbf{T}^S, \mathbf{T}^F$ are partial Cauchy stress tensors, $\hat{\mathbf{p}}$ denotes the source of momentum, and $\mathbf{b}^S, \mathbf{b}^F$ are external mass forces,

3) integrability condition for the deformation of the skeleton

$$\frac{\partial \mathbf{e}^S}{\partial t} = \text{symgrad}\mathbf{v}^S, \tag{5.7}$$

and this is equivalent to the existence of the displacement vector $\mathbf{u}^S$ for the skeleton.

We have to close this system by appropriate constitutive relations. Further in this article we discuss two models: the **Biot's model** and the **simple mixture model**.

The Biot model is based on the following constitutive relations (Biot (1956), see also a full collection of Biot's works on porous materials Tolstoy (1991))

$$\begin{aligned} \mathbf{T}^S &= \mathbf{T}_0^S + \lambda^S \text{tr}\mathbf{e}^S \mathbf{1} + 2\mu^S \mathbf{e}^S + Q\varepsilon \mathbf{1}, \\ \mathbf{T}^F &= -p^F \mathbf{1}, \quad p^F = p_0^F - \kappa \rho_0^F \varepsilon - Q \text{tr}\mathbf{e}^S, \\ \hat{\mathbf{p}} &= \pi \left( \mathbf{v}^F - \mathbf{v}^S \right) - \rho_{12} \left( \frac{\partial \mathbf{v}^F}{\partial t} - \frac{\partial \mathbf{v}^S}{\partial t} \right), \\ n &= n_0 \left( 1 + \delta \text{tr}\mathbf{e}^S + \gamma \left( \text{tr}\mathbf{e}^S - \varepsilon \right) \right), \end{aligned} \tag{5.8}$$

where $\mathbf{T}_0^S, p_0^F$ are the initial partial stress in the skeleton and the initial partial pressure in the fluid, respectively. $\lambda^S, \mu^S$ are effective (i.e. macroscopic, dependent on the initial porosity $n_0$) Lamé constants describing the skeleton, $\kappa$ is the effective compressibility coefficient of the fluid component, $Q$ is the **coupling parameter** between partial stresses, $\rho_{12}$ is the **added mass parameter**, and $\delta, \gamma$ are parameters describing changes of porosity caused by volume changes of both components.

The above set of material parameters is not commonly accepted and the literature is full of its different variations. For instance, Stoll (1989) relies on the following set

$$\begin{aligned} K &= \lambda^S + \frac{2}{3}\mu^S + \rho_0^F \kappa + 2Q, \quad G = \mu^S, \\ C &= \frac{1}{n_0}\left(Q + \rho_0^F \kappa\right), \quad M = \frac{\rho_0^F \kappa}{n_0^2}. \end{aligned} \tag{5.9}$$

In the standard reference book on linear acoustics of porous materials Bourbie et al. (1987), the following set of equations is applied

$$\begin{gathered} \rho \frac{\partial \mathbf{v}^S}{\partial t} + \rho_{uw}\frac{\partial \mathbf{w}}{\partial t} = \operatorname{div}\mathbf{T}, \quad \mathbf{T} := \mathbf{T}^S + \mathbf{T}^F, \\ \rho_{uw}\frac{\partial \mathbf{v}^F}{\partial t} + \rho_w \frac{\partial \mathbf{w}}{\partial t} = -\operatorname{grad}p - \frac{1}{\mathcal{K}}\mathbf{w}, \quad \mathbf{w} := n_0\left(\mathbf{v}^F - \mathbf{v}^S\right), \\ \mathbf{T} = \lambda_f \operatorname{tr}\mathbf{e}^S \mathbf{1} + 2\mu \mathbf{e}^S - \beta M \zeta \mathbf{1}, \quad \mathbf{T}^F = -n_0 p \mathbf{1}, \quad p = M\left(-\beta \operatorname{tr}\mathbf{e}^S + \zeta\right), \end{gathered} \tag{5.10}$$

where $\mathbf{w}$ corresponds to the so-called filter velocity and

$$\begin{gathered} \rho := \rho_0^S + \rho_0^F, \quad \rho_{uw} := \rho_0^{FR}, \quad \rho_w := \frac{1}{n_0^2}\left(\rho_0^F - \rho_{12}\right), \\ \lambda_f := \lambda^S + \kappa \rho_0^F + 2Q, \quad \mu := \mu^S, \quad \beta M := \frac{1}{n_0}\left(Q + \kappa \rho_0^F\right), \quad \frac{1}{\mathcal{K}} := \frac{\pi}{n_0^2}. \end{gathered} \tag{5.11}$$

The quantity $p = \frac{1}{n_0} p^F$ is called the **fluid pore pressure** in contrast to $p^F$ which is the partial pressure.

Certainly, all these relations are equivalent to the original Biot equations.

The added mass is usually related to the **tortuosity** $a$ by the relation

$$\rho_{12} = \rho_0^F (1 - a). \tag{5.12}$$

In addition to the porosity $n$, the tortuosity $a$ describes the morphology of the porous material and, roughly speaking, it is the ratio of the average length of microchannels in the porous material to the average characteristic distance on the microlevel, i.e. for straight channels it is equal to one and otherwise bigger than one: $1 \leq a < \infty$. In the linear model, this quantity is constant.

It can be easily shown that the added mass effect incorporated in the Biot model through the relative acceleration violates the **principle of material objectivity** (Wilmanski (2001a)) and the coupling effect described by the material parameter $Q$ violates the **second law of thermodynamics** (Wilmanski (2002b)). However, one can construct a

**nonlinear model** (e.g. Wilmanski (2003), Wilmanski (2004)) which satisfies these two principles of continuum thermodynamics and, simultaneously, whose linearization leads to the Biot model. In this sense the Biot model can be considered as a thermodynamically sound way of a **linear description** of dynamics of saturated porous materials.

The second model which can be used to describe dynamics of linear poroelastic materials is the so-called **simple mixture model.** This model has been constructed by means of continuum thermodynamics for modelling **nonlinear** processes (e.g. Wilmanski (1998b), Wilmanski (1998a)). For this reason changes of porosity are described by the balance equation rather then the relation $(5.8)_4$. Namely, in the linear model this equation is of the form

$$\frac{\partial\left(n-n_E\right)}{\partial t}+\Phi_0 \operatorname{div}\left(\mathbf{v}^F-\mathbf{v}^S\right)=-\frac{n-n_E}{\tau}, \quad n_E=n_0\left(1+\delta \operatorname{tr} \mathbf{e}^S\right), \tag{5.13}$$

where $\tau$ is the **relaxation time of porosity** and $\Phi_0 = n_0\gamma$.

The coupling parameter $Q$ is equal to zero in the simple mixture model and the tortuosity $a = 1$, i.e there is no influence of relative acceleration. Such a model satisfies both the principle of material objectivity and the second law of thermodynamics.

The questions arise if the simple mixture model reflects properly dynamical features of porous materials and what are the differences between the results obtained by the Biot model and the simple mixture model. Before we proceed to a discussion of surface waves described by a two-component model we address these two questions.

**Objective relative acceleration** In order to appreciate the problem of material objectivity, we consider some kinematic properties of the two-component system which we describe in two different reference systems specified by the following transformation of the point $\mathbf{x}$ in the space of motions (the so-called configuration space)

$$\mathbf{x}^*=\mathbf{O}(t)\,\mathbf{x}+\mathbf{c}(t), \quad \mathbf{O}^{-1}=\mathbf{O}^T, \tag{5.14}$$

where the time dependent orthogonal matrix $\mathbf{O}$ describes the rotation of one reference system with respect to the other while the time dependent vector $\mathbf{c}$ describes the motion of the origin of one system with respect to the other.

This transformation (**isometry**) describes the most general change of reference which preserves the distance between two arbitrary points, say $\mathbf{x}_1, \mathbf{x}_2$:

$$\left|\mathbf{x}_1^*-\mathbf{x}_2^*\right|^2=\left(\mathbf{x}_1-\mathbf{x}_2\right) \cdot \mathbf{O}^T \mathbf{O}\left(\mathbf{x}_1-\mathbf{x}_2\right) \equiv\left|\mathbf{x}_1-\mathbf{x}_2\right|^2. \tag{5.15}$$

In particular, this transformation may describe the change of the inertial reference system into a noninertial one (e.g. the change from the reference with respect to fixed stars to the reference which rotates with a turntable in the laboratory).

It is shown in continuum mechanics that the above transformation yields certain rules of transformation for fields of mechanics. Any scalar quantity does not change under this transformation. We say for this reason that scalars are **objective**. Vectors and tensors may or may not be objective. If a vector $\mathbf{w}$ and a tensor of the second grade $\mathbf{e}$ transform according to the rules

$$\mathbf{w}^*=\mathbf{O}\mathbf{w}, \quad \mathbf{e}^*=\mathbf{O}\mathbf{e}\mathbf{O}^T, \tag{5.16}$$

we call them objective. For instance, the Almansi-Hamel deformation tensor is objective. Certainly, these rules reduce to simple transformations of coordinates if we refer the objects to their coordinates.

Time dependence of the orthogonal tensor $\mathbf{O}$ and of the vector $\mathbf{c}$ yields modifications of these transformation rules if they involve some time differentiation. In the case of a two-component mixture for the basic kinematic quantities we have the rules of the following form

1) partial velocities

$$\mathbf{v}^{S*} = \mathbf{O}\mathbf{v}^S + \dot{\mathbf{O}}\mathbf{x} + \dot{\mathbf{c}}, \quad \mathbf{v}^{F*} = \mathbf{O}\mathbf{v}^F + \dot{\mathbf{O}}\mathbf{x} + \dot{\mathbf{c}}, \tag{5.17}$$

2) partial velocity gradients

$$\begin{aligned} \mathbf{L}^{S*} &= \mathbf{O}\mathbf{L}^S\mathbf{O}^T + \dot{\mathbf{O}}\mathbf{O}^T, \quad \mathbf{L}^{F*} = \mathbf{O}\mathbf{L}^F\mathbf{O}^T + \dot{\mathbf{O}}\mathbf{O}^T, \\ \mathbf{L}^S &\equiv \operatorname{grad}\mathbf{v}^S, \quad \mathbf{L}^F \equiv \operatorname{grad}\mathbf{v}^F, \end{aligned} \tag{5.18}$$

3) accelerations

$$\mathbf{a}^{S*} = \mathbf{O}\mathbf{a}^S + 2\dot{\mathbf{O}}\mathbf{v}^S + \ddot{\mathbf{O}}\mathbf{x} + \ddot{\mathbf{c}}, \quad \mathbf{a}^{F*} = \mathbf{O}\mathbf{a}^F + 2\dot{\mathbf{O}}\mathbf{v}^F + \ddot{\mathbf{O}}\mathbf{x} + \ddot{\mathbf{c}}, \tag{5.19}$$

and the nonobjective contributions lead to the so-called Coriolis, centrifugal, Euler and translational accelerations (e.g. Liu (2002)). When multiplied by the mass densities they may be considered as **apparent mass forces**. It is known that such forces indeed appear in equations of motion in noninertial frames of reference.

It is obvious that the relative acceleration does not transform in an objective manner. We have

$$\mathbf{a}^{F*} - \mathbf{a}^{S*} = \mathbf{O}\left(\mathbf{a}^F - \mathbf{a}^S\right) + 2\dot{\mathbf{O}}\left(\mathbf{v}^F - \mathbf{v}^S\right). \tag{5.20}$$

Hence, the result depends on an arbitrary angular velocity $\dot{\mathbf{O}}$. As the relative acceleration enters the momentum balance equations with the constitutive (material) parameter $\rho_{12}$ solutions of these equations depend on the choice of the reference system in a nontrivial manner. Hence, such contributions violate the principle of material objectivity and they are not admissible in continuum models.

Considering a suspension of bubbles in a fluid a definition of the relative acceleration has been introduced in Drew et al. (1979) which transforms in an objective manner. Such a definition reminds definitions of various objective time derivatives appearing in continuum mechanics (such as Jaumann derivative, Oldroyd derivative, Truesdell derivative, etc.). It is rather easy to show that an **objective relative acceleration** can be also introduced in the nonlinear mechanics of porous materials (see: Wilmanski (2004)). The definition may have the following form

$$\begin{aligned} \mathbf{a}_r &= \frac{\partial}{\partial t}\left(\mathbf{v}^F - \mathbf{v}^S\right) + \left(\mathbf{L}^F - \mathbf{L}^S\right)\mathbf{v}^S - (1 - \mathfrak{z})\,\mathbf{L}^F\left(\mathbf{v}^F - \mathbf{v}^S\right) - \mathfrak{z}\mathbf{L}^S\left(\mathbf{v}^F - \mathbf{v}^S\right), \\ \mathbf{a}_r^* &= \mathbf{O}\mathbf{a}_r, \end{aligned} \tag{5.21}$$

where $\mathfrak{z}$ is an arbitrary scalar. Some properties of a thermodynamical model constructed with a contribution of such a relative acceleration can be found elsewhere (see: Wilmanski

(2004)) but we do not need to go into any details in this work. It is sufficient to state that the thermodynamics admits a linear dependence of the momentum source $\hat{\mathbf{p}}$ on the objective relative acceleration, i.e the following relation for momentum source

$$\hat{\mathbf{p}} = \pi \left(\mathbf{v}^F - \mathbf{v}^S\right) - \rho_{12}\mathbf{a}_r, \tag{5.22}$$

satisfies the principle of material objectivity and the second law of thermodynamics.

It is clear that the Biot contribution follows as a linearization of the above formula. One has $\rho_{12}\mathbf{a}_r \approx \rho_{12}\frac{\partial}{\partial t}\left(\mathbf{v}^F - \mathbf{v}^S\right)$ as required by the relations (5.8) of the Biot model. After such a linearization the model is not objective anymore and, consequently, we cannot transform reference systems within the Biot model. Experiments performed in static conditions cannot be compared with experiments performed in a centrifuge if we process the experimental data with the Biot model.

We return to the problem of added mass further in order to estimate the order of its contributions to the propagation conditions of acoustic waves.

**Gassmann relations** The problem of violation of the second law of thermodynamics by Biot's equations is related to the fact that thermodynamics of mixtures does not admit couplings between components if constitutive relations do not contain higher gradients. This property has been noticed by I. Müller (1973), (see also: Müller (1985)) who has shown that partial quantities describing components of a mixture of fluids such as partial free energies or partial pressures may depend on partial mass densities of all components only in the case when they also depend in a constitutive manner on gradients of mass densities. Otherwise the mixture becomes **simple**, i.e. partial free energies, partial pressures, etc. depend only on their own partial mass densities.

The two-component model of porous materials possesses the same property. The partial stress tensor in the skeleton, say, may depend on volume changes of the fluid only if it depends also on some higher gradients such as the gradient of porosity. Otherwise the second law of thermodynamics yields $Q \equiv 0$.

A thermodynamical correction of the model through the extension of the set of constitutive variables has been introduced in the work Wilmanski (2002b). It has been shown that a dependence on the gradient of porosity grad$n$ leads to the following constitutive relations for partial stresses in the linear model

$$\begin{aligned}
\mathbf{T}^S &= \mathbf{T}_0^S + \lambda^S \mathrm{tr}\mathbf{e}^S \mathbf{1} + 2\mu^S \mathbf{e}^S + Q\varepsilon\mathbf{1} + \beta\left(n - n_E\right)\mathbf{1} - N\left(n - n_0\right)\mathbf{1}, \\
\mathbf{T}^F &= -p^F \mathbf{1}, \quad p^F = p_0^F - \kappa\rho_0^F\varepsilon - Q\mathrm{tr}\mathbf{e}^S + \beta\left(n - n_E\right) - N\left(n - n_0\right),
\end{aligned} \tag{5.23}$$

where the parameter $\beta$ describes the influence of local nonequilibrium changes of porosity and the parameter $N$ stems from the contribution to the source of momentum of the form $N$grad$n$. It has been discussed in earlier works on acoustics of poroelastic materials (e.g. Wilmanski (1999)) that the influence of $\beta$ is negligible in processes with slow relative motion (small filtration velocity). This is the case in geomechanics and we neglect this contribution also in this article, i.e. we assume $\beta \equiv 0$.

However, the influence of the constant $N$ cannot be evaluated so easily and we shall do it in this section by means of some Gedankenexperiments. Their role in geoacoustics is quite fundamental (e.g. White (1983)) and we present here the main results.

One of the main questions in applications of waves to nondestructive testing are relations between macroscopic quantities entering a model describing waves and microscopic quantities which we are trying to test. The typical example of such a quantity is the porosity of porous materials. We present here a brief review of results for granular materials which follow by means of certain static Gedankenexperiments for saturated materials. These experiments proposed by Gassmann (1951) and incorporated into the Biot model by Biot and Willis (1957) describe the behavior of homogeneous samples of fully saturated porous materials in **jacketed drained and undrained** experiments as well as in an **unjacketed** experiment. In such Gedankenexperiments we know constitutive relations between microscopic pressures and microscopic volume changes of both components as well as corresponding macroscopic relations. Microscopic and macroscopic deformations must fulfil kinematic and dynamic compatibility relations which lead to a system of equations for unknown fields which is, in turn, overdetermined. Consequently, we obtain admissibility relations which connect microscopic and macroscopic material parameters.

We assume that the relaxation time of porosity $\tau$ is sufficiently large in comparison with other characteristic quantities (for example, $1/\omega_0$, where $\omega_0$ is the characteristic frequency of monochromatic waves) in order to neglect the relaxation in the porosity balance equation (5.13). As the Biot model does not describe the relaxation of porosity this assumption is necessary for the comparison of the two models.

We skip here the derivation of micro-macro relations referring the reader to the paper Wilmanski (2003) and present the results. It is assumed that microscopic (true) compressibility properties of components are known. It means that we assume the measurability of the true compressibility modulus of the solid component $K_s$, the true compressibility modulus of the fluid component $K_f$, the drained modulus $K_d$ (compare Lancellotta (2005)) and the initial porosity $n_0$. Macroscopic parameters $K, M, C, N$ are then given by four equations (see: Wilmanski (2003))

$$\mathcal{C}_1 := C + \frac{K_f(K - K_s) - N(K - K_V)}{n_0(K_s - K_f)} = 0, \tag{5.24}$$

$$\mathcal{C}_2 := n_0 - \frac{C}{M} - \frac{K_b}{K_s}\frac{1-(1-n_0)\frac{K_s}{K_b}}{1-\frac{1-n_0}{n_0}\frac{NC}{K_bM}}\left\{1 - \frac{N(K - n_0 C)}{n_0 M K_b}\right\} = 0, \tag{5.25}$$

$$\mathcal{C}_3 := K - K_d - C\frac{C-N}{M-\frac{N}{n_0}} = 0. \tag{5.26}$$

$$\mathcal{C}_4 := \left(1 - \frac{K}{K_W}\right)\left(M - C - N\frac{1-n_0}{n_0}\frac{C}{K_s}\right) + \tag{5.27}$$

$$+\left(1 - \frac{C}{K_W}\right)\left(K - C - N\frac{1-n_0}{n_0}\left(1 - \frac{K}{K_s}\right)\right) = 0,$$

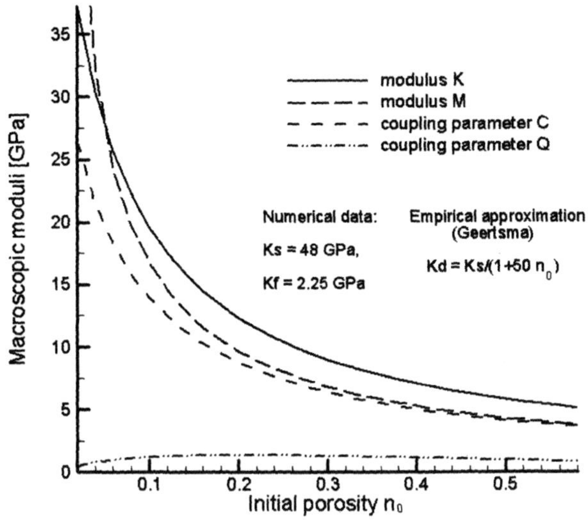


**Figure 22.** Dependence of macroscopic moduli of Biot's model on the porosity $n_0$ following from Gassmann relations.

where

$$K_V = (1-n_0)\,K_s + n_0 K_f, \quad K_b = K - \frac{C^2}{M}, \tag{5.28}$$

$$\frac{1}{K_W} = \frac{1-n_0}{K_s} + \frac{n_0}{K_f}.$$

In general, this set cannot be solved analytically. However, one can show that it possesses two physically admissible solutions. One of them yields a nonzero value of the parameter $N$ but this value is small when compared with other moduli. The other solution contains $N=0$. Then the remaining parameters $K, C, M$ can be found analytically. Their values almost coincide with these obtained in the case of $N \neq 0$. We conclude that the form of relations appearing in the Biot model ($N=0$) is thermodynamically acceptable within reasonable limits.

The analytical solution of the micro-macro transition problem for $N=0$ was obtained for the first time by Gassmann (1951) and it has the following form

$$K = \frac{(K_s - K_d)^2}{\frac{K_s^2}{K_W} - K_d} + K_d, \quad C = \frac{K_s\,(K_s - K_d)}{\frac{K_s^2}{K_W} - K_d}, \quad M = \frac{K_s^2}{\frac{K_s^2}{K_W} - K_d}. \tag{5.29}$$

In Figure 22, we present a numerical example for the above relations. It is clear that the coupling parameter $Q$ is much smaller than the other moduli of the Biot model. In the next Section we investigate its influence on the propagation conditions of bulk waves.

The procedure of micro-macrotransition does not give any information on the shear modulus $\mu^S$. This must be supplemented by some macroscopic measurements. A possible estimate based on values of the Poisson's ratio $\nu$ is given in Albers and Wilmanski (2003). This may have some practical bearing in soil mechanics because the Poisson's ratio seems to vary only a little for soils.

Finally let us mention a result obtained from some geometrical considerations by Berryman for the tortuosity parameter. He has shown (Berryman (1980)) that for granular materials with moderate porosities it may be roughly estimated by the following simple relation

$$a = \frac{1}{2}\left(\frac{1}{n_0} + 1\right).^{6} \tag{5.30}$$

We use this relation in numerical examples presented further in this article.

The governing equations of the Biot model have the form presented in the Box 7.

BOX 7: *Biot's model*

**Fields**: $\mathbf{v}^S$, $\mathbf{v}^F$, $\mathbf{e}^S$, $\varepsilon$ ,

$$\begin{aligned}
\rho_{11}\frac{\partial \mathbf{v}^S}{\partial t} + \rho_{12}\frac{\partial \mathbf{v}^F}{\partial t} &= \lambda^S \operatorname{grad}\operatorname{tr}\mathbf{e}^S + 2\mu^S \operatorname{div}\mathbf{e}^S + Q\operatorname{grad}\varepsilon + \pi\left(\mathbf{v}^F - \mathbf{v}^S\right), \\
\rho_{12}\frac{\partial \mathbf{v}^S}{\partial t} + \rho_{22}\frac{\partial \mathbf{v}^F}{\partial t} &= \rho_0^F \kappa \operatorname{grad}\varepsilon + Q\operatorname{grad}\operatorname{tr}\mathbf{e}^S - \pi\left(\mathbf{v}^F - \mathbf{v}^S\right),
\end{aligned} \tag{5.31}$$

$$\frac{\partial \mathbf{e}^S}{\partial t} = \operatorname{sym}\operatorname{grad}\mathbf{v}^S, \quad \frac{\partial \varepsilon}{\partial t} = \operatorname{div}\mathbf{v}^F,$$

$$\begin{aligned}
\rho_{11} &= \rho_0^S + \rho_{12} \equiv \rho_0^S\left[1 - r(1-a)\right], \quad \rho_{22} = \rho_0^F + \rho_{12} \equiv ra\rho_0^S, \\
\rho_{12} &= r(1-a)\rho_0^S, \quad r := \frac{\rho_0^F}{\rho_0^S},
\end{aligned} \tag{5.32}$$

Material parameters: $\lambda^S, \mu^S, \kappa, \pi, Q, a$.

**For the simple mixture model**: $Q = 0, a = 1$.

[6] Johnson, Plona, Scala, Pasierb and Kojima suggest in the work Johnson et al. (1982) the relation $a = n_0^{-\beta}$, where $\beta = 2/3$ for random array of needles and $\beta = 1/2$ for random array of spheres. Certainly, none of these relations can be true in general as tortuosity is independent of porosity for arbitrary morphology. For instance, two classes of pipe-like channels differing in the area of cross-section on, say, factor two and yielding the same porosity give tortuosities differing also on factor two as the characteristic length in both cases is the same.

The equation for $\varepsilon$ follows easily from the mass balance for the fluid (see (5.5)). In the case of the simple mixture model, we have $Q = 0, \rho_{12} = 0$ (i.e. $a = 1$).

## 5.2 Bulk monochromatic waves in two-component poroelastic materials

We proceed to analyze the propagation of monochromatic waves in two-component (saturated) poroelastic materials. There exists a vast literature on this subject. We mention here only two excellent books: Bourbie et al. (1987), Allard (1993) in which both the wave analysis for Biot's model as well as a comparison with experiments is presented.

We seek solutions of equations (5.31) which have the form of the following monochromatic waves

$$\mathbf{v}^S = \mathbf{V}^S\mathcal{E}, \quad \mathbf{v}^F = \mathbf{V}^F\mathcal{E}, \quad \mathbf{e}^S = \mathbf{E}^S\mathcal{E}, \quad \varepsilon = E^F\mathcal{E}, \tag{5.33}$$

$$\mathcal{E} := \exp\left[i\left(\mathbf{k}\cdot\mathbf{x} - \omega t\right)\right],$$

where $\mathbf{V}^S, \mathbf{V}^F, \mathbf{E}^S, E^F$ are constant amplitudes, $\mathbf{k}$ is the wave vector, $\omega$ real frequency.

By means of the last two equations (5.31) we can eliminate the amplitudes $\mathbf{E}^S, E^F$

$$\mathbf{E}^S = -\frac{1}{2\omega}\left(\mathbf{k}\otimes\mathbf{V}^S + \mathbf{V}^S\otimes\mathbf{k}\right), \quad E^F = -\frac{1}{\omega}\mathbf{V}^F\cdot\mathbf{k}. \tag{5.34}$$

The remaining field equations yield the following compatibility conditions

$$\begin{gathered}\left[\rho_{11}\omega^2\mathbf{1} - \lambda^S\mathbf{k}\otimes\mathbf{k} - \mu^S\left(k^2\mathbf{1} + \mathbf{k}\otimes\mathbf{k}\right) + i\pi\omega\mathbf{1}\right]\mathbf{V}^S + \\ + \left[\rho_{12}\omega^2\mathbf{1} - Q\mathbf{k}\otimes\mathbf{k} - i\pi\omega\mathbf{1}\right]\mathbf{V}^F = 0, \\ \left[\rho_{12}\omega^2\mathbf{1} - Q\mathbf{k}\otimes\mathbf{k} - i\pi\omega\mathbf{1}\right]\mathbf{V}^S + \left[\rho_{22}\omega^2\mathbf{1} - \kappa\rho_0^F\mathbf{k}\otimes\mathbf{k} + i\pi\omega\mathbf{1}\right]\mathbf{V}^F = 0.\end{gathered} \tag{5.35}$$

As usual, the problem of existence of such waves reduces to an eigenvalue problem with the eigenvector $\left[\mathbf{V}^S, \mathbf{V}^F\right]$. As in the case of classical one-component elasticity we split the problem into two parts: in the direction $\mathbf{k}_\perp = \mathbf{k} - \frac{\mathbf{k}\cdot\mathbf{k}}{k^2}\mathbf{k}$, perpendicular to $\mathbf{k}$ (transversal modes) and in the direction of the wave vector $\mathbf{k}$ (longitudinal modes). This yields two dispersion relations $k = k(\omega)$.

In contrast to cases which we have considered before for the elastic single component materials, the above dispersion problem possesses solely complex solutions. For a given real frequency $\omega$, the imaginary part of $k$ arises due to the **dissipation** of energy whose amount is controlled by the permeability coefficient $\pi$ and it is physically caused by the **diffusion**. This energy is transferred as heat to the exterior and cannot be recovered by the system. This makes the physical interpretation of complex solutions of the dispersion relation different from this which we indicated in Section 4.4. on leaky waves. We see further that both leaky and dissipative waves appear in two-component poroelastic materials.

Simultaneously, inspection of the set (5.35) shows that the contribution of permeability $\pi$ enters these equations exactly in the same way as the damping $\eta$ in equation (2.26) – accelerations (second time derivatives) are related to $\omega^2$ and the permeability, similarly to the first time derivative, to $\omega$. Consequently, we could expect the existence of a critical wave number $k$ if this was chosen as the independent variable rather than the frequency $\omega$. This was indeed shown in the work of Edelman (2003). Even though it is an interesting mathematical problem, it does not seem to have much to do with the physical reality as, due to dispersion, one practically does not observe waves of a particular wavelength (i.e. of a particular value of $k$) but rather broad band packages of waves in which the influence of such a critical wave number cannot be spotted. For this reason we use in the analysis of diffusive systems solely the frequency $\omega$ as the independent variable in dispersion relations.

For transversal modes (monochromatic shear waves) we have

$$\begin{aligned}\left[\rho_{11}\omega^2 - \mu^S k^2 + i\pi\omega\right] V_\perp^S + \left[\rho_{12}\omega^2 - i\pi\omega\right] V_\perp^F &= 0, \quad k^2 = \mathbf{k}\cdot\mathbf{k},\\ \left[\rho_{12}\omega^2 - i\pi\omega\right] V_\perp^S + \left[\rho_{22}\omega^2 + i\pi\omega\right] V_\perp^F &= 0,\end{aligned} \tag{5.36}$$

$$V_\perp^S = \mathbf{V}^S\cdot\mathbf{k}_\perp, \quad V_\perp^F = \mathbf{V}^F\cdot\mathbf{k}_\perp.$$

The dispersion relation can be written in this case in the following form

$$\begin{aligned}&\omega\left\{\left(\rho_{11}\rho_{22} - \rho_{12}^2\right)\left(\frac{\omega}{k}\right)^2 - \mu^S\rho_{22}\right\} + \\ &+ i\pi\left\{\left(\rho_{11} + \rho_{22} + 2\rho_{12}\right)\left(\frac{\omega}{k}\right)^2 - \mu^S\right\} = 0,\end{aligned} \tag{5.37}$$

i.e.

$$\left(\frac{\omega}{k}\right)^2 = \frac{\omega r a + i\frac{\pi}{\rho_0^S}}{\omega r\left[a - r(1-a)\right] + i\frac{\pi}{\rho_0^S}(1+r)} c_S^2, \quad c_S^2 := \frac{\mu^S}{\rho_0^S}. \tag{5.38}$$

Consequently, neither the phase velocity $\omega/\mathrm{Re}k$ nor the attenuation $\mathrm{Im}k$ of monochromatic shear waves is dependent on the coupling coefficient $Q$.

In the two limits of frequencies we have then the following solutions

$$\begin{aligned}&\omega\to 0: \quad \lim_{\omega\to 0}\left(\frac{\omega}{\mathrm{Re}k}\right)^2 = \frac{\mu^S}{\rho_0^S + \rho_0^F}, \quad \lim_{\omega\to 0}(\mathrm{Im}k) = 0,\\ &\omega\to\infty: \quad \lim_{\omega\to\infty}\lim\left(\frac{\omega}{\mathrm{Re}k}\right)^2 = \frac{\rho_{22}}{\rho_{11}\rho_{22} - \rho_{12}^2}\mu^S,\\ &\lim_{\omega\to\infty}(\mathrm{Im}k) = \frac{\pi}{2\sqrt{\rho_0^S\mu^S}}\frac{1}{a^2}\sqrt{\frac{a}{a - r(1-a)}}.\end{aligned} \tag{5.39}$$

The first result checks with the results of the classical one-component model commonly used in soil mechanics.

In the simple mixture model, the result for $\omega\to 0$ is, of course, the same. For $\omega\to\infty$ we get

$$\lim_{\omega\to\infty}\left(\frac{\omega}{\mathrm{Re}k}\right)^2 = \frac{\mu^S}{\rho_0^S} \equiv c_S^2, \tag{5.40}$$

and this is identical with the result of the classical elasticity where we used the notation $c_T$ rather than $c_S$ for this limit.

Let us notice that the attenuation in the limit $\omega \to \infty$ is finite and reaches there a maximum. Consequently, for $\pi = 0$ the solution of the dispersion relation would be real. This means that indeed, for the above investigated waves, the attenuation is caused solely by the diffusion.

We demonstrate properties of monochromatic waves on a numerical example in the whole range of frequencies $\omega \in [0, \infty)$. Both for shear waves and for longitudinal waves we use the following numerical data

$$\begin{aligned} c_{P1} &\equiv \sqrt{\frac{\lambda^S + 2\mu^S}{\rho_0^S}} = 2500 \ \frac{\mathrm{m}}{\mathrm{s}}, \quad c_{P2} \equiv \sqrt{\kappa} = 1000 \ \frac{\mathrm{m}}{\mathrm{s}}, \quad c_S \equiv \sqrt{\frac{\mu^S}{\rho_0^S}} = 1500 \ \frac{\mathrm{m}}{\mathrm{s}}, \\ \rho_0^S &= 2500 \ \frac{\mathrm{kg}}{\mathrm{m}^3}, \quad r = 0.1, \quad \pi = 10^8 \ \frac{\mathrm{kg}}{\mathrm{m}^3\mathrm{s}}, \\ Q &= 0.8 \ \mathrm{GPa}, \quad n_0 = 0.4, \quad a = 1.75. \end{aligned} \tag{5.41}$$

Velocities $c_{P1}, c_{P2}, c_S$, the mass density $\rho_0^S$ (i.e $\rho_0^{SR} = 4167 \ \frac{\mathrm{kg}}{\mathrm{m}^3}$ for the porosity $n_0 = 0.4$) and the fraction $r = \rho_0^F / \rho_0^S$ possess values typical for many granular materials under a confining pressure of a few atmospheres and saturated by water. In units standard for soil mechanics the permeability $\pi$ corresponds to app. 0.1 Darcy. The coupling coefficient $Q$ has been estimated by means of the Gassmann relation discussed in Sec. 5.1. The tortuosity coefficient $a = 1.75$ follows from Berryman formula (5.30) for the chosen value of porosity.

Transversal waves described by the relation (5.38) are characterized by the following distribution of velocities and attenuations in function of the frequency (Fig. 23). The solid lines correspond to the solution of Biot's model and the dashed lines to the solution of the simple mixture model.

It is clear that the qualitative behavior of the velocity of propagation is the same in both models. It is a few percent smaller in Biot's model than this in the simple mixture model in the range of high frequencies. A large quantitative difference between these models appears for the attenuation. In the range of higher frequencies it is much smaller in the Biot model, i.e. tortuosity decreases the dissipation of shear waves.

The latter property is illustrated in Fig. 24 where we plot the attenuation of the front of shear waves, i.e. $\lim\limits_{\omega \to \infty} \lim \operatorname{Im} k$, as a function of the tortuosity coefficient $a$. This behavior of attenuation indicates that damping of waves created by the tortuosity, which is connected in the macroscopic model to the relative velocity of components, is not related to scattering of waves on the microstructure. It is rather related to the decrease of the macroscopic diffusion velocity in comparison with the difference of velocities on the microscopic level due to the curvature of channels and volume averaging. Fluctuations are related solely to this averaging and not to temporal deviation from time averages (lack of ergodicity!). Whatever the argument may be the influence of tortuosity on the attenuation of waves seems to be much too strong.

Let us remark in passing that, in order to describe properly the influence of morphology (i.e. a random real geometry of channels, their volume contribution – porosity, their curvature – tortuosity, etc.) on the propagation of waves one would have to account for random scattering of waves on microscopic obstacles. This is, certainly, not done by the added mass coefficient $\rho_{12}$ of Biot's model. However,one way of doing it is the overall

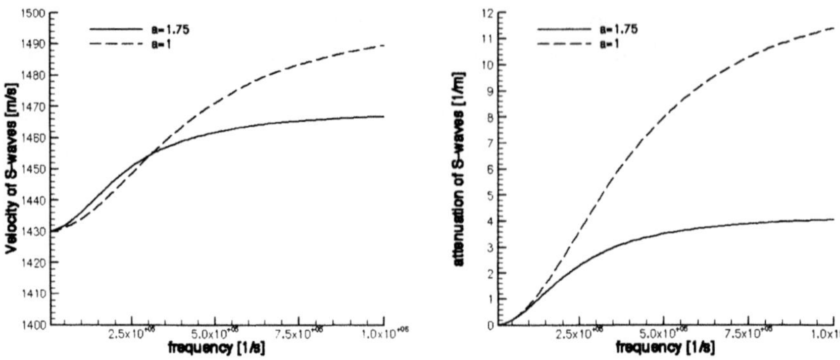


**Figure 23.** Velocity of propagation and attenuation of monochromatic S-waves for two values of the tortuosity coefficient $a$: 1.75 (Biot), 1.00 (simple mixture).

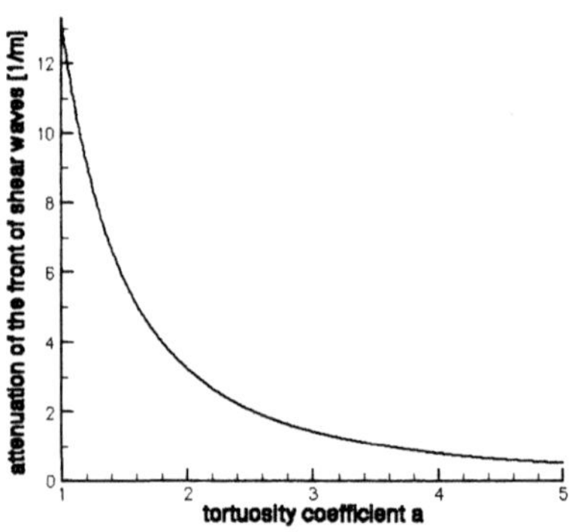


**Figure 24.** Attenuation of the front of shear waves in function of the tortuosity coefficient $a$.

interpretation of the permeability coefficient $\pi$. Usually in the literature on Biot's model, the permeability $\pi$ is attributed to the viscosity of the true fluid component. This is, of course, an unnecessary restriction as curvature microeffects contribute to values of $\pi$ as well and, consequently, a dissipative influence of tortuosity can be included in the model by an effective value of permeability coefficient $\pi$.

Let us return to the second part of the dispersion relation (5.35). For longitudinal waves we obtain

$$\left[\rho_{11}\omega^2 - \left(\lambda^S + 2\mu^S\right)k^2 + i\pi\omega\right]\left[\rho_{22}\omega^2 - \kappa\rho_0^F k^2 + i\pi\omega\right] - \\ - \left(\rho_{12}\omega^2 - Qk^2 - i\pi\omega\right)^2 = 0, \tag{5.42}$$

or, after easy manipulations,

$$\omega\left\{\left[1 - r\left(1-a\right)\right]\left(\frac{\omega}{k}\right)^2 - c_{P1}^2\right\}\left\{a\left(\frac{\omega}{k}\right)^2 - c_{P2}^2\right\} + \\ + \frac{1}{r}i\frac{\pi}{\rho_0^S}\left(\frac{\omega}{k}\right)^2\left\{\left(1+r\right)\left(\frac{\omega}{k}\right)^2 - rc_{P2}^2 - c_{P1}^2 - 2\frac{Q}{\rho_0^S}\right\} - \\ -\frac{1}{r}\omega\left\{r\left(1-a\right)\left(\frac{\omega}{k}\right)^2 - \frac{Q}{\rho_0^S}\right\}^2 = 0, \quad c_{P1}^2 = \frac{\lambda^S + 2\mu^S}{\rho_0^S}, \quad c_{P2}^2 = \kappa. \tag{5.43}$$

Again we check first solutions for two limits of frequencies: $\omega \to 0$, and $\omega \to \infty$.

Let us begin with the second limit which corresponds to the propagation of wave fronts. We have

$$\omega \to \infty : \quad c_\infty := \lim_{\omega\to\infty}\left(\frac{\omega}{\mathrm{Re}k}\right),$$

$$r\left\{\left[1 - r\left(1-a\right)\right]c_\infty^2 - c_{P1}^2\right\}\left\{ac_\infty^2 - c_{P2}^2\right\} - \left\{r\left(1-a\right)c_\infty^2 - \frac{Q}{\rho_0^S}\right\}^2 = 0. \tag{5.44}$$

This is a biquadratic equation for $c_\infty$which yields two nontrivial solutions. These are called the P1-**wave** and the P2-**wave** (or Biot's wave). In contrast to the simple mixture model, in the case of the Biot model these waves are not longitudinal even though they are customarily called so. It can be easily checked by the calculation of corresponding eigenvectors. We shall not do it in this article.

For the simple mixture model, the solution of (5.44) is immediate

$$c_\infty = \begin{cases} c_{P1}, \\ c_{P2}. \end{cases} \tag{5.45}$$

In the classical elasticity, we have $c_{P1} = c_L$and the second longitudinal wave does not exist.

The second, slow wave has been discovered by Frenkel (1944) and then rediscovered by Biot. However, this kind of waves is known since the discovery of the so-called **second sound** in liquid helium by Tisza in 1938. They appear in all hyperbolic models of multicomponent systems – for liquid helium these are the normal fluid and the superfluid. They are very difficult to verify experimentally due to a very high attenuation. However, at least for some porous materials (e.g. sintered glass spheres), they have been indeed observed (e.g. Plona (1980)).

Simultaneously, we obtain the following attenuation in the limit of infinite frequencies

$$\lim_{\omega\to\infty}\left(\mathrm{Im}k\right) = \frac{\pi\Gamma_1}{2\rho_0^S r\Gamma_2}, \tag{5.46}$$

$$\Gamma_1 = c_\infty \left[1 + r - \frac{1}{c_\infty^2}\left(c_{P1}^2 + rc_{P2}^2 + 2\frac{Q}{\rho_0^S}\right)\right],$$

$$\Gamma_2 = c_{P1}^2\left(a - \frac{c_{P2}^2}{c_\infty^2}\right) + c_{P2}^2\left(1 - r(1-a) - \frac{c_{P1}^2}{c_\infty^2}\right) + 2\frac{Q}{\rho_0^S}\left(1 - a - \frac{Q}{r\rho_0^S c_\infty^2}\right),$$

and, for the simple mixture model,

$$\lim_{\omega\to\infty}(\mathrm{Im}k) = \begin{cases} \dfrac{\pi}{2\rho_0^S c_{P1}} & \text{for P1-wave,} \\ \dfrac{\pi}{2\rho_0^F c_{P2}} & \text{for P2-wave.} \end{cases} \tag{5.47}$$

Hence both limits of attenuation for the P1-wave and P2-wave are finite and caused by the permeability $\pi$.

In the case of the low frequency limit we obtain

$$\omega \to 0: \quad c_0 := \lim_{\omega\to 0}\left(\frac{\omega}{\mathrm{Re}k}\right),$$

$$c_0^2\left\{(1+r)c_0^2 - rc_{P2}^2 - c_{P1}^2 + 2\frac{Q}{\rho_0^S}\right\} = 0, \quad \lim_{\omega\to 0}(\mathrm{Im}k) = 0. \tag{5.48}$$

Obviously, we obtain two real solutions of this equation

$$\lim_{\omega\to 0}\left(\frac{\omega}{\mathrm{Re}k}\right)^2\Bigg|_1 : = c_{oP1}^2 = \frac{c_{P1}^2 + rc_{P2}^2 + 2\frac{Q}{\rho_0^S}}{1+r} \equiv \frac{\lambda^S + 2\mu^S + \rho_0^F\kappa + 2\frac{Q}{\rho_0^S}}{\rho_0^S + \rho_0^F}, \tag{5.49}$$

$$\lim_{\omega\to 0}\left(\frac{\omega}{\mathrm{Re}k}\right)^2\Bigg|_2 : = c_{oP2}^2 = 0.$$

These are squares of velocities of propagation of two longitudinal modes in the limit of zero frequency. Clearly, the second mode, P2-wave, does not propagate in this limit. Both limits are independent of tortuosity. The result (5.49) checks with the relation for the velocity of longitudinal waves used in the classical one-component model of soil mechanics provided $Q = 0$.

For the data (5.41) we construct now the numerical example for the P1- and P2-waves discussed above. The solid lines in the following Figures correspond again to Biot's model, the dashed lines to the simple mixture model. In order to show separately the influence of tortuosity $a$ and of the coupling $Q$ we plot as well the solutions with $a = 1$ (dashed dotted lines) and the solutions with $Q = 0$ (dashed double dotted lines).

Even though similar, again, the quantitative differences are much more substantial for P1-waves (Fig. 25). This is primarily the influence of the coupling through partial stresses described by the parameter $Q$. The simple mixture model ($Q = 0, a = 1$) as well as Biot's model with $Q = 0$ yield velocities of these waves which differ only a few percent (lower curves in the left diagram). The coupling $Q$ shifts the curves to higher values and reduces the difference caused by the tortuosity. This result does not seem to be very realistic because the real differences between low frequency and high frequency velocities were measured in soils to be rather as big as indicated by the simple mixture model. This may be an indication that Gassmann relations give much too big values of the coupling parameter $Q$ with respect to those indeed appearing in real granular materials.

Both the tortuosity $a$ and the coupling $Q$ reduce the attenuation quite considerably as indicated in the right Figure.

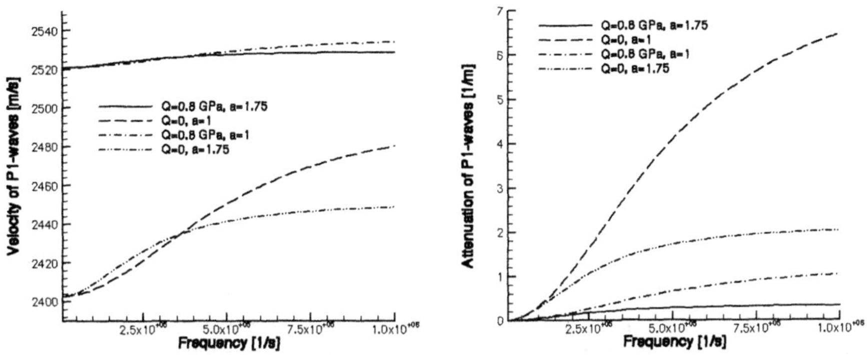


**Figure 25.** Velocity of propagation and attenuation of monochromatic P1-waves for various coupling parameters $Q$ and tortuosity coefficients $a$.

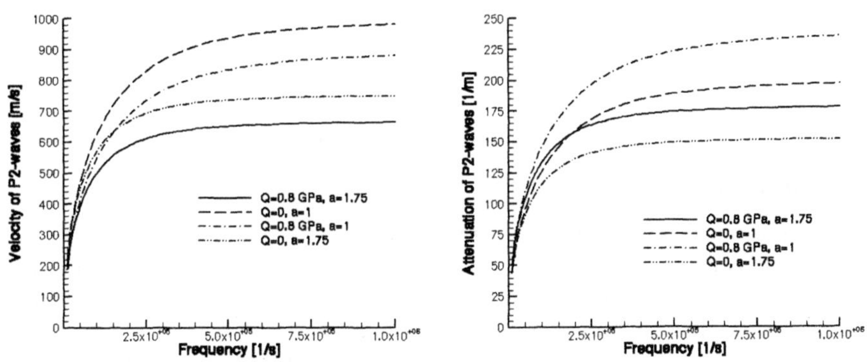


**Figure 26.** Velocity of propagation and attenuation of monochromatic P2-waves for various coupling parameters $Q$ and tortuosity coefficients $a$.

In spite of some claims in the literature the tortuosity $a$ does not influence the existence of the slow (P2-) wave (Fig. 26). Velocities of this wave are again qualitatively similar in Biot's model and in the simple mixture model. The maximum differences appear in the range of high frequencies and reach some 35 percent. The same concerns the attenuation even though quantitative differences are not so big (app. 8 percent).

Let us notice that the lack of coupling through diffusion, $\pi \equiv 0$, yields for both shear waves (5.37) and longitudinal waves (5.43) dispersion relations whose solutions – phase velocities – are independent of frequencies. The waves are nondispersive. Hence, both dispersion and dissipation in the system are caused by the diffusive coupling.

### 5.3 Some remarks on the simple mixture model

It is evident from the above analysis that the simple mixture model yields very substantial technical simplifications in the analysis of wave propagation when compared with the full Biot model. There appear quantitative differences between both models, in some cases substantial. The question arises if Biot's model leads to some qualitative effects which are not appearing in the simple mixture model. The claims that experiments confirm Biot's model are based on quantitative analysis of data rather than on any additional wave effects. Such an analysis may not be very reliable as it is based already on the *a priori* presumption that Biot's model is the only way to describe linear dynamics of porous saturated materials. However, some features of this model like the disastrous diagram in Figure 24 indicate that the model requires corrections.

Simultaneously, an extensive wave analysis for the simple mixture model (e.g. Wilmanski (1998b), Wilmanski (1999), Wilmanski (2001b), Wilmanski (2002a), Wilmanski (2004), Wilmanski and Albers (2003), Albers and Wilmanski (2003)) does not reveal any effects which would not appear as well in Biot's model and, *vice versa*, Biot's model does not indicate any additional effects either. Even such a sophisticated coupling of P1- and P2-waves which yields a local minimum in the velocity of Rayleigh waves (see: Bourbie et al. (1987)) appears in both models. Similar effects are discussed in details in the contribution of B. Albers (2005) to this volume.

As an important consequence of these remarks we proceed with our analysis of surface waves in linear poroelastic materials on the basis of the simple mixture model.

## 6 Surface waves in two-component media

### 6.1 Preliminaries

The theory of surface waves in two-component systems differs qualitatively from such a theory for one-component continua. Such waves are produced in linear models by a combination of bulk waves. In the case of a one-component continuum there are two bulk modes of propagation which yield a single Rayleigh wave. For two-component systems we have three bulk modes: P1-waves, P2-waves and S-waves which produce two surface modes in the case of impermeable boundary. For the permeable boundary, i.e. for the case of an additional system – a fluid in the exterior, there may exists three surface modes, etc. In addition, as all these waves are dissipative, there may exist additional leaky modes similar to these which we have mentioned in Sec. 4.4.

In this Section we consider surface waves in two-component homogeneous poroelastic materials with an impermeable boundary. However, we indicate as well some properties related to the permeable boundary condition. This condition has been proposed by Deresiewicz (1962) (see also: Deresiewicz and Skalak (1963)). The analysis is based on the simple mixture model. We limit the attention solely to high and low frequency ranges. The presentation should be considered to be an introduction to the much more extensive article of B. Albers (2005) contributed to this volume. That article contains also a numerical analysis of dispersion relations in the whole range of frequencies for three types of interfaces: sealed porous medium/vacuum, sealed porous medium/ideal fluid, unsealed porous medium/ideal fluid. We leave entirely open the problem of existence which is at least as complicated as in the case of single component materials.

To the end of this Section we quote a few results obtained within Biot's model. As this model is much more complicated than the simple mixture model results are limited only to some special cases.

### 6.2 Compatibility conditions and dispersion relation

As discussed in Wilmanski (2002a) we seek a solution of the set of fields equations which we obtain from Biot's equations (Box 7) by the substitution $Q = 0, a = 1$. It is convenient to introduce the displacement vector $\mathbf{u}^S$ for the skeleton, and, formally, the displacement vector $\mathbf{u}^F$ for the fluid. The latter is introduced solely for the technical symmetry of considerations and it does not have any physical bearing. Then

$$\begin{aligned} \mathbf{u}^S &= \text{grad}\varphi^S + \text{rot}\boldsymbol{\psi}^S, \quad \mathbf{v}^S = \frac{\partial \mathbf{u}^S}{\partial t}, \quad \mathbf{e}^S = \text{symgrad}\mathbf{u}^S, \\ \mathbf{u}^F &= \text{grad}\varphi^F + \text{rot}\boldsymbol{\psi}^F, \quad \mathbf{v}^F = \frac{\partial \mathbf{u}^F}{\partial t}, \end{aligned} \tag{6.1}$$

where $\varphi^S, \boldsymbol{\psi}^S, \varphi^F, \boldsymbol{\psi}^F$ are two pairs of potentials analogous to those which we were using in the classical elasticity model.

We choose the axes with the downward orientation of the $z$-axis and the $x$-axis in the direction of propagation of the wave. As the problem is assumed to be two-dimensional we make the following ansatz for solutions

$$\begin{aligned} \varphi^S &= A^S(z)\exp\left[i\left(kx-\omega t\right)\right], \quad \varphi^F = A^F(z)\exp\left[i\left(kx-\omega t\right)\right], \\ \psi_y^S &= B^S(z)\exp\left[i\left(kx-\omega t\right)\right], \quad \psi_y^F = B^F(z)\exp\left[i\left(kx-\omega t\right)\right], \\ \psi_x^S &= \psi_z^S = \psi_x^F = \psi_z^F = 0, \end{aligned} \tag{6.2}$$

and

$$\begin{aligned} \rho^S - \rho_0^S &= A_\rho^S(z)\exp\left[i\left(kx-\omega t\right)\right], \quad \rho^F - \rho_0^F = A_\rho^F(z)\exp\left[i\left(kx-\omega t\right)\right], \\ n - n_0 &= A^\Delta \exp\left[i\left(kx-\omega t\right)\right]. \end{aligned} \tag{6.3}$$

Substitution in field equations leads after straightforward calculations to the following compatibility conditions for $z \geq 0$

$$\begin{aligned} B^F &= \frac{i\pi}{\rho_0^F\omega + i\pi}B^S, \quad A^\Delta = -\frac{n_0\omega\tau}{i+\omega\tau}\left(\frac{d^2}{dz^2}-k^2\right)\left(A^F - A^S\right), \\ A_\rho^S &= -\rho_0^S\left(\frac{d^2}{dz^2}-k^2\right)A^S, \quad A_\rho^F = -\rho_0^F\left(\frac{d^2}{dz^2}-k^2\right)A^F, \end{aligned} \tag{6.4}$$

as well as

$$\left[\kappa\left(\frac{d^2}{dz^2}-k^2\right)+\omega^2\right]A^F+\left[\frac{n_0\beta\omega\tau}{\rho_0^F(i+\omega\tau)}\left(\frac{d^2}{dz^2}-k^2\right)+\frac{i\pi}{\rho_0^F}\omega\right]\left(A^F-A^S\right)=0, \quad (6.5)$$

$$\left[\frac{\lambda^S+2\mu^S}{\rho_0^S}\left(\frac{d^2}{dz^2}-k^2\right)+\omega^2\right]A^S-\left[\frac{n_0\beta\omega\tau}{\rho_0^S(i+\omega\tau)}\left(\frac{d^2}{dz^2}-k^2\right)+\frac{i\pi}{\rho_0^S}\omega\right]\left(A^F-A^S\right)=0, \quad (6.6)$$

$$\left[\frac{\mu^S}{\rho_0^S}\left(\frac{d^2}{dz^2}-k^2\right)+\omega^2\right]B^S+\frac{i\pi\rho_0^F}{\rho_0^S\left(\rho_0^F\omega+i\pi\right)}\omega^2B^S=0. \quad (6.7)$$

It is convenient to introduce a dimensionless notation. In order to do so we define the following auxiliary quantities

$$\begin{aligned} c_s &= \frac{c_S}{c_{P1}}<1, \quad c_f=\frac{c_{P2}}{c_{P1}}, \quad \pi'=\frac{\pi\tau}{\rho_0^S}>0, \quad \beta'=\frac{n_0\beta}{\rho_0^S c_{P1}^2}>0, \\ r &= \frac{\rho_0^F}{\rho_0^S}<1, \qquad z'=\frac{z}{c_{P1}\tau}, \quad k'=kc_{P1}\tau, \quad \omega'=\omega\tau. \end{aligned} \quad (6.8)$$

where the velocities $c_{P1}, c_{P2}, c_S$ are defined by (5.41) and, in the simple mixture model, they describe the velocities of fronts of bulk P1-, P2-, and S-wave, respectively. These are, of course, identical with the limits of bulk phase velocities for $\omega \to \infty$. As we neglect processes of relaxation of porosity, the reference time $\tau$ can be chosen arbitrarily.

As we have already mentioned, we neglect further the influence of the nonequilibrium changes of porosity, i.e. $\beta = 0$. In the compatibility relations derived in this Subsection we still keep it in order to show the way in which this influence enters the model if not ignored.

Further we omit the prime for typographical reasons. Substitution of (6.8) in equations (6.5), (6.6), (6.7) yields

$$\begin{aligned} \left[c_f^2\left(\frac{d^2}{dz^2}-k^2\right)+\omega^2\right]A^F+\left[\frac{\beta\omega}{r(i+\omega)}\left(\frac{d^2}{dz^2}-k^2\right)+i\frac{\pi}{r}\omega\right]\left(A^F-A^S\right) &= 0, \\ \left[\left(\frac{d^2}{dz^2}-k^2\right)+\omega^2\right]A^S-\left[\frac{\beta\omega}{i+\omega}\left(\frac{d^2}{dz^2}-k^2\right)+i\pi\omega\right]\left(A^F-A^S\right) &= 0, \\ \left[c_s^2\left(\frac{d^2}{dz^2}-k^2\right)+\omega^2+\frac{i\pi\omega}{\omega+i\frac{\pi}{r}}\right]B^S &= 0. \end{aligned} \quad (6.9)$$

This differential eigenvalue problem can be easily solved because the matrix of coefficients for homogeneous materials is independent of $z$. This is different from the case of waves in heterogeneous materials (compare, for instance, the article of C. G. Lai (2005)). Consequently, we seek solutions in the form

$$A^F=A_f^1e^{\gamma_1 z}+A_f^2e^{\gamma_2 z}, \quad A^S=A_s^1e^{\gamma_1 z}+A_s^2e^{\gamma_2 z}, \quad B^S=B_se^{\zeta z}, \quad (6.10)$$

where, due to the chosen direction of the $z$-axis, the exponents $\gamma_1, \gamma_2, \zeta$ must possess negative real parts. This is the existence requirement for surface waves. Substitution in (6.9) yields them in the form

$$\left(\frac{\zeta}{k}\right)^2 = 1 - \frac{1}{c_s^2}\left(1 + \frac{i\pi}{\omega + i\frac{\pi}{r}}\right)\left(\frac{\omega}{k}\right)^2, \tag{6.11}$$

and

$$\left[c_f^2 + \left(c_f^2 + \frac{1}{r}\right)\frac{\beta\omega}{i+\omega}\right]\left[\left(\frac{\gamma}{k}\right)^2 - 1\right]^2 + \left[1 + \left(1 + \frac{1}{r}\right)\frac{i\pi}{\omega}\right]\left(\frac{\omega}{k}\right)^4$$
$$+\left[1 + c_f^2 + \left(1 + \frac{1}{r}\right)\frac{\beta\omega}{i+\omega} + \left(c_f^2 + \frac{1}{r}\right)\frac{i\pi}{\omega}\right]\left[\left(\frac{\gamma}{k}\right)^2 - 1\right]\left(\frac{\omega}{k}\right)^2 = 0. \tag{6.12}$$

Simultaneously we obtain the following relations for eigenvectors

$$\mathbf{R}^1 = \left(B_s, A_s^1, A_f^1\right)^T, \quad \mathbf{R}^2 = \left(B_s, A_s^2, A_f^2\right)^T, \tag{6.13}$$

where

$$A_f^1 = \delta_f A_s^1, \quad A_s^2 = \delta_s A_f^2, \tag{6.14}$$

$$\delta_f := \frac{1}{r}\frac{\frac{\beta\omega}{i+\omega}\left[\left(\frac{\gamma_1}{k}\right)^2 - 1\right] + \frac{i\pi}{\omega}\frac{\omega^2}{k^2}}{\left(c_f^2 + \frac{1}{r}\frac{\beta\omega}{i+\omega}\right)\left[\left(\frac{\gamma_1}{k}\right)^2 - 1\right] + \left(\frac{\omega}{k}\right)^2 + \frac{i\pi}{\omega r}\frac{\omega^2}{k^2}}, \tag{6.15}$$

$$\delta_s := \frac{\frac{\beta\omega}{i+\omega}\left[\left(\frac{\gamma_2}{k}\right)^2 - 1\right] + \frac{i\pi}{\omega}\frac{\omega^2}{k^2}}{\left(1 + \frac{\beta\omega}{i+\omega}\right)\left[\left(\frac{\gamma_2}{k}\right)^2 - 1\right] + \left(\frac{\omega}{k}\right)^2 + \frac{i\pi}{\omega}\frac{\omega^2}{k^2}}. \tag{6.16}$$

The above solution for the exponents still leaves three unknown constants $B_s, A_f^2, A_s^1$ which must be specified from boundary conditions. This is the subject of the next Subsection.

For technical reasons, we limit the attention only to the limit problems in the range of high and low frequencies.

In the case of **high frequency approximation** we immediately obtain from relations (6.11) and (6.12)

$$\frac{1}{\omega} \ll 1: \qquad \left(\frac{\zeta}{k}\right)^2 = 1 - \frac{1}{c_s^2}\left(\frac{\omega}{k}\right)^2,$$
$$\left(\frac{\gamma_1}{k}\right)^2 = 1 - \left(\frac{\omega}{k}\right)^2, \quad \left(\frac{\gamma_2}{k}\right)^2 = 1 - \frac{1}{c_f^2}\left(\frac{\omega}{k}\right)^2, \tag{6.17}$$

and

$$\delta_f = \delta_s = 0 \quad \Rightarrow \quad \mathbf{R}^1 = \left(B_s, A_s^1, 0\right)^T, \quad \mathbf{R}^2 = \left(B_s, 0, A_f^2\right)^T. \tag{6.18}$$

For the case of **low frequency approximation** the equation (6.12) becomes singular. It can be written in the following form

$$\omega \ll 1: \qquad c_f^2\omega^2\left[\left(\frac{\gamma}{k}\right)^2-1\right]^2+\omega\left[\omega+i\pi\left(1+\frac{1}{r}\right)\right]\left(\frac{\omega}{k}\right)^4+$$
$$+\left[\omega\left(1+c_f^2\right)+i\pi\left(c_f^2+\frac{1}{r}\right)\right]\omega\left[\left(\frac{\gamma}{k}\right)^2-1\right]\left(\frac{\omega}{k}\right)^2=0. \tag{6.19}$$

Making the following substitution

$$W:=\omega\left[\left(\frac{\gamma}{k}\right)^2-1\right]+\omega\left(\frac{\omega}{k}\right)^2\frac{1+c_f^2}{2c_f^2}, \tag{6.20}$$

we obtain a quadratic equation for $W$

$$c_f^2W^2+i\pi\left(c_f^2+\frac{1}{r}\right)\left(\frac{\omega}{k}\right)^2W- \tag{6.21}$$
$$-\left\{i\pi\left[\frac{\left(c_f^2+\frac{1}{r}\right)\left(1+c_f^2\right)}{2c_f^2}-\left(1+\frac{1}{r}\right)\right]\left(\frac{\omega}{k}\right)^4\right\}\omega+O\left(\omega^2\right)=0,$$

which for small $\omega$ can be solved by the regular perturbation method

$$W=W_0+\omega W_1+O\left(\omega^2\right). \tag{6.22}$$

After easy calculations we obtain

$$W=\begin{cases}\left[\frac{1+c_f^2}{2c_f^2}-\frac{r+1}{rc_f^2+1}\right]\left(\frac{\omega}{k}\right)^2\omega, \\ -\left[\frac{1+c_f^2}{2c_f^2}-\frac{r+1}{rc_f^2+1}\right]\left(\frac{\omega}{k}\right)^2\omega-i\pi\frac{rc_f^2+1}{rc_f^2}\left(\frac{\omega}{k}\right)^2\end{cases}. \tag{6.23}$$

Bearing the relation (6.20) in mind we arrive at the following results for the exponents

$$\omega \ll 1: \qquad \left(\frac{\varsigma}{k}\right)^2=1-\frac{r+1}{c_s^2}\left(\frac{\omega}{k}\right)^2,$$
$$\left(\frac{\gamma_1}{k}\right)^2=1-\frac{r+1}{rc_f^2+1}\left(\frac{\omega}{k}\right)^2, \tag{6.24}$$
$$\left(\frac{\gamma_2}{k}\right)^2=1-\frac{rc_f^4+1}{c_f^2\left(rc_f^2+1\right)}\left(\frac{\omega}{k}\right)^2-\frac{i\pi}{\omega}\frac{rc_f^2+1}{rc_f^2}\left(\frac{\omega}{k}\right)^2,$$

and for the coefficients of amplitudes

$$\delta_f=1-\frac{\omega r}{i\pi}\frac{1-c_f^2}{1+rc_f^2}, \quad \delta_s=-rc_f^2\left(1-\frac{\omega r}{i\pi}\frac{1-c_f^2}{1+rc_f^2}\right). \tag{6.25}$$

Obviously due to the singular character of the equation (6.19) the last contribution to $\dfrac{\gamma_2}{k}$ becomes singular for $\omega \to 0$.

## 6.3 Boundary value problems for surface waves

In order to determine surface waves in saturated poroelastic medium we need conditions for $z = 0$. We discuss in some details the problem in which this boundary is **impermeable**, and a poroelastic medium is in contact with **vacuum**. Boundary conditions have then the form

$$T_{13}\big|_{z=0} \equiv T_{13}^S\big|_{z=0} = \mu^S \left( \frac{\partial u_1^S}{\partial z} + \frac{\partial u_3^S}{\partial x} \right)\bigg|_{z=0} = 0, \tag{6.26}$$

$$T_{33}\big|_{z=0} \equiv \left(T_{33}^S - p^F\right)\big|_{z=0} =$$

$$= c_{P1}^2 \rho_0^S \left( \frac{\partial u_1^S}{\partial x} + \frac{\partial u_3^S}{\partial z} \right) - 2c_S^2 \rho_0^S \frac{\partial u_1^S}{\partial x} - c_{P2}^2 \left(\rho^F - \rho_0^F\right)\bigg|_{z=0} = 0, \tag{6.27}$$

$$\frac{\partial}{\partial t}\left(u_3^F - u_3^S\right)\bigg|_{z=0} = 0, \tag{6.28}$$

where the first two conditions mean that the surface $z = 0$ is stress-free (far-field approximation), and the last condition means that there is no transport of fluid mass through this surface (impermeable boundary). $u_1^S, u_3^S$ denote the components of the displacement $\mathbf{u}^S$ in the direction of $x$-axis and $z$-axis, respectively, while $u_3^F$ is the $z$-component of the displacement $\mathbf{u}^F$.

In the case of a permeable boundary neither the condition (6.26) nor the condition (6.28) would hold.

The first condition would have to possess the right-hand side reflecting the external pressure $p_{ext}$ appearing in the fluid outside of the porous material. This change would appear as well in the case of impermeable boundary when we did not have the vacuum outside.

Condition (6.28) which reflects the fact that the impermeable boundary is **material** for both the solid and fluid component would have to describe mass transport through the surface specified by a relation to a driving force. According to the proposition of Deresiewicz and Skalak (1963) such a driving force is proportional to the difference of pore pressures on both sides of the boundary

$$\rho_0^F \frac{\partial}{\partial t}\left(u_3^F - u_3^S\right) - \alpha\left(p^F - n_0 p_{ext}\right)\bigg|_{z=0} = 0, \tag{6.29}$$

where $\alpha$ denotes a surface permeability coefficient and $p_{ext}$ is an external pressure.

The coefficient $\alpha$ is an overall macroscopic description of a boundary layer which is created by the flow of the fluid component from conditions specified by the porous material (i.e. by the permeability $\pi$, porosity $n$, a geometry of the microscopic vicinity of the boundary such as a shape of openings of channels, their average orientation with respect to the surface normal, etc.) to the free space of a pure fluid. It is clear that the limit $\alpha \to 0$ corresponds to the impermeable (sealed) boundary, and the limit $\alpha \to \infty$ corresponds to the continuity of pressure in the fluid: $p^F = n_0 p_{ext}$. Such a boundary condition is used, for instance, in theories of porous materials with a rigid skeleton which are used in the description of various geotechnical diffusion and seepage processes.

In addition, for the permeable boundary we have to account for the continuity of the mass flux through the boundary. This additional boundary condition is necessary with respect to the existence of an additional constant in the solution for the exterior (in the range $z < 0$).

Substitution of results of the previous Subsection in boundary conditions (6.26)-(6.28) yields the following equations for three unknown constants $B_s, A_f^2, A_s^1$

$$\mathbf{AX} = \mathbf{0}, \tag{6.30}$$

where

$$\mathbf{A} := \begin{pmatrix} \left(\frac{\varsigma}{k}\right)^2 + 1 & 2i\frac{\gamma_2}{k}\delta_s & 2i\frac{\gamma_1}{k} \\ -2ic_s^2\frac{\varsigma}{k} & \begin{array}{c}\left[\left(\frac{\gamma_2}{k}\right)^2 - 1 + 2c_s^2\right]\delta_s + \\ +rc_f^2\left[\left(\frac{\gamma_2}{k}\right)^2 - 1\right]\end{array} & \begin{array}{c}\left(\frac{\gamma_1}{k}\right)^2 - 1 + 2c_s^2 + \\ +rc_f^2\left[\left(\frac{\gamma_1}{k}\right)^2 - 1\right]\delta_f\end{array} \\ i\frac{r\omega}{r\omega + i\pi} & -\left(\delta_s - 1\right)\frac{\gamma_2}{k} & \left(\delta_f - 1\right)\frac{\gamma_1}{k} \end{pmatrix}, \tag{6.31}$$

$$\mathbf{X} := \left( B_s, \; A_f^2, \; A_s^1 \right)^T .$$

This homogeneous set yields the **dispersion relation:** $\det \mathbf{A} = 0$ determining the $\omega - k$ relation. We investigate separately solutions of this equation for high and low frequencies.

## 6.4 High frequency approximation

In the case of high frequencies $\frac{1}{\omega} \ll 1$ we have $\delta_s = \delta_f = 0$ and the dispersion relation follows in the form

$$\mathcal{P}_R\sqrt{1 - \frac{1}{c_f^2}\left(\frac{\omega}{k}\right)^2} + \frac{r}{c_s^4}\left(\frac{\omega}{k}\right)^4\sqrt{1 - \left(\frac{\omega}{k}\right)^2} = 0, \tag{6.32}$$

where

$$\mathcal{P}_R := \left(2 - \frac{1}{c_s^2}\left(\frac{\omega}{k}\right)^2\right)^2 - 4\sqrt{1 - \left(\frac{\omega}{k}\right)^2}\sqrt{1 - \frac{1}{c_s^2}\left(\frac{\omega}{k}\right)^2}. \tag{6.33}$$

Hence for $r = 0$ the relation (6.32) reduces to $\mathcal{P}_R = 0$ which is the Rayleigh dispersion relation for single component continua. Otherwise we obtain the relation identical with this analyzed by Edelman and Wilmanski (2002) in the limit of short waves (i.e. $\frac{1}{k} \ll 1$). Consequently, the conclusions for this case are the same as well. As shown in the paper Edelman and Wilmanski (2002) the equation (6.32) possesses two roots defining two surface waves: a true Stoneley wave which propagates with the finite attenuation and with the velocity a bit smaller than $c_f$ as well as a generalized Rayleigh wave which is leaky (i.e. it radiates the energy to the P2-wave) and propagates with the velocity $c_R$: $c_f < c_R < c_s$. The Rayleigh wave is leaky because its attenuation is unbounded, i.e. such a wave cannot exist in the range of high frequencies. Immediately after the initiation, it transforms into bulk waves.

These results are not very surpprising because the dispersion relation (6.32) is identical with the dispersion relation in Box 6 for the Stoneley-Scholte wave. The only difference is that the real Stoneley-Scholte wave propagates on both sides of the interface and the above presented wave propagates solely below the boundary ($z > 0$) in the porous medium.

The detailed description of these waves can be found in Albers (2005).

## 6.5 Low frequency approximation

If we account for the relations (6.24) and (6.25) in the condition det $\mathbf{A}$ =0 then we obtain the dispersion relation reflecting a dependence of $\frac{\omega}{k}$ on $\omega$. The expansion with respect to $\sqrt{\omega}$ yields the identity in the zeroth order and the following relation for the higher order

$$\left(\frac{\omega}{k}\right)\left\{\left(2-\frac{r+1}{c_s^2}\left(\frac{\omega}{k}\right)^2\right)^2-4\sqrt{1-\frac{r+1}{c_s^2}\left(\frac{\omega}{k}\right)^2}\sqrt{1-\frac{r+1}{rc_f^2+1}\left(\frac{\omega}{k}\right)^2}\right\}+ \tag{6.34}$$

$$+O\left(\sqrt{\omega}\right)=0.$$

Clearly we obtain two solutions:

1. Rayleigh wave whose velocity is different from zero in the limit $\omega \longrightarrow 0$ and whose attenuation is of the order $O\left(\sqrt{\omega}\right)$. The relation for the velocity reminds the relation (6.33) with the velocities of bulk waves replaced by the low frequency limits. Namely we have

$$\frac{r+1}{c_s^2}=c_{P1}^2\frac{\rho_0^S+\rho_0^F}{\mu^S}\equiv\frac{c_{P1}^2}{c_{oS}^2},\quad \frac{r+1}{rc_f^2+1}=c_{P1}^2\frac{\rho_0^S+\rho_0^F}{\lambda^S+2\mu^S+\rho_0^F\kappa}\equiv\frac{c_{P1}^2}{c_{oP1}^2}. \tag{6.35}$$

   Consequently

$$\left(2-\frac{c_{P1}^2}{c_{oS}^2}\left(\frac{\omega}{k}\right)^2\right)^2-4\sqrt{1-\frac{c_{P1}^2}{c_{oS}^2}\left(\frac{\omega}{k}\right)^2}\sqrt{1-\frac{c_{P1}^2}{c_{oP1}^2}\left(\frac{\omega}{k}\right)^2}=0. \tag{6.36}$$

2. The Stoneley wave has the velocity of propagation of the order $O\left(\sqrt{\omega}\right)$. Hence, it goes to zero in the same way as the velocity of propagation of the P2-wave.

BOX 8: *Surface waves on an impermeable boundary of porous materials – dispersion relations*

High frequency approximation: $\omega \to \infty$

$$\mathcal{P}_R \sqrt{1 - \frac{1}{c_{P2}^2}\left(\frac{\omega}{k}\right)^2} + \frac{r}{c_S^4}\left(\frac{\omega}{k}\right)^4 \sqrt{1 - \frac{1}{c_{P1}^2}\left(\frac{\omega}{k}\right)^2} = 0,$$

$$\mathcal{P}_R := \left(2 - \frac{1}{c_S^2}\left(\frac{\omega}{k}\right)^2\right)^2 - 4\sqrt{1 - \frac{1}{c_{P1}^2}\left(\frac{\omega}{k}\right)^2}\sqrt{1 - \frac{1}{c_S^2}\left(\frac{\omega}{k}\right)^2},$$

Low frequency approximation: $\omega \to 0$

$$\left(2 - \frac{1}{c_{oS}^2}\left(\frac{\omega}{k}\right)^2\right)^2 - 4\sqrt{1 - \frac{1}{c_{oS}^2}\left(\frac{\omega}{k}\right)^2}\sqrt{1 - \frac{1}{c_{oP1}^2}\left(\frac{\omega}{k}\right)^2} = 0,$$

where

$$c_{P1} = \sqrt{\frac{\lambda^S + 2\mu^S}{\rho_0^S}}, \quad c_{P2} = \sqrt{\kappa}, \quad c_S = \sqrt{\frac{\mu^S}{\rho_0^S}},$$

$$c_{oP1} = \sqrt{\frac{\lambda^S + 2\mu^S + \rho_0^F \kappa}{\rho_0^S + \rho_0^F}}, \quad c_{oS} = \sqrt{\frac{\mu^S}{\rho_0^S + \rho_0^F}}, \quad r = \frac{\rho_0^F}{\rho_0^S}.$$

## 6.6 Remarks on modelling surface waves by Biot's model and the simple mixture model

The results for a two-component model of porous solid-fluid mixtures presented in this Section should be compared with those obtained by means of the Biot model and with experimental observations. We shall not go into details of such a comparison in this work. However, as mentioned before, there is a very good qualitative agreement of both models as far as propagation of acoustic waves is concerned. Velocities of bulk waves are influenced by the coupling parameter $Q$ but this influence reflected in high frequency – low frequency relations seems to be too strong for values of this parameter following from the classical Gassmann relations. The influence of tortuosity $a$ on the velocities of propagation is rather small and, simultaneously, essential changes of attenuation do not correspond to the physical inside, particularly to mechanisms of scattering of waves, of the morphology of porous materials in spite of some claims in the literature (e.g. compare Johnson et al. (1982)).

Neither bulk waves nor surface waves reveal any qualitative differences between Biot's model and the simple mixture model. It can be expected that ranges of existence of different surface modes are different for both models but results of analysis of this problem are not yet available.

It should be mentioned that results for surface waves within the Biot model are often limited to high frequency limits (e.g. Feng and Johnson (1983)). Some early results of Deresiewicz for low frequencies (Deresiewicz (1961), Deresiewicz (1962), Deresiewicz (1964a), Deresiewicz (1964b)) do not depart from those obtained within the simple mixture model. There exists a number of works in which solutions similar to those of Deresiewicz are obtained for more complex systems (e.g. Kumar and Miglani (1996), Kumar et al. (2002), Sharma et al. (1990), Sharma et al. (1991)). For instance, in the work Kumar et al. (2002) a porous semiinfinite medium described by the Biot model is considered to be in contact with a layer of an ideal homogeneous fluid on which there is another layer of ideal inhomogeneous fluid with a linear dependence of material parameters on the $z$-coordinate. Surface waves are sougth under the following simplifying assumptions. In the Deresiewicz-Skalak boundary condition (6.29), it is assumed that the boundary permeability $\alpha \to \infty$. The second assumption concerns the diffusion. Authors assume lack of dissipation which means that the partial velocities of components are equal. This happens approximately indeed in the range of low frequencies where the components move almost synchronized but not in the range of high frequencies. However, both limits are investigated numerically in the work. In these calculations, the data are chosen in such a way that the tortuosity $a = 1.01$ which is, of course almost eliminating this effect from Biot's model. It is rather natural that results practically do not differ from those for single component materials.

The most interesting work with a very extensive comparison of different models has been done by A. Norris (1989) for the propagation of Stoneley waves on the permeable boundary of the porous material. He investigates two problems. A low frequency regime is first described by the model of pore-pressure diffusion in a compressible frame. In this range, the Stoneley wave in a borehole is called a **tube wave** and Norris presents a comparison of various so-called quasi-static models. He is using the full Deresiewicz-Skalak boundary condition. The most extensive part of the work concerns the application of Biot's model for which the frequency variations of the velocity and attenuation of the Stoneley wave are discussed in details for the range app. 0-50 kHz. For the lack of space in this article we have to refer to the original work for details. It would be, certainly, interesting to compare these results based on Biot's model with those which follow from the simple mixture model.

Further details concerning the comparison of both models and a full account of numerical results for surface waves described by the simple mixture model can be found in the article of B. Albers (2005). Certainly, it should be born in mind that the simple mixture model must be quantitatively a worse approximation than Biot's model because it does not contain natural physical couplings. However, it is known from the theory of mixture of fluids that many results of the simple mixture theory are good enough for some practical purposes. This seems to be also the case for porous materials as far as a qualitative analysis of acoustic waves is concerned.

## Bibliography

J. D. Achenbach. *Wave propagation in elastic solids.* North-Holland Publ. Comp., Amsterdam, 1973.

K. Aki and P. G. Richards. *Quantitative seismology.* University Science Books, Sansalito, second edition edition, 2002.

B. Albers. Modelling of surface waves in poroelastic saturated materials by means of a two-component continuum. In C. Lai and K. Wilmanski, editors, *Surface Waves in Geomechanics, Direct and Inverse Modelling for Soils and Rocks*, CISM Courses and Lectures. Springer Verlag, Wien-New York, 2005.

B. Albers and K. Wilmanski. On modeling acoustic waves in saturated poroelastic media. Preprint 874, WIAS, 2003. to appear in: J. Engn. Mech., 2004.

J. F. Allard. *Propagation of Sound in Porous Media. Modelling Sound Absorbing Materials.* Elsevier, Essex, 1993.

J. H. Ansell. The roots of the stoneley wave equation for solid-liquid interfaces. *Pure Appl. Geophys.*, 94:172–188, 1972.

E. A. Ash and E. G. S. Paige. *Rayleigh-Wave theory and application.* Springer, Berlin, 1985.

A. Ben-Menahem and S. J. Singh. *Seismic waves and sources.* Springer-Verlag, Dover, New York, second edition, 2000.

J. G. Berryman. Confirmation of Biot's theory. *Appl. Phys. Lett.*, 37:382–384, 1980.

M. A. Biot. Theory of propagation of elastic waves in a fluid-saturated porous solid. I. Low-frequency range. *J. Acoust. Soc. Am.*, 28:168–178, 1956.

M. A. Biot and D. G. Willis. The elastic coefficients of the theory of consolidation. *J. Appl. Mech.*, 24:594–601, 1957.

T. Bourbie, O. Coussy, and B. Zinszner. *Acoustics of porous media.* Éditions Technip, Paris, 1987.

L. M. Brekhovskikh and O. A. Godin. *Acoustics of layered media. I. Plane and quasi-plane waves.* Springer, Berlin, 1998.

L. M. Brekhovskikh and V. Goncharov. *Mechanics of continua and wave dynamics.* Springer, Berlin, second edition, 1994.

L. Cagniard. *Reflection and refraction of progressive waves.* McGraw-Hill Book Co., 1962.

C. A. Coulson and A. Jeffrey. *Waves; A mathematical approach to the common types of wave motion.* Longman, London, second edition, 1977.

H. Deresiewicz. The effect of boundaries on wave propagation in a liquid-filled porous solid: II. Love waves in a porous layer. *Bull. Seismol. Soc. Am.*, 51(1):51–59, 1961.

H. Deresiewicz. The effect of boundaries on wave propagation in a liquid-filled porous solid: IV. Surface waves in a half-space. *Bull. Seismol. Soc. Am.*, 52(3):627–638, 1962.

H. Deresiewicz. The effect of boundaries on wave propagation in a liquid-filled porous solid: VI. Love waves in a double surface layer,. *Bull. Seismol. Soc. Am.*, 54:417–423, 1964a.

H. Deresiewicz. The effect of boundaries on wave propagation in a liquid filled porous solid. VII. Surface waves in a half space in the presence of a liquid layer. *Bull. Seismol. Soc. Am.*, 54(1):425–430, 1964b.

H. Deresiewicz and R. Skalak. On uniqueness in dynamic poroelasticity. *Bull. Seismol. Soc. Am.*, 53:783–788, 1963.

D. Drew, L. Cheng, and R. T. Lahey Jr. The analysis of virtual mass effects in two-phase flow. *Int. J. Multiphase Flow*, 5:233–242, 1979.

I. Edelman. Bifurcation of the Biot slow wave in a porous medium. *J. Acoust. Soc. Am.*, 114(1):90–97, 2003.

I. Edelman and K. Wilmanski. Asymptotic analysis of surface waves at vacuum/porous medium and liquid/porous medium interfaces. *Contin. Mech. Thermodyn.*, 14(1): 25–44, 2002.

W. C. Elmore and M. A. Heald. *Physics of waves.* Dover, New York, 1985.

S. Feng and D. L. Johnson. High-frequency acoustic properties of a fluid/porous solid interface. I. New surface mode. *J. Acoust. Soc. Am.*, 74(3):906–914, 1983.

S. Foti. Surface wave testing for geotechnical characterization of a real site. In C. Lai and K. Wilmanski, editors, *Surface Waves in Geomechanics, Direct and Inverse Modelling for Soils and Rocks*, CISM Courses and Lectures. Springer Verlag, Wien-New York, 2005.

Ya. Frenkel. On the theory of seismic and seismoelectric phenomena in moist soils. *J. Phys.*, 8(4):230–241, 1944.

F. Gassmann. Über die Elastizität poröser Medien. *Vierteljahresschrift der Naturforschenden Gesellschaft in Zürich*, 96:1–23, 1951.

A. S. Ginzbarg and E Strick. Stoneley-wave velocities for a solid-solid interface. *Bull. Seismol. Soc. Am.*, 48:51–63, 1958.

P. Hess. Surface acousic waves in materials science. *Physics Today*, March:42–47, 2002.

D. L. Johnson, T. J. Plona, C. Scala, F. Pasierb, and H. Kojima. Tortuosity and acoustic slow waves. *Phys. Rev. Lett.*, 49:1840–1844, 1982.

E. Kausel. Do complex rayleigh modes really exist? private communication, 2004.

E. Kausel. Numerical technics in eigenvalue problems for surface waves. In C. Lai and K. Wilmanski, editors, *Surface Waves in Geomechanics, Direct and Inverse Modelling for Soils and Rocks*, CISM Courses and Lectures. Springer Verlag, Wien-New York, 2005.

H. Kolsky. *Stress waves in solids.* Clarendon Press, Oxford, 1953.

R. Kumar and A. Miglani. Effect of pore alignment on surface wave propagation in a liquid-saturated porous layer over a liquid-saturated porous half-space with loosely bonded interface. *J. Phys. Earth*, 44:153–172, 1996.

R. Kumar, A. Miglani, and N. R. Garg. Surface wave propagation in a double liquid layer over a liquid-saturated porous half-space. *Sādhanā*, 27(6):643–655, 2002.

C. G. Lai. *Simultaneous inversion of Rayleigh phase velocity and attenuation for near-surface site characterization.* PhD thesis, Georgia Institute of Technology, 1998.

C. G. Lai. Surface waves in dissipative media: Forward and inverse modelling. In C. Lai and K. Wilmanski, editors, *Surface Waves in Geomechanics, Direct and Inverse Modelling for Soils and Rocks*, CISM Courses and Lectures. Springer Verlag, Wien-New York, 2005.

R. Lancellotta. Experimental soil behavior, its testing by waves, engineering applications. In C. Lai and K. Wilmanski, editors, *Surface Waves in Geomechanics, Direct and Inverse Modelling for Soils and Rocks*, CISM Courses and Lectures. Springer Verlag, Wien-New York, 2005.

G. Liu and J. Qu. Guided circumferential waves in a circular annulus. *J. Appl. Mech.*, 65:424–430, 1998.

I-Shih Liu. *Continuum Mechanics.* Springer, Berlin, 2002.

A. E. H. Love. *Some problems of geodynamics.* Cambridge University Press, 1911. reprinted by Dover, N.Y., 1967.

G. Maugin. Theory of nonlinear surface waves and solitons. In C. Lai and K. Wilmanski, editors, *Surface Waves in Geomechanics, Direct and Inverse Modelling for Soils and Rocks*, CISM Courses and Lectures. Springer Verlag, Wien-New York, 2005.

I. Müller. A new approach to thermodynamics of simple mixtures. *Zeitschrift für Naturforschung*, 28a, 1973.

I. Müller. *Thermodynamics.* Pitman, Boston, 1985.

E. G. Nesvijski. On a possibility of Rayleigh transformed sub-surface waves propagation. *NDT.net*, 5(09), September 2000.

A. N. Norris. Back reflection of ultrasonic waves from a liquid-solid interface. *J. Acoust. Soc. Am.*, 73(2):427–434, 1983.

A. N. Norris. Stoneley-wave attenuation and dispersion in permeable formations. *Geophysics*, 54(3):330–341, 1989.

A. N. Norris and B. K. Sinha. The speed of a wave along a fluid/solid interface in the presence of anisotropy and prestress. *J. Acoust. Soc. Am*, 98(2):1147–1154, 1995.

D. F. Parker and G. A. Maugin. *Recent developments in surface acoustic waves.* Springer, Berlin, 1988.

R. A. Phinney. Propagation of leaking interface waves. *Bull. Seismol. Soc. Am.*, 51(4): 527–555, 1961.

T. J. Plona. Observation of a second bulk compressional wave in a porous medium at ultrasonic frequencies. *Appl. Phys. Lett.*, 36:259–261, 1980.

J. W. (Strutt) Rayleigh. On waves propagated along the plane surface of an elastic solid. *Proc. London Math. Soc.*, 17:4–11, 1887.

G. J. Rix. Surface testing for near-surface site characterization. In C. Lai and K. Wilmanski, editors, *Surface Waves in Geomechanics, Direct and Inverse Modelling for Soils and Rocks*, CISM Courses and Lectures. Springer Verlag, Wien-New York, 2005.

G. J. Rix, C. G. Lai, and S. Foti. Simultaneous measurement of surface wave dispersion and attenuation curves. *Geotechnical Testing Journal*, 24(4):350–358, 2001.

H. Sato and M. Fehler. *Seismic wave propagation and scattering in the heterogeneous earth.* Springer, New York, 1998.

J. G. Scholte. On seismic waves in a spherical earth. *Koninkl. Ned. Meteorol. Inst.*, pages 122–165, 1947a.

J. G. Scholte. The range of existence of Rayleigh and Stoneley waves. *Monthly Notices of the Royal Astronomical Society, Geophysical Supplement*, 5:120–126, 1947b.

C. T. Schroeder and W. R. Scott. A finite-difference model to study the elastic-wave interactions with buried land mines. *IEEE Trans. Geosci. Remote Sensing*, 38(4): 1505–1512, 2000.

C. T. Schroeder and W. R. Scott. On the complex conjugate roots of the Rayleigh equation: The leaky surface wave. *J. Acoust. Soc. Am.*, 110(6):2867–2877, 2001.

M. D. Sharma, R. Kumar, and M. L. Gogna. Surface wave propagation in a transversely isotropic elastic layer overlying a liquid-saturated porous solid half-space and lying under a uniform layer of liquid. *Pure Appl. Geophys.*, 133:523–540, 1990.

M. D. Sharma, R. Kumar, and M. L. Gogna. Surface wave propagation in a liquid-saturated porous layer overlying a homogeneous transversely isotropic half-space and lying under a uniform layer of liquid. *Int. J. Solids Struct.*, 27:1255–1267, 1991.

R. D. Stoll. *Sediment Acoustics.* Number 26 in Lecture Notes in Earth Sciences. Springer, New York, 1989.

R. Stoneley. Elastic waves at the surface of separation of two solids. *Proc. Roy. Soc. London, Ser. A*, 106:416–428, 1924.

E. Strick. The pseudo-Rayleigh wave. In W. L. Roever, T. F. Vining, and E. Strick, editors, *On the propagation of elastic wave motion*, volume 251 of *Phil. Trans. Roy. Soc. (London), Ser. A*, pages 488–509. Royal Society, 1959.

E. Strick and A. S. Ginzbarg. Stoneley-wave velocities for a fluid-solid interface. *Bull. Seismol. Soc. Am.*, 46:281–292, 1956.

I. Tolstoy, editor. *Acoustics, Elasticity and Thermodynamics of Porous Media: Twenty-one papers by M. A. Biot.* Acoustical Society of America, 1991.

J. Trefil. *A scientist at the seashore.* Macmillan Publ. Co., New York, 1984.

A. Udias. *Principles of Seismology.* Cambridge University Press, 1999.

C. Valle, J. Qu, and L. J. Jacobs. Guided circumferential waves in layered cylinders. *Int. J. Engn. Sci.*, 37:1369–1387, 1999.

I. A. Viktorov. Rayleigh-type waves on curved surfaces. *J. Acoust. Soc. Am.*, 4:131–136, 1958a.

I. A. Viktorov. Wolny tipa relejewskich na cilindricheskich powierchnostiach. *Akust. Zurnal*, 4(2):131–136, 1958b. (in Russian).

I. A. Viktorov. *Rayleigh and Lamb waves. Physical theory and applications.* Plenum Press, New York, 1967.

I. A. Viktorov. *Zwukovyje powierchnostnyje wolny w twiordych telach.* Nauka, Moskwa, 1981. (in Russian).

J. E. White. *Underground sound, application of seismic waves.* Elsevier Sci. Publ., New York, 1983.

R. M. White. Surface elastic waves. *Proc. IEEE*, 58(8):1238–1276, 1970.

G. B. Whitham. *Linear and nonlinear waves.* John Wiley & Sons, New York, 1974.

K. Wilmanski. A thermodynamic model of compressible porous materials with the balance equation of porosity. *Transp. Porous Media*, 32:21–47, 1998a.

K. Wilmanski. *Thermomechanics of continua.* Springer, Berlin, 1998b.

K. Wilmanski. Waves in porous and granular materials. In K. Hutter and K. Wilmanski, editors, *Kinetic and continuum theories of granular and porous media*, number 400 in CISM Courses and Lectures, pages 131–186. Springer, Wien New York, 1999.

K. Wilmanski. Some questions on material objectivity arising in models of porous materials. In P. Podio-Guidugli and M. Brocato, editors, *Rational Continua, Classical and New. A collection of 7 papers dedicated to G. Capriz*, pages 149–161. Springer-Italy, Milan, 2001a.

K. Wilmanski. Thermodynamics of multicomponent continua. In R. Teisseyre and E. Majewski, editors, *Earthquake Thermodynamics and Phase Transformations in the Earth's Interior*, volume 76 of *International Geophysics Series*, chapter 25, pages 567–656. Academic Press, San Diego, 2001b.

K. Wilmanski. Propagation of sound and surface waves in porous materials. In B. T. Maruszewski, editor, *Structured Media*, pages 312–326. Publishing House of Poznan University of Technology, 2002a.

K. Wilmanski. Thermodynamical admissibility of Biot's model of poroelastic saturated materials. *Arch. Mech.*, 54(5-6):709–736, 2002b.

K. Wilmanski. On a micro-macro transition for poroelastic Biot's model and corresponding Gassmann-type relations. Preprint 868, WIAS, 2003. to appear in: Géotechnique, 2004.

K. Wilmanski. Tortuosity and objective relative acceleration in the theory of porous materials. Preprint 922, WIAS, 2004. to appear in: Proc. Royal Soc. (London), Ser. A, 2004.

K. Wilmanski and B. Albers. Acoustic waves in porous solid-fluid mixtures. In K. Hutter and N. Kirchner, editors, *Dynamic response of granular and porous materials under large and catastrophic deformations*, volume 11 of *Lecture Notes in Applied and Computational Mechanics*, pages 283–313. Springer, Berlin, 2003.

Yan Qing Zeng and Qing Huo Liu. Acoustic detection of buried objects in 3-D fluid saturated porous media: Numerical modeling. *IEEE Transactions on Geoscience and Remote Sensing*, 39(6):1165–1173, 2001a.

Yan Qing Zeng and Qing Huo Liu. A staggered-grid finite-difference method with perfectly matched layers for poroelastic wave equations. *J. Acoust. Soc. Am.*, 109(6): 2571–2580, 2001b.

# Modelling of surface waves in poroelastic saturated materials by means of a two component continuum

Bettina Albers
Weierstrass Institute for Applied Analysis and Stochastics
Mohrenstr. 39, D-10117 Berlin, Germany

**Abstract** These lecture notes are devoted to an overview of the modelling and the numerical analysis of surface waves in two-component saturated poroelastic media. This is an extension to the part of this book by K. Wilmanski which is primarily concerned with the classical surface waves in single component media. We use the "simple mixture model" which is a simplification of the classical Biot model for poroelastic media. Two interfaces are considered here: firstly the interface between a porous half space and a vacuum and secondly the interface between a porous halfspace and a fluid halfspace. For both problems we show how a solution can be constructed and a numerical solution of the dispersion relation can be found. We discuss the results for phase and group velocities and attenuations, and compare some of them to the high and low frequency approximations ($\omega \to \infty$, $\omega \to 0$, respectively).

For the interface porous medium/vacuum there exist in the whole range of frequencies two surface waves – a leaky Rayleigh wave and a true Stoneley wave. For the interface porous medium/fluid one more surface wave appears – a leaky Stoneley wave. For this boundary velocities and attenuations of the waves are shown in dependence on the surface permeability. The true Stoneley wave exists only in a limited range of this parameter. At the end we have a look on some results of other authors and a glance on a logical continuation of this work, namely the description of the structure and the acoustic behavior of partially saturated porous media.

## Contents

**1 Introduction** **278**
1.1 Practical aims . . . . . . . . . . 279
1.2 Theoretical background . . . . . . . . . . 280
- Biot's model
- Simple mixture model
- Boundary conditions

**2 Analysis by means of the "simple mixture model" – boundary porous medium/vacuum** **284**

2.1 Construction of solution for a semiinfinite medium . . . . . . . . . . . . . 285
- Compatibility with field equations
- Dimensionless notation
- Ansatz
- Dispersion relation
- High and low frequency approximations
2.2 Numerical procedure . . . . . . . . . . . . . . . . . . . . . . . . . . . 291
2.3 Parameters . . . . . . . . . . . . . . . . . . . . . . . . . . . . . . . 292
2.4 Numerical results . . . . . . . . . . . . . . . . . . . . . . . . . . . . 292
- Phase velocities of Rayleigh and Stoneley waves
- Group velocities of Rayleigh and Stoneley waves
- Attenuation of Rayleigh and Stoneley waves
2.5 Summary of results for boundary porous medium/vacuum . . . . . . . . . 300

**3 Analysis by means of the "simple mixture model" – boundary porous medium/fluid 302**
3.1 Construction of solution . . . . . . . . . . . . . . . . . . . . . . . . . 302
- Compatibility with field equations
- Insertion into boundary conditions
- Dispersion relation
- High and low frequency approximations
3.2 Numerical results . . . . . . . . . . . . . . . . . . . . . . . . . . . . 308
- Procedure and parameters
- Dependence of phase velocities and attenuations on the frequency
- Dependence of velocities/attenuations on the surface permeability
- Group velocities of the three surface waves
- Summary of results for boundary porous medium/fluid

**4 Comparison to results of other authors 314**
4.1 Deresiewicz . . . . . . . . . . . . . . . . . . . . . . . . . . . . . . . 314
4.2 Feng & Johnson . . . . . . . . . . . . . . . . . . . . . . . . . . . . . 315

**5 Glance on three-component porous media 317**

**Bibliography 320**

# 1 Introduction

The classical approach to the propagation of surface waves is based on one-component elastic models (for a detailed presentation of solutions for some of such problems see e.g. Wilmanski (2005)). At a plane boundary of a homogeneous linear elastic material solely a true Rayleigh wave appears, while at the interface of a semiinfinite elastic solid and a semiinfinite ideal fluid the Rayleigh wave is a leaky surface wave and an additional true Stoneley wave emerges. A true surface wave propagates along the surface direction and decays exponentially with depth, while leaky surface waves are attenuated in the surface direction and radiate energy into bulk or other surface waves. It is customary in the literature on waves in porous materials to call leaky surface waves pseudosurface waves (e.g. Feng and Johnson (1983a), Chao et al. (2004), first used in Phinney (1961)). For instance, there exist pseudo-Rayleigh, pseudo-Stoneley etc. waves. In this article we prefer to call these waves *leaky surface waves* (as e.g. done in Schröder and Scott (2001)). However, the two terms possess the same meaning: *A leaky surface wave is identical with a pseudosurface wave.* It is known that surface waves in linear models result from the combination of bulk waves (see: Viktorov (1967)). Physically, this means that at any point of the boundary bulk longitudinal and shear waves combine into the surface wave which must be slower than all bulk waves. In a one-component medium we have one longitudinal wave $P$ (primary) and one transversal wave $S$ (secondary or shear). The additional component in saturated porous media yields – compared to the one-component body – the existence of an additional bulk wave, the $P2$-wave, and thus also to an additional surface wave which, in dependence on the interface, may be of true or leaky type. For instance, on a permeable boundary, dividing the system into a saturated poroelastic material and an ideal fluid, three surface waves appear. This is due to the fact that the system of bulk waves consists of four waves: $P1$-, $P2$-, $S$-waves in the poroelastic material and a $P$-wave in the fluid. The third surface wave is carried primarily by the fluid outside of the porous material and, consequently, its velocity is bigger than this of the $P2$-wave but smaller than the velocity of the $P$-wave in the fluid. Such waves are commonly observed in laboratory (e.g. Nagy (1999)) and borehole experiments (e.g. Winkler et al. (1989)), their theoretical investigation within Biot's model was carried, for instance, by Norris (1989).

In application to soils, the underground consists of several layers. Thus an extension of the theoretical description of wave propagation is necessary. Up to now it is done for single component materials (e.g. Lai (1998), Rix et al. (2001)) in which the soil is considered to be heterogeneous. Then, dispersion on heterogeneities is responsible for the existence of several modes of propagation each travelling at a different phase and group velocity (modal velocities). In the approximation by a layered structure dispersion arises from constructive interference phenomena (see: Udias (1999)) occurring in media which are either bounded (e.g. rods plates, and other types of waveguides) or heterogeneous.

### 1.1 Practical aims

The theoretical investigation of surface waves in porous materials which is performed in this work is motivated by the possible construction of a method of nondestructive testing of materials. The term *nondestructive testing* denotes inspection methods which are used to investigate either the structure of a body or to search for the presence of defects without causing any effect on the properties of the medium. The notion "nondestructive testing" stands for a number of different methods like acoustics, radiography, dye penetration or magnetic particle inspection (for an introduction and examples see e.g. Meyendorf et al. (2004)). Nondestructive testing methods are used for many reasons such as to ensure product integrity and reliability, to prevent failure, due to often lower costs in comparison to conventional methods etc. Thus it is, for instance, possible to analyze porosity, voids, inclusions, cracks etc. Nondestructive testing is used in various areas such as, for example,

- geophysics, i.e. testing of soils and rocks,
- civil engineering, i.e. testing of concrete and other porous construction materials,
- road construction, i.e. testing of the road surface and the pavement,
- medicine, i.e. testing of bones and soft tissues,
- electronic industry, e.g. testing of surface coatings by nanomaterials.

Different operational areas require very different regions of frequencies of the waves: for field measurements in soil mechanics which are required in practical applications the range of applicable frequencies lies between 1 Hz and 100 Hz. However, this value differs considerably from laboratory measurements on such materials (White (1983)). For soil samples the frequency region in such tests lies around some kHz. This range can be extended to some MHz for stones or concrete by the application of a laser technique (Owino and Jacobs (1999), Hess (2002)). Medical applications (see e.g. Kutruff (1991)) allow for frequencies up to approx. 3 MHz while testing of nanomaterials requires frequencies of approx. 100 MHz (see Chapter 11 of Meyendorf et al. (2004)).

Here, we want to focus our attention on the nondestructive testing of geomaterials by the analysis of surface waves. For this analysis one can choose two different ways in which a dynamic disturbance is excited: harmonic or impact excitation. In contrast to the far field approximation of seismic waves which is usually based on the frequency analysis, engineering applications were primarily concerned with waves initiated by chopping or explosions. This is not the case any more. Numerous experiments and measurements are made nowadays by devices producing harmonic vibrations (e.g. Rix et al. (2001)). They show that the investigation of a problem in dependence on a real frequency $\omega$ is more adequate than a comparable analysis with a real wavelength $\frac{2\pi}{k}$ as control variable.

## 1.2 Theoretical background

### Biot's model

The most popular and widespread model for the theoretical description of processes in fluid-saturated porous media is the *Biot model.* It has been introduced by Maurice A. Biot in 1941 for a consolidation problem and was the basis for the analysis of the propagation of body waves in a two-component medium since 1956 (see: Tolstoy (1992)). In the notation of the present article Biot's momentum balances have the following form

$$\begin{aligned} \rho_0^F \frac{\partial \mathbf{v}^F}{\partial t} &= \kappa\rho_0^F \operatorname{grad}\varepsilon + Q \operatorname{grad} \operatorname{tr}\mathbf{e}^S - \pi\left(\mathbf{v}^F - \mathbf{v}^S\right) + \rho_{12}\left(\frac{\partial \mathbf{v}^F}{\partial t} - \frac{\partial \mathbf{v}^S}{\partial t}\right), \qquad (1.1) \\ \rho_0^S \frac{\partial \mathbf{v}^S}{\partial t} &= \lambda^S \operatorname{grad} \operatorname{tr}\mathbf{e}^S + 2\mu^S \operatorname{div}\mathbf{e}^S + Q \operatorname{grad}\varepsilon + \pi\left(\mathbf{v}^F - \mathbf{v}^S\right) - \rho_{12}\left(\frac{\partial \mathbf{v}^F}{\partial t} - \frac{\partial \mathbf{v}^S}{\partial t}\right), \end{aligned}$$

where $\mathbf{v}^F, \mathbf{v}^S$ denote the macroscopic (average) partial velocities of the fluid and of the skeleton, respectively, $\mathbf{e}^S$ is the symmetric Almansi-Hamel tensor of small deformations of the skeleton, $\varepsilon$ is the volume change of the fluid. These quantities fulfil the following relations

$$\frac{\partial \mathbf{e}^S}{\partial t} = \operatorname{sym} \operatorname{grad} \mathbf{v}^S, \quad \frac{\partial \varepsilon}{\partial t} = \operatorname{div} \mathbf{v}^F,$$
$$\varepsilon := \frac{\rho_0^F - \rho^F}{\rho_0^F}, \quad \zeta := n_0(\operatorname{tr} \mathbf{e}^S - \varepsilon); \tag{1.2}$$

the latter is the change of fluid contents commonly used as a field in Biot's model instead of $\varepsilon$. Relations (1.2) yield the existence of partial displacement fields $\mathbf{u}^S \equiv \mathbf{u}$, $\mathbf{u}^F \equiv \mathbf{U}$, with

$$\mathbf{e}^S = \operatorname{sym} \operatorname{grad} \mathbf{u}, \quad \mathbf{v}^S = \frac{\partial \mathbf{u}}{\partial t}, \quad \varepsilon = \operatorname{div} \mathbf{U}, \quad \mathbf{v}^F = \frac{\partial \mathbf{U}}{\partial t}. \tag{1.3}$$

The quantities $\rho_0^F = n_0 \rho_0^{FR}$, $\rho_0^S = (1 - n_0)\, \rho_0^{SR}$ denote constant (initial) mass densities, $\rho_0^{FR}, \rho_0^{SR}$ are the initial true mass densities, $n_0$ is the initial porosity. The material parameters $\lambda^S, \mu^S, \kappa, Q, \pi, \rho_{12}$ are assumed to be constant. The parameter $Q$ describes the coupling of partial stresses in the Biot model, while $\rho_{12}$ is attributed to the tortuosity $a$. A typical simplified relation among those quantities was proposed by Berryman in 1980 (e.g. Bourbié et al. (1987)) and has the form

$$\rho_{12} = \rho^F (1 - a), \quad a = \frac{1}{2}\left(\frac{1}{n_0} + 1\right). \tag{1.4}$$

The permeability coefficient $\pi$ contains an overall information on the morphology of the porous material which includes porosity, tortuosity, viscosity of the real pore fluid etc. The form of this dependence is not needed in this work.

**Simple mixture model**

In our analysis we rely on a simpler model than this of Biot. We neglect two effects:

- the added mass effect reflected in the Biot model by off-diagonal contributions to the matrix of partial mass densities (the parameter $\rho_{12}$),
- the static coupling effect between partial stresses (the parameter $Q$).

The first contribution is neglected because it yields a non-objectivity of Biot's equations (Wilmanski (2001)). This does not mean that the tortuosity may not have an influence on the propagation of surface waves but solely that it is not well modelled by Biot. For details concerning this question see Wilmanski (2004).

The second contribution – coupling of partial stresses $Q$ – is neglected because it yields solely *quantitative* corrections without changing the *qualitative* behavior of the system, at least in the range of a relatively high stiffness of the skeleton in comparison with the fluid. This has been analyzed for bulk waves in Albers and Wilmanski (2003b).

Bearing the above remarks in mind it seems to be appropriate to rely on the simplified model ("simple mixture model" in which $Q = 0$, $\rho_{12} = 0$) which, in turn, reduces essentially technical difficulties.

We present here the linear form of the "simple mixture" model for a two-component poroelastic saturated medium (for details see: Wilmanski (1999)).

Within this model the process is described by the *macroscopic* fields $\rho^F(\mathbf{x},t)$ – partial mass density of the fluid, $\mathbf{v}^F(\mathbf{x},t)$ – velocity of the fluid, $\mathbf{v}^S(\mathbf{x},t)$ – velocity of the skeleton, $\mathbf{e}^S(\mathbf{x},t)$ – symmetric tensor of small deformations of the skeleton and the porosity $n$. These fields satisfy the following set of linear equations

$$\begin{aligned}
&\frac{\partial \rho^F}{\partial t} + \rho_0^F \operatorname{div} \mathbf{v}^F = 0, \quad \left|\frac{\rho^F - \rho_0^F}{\rho_0^F}\right| \ll 1, \\
&\rho_0^F \frac{\partial \mathbf{v}^F}{\partial t} + \kappa \operatorname{grad} \rho^F + \beta \operatorname{grad}(n - n_E) + \hat{\mathbf{p}} = 0, \quad \hat{\mathbf{p}} := \pi\left(\mathbf{v}^F - \mathbf{v}^S\right), \\
&\rho_0^S \frac{\partial \mathbf{v}^S}{\partial t} - \operatorname{div}\left(\lambda^S\left(\operatorname{tr} \mathbf{e}^S\right)\mathbf{1} + 2\mu \mathbf{e}^S + \beta(n - n_E)\mathbf{1}\right) - \hat{\mathbf{p}} = 0, \\
&\frac{\partial \mathbf{e}^S}{\partial t} = \operatorname{sym} \operatorname{grad} \mathbf{v}^S, \quad \left\|\mathbf{e}^S\right\| \ll 1, \quad n_E := n_0\left(1 + \delta \operatorname{tr} \mathbf{e}^S\right), \\
&\frac{\partial(n - n_E)}{\partial t} + \Phi \operatorname{div}\left(\mathbf{v}^F - \mathbf{v}^S\right) + \frac{n - n_E}{\tau_n} = 0, \quad \left|\frac{n - n_0}{n_0}\right| \ll 1.
\end{aligned} \tag{1.5}$$

As before $\rho_0^F, \rho_0^S, n_0$ denote constant reference values of partial mass densities, and porosity, respectively, and $\kappa, \lambda^S, \mu^S, \beta, \pi, \tau_n, \delta, \Phi$ are constant material parameters. The first one describes the macroscopic compressibility of the fluid component, the next two are macroscopic elastic constants of the skeleton, $\beta$ is the coupling constant, $\pi$ is the coefficient of bulk permeability, $\tau_n$ is the relaxation time of porosity and $\delta$, $\Phi$ describe equilibrium and nonequilibrium changes of porosity, respectively. For the purpose of this work we assume $\beta = 0$. Then the problem of evolution of porosity described by equation $(1.5)_5$ can be solved separately from the rest of the problem and does not influence the propagation of acoustic waves in the medium. The norm of the deformation tensor $\|\mathbf{e}^S\|$ is understood as the maximum absolute value of eigenvalues of $\mathbf{e}^S$ (the maximum stretch).

In this article we present the *simple mixture model* in order to investigate the dispersion relation for surface waves on impermeable and permeable boundaries of a fluid saturated poroelastic medium in the *whole range* of frequencies. The problem of the frequency dependence for intermediate frequencies has been widely ignored in the literature due to numerical difficulties within the Biot model. Many numerical results were not published yet elsewhere.

**Boundary conditions**

In order to determine surface waves in a saturated poroelastic medium we need conditions for $z = 0$. The most general case which we are going to consider is the interface between a saturated porous material and an ideal fluid. Boundary conditions for such an interface were formulated by Deresiewicz and Skalak (1963). We quote them here in a slightly modified form. If we denote quantities outside of the porous medium by a "+" sign the boundary conditions have the form

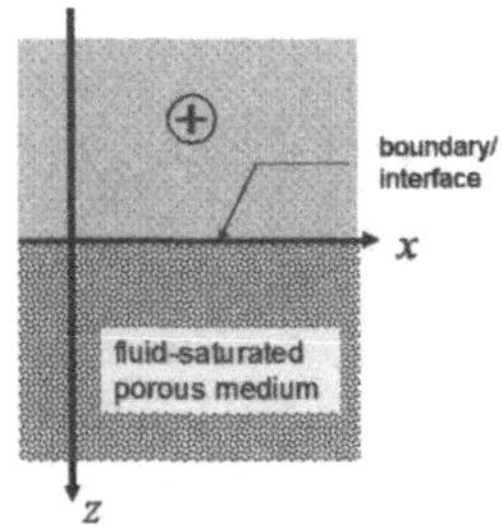


**Figure 1.** Contact of a porous halfspace with an exterior.

$$
\begin{aligned}
\bullet \quad & \left(T_{13} - T_{13}^{+}\right)\Big|_{z=0} \equiv T_{13}^{S}\Big|_{z=0} = c_S^2 \rho_0^S \left(\frac{\partial u_1^S}{\partial z} + \frac{\partial u_3^S}{\partial x}\right)\Bigg|_{z=0} = 0, \\
\bullet \quad & \left(T_{33} - T_{33}^{+}\right)\Big|_{z=0} \equiv \left(T_{33}^{S} + p^{F+} - p^{F}\right)\Big|_{z=0} = \\
& = c_{P1}^2 \rho_0^S \left(\frac{\partial u_1^S}{\partial x} + \frac{\partial u_3^S}{\partial z}\right) - 2 c_S^2 \rho_0^S \frac{\partial u_1^S}{\partial x} + \\
& + c_{+}^2 \left(\rho^{F+} - \rho_0^{F+}\right) - c_{P2}^2 \left(\rho^{F} - \rho_0^{F}\right)\Big|_{z=0} = 0, \qquad (1.6) \\
\bullet \quad & \rho_0^F \frac{\partial}{\partial t}\left(u_3^F - u_3^S\right)\Bigg|_{z=0} = \rho_0^{F+} \frac{\partial}{\partial t}\left(u_3^{F+} - u_3^S\right)\Bigg|_{z=0}, \\
\bullet \quad & \rho_0^F \frac{\partial}{\partial t}\left(u_3^F - u_3^S\right) + \alpha\left(p^F - n_0 p^{F+}\right)\Bigg|_{z=0} = 0,
\end{aligned}
$$

where $u_1^S, u_3^S$ are $x$-, and $z$-components of the displacement $\mathbf{u}^S$, respectively, and $u_3^F, u_3^{F+}$ are $z$-components of the displacements $\mathbf{u}^F$ and $\mathbf{u}^{F+}$, respectively. Simultaneously,

$$
c_{P1}^2 := \frac{\lambda^S + 2\mu^S}{\rho_0^S}, \quad c_S^2 := \frac{\mu^S}{\rho_0^S}, \quad c_{P2}^2 := \kappa, \quad c_{+}^2 := \kappa^{+}, \qquad (1.7)
$$

are squares of the front velocities of the bulk waves in the porous material: $P1$ (fast wave), $S$ (shear wave), $P2$ (slow wave, also called Biot's wave), and of the $P$-wave in the fluid, respectively. In the case of Biot's model there would be an additive contribution in the numerator of $c_{P1}$ of the coupling parameter $Q$ which is of the order of a few percent of $\lambda^S$ (see Albers and Wilmanski (2003b) for a detailed analysis).

The first two conditions of (1.6) describe the continuity of the full traction, $\mathbf{t} := \left(\mathbf{T}^S + \mathbf{T}^F\right)\mathbf{n}$, $\mathbf{n} = (0,0,1)^T$, on the boundary (i.e. momentum balance); the third condition is the continuity of the fluid mass flux (i.e. mass balance), and the last condition specifies the mass transport through the surface. The in- and outflow through the boundary is assumed to be proportional to the difference of the pore pressures on both sides of the boundary. In this condition $\alpha$ denotes a surface permeability coefficient (it corresponds to $\frac{1}{T}$ in the works Feng and Johnson (1983a) and Feng and Johnson (1983b)) and $p^{F+}$ is the external pressure. This condition relies on the assumption that the pore pressure $p$ and the partial pressure $p^F$ satisfy the relation $p^F \approx n_0 p$ at least in a small vicinity of the surface. For a justification of this condition it would be necessary to consider a boundary layer in the limit of zero thickness.

Some words are appropriate to explain the notion of *surface permeability*: As we see in the neighbouring figure the flow of the fluid is "straight" both outside the porous medium and inside the channels of the porous medium. However, at the entrance to the porous medium the flow is disturbed: the fluid has to find its way through the voids between the solid particles. In principle we consider a flow which is perpendicular to the boundary but in a small boundary layer it deviates from this direction. The more complicated the structure of the boundary the bigger is the resistance against the inflow and the smaller is the surface permeability parameter $\alpha$. The value $\alpha = 0$ corresponds to a sealed pore situation, for $\alpha = \infty$ the boundary is completely open.

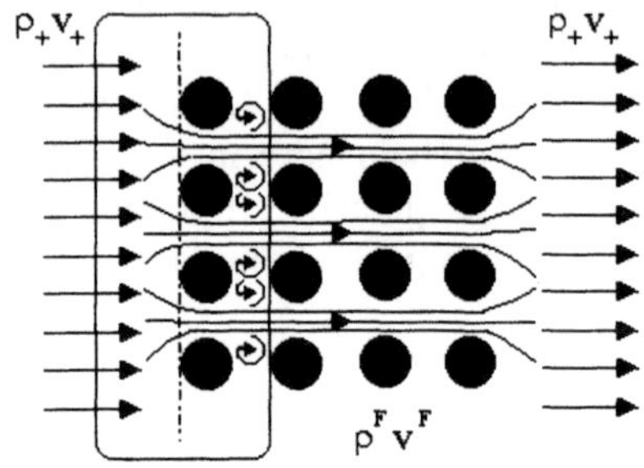


**Figure 2.** Sketch of the flow into and out of a porous medium.

For the impermeable boundary with vacuum ($\alpha = 0$ in our notation, $T \to \infty$ in the work Feng and Johnson (1983a)), the above boundary conditions simplify to the following ones

- $$T_{13}\big|_{z=0} \equiv T_{13}^S\big|_{z=0} = c_S^2 \rho_0^S \left( \frac{\partial u_1^S}{\partial z} + \frac{\partial u_3^S}{\partial x} \right)\bigg|_{z=0} = 0, \tag{1.8}$$
- $$T_{33}\big|_{z=0} \equiv \left(T_{33}^S - p^F\right)\big|_{z=0} = \\ = c_{P1}^2 \rho_0^S \left( \frac{\partial u_1^S}{\partial x} + \frac{\partial u_3^S}{\partial z} \right) - 2c_S^2 \rho_0^S \frac{\partial u_1^S}{\partial x} + \\ - c_{P2}^2 \left(\rho^F - \rho_0^F\right)\big|_{z=0} = 0, \tag{1.9}$$
- $$\frac{\partial}{\partial t}\left(u_3^F - u_3^S\right)\bigg|_{z=0} = 0, \tag{1.10}$$

due to the fact that the quantities outside of the porous medium are equal to zero. Then the third and the fourth boundary conditions are identical.

The sealed boundary between the porous medium and a fluid yields three conditions as well but in this case there is a right hand side in relation for normal stresses.

## 2 Analysis by means of the "simple mixture model" – boundary porous medium/vacuum

The procedure and the numerical results for phase velocities and attenuations for this boundary are also presented in Albers and Wilmanski (2003a).

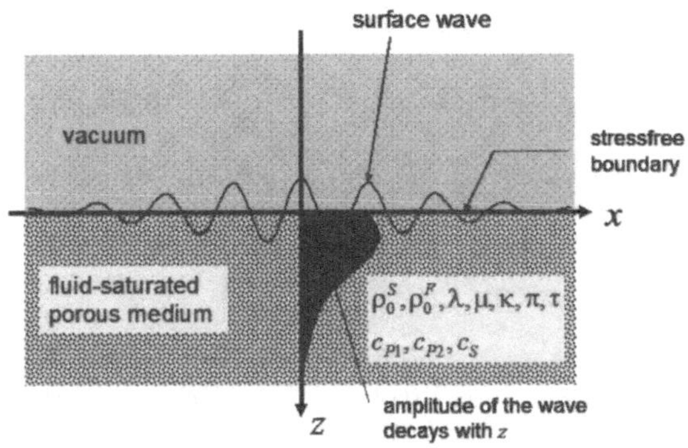


**Figure 3.** Geometry of the boundary porous medium/vacuum.

### 2.1 Construction of solution for a semiinfinite medium

We follow here the same procedure of construction of solutions as in the works Wilmanski (2002), Edelman and Wilmanski (2002) and Wilmanski and Albers (2003). While in Edelman and Wilmanski (2002) an asymptotic analysis of the high-frequency properties of surface waves in function of the wavelength $1/k$ ($k$ – wave number) within the simple mixture model was carried out, here, we consider monochromatic waves with a given *real frequency* $\omega$. This may be considered either as a Fourier component of the expansion of the solution in time space or as a far field boundary value problem with a harmonic surface source of waves.

This choice of the description seems to be better justified from the physical point of view than the choice of the wave number, $k$, or, equivalently, of the wave length, $\frac{1}{k}$. In contrast to the latter case, a monochromatic wave of a given frequency, $\omega$, can be easily induced in real field experiments. It should be mentioned, however, that these two approaches are not equivalent from the formal point of view. For an explanation of this statement see the example to the linear wave equation with damping describing longitudinal vibrations of a rod in the chapter of K. Wilmanski in this book (Wilmanski (2005)). This example is simpler but has the same structure as the problem considered in this article. The analysis in terms of real $k$ yields, indeed, a critical behavior as shown in this example. This has been observed for Biot's wave ($P2$-wave) by Edelman (2003). However, according to the above remarks this observation seems to be solely of an academic interest.

#### Compatibility with field equations

We introduce the displacement vector $\mathbf{u}^S$ for the skeleton, and formally the displacement vector $\mathbf{u}^F$ for the fluid. The latter is introduced solely for the technical symmetry of considerations and does not have any direct physical bearing. Then we have according to Helmholtz's Theorem

$$\begin{aligned} \mathbf{u}^S &= \operatorname{grad}\varphi^S + \operatorname{rot}\boldsymbol{\psi}^S, \quad \mathbf{v}^S = \frac{\partial \mathbf{u}^S}{\partial t}, \quad \mathbf{e}^S = \operatorname{sym}\operatorname{grad}\mathbf{u}^S, \\ \mathbf{u}^F &= \operatorname{grad}\varphi^F + \operatorname{rot}\boldsymbol{\psi}^F, \quad \mathbf{v}^F = \frac{\partial \mathbf{u}^F}{\partial t}. \end{aligned} \tag{2.1}$$

This means in coordinates

$$u_i^F = \frac{\partial \varphi^F}{\partial x^i} + \varepsilon_{ijk}\frac{\partial \psi_k^F}{\partial x^j}, \quad u_i^S = \frac{\partial \varphi^S}{\partial x^i} + \varepsilon_{ijk}\frac{\partial \psi_k^S}{\partial x^j}, \tag{2.2}$$

where $\varepsilon_{ijk}$ is the permutation symbol

$$\varepsilon_{ijk} = \begin{cases} 0 & \text{for } i=j, j=k, k=i \\ 1 & \text{for } (i,j,k) \in \{(1,2,3),(2,3,1),(3,1,2)\} \\ -1 & \text{for } (i,j,k) \in \{(1,3,2),(3,2,1),(2,1,3)\} \end{cases}.$$

For the two-dimensional case[1] we make the following ansatz for monochromatic wave solutions in the $x$-direction

$$\begin{aligned} \varphi^S &= A^S(z)\exp[i(kx-\omega t)], \quad \varphi^F = A^F(z)\exp[i(kx-\omega t)], \\ \psi_y^S &= B^S(z)\exp[i(kx-\omega t)], \quad \psi_y^F = B^F(z)\exp[i(kx-\omega t)], \\ \psi_x^S &= \psi_z^S = \psi_x^F = \psi_z^F = 0, \end{aligned} \tag{2.3}$$

and

$$\rho^F - \rho_0^F = A_\rho^F(z)\exp[i(kx-\omega t)]. \tag{2.4}$$

We investigate the approximation of a very large relaxation time of porosity ($\tau_n \to \infty$) compared to the inverse of characteristic frequencies. This is justified in applications to soils where viscous effects related to porosity do not seem to appear. Then the non-equilibrium porosity $n$ is determined by volume changes of the skeleton and of the fluid: $n = n_E + \Phi\left(\mathrm{tr}\mathbf{e}^S - \varepsilon\right)$. Consequently, they do not give any independent contribution to the wave problem.

Substitution in field equations (see: (1.5))

$$\begin{aligned} &\frac{\partial \rho^F}{\partial t} + \rho_0^F \operatorname{div}\left(\frac{\partial \mathbf{u}^F}{\partial t}\right) = 0, \qquad \mathbf{e}^S = \operatorname{sym}\operatorname{grad}\mathbf{u}^S, \\ &\rho_0^F \frac{\partial^2 \mathbf{u}^F}{\partial t^2} + \kappa \operatorname{grad}\rho^F + \pi\left(\frac{\partial \mathbf{u}^F}{\partial t} - \frac{\partial \mathbf{u}^S}{\partial t}\right) = 0, \\ &\rho_0^S \frac{\partial^2 \mathbf{u}^S}{\partial t^2} - \operatorname{div}\left(\lambda^S\left(\operatorname{tr}\mathbf{e}^S\right)\mathbf{1} + 2\mu\mathbf{e}^S\right) - \pi\left(\frac{\partial \mathbf{u}^F}{\partial t} - \frac{\partial \mathbf{u}^S}{\partial t}\right) = 0. \end{aligned} \tag{2.5}$$

[1]Under this assumption of plane wave motion we are not able to describe the geometrical dispersion of waves which is the main practical motivation for applications of surface waves rather than bulk waves.

leads after straightforward calculations[2] to the following compatibility conditions

$$B^F = \frac{i\pi}{\rho_0^F \omega + i\pi} B^S, \qquad A_\rho^F = -\rho_0^F \left( \frac{d^2}{dz^2} - k^2 \right) A^F, \tag{2.6}$$

from momentum balance for the fluid in $z$-direction — from mass balance for the fluid

as well as

$$\left[ \kappa \left( \frac{d^2}{dz^2} - k^2 \right) + \omega^2 \right] A^F + \frac{i\pi}{\rho_0^F} \omega \left( A^F - A^S \right) = 0, \tag{2.7}$$

from momentum balance for the fluid in $x$-direction

$$\left[ \frac{\lambda^S + 2\mu^S}{\rho_0^S} \left( \frac{d^2}{dz^2} - k^2 \right) + \omega^2 \right] A^S - \frac{i\pi}{\rho_0^S} \omega \left( A^F - A^S \right) = 0, \tag{2.8}$$

from momentum balance for the skeleton in $x$-direction

$$\left[ \frac{\mu^S}{\rho_0^S} \left( \frac{d^2}{dz^2} - k^2 \right) + \omega^2 + \frac{i\pi \rho_0^F}{\rho_0^S \left( \rho_0^F \omega + i\pi \right)} \omega^2 \right] B^S = 0. \tag{2.9}$$

from momentum balance for the skeleton in $z$-direction

**Dimensionless notation**

It is convenient to introduce a dimensionless notation. Therefore we define the following dimensionless quantities

$$\begin{aligned} c_s &:= \frac{c_S}{c_{P1}} < 1, \quad c_f := \frac{c_{P2}}{c_{P1}}, \quad \pi' := \frac{\pi\tau}{\rho_0^S} > 0, \\ r &:= \frac{\rho_0^F}{\rho_0^S} < 1, \quad z' := \frac{z}{c_{P1}\tau}, \quad k' := k c_{P1} \tau, \quad \omega' := \omega\tau, \end{aligned} \tag{2.10}$$

where $\tau$ is an arbitrary reference time. It may be chosen as $\tau = \frac{\rho_0^S}{\pi}$ which would lead to $\pi' = 1$, or it may be identical with the relaxation time of porosity $\tau_n$. Further, we make an arbitrary choice of this normalization parameter.

**Ansatz**

For simplicity we further omit the prime in (2.10). Substitution of these quantities in equations (2.7), (2.8), (2.9) yields

$$\begin{aligned} \left[ c_f^2 \left( \frac{d^2}{dz^2} - k^2 \right) + \omega^2 \right] A^F + i\frac{\pi}{r} \omega \left( A^F - A^S \right) &= 0, \\ \left[ \left( \frac{d^2}{dz^2} - k^2 \right) + \omega^2 \right] A^S - i\pi\omega \left( A^F - A^S \right) &= 0, \\ \left[ c_s^2 \left( \frac{d^2}{dz^2} - k^2 \right) + \omega^2 + \frac{i\pi\omega^2}{\omega + i\frac{\pi}{r}} \right] B^S &= 0. \end{aligned} \tag{2.11}$$

2 $\mathrm{div}\mathbf{u}_{\mathrm{transversal}} \equiv \mathrm{div}\,\mathrm{rot}\psi \equiv 0$,
$\mathrm{rot}\mathbf{u}_{\mathrm{longitudinal}} \equiv \mathrm{rot}\,\mathrm{grad}\varphi \equiv 0$.

The matrix of coefficients for homogeneous materials is independent of $z$. Hence the differential eigenvalue problem can be easily solved. Consequently, we seek solutions in the form

$$A^F = A_f^1 e^{\gamma_1 z} + A_f^2 e^{\gamma_2 z}, \quad A^S = A_s^1 e^{\gamma_1 z} + A_s^2 e^{\gamma_2 z}, \quad B^S = B_s e^{\zeta z}. \tag{2.12}$$

Substitution in (2.11) yields relations for the exponents in the form

$$\left(\frac{\zeta}{k}\right)^2 = 1 - \frac{1}{c_s^2}\left(1 + \frac{i\pi}{\omega + i\frac{\pi}{r}}\right)\left(\frac{\omega}{k}\right)^2, \tag{2.13}$$

$$\Rightarrow \quad \boxed{\frac{\zeta_{1,2}}{k} = \pm\sqrt{1 - \frac{1}{c_s^2}\left(1 + \frac{i\pi}{\omega + i\frac{\pi}{r}}\right)\left(\frac{\omega}{k}\right)^2},}$$

and

$$\begin{gathered} c_f^2\left[\left(\frac{\gamma}{k}\right)^2 - 1\right]^2 + \left[1 + \left(1 + \frac{1}{r}\right)\frac{i\pi}{\omega}\right]\left(\frac{\omega}{k}\right)^4 + \\ + \left[1 + c_f^2 + \left(c_f^2 + \frac{1}{r}\right)\frac{i\pi}{\omega}\right]\left[\left(\frac{\gamma}{k}\right)^2 - 1\right]\left(\frac{\omega}{k}\right)^2 = 0, \end{gathered} \tag{2.14}$$

$$\Rightarrow \quad \boxed{\begin{aligned} \frac{\gamma_{1,2,3,4}}{k} = \pm\Bigg\{ & 1 - \frac{1}{2c_f^2}\left[\left(1 + c_f^2\right)\left(\frac{\omega}{k}\right)^2 + \left(c_f^2 + \frac{1}{r}\right)\frac{i\pi\omega}{k^2}\right] \pm \\ & \pm\Bigg(\left(\frac{\omega}{k}\right)^4\left[\frac{1}{4c_f^4}\left(1 + c_f^2\right)^2 - \frac{1}{c_f^2}\right] + \frac{i\pi}{k}\left(\frac{\omega}{k}\right)^3\left[\frac{1}{2c_f^4}\left(1 + c_f^2\right)\left(c_f^2 + \frac{1}{r}\right) - \right. \\ & \left. - \frac{1}{c_f^2}\left(1 + \frac{1}{r}\right)\right] - \frac{\pi^2\omega^2}{4c_f^4 k^4}\left(c_f^2 + \frac{1}{r}\right)^2\Bigg)^{1/2}\Bigg\}^{1/2}. \end{aligned}}$$

Simultaneously, we obtain for the eigenvectors the following relations

$$\mathbf{R}^1 = \left(B_s, A_s^1, A_f^1\right)^T, \quad \mathbf{R}^2 = \left(B_s, A_s^2, A_f^2\right)^T, \tag{2.15}$$

where

$$A_f^1 = \delta_f A_s^1, \quad A_s^2 = \delta_s A_f^2, \tag{2.16}$$

$$\delta_f := \frac{1}{r}\frac{\frac{i\pi}{\omega}\frac{\omega^2}{k^2}}{c_f^2\left[\left(\frac{\gamma_1}{k}\right)^2 - 1\right] + \left(\frac{\omega}{k}\right)^2 + \frac{i\pi}{\omega r}\frac{\omega^2}{k^2}}, \tag{2.17}$$

$$\delta_s := \frac{\dfrac{i\pi}{\omega}\dfrac{\omega^2}{k^2}}{\left(\dfrac{\gamma_2}{k}\right)^2 - 1 + \left(\dfrac{\omega}{k}\right)^2 + \dfrac{i\pi}{\omega}\dfrac{\omega^2}{k^2}}. \tag{2.18}$$

The above solution for the exponents still leaves three unknown constants $B_s, A_f^2, A_s^1$ which must be specified from boundary conditions.

Complex values of $\zeta_{1,2}$ and $\gamma_{1,2,3,4}$ result from the dissipation caused by the relative motion, i.e. the influence of the permeability $\pi$. As a consequence, solutions decay in $z$-direction but, simultaneously, they vibrate (see: Fig. 4).

**Dispersion relation**

Substitution of the above results in the boundary conditions (1.8)-(1.10) yields the following equations for the three unknown constants $B_s, A_f^2$ and $A_s^1$

$$\mathbf{AX} = \mathbf{0}, \tag{2.19}$$

where

$$\mathbf{A} := \begin{pmatrix} \left(\frac{\zeta}{k}\right)^2 + 1 & 2i\frac{\gamma_2}{k}\delta_s & 2i\frac{\gamma_1}{k} \\ -2ic_s^2\frac{\zeta}{k} & \begin{array}{c}\left[\left(\frac{\gamma_2}{k}\right)^2 - 1 + 2c_s^2\right]\delta_s + \\ +rc_f^2\left[\left(\frac{\gamma_2}{k}\right)^2 - 1\right]\end{array} & \begin{array}{c}\left(\frac{\gamma_1}{k}\right)^2 - 1 + 2c_s^2 + \\ +rc_f^2\left[\left(\frac{\gamma_1}{k}\right)^2 - 1\right]\delta_f\end{array} \\ i\frac{r\omega}{r\omega + i\pi} & -\left(\delta_s - 1\right)\frac{\gamma_2}{k} & \left(\delta_f - 1\right)\frac{\gamma_1}{k} \end{pmatrix}, \tag{2.20}$$

$$\mathbf{X} := \left( B_s, \ A_f^2, \ A_s^1 \right)^T.$$

This homogeneous set yields the *dispersion relation:* $\det \mathbf{A} = 0$ determining the $\omega - k$ relation. The wave number $k$ is complex for two reasons: the first one is the dissipation due to diffusion which is always present for $\pi \neq 0$ and the second is the loss of energy in leaky waves independently of $\pi$.

We investigate the numerical solution of the dispersion relation and compare the results for high and low frequencies with approximations shown in Wilmanski and Albers (2003). These are briefly summarized in the next subsection.

**High and low frequency approximations**

**High frequencies** In the limit $\omega \to \infty$ we immediately obtain from relations (2.13) and (2.14)

$$\left(\frac{\zeta}{k}\right)^2 = 1 - \frac{1}{c_s^2}\left(\frac{\omega}{k}\right)^2,$$

$$\left(\frac{\gamma_1}{k}\right)^2 = 1 - \left(\frac{\omega}{k}\right)^2, \quad \left(\frac{\gamma_2}{k}\right)^2 = 1 - \frac{1}{c_f^2}\left(\frac{\omega}{k}\right)^2, \tag{2.21}$$

and

$$\delta_f = \delta_s = 0 \quad \Rightarrow \quad \mathbf{R}^1 = \left(B_s, A_s^1, 0\right)^T, \quad \mathbf{R}^2 = \left(B_s, 0, A_f^2\right)^T. \tag{2.22}$$

Hence, the exponents are *real* in this limit, i.e. there appear no vibrations in $z$-direction.

The dispersion relation follows in the form

$$\mathcal{P}_R\sqrt{1 - c_f^2\left(\frac{\omega}{k}\right)^2} + \frac{r}{c_s^4}\left(\frac{\omega}{k}\right)^4\sqrt{1 - \left(\frac{\omega}{k}\right)^2} = 0, \tag{2.23}$$

where

$$\mathcal{P}_R := \left(2 - \frac{1}{c_s^2}\left(\frac{\omega}{k}\right)^2\right)^2 - 4\sqrt{1 - \left(\frac{\omega}{k}\right)^2}\sqrt{1 - \frac{1}{c_s^2}\left(\frac{\omega}{k}\right)^2}. \tag{2.24}$$

For $r = 0$ the relation (2.23) reduces to $\mathcal{P}_R = 0$ which is the Rayleigh dispersion relation for single component continua. The high frequency limit, described by (2.23), exists for any choice of material parameters in contrast to results of Feng and Johnson (1983a). The range of nonexistence of this wave related to small values of the longitudinal frame modulus, corresponding to $\mu^S < \rho^S \kappa$ in our notation, has not been investigated within the simple mixture model.

**Low frequencies** For the limit $\omega \to 0$ one arrives at the following results for the first order approximation of exponents (notice a singularity of the last contribution to $\frac{\gamma_2}{k}$!)

$$\left(\frac{\zeta}{k}\right)^2 = 1 - \frac{r+1}{c_s^2}\left(\frac{\omega}{k}\right)^2, \qquad \left(\frac{\gamma_1}{k}\right)^2 = 1 - \frac{r+1}{rc_f^2+1}\left(\frac{\omega}{k}\right)^2,$$
$$\left(\frac{\gamma_2}{k}\right)^2 = 1 - \frac{rc_f^4+1}{c_f^2\left(rc_f^2+1\right)}\left(\frac{\omega}{k}\right)^2 - \frac{i\pi}{\omega}\frac{rc_f^2+1}{rc_f^2}\left(\frac{\omega}{k}\right)^2, \tag{2.25}$$

and for the coefficients of amplitudes

$$\delta_f = 1 - \frac{\omega r}{i\pi}\frac{1-c_f^2}{1+rc_f^2}, \qquad \delta_s = -rc_f^2\left(1 - \frac{\omega r}{i\pi}\frac{1-c_f^2}{1+rc_f^2}\right). \tag{2.26}$$

These coefficients of amplitudes for low frequencies are complex while we have seen that those of the high frequency limit are real. This means that here vibrations in $z$-direction appear. We show schematically the behavior of the amplitudes in the neighbouring figure. The dashed lines show the decreasing of the amplitude for real coefficients of amplitudes, the solid line shows it for complex coefficients where vibrations appear. If we account for the relations (2.25) and (2.26) in the condition det $\mathbf{A}$ =0 then we obtain a relation specifying $\frac{\omega}{k}$. From this we get as a solution a Rayleigh wave whose speed is given by a relation analogous to (2.24) in which the speeds of bulk waves (1.7) are replaced by their low frequency counterparts $c_{oP1}$ and $c_{oS}$:

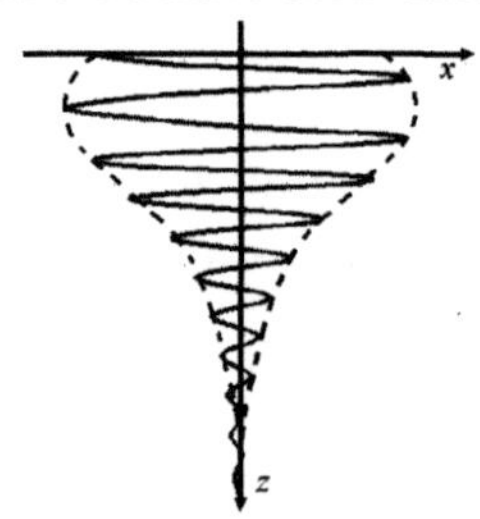


**Figure 4.** Behavior of amplitudes of surface waves.

$$\left(2-\frac{c_{P1}^2}{c_{oS}^2}\left(\frac{\omega}{k}\right)^2\right)^2-4\sqrt{1-\frac{c_{P1}^2}{c_{oS}^2}\left(\frac{\omega}{k}\right)^2}\sqrt{1-\frac{c_{P1}^2}{c_{oP1}^2}\left(\frac{\omega}{k}\right)^2}=0,$$
$$c_{oP1}:=\sqrt{\frac{\lambda^S+2\mu^S+\rho_0^F\kappa}{\rho_0^S+\rho_0^F}},\qquad c_{oS}:=\sqrt{\frac{\mu^S}{\rho_0^S+\rho_0^F}}. \tag{2.27}$$

Both components move in a synchronized way, thus the influence of diffusion is not visible. This shows that it is often sufficient to use one-component models for the analysis of surface waves in the low frequency range which is the proper region in geophysics.

In contrast to the high frequency approximation this wave is neither dispersive nor dissipative (no attenuation). The Stoneley wave does not exist in this approximation which means, as we see further, that its velocity of propagation is zero in the limit $\omega \to 0$.

## 2.2 Numerical procedure

The problem det $\mathbf{A}$ =0 has been solved for the complex wave number, $k$, using the two computing packages MATLAB 5.3 and MAPLE V Release 5.1. The attempt to execute the calculation with the third package, MATHEMATICA 5, failed. In both successful packages, in principle, it is possible to use the existing equation solvers although they need for the calculations with complex variables a very extensive main memory. It has been observed that the package MAPLE calculates solely *one of the solutions* for $k$ for any choice of sign combinations of exponents $\gamma_1, \gamma_2$ and $\zeta$, changing between branches of solution by the variation of exponents without any apparent reason. Sometimes it was the Rayleigh solution which was calculated and sometimes the Stoneley solution. On the other hand, MATLAB revealed *all solutions* independently of a chosen combination of signs of exponents and this required testing any solution in order to find a corresponding combination of signs. Of course, the ascertained values agreed in both packages.

Inspection of the above solutions shows that they contain roots of quantities which are either already complex or may change the sign. As well known, this means that there exist many Riemann surfaces. It has been seen already in the case of classical Rayleigh waves (Schröder and Scott (2001), Wilmanski (2005)) that, in order to obtain a true surface wave one has to choose a proper Riemann plane otherwise one obtains leaky surface waves. This is, of course, also visible in the numerical procedure which yields solutions on both Riemann surfaces which are related to the true surface wave as well as to solutions for leaky waves. An additional problem arises due to the fact that complex solutions in the present case follow not only from the choice of the Riemann surface but also from the attenuation through dissipation. Consequently, a numerical analysis has to be done with a particular care.

The duration of the calculation for one value of $\omega$ was about 90s on a 1000 MHz machine.

From the complex results for $k$ we are able to determine the normalized velocities of the Rayleigh and Stoneley modes $c'_{Ra} = \frac{\omega}{\mathrm{Re}k_1}$, $c'_{St} = \frac{\omega}{\mathrm{Re}k_2}$, respectively, and the corresponding normalized attenuations $\mathrm{Im}k_1$ for the Rayleigh wave and $\mathrm{Im}k_2$ for the Stoneley wave.

### 2.3 Parameters

The results have been obtained for the following numerical data

$$\begin{array}{llll} \beta = 0, & c_{P1} = 2500\ \dfrac{\mathrm{m}}{\mathrm{s}}, & c_{P2} = 1000\ \dfrac{\mathrm{m}}{\mathrm{s}}, & c_S = 1500\ \dfrac{\mathrm{m}}{\mathrm{s}}, \\ \rho_0^S = 2500\ \dfrac{\mathrm{kg}}{\mathrm{m}^3}, & \rho_0^F = 250\ \dfrac{\mathrm{kg}}{\mathrm{m}^3}, & c_f = \dfrac{c_{P2}}{c_{P1}} = 0.4, & c_s = \dfrac{c_S}{c_{P1}} = 0.6, \\ r = \dfrac{\rho_0^F}{\rho_0^S} = 0.1, & \pi = \begin{cases} 10^7\ \frac{\mathrm{kg}}{\mathrm{m}^3\mathrm{s}} \text{ or} \\ \text{variable} \end{cases}, & \tau = 10^{-6}\ \mathrm{s}, & \pi' := \dfrac{\pi\tau}{\rho_0^S} = \begin{cases} 0.004 \text{ or} \\ \text{variable} \end{cases}. \end{array} \tag{2.28}$$

These data correspond approximately to, for instance, either marls or porous and saturated sandstones (Bourbié et al. (1987)).

### 2.4 Numerical results

In the whole range of frequencies there exist two surface modes of propagation corresponding to the classical Rayleigh and Stoneley waves.

Results are shown for different values of the bulk permeability coefficient, $\pi$. This parameter describes the resistance of the porous medium against the flow of the fluid.

The classical form of the Darcy law refers to simple seepage experiments and describes the seepage velocity $v_{seep}$ in terms of the pressure gradient

$$v_{seep} = -\frac{K}{\rho g}\frac{dp}{dx},$$

where $\rho$ – mass density of the fluid, $g$ – earth acceleration, and $K$ – hydraulic conductivity. This corresponds to the quasistatic form of the relation $(1)_2$

$$\frac{dp}{dx} + \pi\left(v^F - v^S\right) = 0,$$

if we require the relative and seepage velocities to be identical $v_{seep} = v^F - v^S$. Consequently, the relation between permeability coefficients in these two approaches is as follows

$$\frac{K}{\rho g} \sim \frac{1}{\pi}.$$

For instance, for water saturated sands $K \sim 10^{-2} \div 10^{-3}\ \frac{\mathrm{m}}{\mathrm{s}}$, $\rho \sim 10^3\ \frac{\mathrm{kg}}{\mathrm{m}^3}$ and $g \sim 10\ \frac{\mathrm{m}}{\mathrm{s}^2}$ $\Longrightarrow \pi \sim 10^6 \div 10^7\ \frac{\mathrm{kg}}{\mathrm{m}^3\mathrm{s}}$ (see e.g. Bear (1988)). In standard units of permeability this corresponds to app. $1 \div 0.1$ darcy.

Numerical results for velocities and attenuations are partially normalized. Namely values of velocities are divided by the high frequency velocity of the $P1$-wave $c_{P1}$ given by (2.28) and these of attenuations – by the product $c_{P1}\tau$.

Consequently, according to relations (2.10) the normalization of the independent variable $\omega$ by the same constant $\tau$ would lead to identical results for different values of the permeability coefficient $\pi$. This selfsimilarity is lost in the figures presented in the paper due to the fact that we do not normalize the frequency $\omega$. It appears in all figures in real physical units (Hz or MHz).

**Phase velocities of Rayleigh and Stoneley waves**

Fig. 5 shows the velocity of the Rayleigh wave normalized by the $P1$-velocity in dependence on the frequency (see: (2.10)). The velocity is given for different values of the bulk permeability parameter $\pi$. The left and the right figure differ in the range of frequencies. In the left one it varies between zero and the very large value of 100 MHz while the range in the right one is smaller. As in the classical case velocities of all monochromatic Rayleigh waves are smaller than the velocity of the bulk shear wave, $c_S$, whose normalized value is $c_s \equiv \frac{c_S}{c_{P1}} = 0.6$. The high and low frequency limits which are indicated in the left figure are different and correspond to different limits of bulk waves. For our data these dimensionless limits are approx. 0.5493 and 0.5237, respectively.

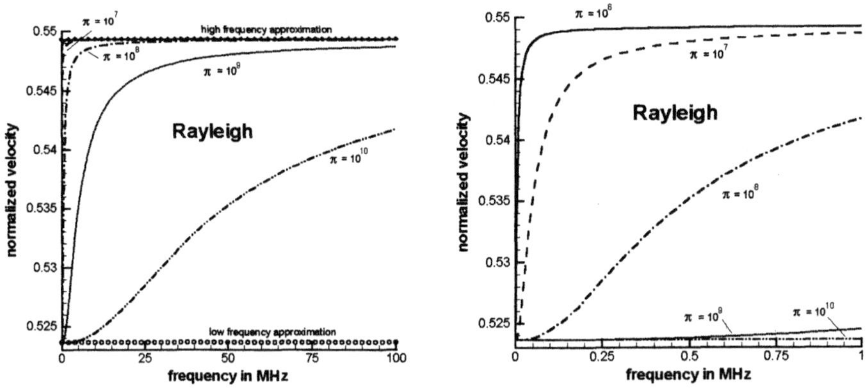


**Figure 5.** Normalized velocities of the Rayleigh wave $c'_{Ra} \equiv \frac{c_{Ra}}{c_{P1}}$ for different values of the permeability coefficient $\pi$ in units $\left[\frac{\text{kg}}{\text{m}^3\text{s}}\right]$ and in different ranges of frequencies.

The velocity of the Rayleigh wave possesses in the range of relatively small frequencies an interesting feature. As we see in the next figure the dependence on the frequency is not monotonous. However, the nonmonotonicity is so weak that in the range of low frequencies applied in geotechnics the Rayleigh velocity may be assumed to be constant. The nonmonotonicity can be observed in a blow-up presented in Fig. 6. The Rayleigh velocity decays first a little and then it becomes growing to its limit value for $\omega \to \infty$. The maximum decay is very little indeed – approx. 0.025% of the difference of limit values for $\omega = 0$ and $\omega \to \infty$. Interestingly, the minimum value remains constant for the different values of $\pi$. This means that the decay is not driven by the diffusion. Details are shown in the right figure for the permeability $\pi = 10^7 \frac{\text{kg}}{\text{m}^3\text{s}}$.

Bourbié et al. (1987) prescribe this effect to an influence of the $P2$-wave, whose velocity goes very rapidly to zero as $\omega \to 0$ and becomes already very small in this range of frequencies. Consequently, it has a bigger influence than in the range of higher frequencies on the behavior of Rayleigh waves. It is interesting that this coupling is present in the simplified model where the coupling term in stresses is absent and bulk modes couple only through boundary conditions. Due to the latter property of the model the size of the effect is very small indeed, even though within the Biot model it is not very large either (compare: Fig. 6.11 in Bourbié et al. (1987), where the maximum corresponds to approx. 0.3%).

In any case this region of the Rayleigh velocity has been investigated particularly carefully to eliminate the possibility of numerical artefacts.

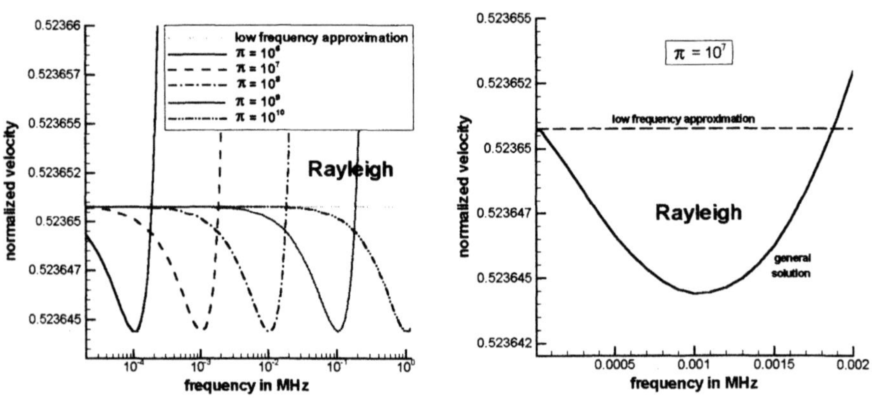


**Figure 6.** Decay of the Rayleigh velocity $\frac{c_{Ra}}{c_{P1}}$ in the range of relatively small frequencies. Left: for different values of the permeability coefficient $\pi$ in units $\left[\frac{\text{kg}}{\text{m}^3\text{s}}\right]$, right: detail for $\pi = 10^7 \frac{\text{kg}}{\text{m}^3\text{s}}$.

In Fig. 7 we present the velocity of the second surface wave possible for the boundary porous medium/vacuum – the Stoneley wave. It is normalized in the same way as the Rayleigh velocity. Also in these figures we see the curves for several values of $\pi$. The velocity increases from the zero value for $\omega = 0$. This property was indicated earlier as a nonexistence of the Stoneley wave. The growth is faster than the growth of the Rayleigh velocity but the maximum value is smaller. It lies always below the normalized velocity of the fluid $c_f \equiv \frac{c_{P2}}{c_{P1}} = 0.4$. This happens for all values of $\pi$. The maximum value of the Stoneley velocity appearing for $\omega \to \infty$ is approximately 0.15% smaller than the velocity of the fluid. One should point out that the Stoneley velocity behaves regularly in the whole range of frequencies and it ceases to exist only for $\omega = 0$. In the vicinity of this point the Stoneley velocity possesses a similar feature to the $P2$-wave: it decays to zero as $\sqrt{\omega}$.

In order to be more specific, in Fig. 8 we consider a selected case which may appear in geotechnics and shows the normalized velocities of both Rayleigh and Stoneley waves for

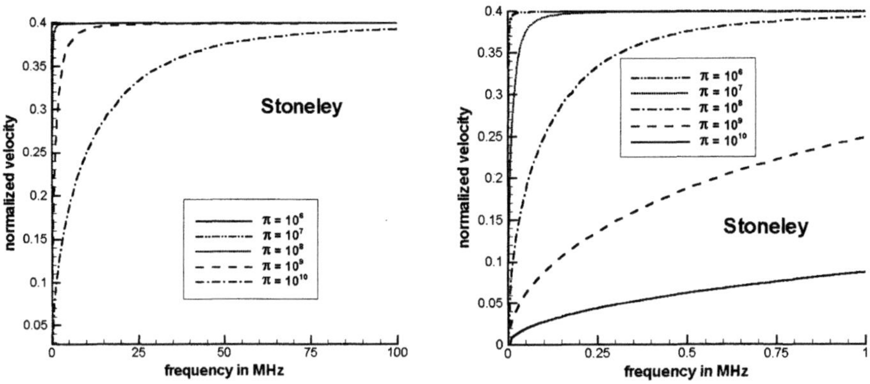


**Figure 7.** Normalized velocities of the Stoneley wave $c'_{St} \equiv \frac{c_{St}}{c_{P1}}$ for different values of the permeability coefficient $\pi$ in units $\left[\frac{\mathrm{kg}}{\mathrm{m}^3\mathrm{s}}\right]$ in different ranges of frequencies.

a permeability coefficient $\pi = 10^7 \frac{\mathrm{kg}}{\mathrm{m}^3\mathrm{s}}$. This corresponds, as shown above, to sandstone saturated with water.

We see the velocities of both surface modes in different ranges of frequencies. The first graph shows properties up to 0.5 MHz, and the second one – up to 10 Hz. Each wave attends a finite asymptotic value as $\omega \to \infty$. This value is bigger for the Rayleigh wave than for the Stoneley wave. While the Stoneley wave velocity starts from zero for $\omega = 0$ the Rayleigh wave yields a low frequency limit unequal to zero. With growing frequency the Rayleigh wave remains at first nearly constant and then increases little until it reaches the high frequency limit. On the other hand the growth of the Stoneley wave velocity is much steeper.

In order to demonstrate the existence of the Stoneley wave in the range of very small frequencies we show in the right figure some calculated points which make obvious that there do not appear any numerical problems to calculate the Stoneley velocity in the limit $\omega \to 0$. In the same range of frequencies, usual in geophysics, the Rayleigh velocity remains nearly constant. Notice that the velocity axis is broken in order to show the behavior of both velocities in the same units.

### Group velocities of Rayleigh and Stoneley waves

The figures for phase velocities of Rayleigh and Stoneley waves show that both of them depend on the frequency $\omega$. In inhomogeneous media waves of different frequency (or wavelength) in general propagate with different phase velocities. This phenomenon is known as *dispersion*. The dispersion in heterogeneous materials appears in a nondissipative manner which is not the case in systems with diffusion. It is easy to see that dissipative waves considered in this article become nondispersive in the nondissipative limit $\pi \to 0$. Compare the chapter of K. Wilmanski in this book (Wilmanski (2005)) where it is shown that in homogeneous media waves propagate undispersed.

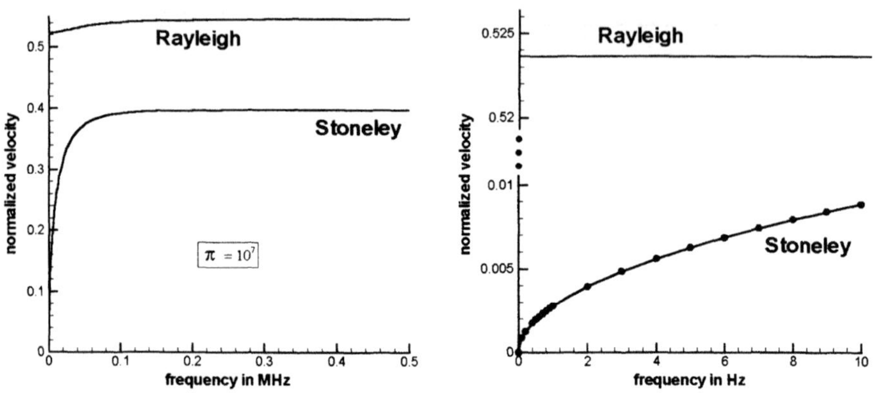


**Figure 8.** Comparison of the behavior of Rayleigh and Stoneley wave velocities in high and low frequency ranges for a permeability coefficient $\pi = 10^7 \frac{\mathrm{kg}}{\mathrm{m}^3\mathrm{s}}$.

A monochromatic wave as investigated in this section is an idealization which is never strictly realized in nature. Most sources emit signals with a continuous spectrum over a limited frequency band. The group velocity $c_g$ (for details in the case of real $k$ see e.g. Aki and Richards (1980), Ben-Menahem and Singh (1981) or Whitham (1974)) for a given frequency $\omega$ is the velocity of transport of a wave package consisting of contributions from a band of frequencies around $\omega$. Then, accounting for the fact that the real wavelength $k$ and the phase velocity $c_{ph}$ depend on the frequency $\left(k = \frac{\omega}{c_{ph}}\right)$

$$\frac{dk}{d\omega} = \frac{1}{c_g} = \frac{1}{c_{ph}} - \frac{\omega}{c_{ph}^2}\frac{dc_{ph}}{d\omega}, \quad \Rightarrow c_g = \frac{c_{ph}}{1 - \frac{\omega}{c_{ph}}\frac{dc_{ph}}{d\omega}}. \tag{2.29}$$

However, in our case the wave number is complex ($k = k_R(\omega) + ik_I(\omega)$) and a relation similar to (2.29) follows under some simplifying assumptions. We consider the wave consisting of a narrow band of frequencies near the middle frequency $\omega_0$. The solution for the amplitude $A$ can be described by a Fourier integral which accounts for all frequencies entering the band

$$A(x,t) = \frac{1}{2\pi}\int_{-\infty}^{\infty} A(\omega)\, \mathrm{e}^{-k_I x} \mathrm{e}^{i(k_R x - \omega t)} d\omega. \tag{2.30}$$

Under the assumptions of small changes of the amplitude and small changes of damping

$k_I$ in the range $\omega_0 \leq \omega \leq \omega_0 + \Delta\omega$, this is approximately

$$\begin{aligned}
A(x,t) &\cong \frac{1}{2\pi} A(\omega_0)\, e^{-k_I^0 x} e^{i\left(k_R^0 x - \omega_0 t\right)} \int_{\omega_0}^{\omega_0+\Delta\omega} e^{i\left(\frac{x}{c_g}-t\right)(\omega-\omega_0)} d\omega \cong \\
&\cong \frac{1}{2\pi} A(\omega_0)\, e^{i(k_0 x - \omega_0 t)} \int_0^{\Delta\omega} e^{i\left(\frac{x}{c_g}-t\right)\xi} d\xi = \\
&= \frac{1}{2\pi} A(\omega_0)\, e^{i(k_0 x - \omega_0 t)} \int_0^{\Delta\omega} \left[\cos\left(\tfrac{x}{c_g}-t\right)\xi + i \sin\left(\tfrac{x}{c_g}-t\right)\xi\right] d\xi = \qquad (2.31) \\
&= \frac{1}{2\pi} A(\omega_0)\, e^{i(k_0 x - \omega_0 t)} \frac{1}{\frac{x}{c_g}-t} \left[\sin\left(\tfrac{x}{c_g}-t\right)\Delta\omega - i\underbrace{\left(\cos\left(\tfrac{x}{c_g}-t\right)\Delta\omega - 1\right)}_{\approx 0}\right] = \\
&= \frac{1}{2\pi} A(\omega_0) \underbrace{e^{ik_0(x-c_{ph}t)}}_{\text{carrier}} \underbrace{\left(\frac{\sin\left[\left(\frac{x}{c_g}-t\right)\Delta\omega\right]}{\left(\frac{x}{c_g}-t\right)\Delta\omega}\right)}_{\text{modulator}} \Delta\omega,
\end{aligned}$$

with $\boxed{c_g := \left(\dfrac{dk_R}{d\omega}\right)^{-1}_{\omega=\omega_0} = \dfrac{c_{ph}}{1-\frac{\omega}{c_{ph}}\frac{dc_{ph}}{d\omega}}}$ and $\left|\dfrac{\Delta\omega}{\omega_0}\right| \ll 1, \quad c_{ph} = \dfrac{\omega}{k_R}$

$$k_0 = k_R^0 + ik_I^0 = k(\omega_0).$$

In the following figure we show both the phase velocities and the group velocities of both surface waves at the boundary porous medium/vacuum for the permeability coefficient $\pi = 10^7\ \frac{\text{kg}}{\text{m}^3\text{s}}$. The derivative $\frac{dc_{ph}}{d\omega}$ has been calculated as central difference.

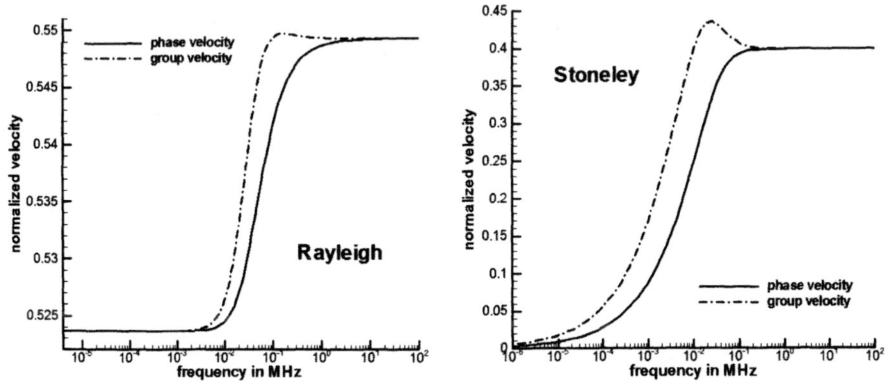


**Figure 9.** Rayleigh and Stoneley wave phase and group velocities in the whole frequency range for a permeability coefficient $\pi = 10^7\ \frac{\text{kg}}{\text{m}^3\text{s}}$.

We see that both group velocities have a maximum while these maxima do not appear in the phase velocity curves. Consequently, even though we have the inverse behavior to this typical for classical elastic systems (phase speeds grow rather than decay with the frequency $\omega$), we may expect sharp changes in the amplitude of arriving waves. These correspond to the classical Airy phase, namely amplitudes at the time in a seismogram at which sharp pulses arrive (see the chapter of Wilmanski in this book).

### Attenuation of Rayleigh and Stoneley waves

This section is devoted to the behavior of the attenuation of the Rayleigh and Stoneley waves. Imaginary parts of the wave number $k$ determine the damping of waves. It is normalized by the product with the $P1$-velocity and the relaxation time (see: (2.10)). This means for our parameters that the values presented in the figures are 400 times smaller ($(2500 \times 10^{-6})^{-1}$) than in real physical units $\left[\frac{1}{\mathrm{m}}\right]$.

Let us first turn our attention to the Rayleigh wave. Fig. 10 shows the attenuation of this wave in different ranges of frequency. The left hand side conveys the impression that the attenuation would not start from zero with zero frequency. This stems from the double-logarithmic scale. However, in the right figure (at least at the frequency-axis) the right physical behavior is obvious: the attenuation for all values of $\pi$ starts from zero. Although distorting the $\omega - \mathrm{Im}k$ behavior, we use on the left hand side the double-logarithmic scale to expose two characteristic features: intersections of curves for different permeabilities $\pi$ as well as the common asymptotics for $\omega \to \infty$. In contrast to all other waves whose attenuation goes to a finite limit as $\omega \to \infty$ – a property which, incidentally, was not clearly stated in earlier works – the attenuation of Rayleigh waves goes to $\infty$ as $\omega \to \infty$. We demonstrate this point separately again in Fig. 13 but we want to stress here that this singular behavior of the Rayleigh wave in relation to the attenuation is the feature of a *leaky wave*. In contrast to the attenuation of other waves which results solely from the dissipation, for a leaky wave the infinite limit of attenuation (asymptote) is independent of the permeability $\pi$, provided $\pi \neq 0$. The left hand side of Fig. 10 also clarifies that the attenuation is in the same manner selfsimilar as the velocity. The existence of intersection points of curves for different $\pi$ is also characteristic for bulk $P1$-waves and indicates the decay of curvature of these curves as functions of the physical frequency $\omega$ with the growing permeability $\pi$.

In order to expose a region of very small frequencies important in geotechnical applications, we present in Fig. 11 the attenuation of both surface waves and two bulk waves: $P1$ and $P2$ in the range of frequencies up to 1000 Hz. Clearly, in this range, the Rayleigh wave is attenuated stronger than the $P1$ wave but still weaker than $P2$.

Inspection of Fig. 12 shows that also the normalized attenuation of the Stoneley wave starts from the zero value for $\omega = 0$. But, in contrast to the Rayleigh wave attenuation for small frequencies, it increases much faster and then approaches a horizontal asymptotic value for larger values of the frequency. This means that the limit $\omega \to \infty$ is finite and dependent on the permeability coefficient $\pi$.

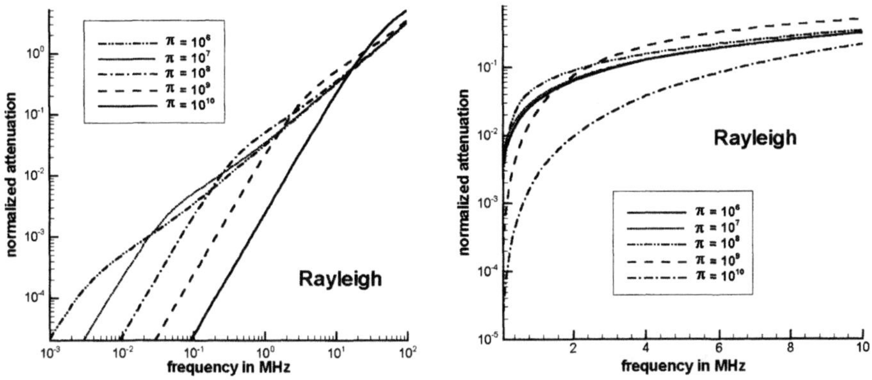


**Figure 10.** Attenuation of the Rayleigh wave for different values of the permeability coefficient $\pi$ in units $\left[\frac{\text{kg}}{\text{m}^3\text{s}}\right]$ in different ranges of frequencies.

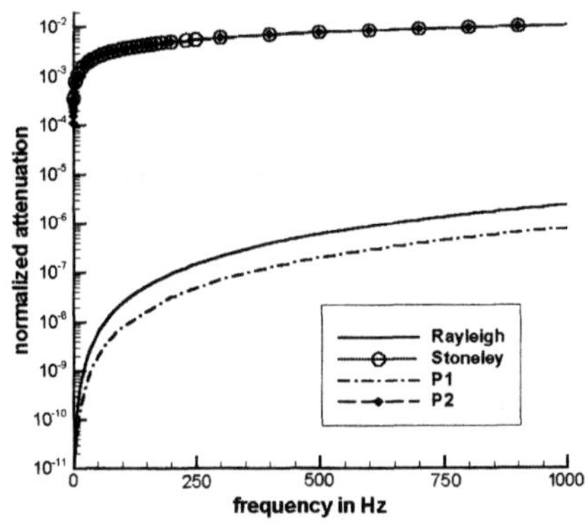


**Figure 11.** Attenuation of Rayleigh, Stoneley, $P1$ and $P2$-wave, for $\pi = 10^7 \ \frac{\text{kg}}{\text{m}^3\text{s}}$ in a range of small frequencies.

This property does not coincide with the statement in Edelman and Wilmanski (2002) in which it is claimed that the Stoneley wave propagates "almost without attenuation" (p. 38). This feature concerns the quantity $\text{Im}\tilde{\omega} \equiv \frac{\text{Im}\omega}{k}$ in terms of the quoted paper, and means solely that $\lim_{k\to\infty} \text{Im}\omega < \infty$. In the frequency space considered in the present work we have similarly $\lim_{\omega\to\infty} \text{Im}k < \infty$.

We can reinterpret those results in terms of the *quality factor* used frequently in works on acoustics. As indicated, for instance, in Aki and Richards (1980), Bourbié et al. (1987) it may be defined as

$$Q(\omega) := \left|\frac{\text{Re}k(\omega)}{\text{Im}k(\omega)}\right| \equiv \left|\frac{\text{Re}k(\omega)}{\omega}\frac{\omega}{\text{Im}k(\omega)}\right| = \frac{1}{c_{ph}}\left|\frac{\omega}{\text{Im}k(\omega)}\right|, \quad c_{ph} := \frac{\omega}{\text{Re}k(\omega)}. \tag{2.32}$$

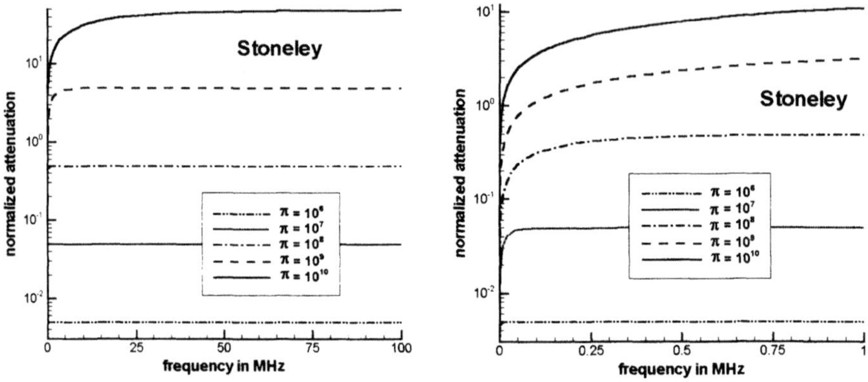


**Figure 12.** Attenuation of the Stoneley wave for different values of the permeability coefficient $\pi$ in units $\left[\frac{\mathrm{kg}}{\mathrm{m}^3\mathrm{s}}\right]$ in different ranges of frequencies.

Clearly, for $|\mathrm{Im}k(\omega)| < \infty$, we have the limit $\lim_{\omega\to\infty} Q(\omega) = \infty$, which is qualified in these works as a lack of dissipation. The other limit $Q = 0$ would mean an infinitely attenuating medium.
We do not use the notion of the quality factor in this article because neither its definition (2.32) is universally accepted, particularly for the limit $\omega \to 0$, nor it possesses such a clear physical interpretation as $\mathrm{Im}k$.

In Fig. 13 we show the normalized attenuation of the surface and bulk waves. For low frequencies the attenuation of the Stoneley wave is much higher than this of the Rayleigh wave. Both attenuations are starting from zero for $\omega = 0$. The Stoneley wave attenuation increases rapidly until it reaches a certain value which depends on the permeability coefficient $\pi$, in the case under consideration – approx. $0.0496\times(c_{P1}\tau)^{-1} \simeq 19.84\frac{1}{\mathrm{m}}$. After reaching this value – which happens in the low frequency range – it remains almost constant. The Rayleigh wave attenuation, however, does not have a finite value for $\omega \to \infty$. As we have already mentioned, the Rayleigh wave is for this reason a leaky wave. Generally, the Rayleigh attenuation increases linearly with growing $\omega$ (i.e. the corresponding quality factor $\lim_{\omega\to\infty} Q(\omega) < \infty$), only for very low frequencies the growth is a little bit faster. Consequently, there appears an intersection of the attenuation curves of both waves. This point lies in the range of high frequencies.

### 2.5 Summary of results for boundary porous medium/vacuum

In the whole range of frequencies there exist two modes of surface waves: a leaky Rayleigh wave and a Stoneley wave. We have shown numerical results for the normalized velocities $\left(\frac{1}{c_{P1}}\frac{\omega}{\mathrm{Re}k(\omega)}\right)$ and attenuations $(\mathrm{Im}k(\omega)\,c_{P1}\tau)$ of these waves for different values of the bulk permeability coefficient, $\pi$, in different ranges of frequencies, $\omega$.

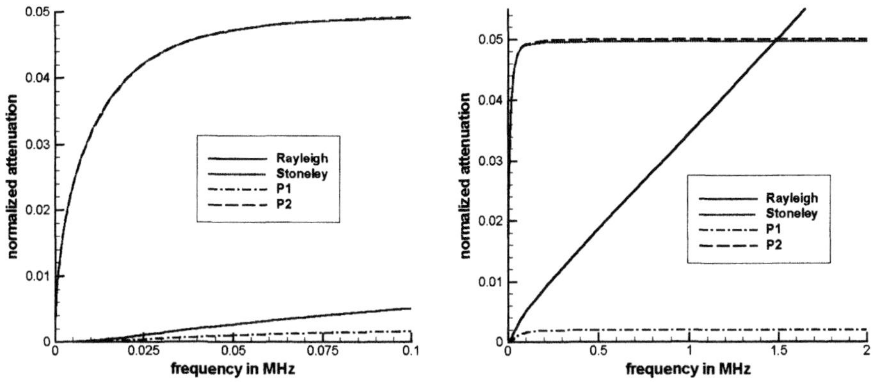


**Figure 13.** Normalized attenuation of Rayleigh, Stoneley, $P1$ and $P2$-wave, for $\pi = 10^7$ $\frac{\text{kg}}{\text{m}^3\text{s}}$ in different ranges of frequencies.

### Leaky Rayleigh

- the velocity of propagation of this wave lies in the interval determined by the limits $\omega \to 0$ (following from (2.27)) and $\omega \to \infty$ (following from (2.23)). The high frequency limit is approx. 4.7% higher than the low frequency limit. The velocity is always smaller than $c_S$, i.e. slower than the $S$-wave. As a function of $\omega$ it possesses an inflection point and it is slightly nonmonotonous,
- this nonmonotonicity appears in the range of small frequencies. The velocity possesses in this range a minimum whose size is very small (approx. 0.025% of the difference of limit values for $\omega = 0$ and $\omega \to \infty$). Interestingly, the minimum value remains constant for the different values of $\pi$. This means that the decay is not driven by diffusion. Such a behavior is also observed within Biot's model. It results from a coupling of $P1$- and $P2$-waves through boundary conditions;
- the attenuation of this wave grows from zero for $\omega = 0$ to infinity as $\omega \to \infty$. In the range of large frequencies it is linear (i.e. the quality factor is a constant different from zero) and independent of the permeability $\pi$. This means that it is a leaky wave.

### Stoneley

- the velocity of this wave grows monotonically from the zero value for $\omega = 0$ to a finite limit which is slightly smaller (approx. 0.15%) than the velocity $c_{P2}$ of the $P2$-wave. The growth of the velocity of this wave in the range of low frequencies is much steeper than this of Rayleigh waves similarly to the growth of the $P2$-velocity;
- both the velocity and attenuation of the Stoneley wave approach zero as $\sqrt{\omega}$,
- the attenuation of the Stoneley wave grows monotonically to a finite limit for $\omega \to \infty$. It is slightly smaller than the attenuation of $P2$-waves. Consequently, in

contrast to the claims in the literature, the Stoneley wave is attenuated. Solely its quality factor goes to zero as $\omega \to \infty$.

Results for different values of the permeability coefficient $\pi$ are selfsimilar, i.e. a change of $\pi$ yields a corresponding change in the scale of the frequency axis for velocities, and of both axes for attenuations. Otherwise the qualitative behavior remains unchanged.

# 3 Analysis by means of the "simple mixture model" – boundary porous medium/fluid

In the previous part we have introduced the equations for the porous medium. Investigating the interface between a porous medium and a liquid we have to consider additional equations for the liquid outside of the porous medium. We distinguish this part of the system by the sign "+".

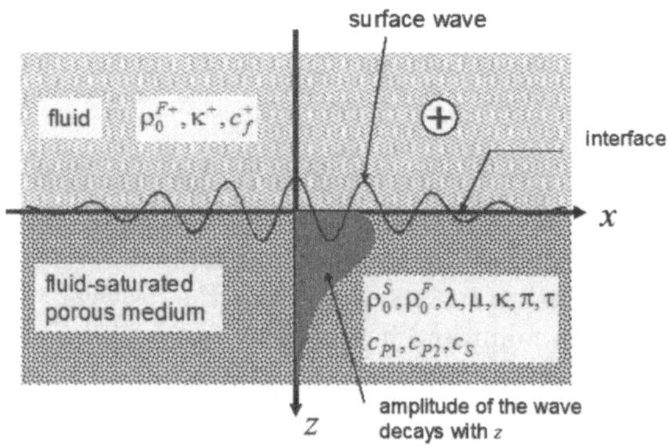


**Figure 14.** Geometry of the interface porous medium/fluid.

## 3.1 Construction of solution

The additional equations for the fluid in the exterior of the porous material read

$$\frac{\partial \rho^{F+}}{\partial t} + \rho_0^{F+} \operatorname{div} \mathbf{v}^{F+} = 0,$$
$$\rho_0^{F+} \frac{\partial \mathbf{v}^{F+}}{\partial t} + \kappa^+ \operatorname{grad} \rho^{F+} = 0. \tag{3.1}$$

Here $\rho^{F+}$ denotes the partial mass density of the fluid in the +-region and $\rho_0^{F+}$ is its constant reference value. $\kappa^+$ describes the compressibility of the fluid. The identification of compressibility coefficients in porous materials can be done in many ways and they do not necessarily give the same results. To make this issue clear, we present here two approaches.

One of them is commonly used in micro-macro transitions for granular materials. Such a transition is described by relations $\rho^F = n\rho^{F+}$, $p^F = np^{F+}$, $p^F - p_0^F = \kappa\left(\rho^F - \rho_0^F\right)$, $p^{F+} - p_0^{F+} = \kappa^+\left(\rho^{F+} - \rho_0^{F+}\right)$, and, under the condition of constant porosity $n = n_0$, we have $\kappa = \kappa^+$.

However, a simple wave analysis shows that this relation cannot hold as the speeds of waves carried by the fluid component are, respectively, $c_+ = \sqrt{\kappa^+}$, $c_{P2} = \sqrt{\kappa}$, and these are, of course, different. The situation does not improve essentially if we introduce the Biot's coupling. This means that linear models used in the description of waves cannot be based on the above quoted simple micro-macro relations. They possess their own macroscopic status and their effective material parameters account for such microscopic effects as scattering of waves. Different macroscopic compressibilities for the same material inside and outside of the porous medium are caused by the channels of the porous medium. While a wave can take a free "path" in the fluid outside, inside the porous medium waves are slower due to reflection on the boundaries of the channels. This is, certainly, not present in simple micro-macro transitions.

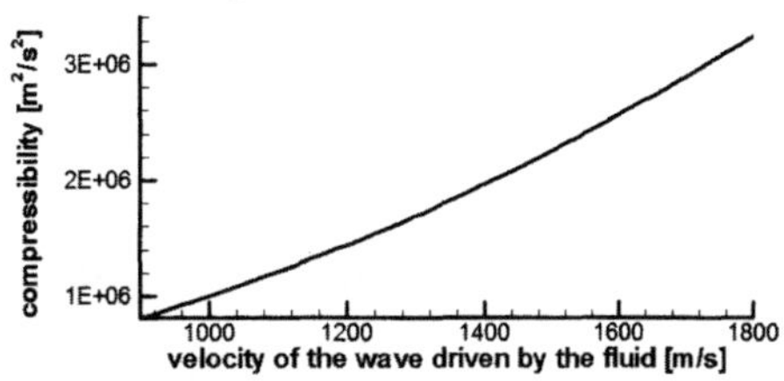


**Figure 15.** Compressibilities for different fluid velocities.

We assume the fluid to be water. A wave in water travels with the velocity of around $c_+ \approx 1500\ \frac{\mathrm{m}}{\mathrm{s}}$ and the corresponding compressibility $\kappa^+ = 2.25 \cdot 10^6\ \frac{\mathrm{m}^2}{\mathrm{s}^2}$. Inside the porous medium the velocity of the $P2$-wave is assumed to be $c_{P2} = 1000\ \frac{\mathrm{m}}{\mathrm{s}}$ to which corresponds a compressibility $\kappa = c_{P2}^2 = 1 \cdot 10^6\ \frac{\mathrm{m}^2}{\mathrm{s}^2}$ (see Fig. 15).

**Compatibility with field equations**

As also done in the equations for the porous medium we introduce the displacement vector $\mathbf{u}^{F+}$. But it also is introduced solely for technical symmetry of the considerations and it does not have any physical bearing. Then

$$\mathbf{u}^{F+} = \operatorname{grad}\varphi^{F+}, \quad \mathbf{v}^{F+} = \frac{\partial \mathbf{u}^{F+}}{\partial t}. \tag{3.2}$$

Again, we make an ansatz for solutions harmonic in the $x$-direction

$$\begin{aligned} \varphi^{F+} &= A^{F+}(z)\exp\left[i\left(kx - \omega t\right)\right], \\ \rho^{F+} - \rho_0^{F+} &= A_\rho^{F+}(z)\exp\left[i\left(kx - \omega t\right)\right]. \end{aligned} \tag{3.3}$$

Substitution in (3.1) leads to the following compatibility conditions

$$A_\rho^{F+} = -\rho_0^{F+}\left(\frac{d^2}{dz^2} - k^2\right)A^{F+}, \qquad \left[\kappa^+\left(\frac{d^2}{dz^2} - k^2\right) + \omega^2\right]A^{F+} = 0. \tag{3.4}$$

We introduce dimensionless quantities

$$c_f^+ := \frac{\sqrt{\kappa^+}}{c_{P1}}, \quad r^+ := \frac{\rho_0^{F+}}{\rho_0^S} < 1, \quad \alpha' := \alpha c_{P1}. \tag{3.5}$$

Further we omit the prime. Then substitution of (2.10) and (3.5) in equations (3.4) yields

$$\left[c_f^{+2}\left(\frac{d^2}{dz^2}-k^2\right)+\omega^2\right]A^{F+}=0. \tag{3.6}$$

We seek a solution in the form

$$A^{F+}=A_f^+e^{\gamma^+z}, \tag{3.7}$$

where the exponent $\gamma^+$ must possess a positive real part to describe a surface wave. Substitution in (3.6) yields it in the form

$$\left(\frac{\gamma^+}{k}\right)^2=1-\frac{1}{c_f^{+2}}\left(\frac{\omega}{k}\right)^2 \quad\Rightarrow\quad \boxed{\frac{\gamma_{1,2}^+}{k}=\pm\sqrt{1-\frac{1}{c_f^{+2}}\left(\frac{\omega}{k}\right)^2}.} \tag{3.8}$$

**Insertion into boundary conditions**

Using (2.1) and (3.2) we get (not yet dimensionless)

$$\begin{aligned}
&\bullet\quad \left.\left(2\frac{\partial^2\varphi^S}{\partial x\partial z}+\frac{\partial^2\psi_y^S}{\partial x^2}-\frac{\partial^2\psi_y^S}{\partial z^2}\right)\right|_{z=0}=0,\\
&\bullet\quad c_{P1}^2\rho_0^S\left(\frac{\partial^2\varphi^S}{\partial x^2}+\frac{\partial^2\varphi^S}{\partial z^2}\right)-2c_S^2\rho_0^S\left(\frac{\partial^2\varphi^S}{\partial x^2}-\frac{\partial^2\psi_y^S}{\partial x\partial z}\right)+\\
&\qquad\qquad +c_+^2\left(\rho^{F+}-\rho_0^{F+}\right)-c_{P2}^2\left(\rho^F-\rho_0^F\right)\big|_{z=0}=0,\\
&\bullet\quad \rho_0^F\frac{\partial}{\partial t}\left.\left(\frac{\partial\varphi^F}{\partial z}+\frac{\partial\psi_y^F}{\partial x}-\frac{\partial\varphi^S}{\partial z}-\frac{\partial\psi_y^S}{\partial x}\right)\right|_{z=0}=\\
&\qquad\qquad =\rho_0^{F+}\frac{\partial}{\partial t}\left.\left(\frac{\partial\varphi^{F+}}{\partial z}-\frac{\partial\varphi^S}{\partial z}-\frac{\partial\psi_y^S}{\partial x}\right)\right|_{z=0},\\
&\bullet\quad \rho_0^F\frac{\partial}{\partial t}\left.\left(\frac{\partial\varphi^F}{\partial z}+\frac{\partial\psi_y^F}{\partial x}-\frac{\partial\varphi^S}{\partial z}-\frac{\partial\psi_y^S}{\partial x}\right)\right|_{z=0}+\alpha\left\{p_0^F+c_{P2}^2\left(\rho^F-\rho_0^F\right)-\right.\\
&\qquad\qquad \left.-n_0\left[p_0^{F+}+c_+^2\left(\rho^{F+}-\rho_0^{F+}\right)\right]\right\}\big|_{z=0}=0,
\end{aligned} \tag{3.9}$$

where we have used the constitutive relations

$$\begin{aligned}
p^F=p_0^F+\kappa\left(\rho^F-\rho_0^F\right),\quad p^{F+}=p_0^{F+}+\kappa^+\left(\rho^{F+}-\rho_0^{F+}\right),\\
c_{P2}^2=\kappa,\quad c_+^2=\kappa^+.
\end{aligned} \tag{3.10}$$

Making use of (2.3), (2.4), (2.6), (3.3) and (3.4) we obtain

$$\begin{aligned}
&\bullet \quad \left(-k^2 B^S - \frac{d^2 B^S}{dz^2} + 2ik\frac{dA^S}{\partial z}\right)\bigg|_{z=0} = 0,\\
&\bullet \quad c_{P1}^2\rho_0^S\left(-k^2A^S + \frac{d^2A^S}{dz^2}\right) - 2c_S^2\rho_0^S\left(-k^2A^S - ik\frac{dB^S}{dz}\right) +\\
&\qquad + c_{P2}^2\rho_0^F\left(\frac{d^2A^F}{dz^2} - k^2A^F\right) - c_+^2\rho_0^{F+}\left(\frac{d^2A^{F+}}{dz^2} - k^2A^{F+}\right)\bigg|_{z=0} = 0,\\
&\bullet \quad -i\omega\rho_0^F\left(\frac{dA^F}{\partial z} + ikB^F - \frac{dA^S}{\partial z} - ikB^S\right)\bigg|_{z=0} =\\
&\qquad = -i\omega\rho_0^{F+}\left(\frac{dA^{F+}}{\partial z} - \frac{dA^S}{\partial z} - ikB^S\right)\bigg|_{z=0},\\
&\bullet \quad i\omega\rho_0^F\left(\frac{dA^F}{\partial z} + ikB^F - \frac{dA^S}{\partial z} - ikB^S\right)\bigg|_{z=0} = \alpha\left\{-c_{P2}^2\rho_0^F\left(\frac{d^2A^F}{dz^2} - k^2A^F\right) +\right.\\
&\qquad \left. + n_0c_+^2\rho_0^{F+}\left(\frac{d^2A^{F+}}{dz^2} - k^2A^{F+}\right)\right\},
\end{aligned} \tag{3.11}$$

where it was assumed that

$$p_0^{F+} = \frac{p_0^F}{n_0}. \tag{3.12}$$

It means that the initial external pressure is equal to the initial pore pressure.

Eqs. (3.11) read in dimensionless form (again omitting the primes)

$$\begin{aligned}
&\bullet \quad \left(-k^2 B^S - \frac{d^2 B^S}{dz^2} + 2ik\frac{dA^S}{\partial z}\right)\bigg|_{z=0} = 0,\\
&\bullet \quad -k^2A^S + \frac{d^2A^S}{dz^2} + 2c_s^2\left(k^2A^S + ik\frac{dB^S}{dz}\right) +\\
&\qquad + c_f^2 r\left(\frac{d^2A^F}{dz^2} - k^2A^F\right) - c_f^{+2}r^+\left(\frac{d^2A^{F+}}{dz^2} - k^2A^{F+}\right)\bigg|_{z=0} = 0,\\
&\bullet \quad r\left(\frac{dA^F}{\partial z} + ikB^F - \frac{dA^S}{\partial z} - ikB^S\right)\bigg|_{z=0} = r^+\left(\frac{dA^{F+}}{\partial z} - \frac{dA^S}{\partial z} - ikB^S\right)\bigg|_{z=0},\\
&\bullet \quad -i\omega\left(\frac{dA^F}{\partial z} + ikB^F - \frac{dA^S}{\partial z} - ikB^S\right)\bigg|_{z=0} = \alpha\left\{c_f^2\left(\frac{d^2A^F}{dz^2} - k^2A^F\right) -\right.\\
&\qquad \left. - \underbrace{n_0\frac{r^+}{r}}_{=1} c_f^{+2}\left(\frac{d^2A^{F+}}{dz^2} - k^2A^{F+}\right)\right\}.
\end{aligned} \tag{3.13}$$

where we have used the relation $n_0 = \frac{r}{r^+}$ if $r^+ \neq 0$.

**Dispersion relation**

Inserting (2.6), (2.12) and (3.7) in these boundary conditions we obtain

- $-k^2 B_s - \zeta^2 B_s + 2ik\left(\gamma_1 A_s^1 + \gamma_2 A_s^2\right) = 0,$
- $-k^2\left(A_s^1 + A_s^2\right) + \gamma_1^2 A_s^1 + \gamma_2^2 A_s^2 + 2c_s^2\left[k^2\left(A_s^1 + A_s^2\right) + ik\zeta B_s\right] +$
  $+c_f^2 r\left(\gamma_1^2 A_f^1 + \gamma_2^2 A_f^2 - k^2 A_f^1 - k^2 A_f^2\right) - c_f^{+2} r^+\left(\gamma^{+2} A_f^+ - k^2 A_f^+\right) = 0,$
- $\gamma_1 A_f^1 + \gamma_2 A_f^2 - \dfrac{k\pi}{r\omega + i\pi} B_s + \left(\dfrac{r^+}{r} - 1\right)\left(\gamma_1 A_s^1 + \gamma_2 A_s^2 + ikB_s\right) - \dfrac{r^+}{r}\gamma^+ A_f^+ = 0,$
- $$\begin{aligned} &-i\omega\left(\gamma_1 A_f^1 + \gamma_2 A_f^2 - \frac{k\pi}{r\omega + i\pi} B_s - \gamma_1 A_s^1 - \gamma_2 A_s^2 - ikB_s\right) = \\ &\alpha\left\{c_f^2\left(\gamma_1^2 A_f^1 + \gamma_2^2 A_f^2 - k^2 A_f^1 - k^2 A_f^2\right) - c_f^{+2}\left(\gamma^{+2} A_f^+ - k^2 A_f^+\right)\right\}. \end{aligned} \tag{3.14}$$

Now (2.16) yields the following equations for the four unknown constants $B_s$, $A_s^1$, $A_f^2$, $A_f^+$

$$\mathbf{AX} = \mathbf{0}, \tag{3.15}$$

where

$$\mathbf{A} := \begin{pmatrix} \left(\frac{\zeta}{k}\right)^2 + 1 & -2i\frac{\gamma_1}{k} & -2i\frac{\gamma_2}{k}\delta_s & 0 \\ 2ic_s^2\frac{\zeta}{k} & \begin{matrix}\left(\frac{\gamma_1}{k}\right)^2 - 1 + 2c_s^2 + \\ +rc_f^2\left[\left(\frac{\gamma_1}{k}\right)^2 - 1\right]\delta_f\end{matrix} & \begin{matrix}\left[\left(\frac{\gamma_2}{k}\right)^2 - 1 + 2c_s^2\right]\delta_s + \\ +rc_f^2\left[\left(\frac{\gamma_2}{k}\right)^2 - 1\right]\end{matrix} & -r^+ c_f^{+2}\left[\left(\frac{\gamma^+}{k}\right)^2 - 1\right] \\ -\frac{\pi}{r\omega+i\pi} + i\left(\frac{r^+}{r} - 1\right) & \frac{\gamma_1}{k}\left[\delta_f + \left(\frac{r^+}{r} - 1\right)\right] & \frac{\gamma_2}{k}\left[1 + \delta_s\left(\frac{r^+}{r} - 1\right)\right] & -\frac{r^+}{r}\frac{\gamma^+}{k} \\ \frac{r\omega}{r\omega+i\pi}\frac{\omega}{k} & \begin{matrix}i\frac{\omega}{k}\frac{\gamma_1}{k}\left(\delta_f - 1\right) + \\ +\alpha c_f^2\left[\left(\frac{\gamma_1}{k}\right)^2 - 1\right]\delta_f\end{matrix} & \begin{matrix}i\frac{\omega}{k}\frac{\gamma_2}{k}\left(1 - \delta_s\right) + \\ +\alpha c_f^2\left[\left(\frac{\gamma_2}{k}\right)^2 - 1\right]\end{matrix} & -\alpha c_f^{+2}\left[\left(\frac{\gamma_+}{k}\right)^2 - 1\right] \end{pmatrix}, \tag{3.16}$$

$$\mathbf{X} := \left(\ B_s,\ \ A_s^1,\ \ A_f^2,\ \ A_f^+\ \right)^T. \tag{3.17}$$

This homogeneous set yields the *dispersion relation:* $\det \mathbf{A} = 0$ determining the $\omega - k$ relation. We investigate also here the solutions of this equation for high and low frequencies.

**High and low frequency approximations**

**High frequencies** In the case of high frequencies (physical dimensions) $\frac{1}{\omega\tau} \ll 1$ we have $\delta_s = \delta_f = 0$, the exponents $\left(\frac{\gamma_1}{k}\right)^2 = 1 - \left(\frac{\omega}{k}\right)^2$, $\left(\frac{\gamma_2}{k}\right)^2 = 1 - \frac{1}{c_f^2}\left(\frac{\omega}{k}\right)^2$, $\left(\frac{\zeta}{k}\right)^2 = 1 - \frac{1}{c_s^2}\left(\frac{\omega}{k}\right)^2$ and $\left(\frac{\gamma^+}{k}\right)^2 = 1 - \frac{1}{c_f^{+2}}\left(\frac{\omega}{k}\right)^2$, and the dispersion relation follows in the form

$$\begin{aligned}\alpha \frac{r}{c_s^4}\frac{r}{r^+}\left(\frac{r^+}{r} - 1\right)^2 \frac{\gamma_1}{k}\left(\frac{\omega}{k}\right)^5 + i\frac{r}{c_s^4}\left(\frac{\gamma^+}{k} - \frac{r^+}{r}\frac{\gamma_2}{k}\right)\frac{\gamma_1}{k}\left(\frac{\omega}{k}\right)^4 + \\ +\alpha \mathcal{P}_R\left(\frac{r}{r^+}\frac{\gamma_2}{k} - \frac{\gamma^+}{k}\right)\frac{\omega}{k} + i\frac{\gamma_2}{k}\frac{\gamma_+}{k}\mathcal{P}_R = 0,\end{aligned} \tag{3.18}$$

where

$$\mathcal{P}_R := \left[2 - \frac{1}{c_s^2}\left(\frac{\omega}{k}\right)^2\right]^2 - 4\frac{\gamma_1}{k}\frac{\zeta}{k}. \tag{3.19}$$

We consider two cases of the dispersion relation (3.18):

1. $\underline{\alpha = 0}$ (impermeable boundary; i.e. sealed porous medium in contact with an external fluid)

   We get from (3.18)

$$\frac{\gamma_+}{k}\frac{\gamma_2}{k}\mathcal{P}_R + \left(\frac{\gamma^+}{k} - \frac{r^+}{r}\frac{\gamma_2}{k}\right)\frac{r}{c_s^4}\frac{\gamma_1}{k}\left(\frac{\omega}{k}\right)^4 = 0. \tag{3.20}$$

   The case $\alpha = 0$ does not correspond to the case of the boundary porous medium/vacuum which we have considered in the last section because the fluid outside yields a pressure on the boundary. However, if additional to $\alpha = 0$ also $r^+ = 0$ we have the same conditions and (3.20) must be identical to the Rayleigh dispersion relation for the boundary porous medium/vacuum (2.23) derived in the last section (see also: Wilmanski and Albers (2003))

$$\mathcal{P}_V = \mathcal{P}_R\frac{\gamma_2}{k} + \frac{r}{c_s^4}\frac{\gamma_1}{k}\left(\frac{\omega}{k}\right)^4 = 0. \tag{3.21}$$

   This is, indeed, true if we cancel $\frac{\gamma^+}{k}$ on both sides after setting $r^+$ equal to zero.

2. $\underline{\alpha \to \infty}$
   Here, the other two terms of (3.18) remain and we obtain after division by $\alpha$

$$\mathcal{P}_R\left(\frac{\gamma_2}{k} - \frac{r^+}{r}\frac{\gamma^+}{k}\right) + \left(\frac{r^+}{r} - 1\right)^2\frac{r}{c_s^4}\frac{\gamma_1}{k}\left(\frac{\omega}{k}\right)^4 = 0. \tag{3.22}$$

Both equations (3.20) and (3.22) remind equation (3.21) but they are both modified by the influence of the fluid outside.

**Low frequencies** In contrast to the impermeable boundary with vacuum analytical calculations in this case become very complicated. Therefore, we investigate this case solely on a numerical example.

### 3.2 Numerical results

**Procedure and parameters**

In principle the numerical procedure of solution of the problem $\det \mathbf{A} = 0$ for the complex wave number $k$ is the same as described in Sect. 2.2. However, in this more complicated case MATLAB revealed no result. Thus the results of the programs MAPLE 7 and MAPLE V Release 5.1 were the sole attainable values. Interestingly, MAPLE for some sign combination of exponents produces two solutions which are close to each other but, for most frequencies, not the same.

In order to expose better some properties, we change the input data for $c_S$. We need as well additional quantities in the exterior:

$$\begin{aligned}
&c_{P1} = 2500\,\frac{\mathrm{m}}{\mathrm{s}}, \quad && c_{P2} = 1000\,\frac{\mathrm{m}}{\mathrm{s}}, \quad && c_S = 1250\,\frac{\mathrm{m}}{\mathrm{s}}, \quad && c_+ = 1500\,\frac{\mathrm{m}}{\mathrm{s}}, \\
&\rho_0^S = 2500\,\frac{\mathrm{kg}}{\mathrm{m}^3}, && \rho_0^F = 250\,\frac{\mathrm{kg}}{\mathrm{m}^3}, && \rho_0^{F+} = 1000\,\frac{\mathrm{kg}}{\mathrm{m}^3}, && r = \frac{\rho_0^F}{\rho_0^S} = 0.1, \\
&c_f = \frac{c_{P2}}{c_{P1}} = 0.4, && c_s = \frac{c_S}{c_{P1}} = 0.5, && c_f^+ = \frac{c_+}{c_{P1}} = 0.6, && r^+ = \frac{\rho_0^{F+}}{\rho_0^S} = 0.4.
\end{aligned} \tag{3.23}$$

While some of the results for the boundary porous medium/vacuum have been shown for the varying bulk permeability parameter, $\pi$, this coefficient is constant here, namely $\pi = 10^7\ \frac{\mathrm{kg}}{\mathrm{m}^3\mathrm{s}}$. However, instead of this we show for this boundary the influence of the surface permeability, $\alpha$. We demonstrate that two surface waves, both of them leaky, exist in the whole range of frequencies for each choice of $\alpha$. They correspond to the classical Rayleigh wave, and to a Stoneley wave which is produced due to presence of the fluid outside the porous medium, respectively. The latter can be supported by the fact that it appears also for the sealed pore situation if there is a fluid outside. Moreover we obtain a third type of wave which appeared also on the boundary porous medium/vacuum, namely a true Stoneley wave. However, this wave exists only for small values of $\alpha$.

Numerical results for velocities and attenuations are normalized in the same way as indicated in Sect. 2.4.

**Dependence of phase velocities and attenuations on the frequency**

Fig. 16 shows both the phase velocities and the attenuations of all three surface waves appearing at the interface between a porous halfspace and a fluid halfspace. On the left hand side the velocities are plotted and figures on the right hand side show the attenuations. Both quantities are given for a wide range of frequencies between 1 Hz and 100

MHz. The different curves correspond to various values of the surface permeability parameter $\alpha$. As explained above, $\alpha = 0$ means that the surface is completely impermeable while $\alpha = \infty$ corresponds to an open pore situation.

In Fig. 16 both the frequency and the attenuations are shown in logarithmic scale while the velocity is presented in normal scale.

**Phase velocities** Again, the velocity of the Rayleigh wave lies under the velocity of the bulk shear wave, $c_S$, whose normalized value here is $c_s \equiv \frac{c_S}{c_{P1}} = 0.5$. Curves for different values of $\alpha$ have each a low and a high frequency limit which is unequal to zero. While for small frequencies the velocity is the same independently of $\alpha$, the high frequency limit decreases with increasing $\alpha$. For the open pore situation the difference between high and low frequency limits is approximately one half of the difference for a close boundary. Inbetween there is a steep increase of the Rayleigh velocity. The range of frequencies for this increase is smaller for small values of $\alpha$ (10-100 kHz) than for large values of $\alpha$ (1-500 kHz). Moreover for the latter there appears a small plateau in this zone. This may be an indication of the change of the Riemann surface which is, however, much better pronounced in the attenuations.

The velocity of the leaky Stoneley wave behaves similar to the Rayleigh wave. Also this wave possesses high an low frequency limits unequal to zero and the steep increase inbetween appears in the same frequency region. However, in contrast to the Rayleigh wave for this wave the high frequency limit is larger for bigger values of $\alpha$ than for smaller ones. The frequency behavior of this wave is – at least for small $\alpha$ – not monotonous. A maximum value appears in the region of order 100 kHz. Interestingly, the velocity of this wave is smaller than this of the Rayleigh wave although it is driven by the fluid outside of the porous medium whose longitudinal bulk wave is faster than that of the fluid inside the porous medium ($c_f^+ = 0.6$, $c_f = 0.4$). This result has been obtained also by Feng&Johnson Feng and Johnson (1983a) within Biot's model. As we will see in the next subsection the true Stoneley wave exists only for small values of $\alpha$, and, therefore, we show its behavior only for two values of $\alpha$. For these the velocities do not differ substantially. They start form zero at $\omega = 0$ and increase until around 100 kHz where they reach a high frequency limit which is a little bit smaller than the velocity of the $P2$ wave, $c_f \equiv \frac{c_{P2}}{c_{P1}} = 0.4$.

**Attenuations** Let us start with the attenuation of the true Stoneley wave. This has the same appearance as this obtained for the boundary porous medium/vacuum. We show a log-log-plot of this attenuation. Thus the similarity to Fig. 12 is not directly obvious. The attenuation starts from zero as $\omega = 0$ and reaches a horizontal asymptote at around 100 kHz. The only amazing point is, that in the mapped region of frequencies, starting from 1 Hz the value for $\alpha = 10^{-4}\ \frac{\mathrm{s}}{\mathrm{m}}$ is much smaller than for $\alpha = 0$ even though the difference in velocities is small. It is surprising because for the other attenuations the value for $\alpha = 0$ lies under all other values for different $\alpha$.

In contrast to the true surface wave the remaining leaky waves possess singularities in the attenuation for two intermediate frequencies. These frequencies seem to be related

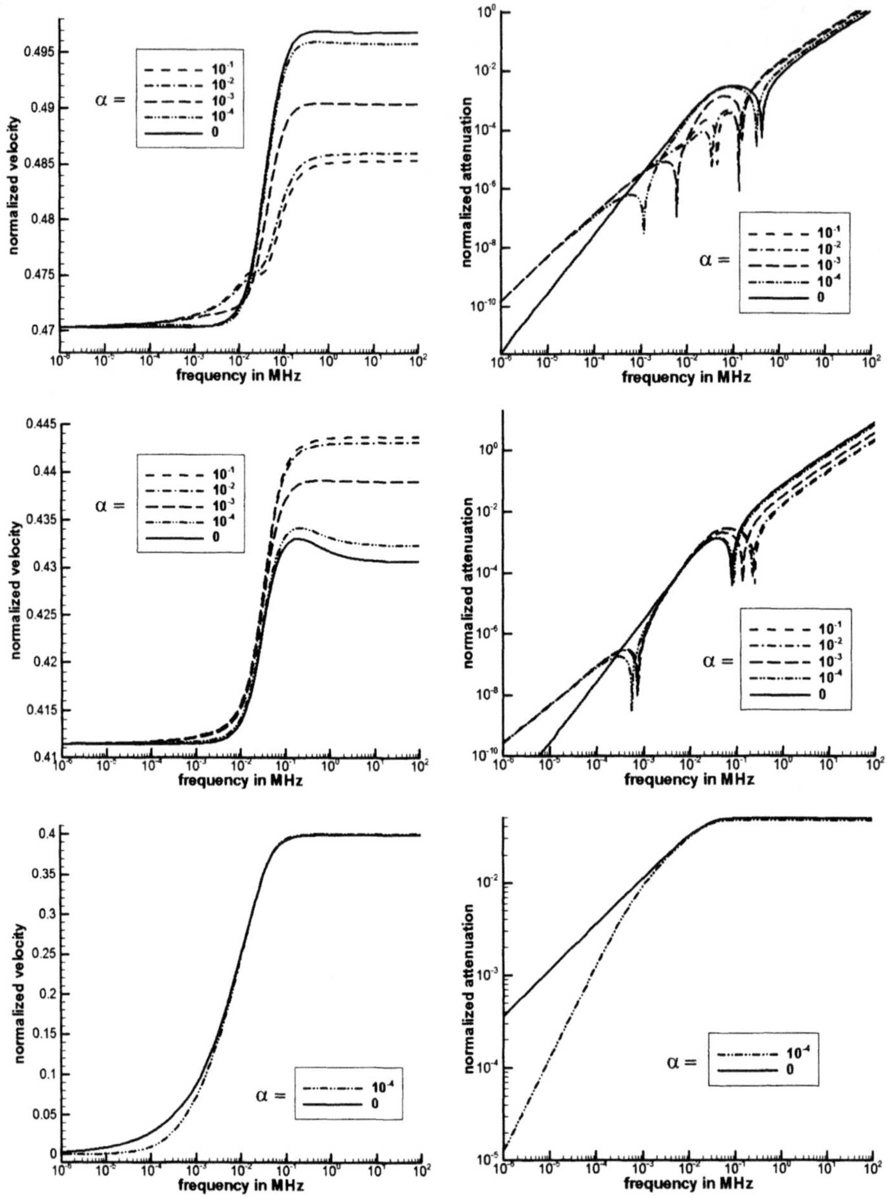

**Figure 16.** Normalized phase velocities and attenuations of the leaky Rayleigh (first row), leaky Stoneley (second row) and the Stoneley wave (third row) in dependence on the frequency. Different curves correspond to different values of the surface permeability $\alpha$ in units $\left[\frac{\mathrm{s}}{\mathrm{m}}\right]$). The smaller $\alpha$ the denser the boundary.

with characteristic frequencies $\frac{\pi}{2\rho_0^S}$ and $\frac{\pi}{2\rho_0^F}$ which have already appeared in the stability analysis of adsorption processes (Albers (2003)). However, there exists an influence of the parameter $\alpha$, responsible for dissipation, and, simultaneously, the location of the singularities changes with the variation of this coefficient. Consequently, as indicated also in some papers on Biot's model, the diffusion-driven resonances appear also in the surface waves. Their existence seems to be confirmed experimentally (for results of measurements see in particular Chao et al. (2004), Wisse et al. (2002)). In Fig. 17 we reproduce one figure of the paper by Chao et al. (2004). It shows the measured damping coefficients of the pseudo-Stoneley wave (in our terminology: leaky Stoneley wave) in a shock-induced borehole experiment. The formation is a Berea sandstone. The damping coefficient is given in a frequency range up to 50 kHz and we see that there appear also some well pronounced singularities.

Little is known about their mathematical origin. However, the numerical analysis indicates that they appear due to the change of the Riemann surface. If one would ignore the singularities and look only at the connecting line between the maxima the curves would look similar to the curve for the boundary porous medium/vacuum.
In any case, it is obvious that the curves show the leaky character as observed already for the Rayleigh wave for this boundary: for high frequencies the attenuation grows linearly and unbounded.

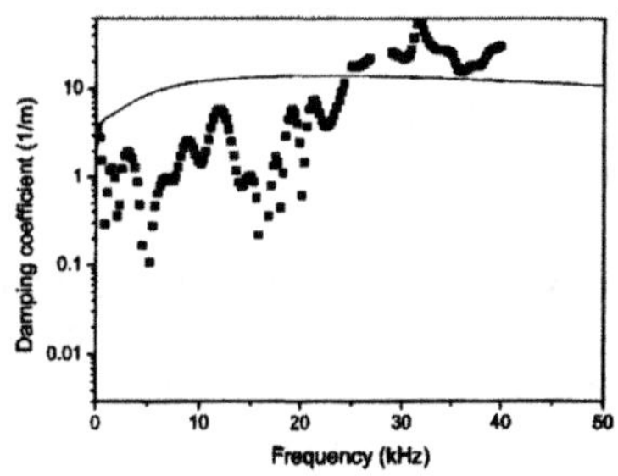


**Figure 17.** Experimental damping coefficients of the pseudo-Stoneley wave. From Chao et al. (2004)

**Dependence of phase velocities and attenuations on the surface permeability**

In the next figure we illustrate the behavior of the three surface waves appearing at the boundary porous medium/fluid for the frequency limits $\omega \to \infty$, $\omega \to 0$ but in dependence on the surface permeability parameter $\alpha$. In the first row of Fig. 18 we show normalized velocities and attenuations for the same material parameters as used in the last subsection. In the bottom row the same quantities are shown for another choice of material parameters: with unchanged mass densities and porosity and different velocities for two bulk waves ($c_{P1} = 3500\ \frac{\text{m}}{\text{s}}$, $c_S = 1750\ \frac{\text{m}}{\text{s}}$, $c_{P2}, c_{P+}$ unchanged).

The upper left figure shows the phase velocities of the surface waves for the two limits of frequencies ($\omega \to \infty$ – solid lines, $\omega \to 0$ – dashed lines). Additional to the shown range of surface permeability parameters between $10^{-6}\ \frac{\text{s}}{\text{m}}$ and $10^{-1}\ \frac{\text{s}}{\text{m}}$ we show on the left and right hand side points corresponding to the limit cases $\alpha = 0$ and $\alpha = \infty$. It is obvious that the values for $10^{-6}$ and $10^{-1}$ correspond already well to the limit values. While both leaky surface waves exist in the whole range of surface permeabilities, the true surface wave appears only for small values of the surface permeability (approx. in the interval $0 \leq \alpha \leq 10^{-3.9}\ \frac{\text{s}}{\text{m}}$). Thus for a relatively open boundary (large $\alpha$) no real surface wave exists. It is obvious that for each wave the low frequency value of the velocity (dashed lines) is independent of $\alpha$. For the true surface wave this value is zero while it is bigger than zero for the leaky waves. Both the high frequency limit of the leaky Rayleigh and of the leaky Stoneley wave change monotonously from the limit $\alpha = 0$ to the limit $\alpha = \infty$. However, the Rayleigh wave velocity is bigger for a dense boundary, while for the leaky Stoneley wave the limit for the open boundary is bigger. The velocity behavior of the waves for the other choice of material parameters ($c_S > c_{f+}$, see: bottom left figure) does not change substantially. Again, $c_R < c_s$, $c_{St} < c_f$ and $c_{leakySt} < c_{f+}$. The latter is a little bit more obvious for this choice of material parameters than for the other choice. We will see in Sect. 4.2 that the last choice corresponds to a higher value of the modulus of the skeleton frame (5.53 instead of 3.95). For this stiffer medium the true Stoneley wave exists only for a still denser boundary, namely for approx. $0 \leq \alpha \leq 10^{-4.8}$ $\frac{\mathrm{s}}{\mathrm{m}}$).

The figures on the right hand side show the normalized attenuation of the three surface waves for a chosen frequency. As already mentioned, the true surface wave – the Stoneley wave – ceases to exist in the range of high surface permeabilities $\alpha$. In the limit value, its attenuation becomes infinite. The remaining two leaky waves possess finite attenuation in the whole range of $\alpha$.

**Group velocities of the three surface waves**
For the impermeable boundary ($\alpha = 0$) between a porous material and a fluid we present the group velocities of the surface waves. The group velocity of the leaky Stoneley wave behaves differently from the both waves which also appeared for the boundary porous medium/vacuum. While they only possess a maximum for a certain frequency, the leaky Stoneley wave exhibits first a strong maximum and then a slight minimum. The mathematical reason is obvious: the growth in phase velocities of Rayleigh and Stoneley waves is almost monotonous but for the leaky Stoneley wave the phase velocity possesses a clear maximum (compare: Fig. 16), and, as indicated in the box of (2.31) the group velocity depends on the slope of the phase velocity.

**Summary of results for boundary porous medium/fluid**
In the whole range of frequencies there exist two leaky surface waves: a leaky Rayleigh wave and a leaky Stoneley wave. In the range of very small values of the surface permeability parameter $\alpha$, there exists a third true surface wave – a Stoneley wave. The behavior of the Rayleigh wave and the true Stoneley wave in dependence on the frequency is the same as described in Sect. 2.5. Here, we summarize only some additional comments on the $\alpha$-dependence and on the additional wave appearing at this boundary, the leaky Stoneley wave:

**Leaky Rayleigh**

- for low frequencies the phase velocity for different values of the surface permeability $\alpha$ remains almost constant. For high frequencies smaller values of $\alpha$ yield bigger velocities; for the open pore case the difference between high and low frequency limits is approx. one half of the difference for a close boundary;
- the attenuation grows linearly and unbounded, there appear singularities which depend on $\alpha$ and seem to be related to the characteristic frequencies $\frac{\pi}{2\rho_0^S}$ and $\frac{\pi}{2\rho_0^F}$;

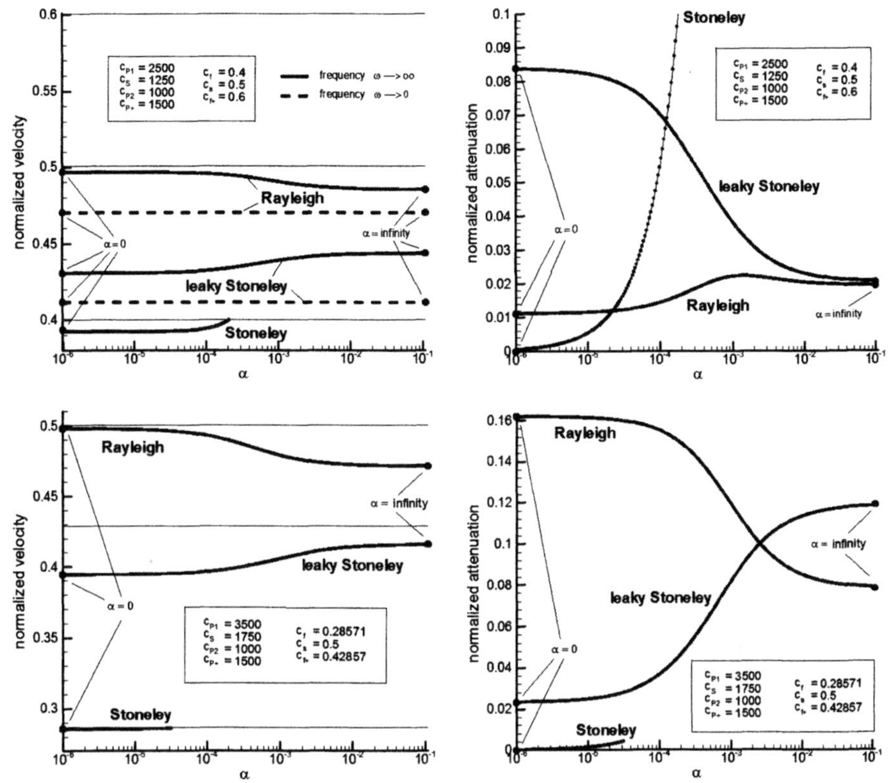


**Figure 18.** Normalized phase velocities and attenuations of the three surface waves in dependence on the surface permeability parameter, $\alpha$, for different material parameters.

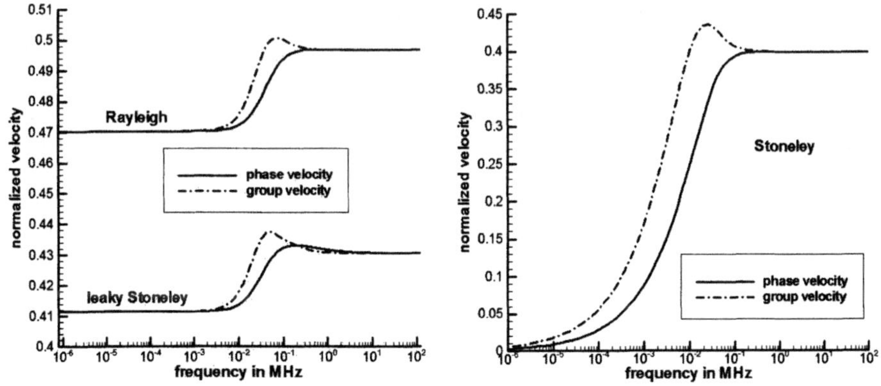


**Figure 19.** Normalized phase and group velocities of the three surface waves.

**Leaky Stoneley**

- the phase velocity of this wave behaves similarly to this of the leaky Rayleigh wave; however, the high frequency limit is larger for bigger values of $\alpha$ than for smaller ones; a maximum value appears in the region of order 100 kHz; the velocity of the leaky Stoneley wave is for each pair $(\omega, \alpha)$ smaller than this of the leaky Rayleigh wave.
- also the attenuation behaves similar to this of the leaky Rayleigh wave; however, the singularities are weaker dependent on $\alpha$;

**Stoneley**

- it exists only for small values of the surface permeability $\alpha$; for different values of $\alpha$ the velocity is nearly the same; it starts from zero and approaches a horizontal asymptote;
- the attenuation of the Stoneley wave grows monotonically to a finite limit for $\omega \to \infty$.

# 4 Comparison to results of other authors

## 4.1 Deresiewicz

One of the first attempts to investigate surface waves in two-component porous materials stems from (Deresiewicz (1962), Deresiewicz (1964)). In the first of these papers he has studied the boundary porous medium/vacuum using the Biot's equations. He has calculated Rayleigh wave velocities and attenuations. Similar to the procedure of the present article he has found analytical approximations for high and low frequencies and carried out numerical calculations for a kerosene-saturated sandstone. He wrote: "Because of its complexity, the secular (comment by B. A.: dispersion) equation does not lend itself to analytical study for intermediate values of the frequency. Accordingly, a numerical study was undertaken, of the variation of velocity and dissipation per cycle with frequency, for a material whose elastic and dynamical coefficients were available, with several curious results." One of these "curious" results is a minimum of the phase velocity in the region of small frequencies like we found for the Rayleigh wave. However, Deresiewicz located also a second minimum in the range of high frequencies which we did not ascertain. The results for the attenuation, in Deresiewicz's work given as the specific energy loss, are hardly comparable to our results. It is impressing how much the author did know about the Rayleigh wave already in the 60ies. However, it is strange that he did not get the Stoneley solution from the dispersion equation. The second paper concerns surface wave in the presence of a liquid layer. This work is not comparable to our case of a boundary between porous medium/fluid because the liquid layer has a given thickness. Moreover, for this case Deresiewicz only considered low frequencies and did not attain numerical results.

### 4.2 Feng & Johnson

Feng and Johnson (1983a) use Biot's theory to search numerically primarily the velocities of various surface waves. They distinguish between the true slow surface wave and the leaky surface waves and calculate their velocities at an interface between a fluid half-space and a half space of a fluid-saturated porous medium like we did in the last section.

They rely on a somewhat different approach to obtain results for the surface waves. Namely, they assume a surface wave to have the characteristic form of being a linear combination of all bulk waves. Under this assumption the displacements can be expressed as

$$\mathbf{u} = \sum_J C_j \widetilde{\mathbf{u}}_j e^{-\gamma_j |z|} e^{i(kx-\omega t)}, \qquad \mathbf{U} = \sum_J C_j \widetilde{\mathbf{U}}_j e^{-\gamma_j |z|} e^{i(kx-\omega t)}, \tag{4.1}$$

where $\mathbf{u}, \mathbf{U}$ are the displacement vectors of the solid part and the fluid part, respectively. The summation is over the three bulk modes (fast, slow, shear) and $\gamma_j \equiv \left(k^2 - \omega^2/V_j^2\right)^{1/2}$ ($V_j$ are bulk wave speeds). The quantities $C_j$ are the amplitudes of the bulk modes. The polarization vectors for each bulk wave $\left(\widetilde{\mathbf{U}}_j, \widetilde{\mathbf{u}}_j\right)$ are related to each other and to the complex wave vector by requiring the displacements to satisfy the bulk differential equations.

However, in contrast to our approach, Feng &Johnson focus only on the high-frequency range. They also use the boundary conditions introduced by Deresiewicz and Skalak butthey limit their attention to the open-pore and to the sealed pore situation (i.e. in our notation: $\alpha = \infty$ and $\alpha = 0$, respectively, and in their notation: $T = 0$ and $T = \infty$). In dependence on the stiffness of the skeleton (in their notation: longitudinal modulus of the skeleton frame $\left(K_b + \frac{4}{3}N\right)^{1/2}$, where $K_b$ : bulk modulus of porous drained solid, $N$ : shear modulus of both the drained porous solid and of the composite) and e.g. on the coefficient of added mass ("tortuosity", in their notation: $\alpha$) they have investigated the existence of the surface modes. We show here some of their numerical results for material parameters which fit best to ours, namely for the case: Water as the fluid (in both $z < 0$ and $z > 0$ regions) and fused glass beads as the porous medium. In order to compare the results we calculate the stiffness of the sandstone of our example in their notation: With the formula for the longitudinal velocity in the dry sample

$$V_L = \sqrt{\frac{K_b + \frac{4}{3}N}{(1-\phi)\,\rho^{SR}}}, \tag{4.2}$$

where $\phi$ : porosity, $\rho^{SR}$ : realistic mass density of the solid, we are able to calculate the quantity appearing as the variable on the horizontal axis

$$\begin{aligned} \sqrt{K_b + \frac{4}{3}N} &= V_L\sqrt{(1-\phi)\,\rho^{SR}} = c_{P1}\sqrt{\rho_0^S} = \\ &= 2500\sqrt{2500}\frac{\sqrt{\mathrm{N}}}{\mathrm{m}} = 125000\frac{\sqrt{\mathrm{N}}}{\mathrm{m}} \approx \\ &\approx 125000\frac{316.23\sqrt{\mathrm{dynes}}}{100\,\mathrm{cm}} \approx 3.9528 \cdot 10^5\,\frac{\sqrt{\mathrm{dynes}}}{\mathrm{cm}}. \end{aligned} \tag{4.3}$$

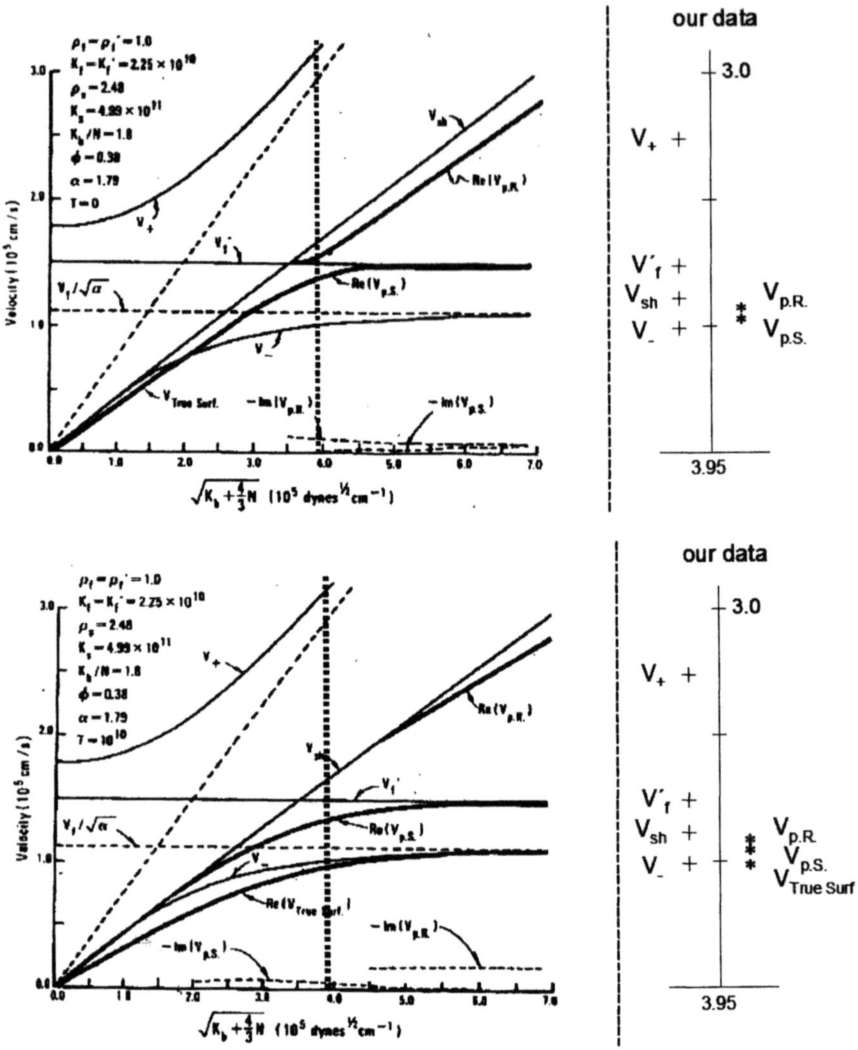


**Figure 20.** Figures of the paper by Feng and Johnson (1983a) for the case: Water as the fluid (in both $z < 0$ and $z > 0$ regions) and fused glass beads as the porous medium extended by a sketch of our results (sandstone instead of glass beads as the porous medium). The points "+" and "$\star$" mark the location of corresponding quantities.
At the top: open pore situation, at the bottom: sealed pore situation.

Feng & Johnson found that for an open-pore surface situation, the true surface wave exists for a limited range of material parameters and changes continuously into a slightly leaky-Stoneley wave as its velocity crosses over the slowest bulk wave velocity. For the sealed pore situation there exist simultaneously a true surface wave (for all values of material parameters) and a leaky-Stoneley wave. The leaky-Rayleigh wave has features similar to those of the leaky-Rayleigh wave for a fluid/nonporous solid case.

These results are demonstrated in juxtaposition in the following table.

| vacuum / solid | fluid / solid | fluid / porous medium | |
|---|---|---|---|
| **Rayleigh** | **leaky-Rayleigh**<br>**Stoneley** | **leaky-Rayleigh**<br>possibly **Stoneley**<br>possibly **leaky-Stoneley** | } or both |

For the leaky waves the authors claim that "they are roots of the same equation on the wrong Riemann sheet". They explain it as follows: "The pseudo-Stoneley mode has a velocity faster than the slowest bulk mode but slower than the rest of the bulk modes. The pseudo-Rayleigh mode has a velocity faster than the slowest bulk mode and the bulk fluid mode (in the region $z > 0$). These relations imply that these surface modes radiate into the slowest bulk mode (for pseudo-Rayleigh mode: both the slowest and the fluid bulk mode) as they propagate along the surface (i.e. $\mathrm{Im}\,(\gamma_j) \neq 0$). Furthermore, the radiation condition requires the radiation of bulk modes to have an exponentially growing factor in the $z$ direction instead of a decaying one; in addition, these pseudosurface modes are attenuated in the $x$ direction, i.e., $\mathrm{Im}\,(V_{\text{ps. surf.}}) < 0$. These two physical requirements imply that both $\mathrm{Re}\,(\gamma_{\text{rad. bulk modes}})$ and $\mathrm{Im}\,(\gamma_{\text{rad. bulk modes}})$ are negative. Usually the condition $|\mathrm{Re}\,(\gamma_{\text{rad. bulk modes}})| \ll |\mathrm{Im}\,(\gamma_{\text{rad. bulk modes}})|$ (or $|\mathrm{Im}\,(V_{\text{ps. surf.}})| \ll |\mathrm{Re}\,(V_{\text{ps. surf.}})|$) is satisfied; this corresponds to the situation of "slightly" leaky pseudosurface modes. So in most cases, one can incorporate conditions on the square roots in $\gamma_j$'s for both the true surface mode and the pseudosurface modes by requiring $\mathrm{Re}\,(\gamma_j) - \mathrm{Im}\,(\gamma_j) \geq 0$. This treatment of the pseudosurface modes is analogous to the treatment of the pseudo-Rayleigh mode for the fluid/solid case about which there is some controversy (Ansell (1972), Strick (1959), Phinney (1961))."

## 5 Glance on three-component porous media

Classical soil mechanics and also the classical theory of porous materials are limited to the description of soils and other porous materials which are fully saturated with a fluid (e.g. water or oil). However, there are many practically relevant examples of materials for which such a condition is not fulfilled: their voids are filled partially with a fluid, and partially with a gas – often with vapor – and these possess different partial pressures. Such materials are called "unsaturated" or "partially saturated". In the last 40 years numerous endeavors have been carried out to examine their mechanical behavior (see e.g. Fredlund and Rahardjo (1993), Pietruszczak and Pande (1996), Schrefler et al. (1990), Wheeler (1988), Wood (1979)) and also the wave propagation in such materials (e.g. Anderson and Hampton (1980), Bedford and Stern (1983), Berryman et al. (1988), Santamarina (2001), Smeulders et al. (1992)).

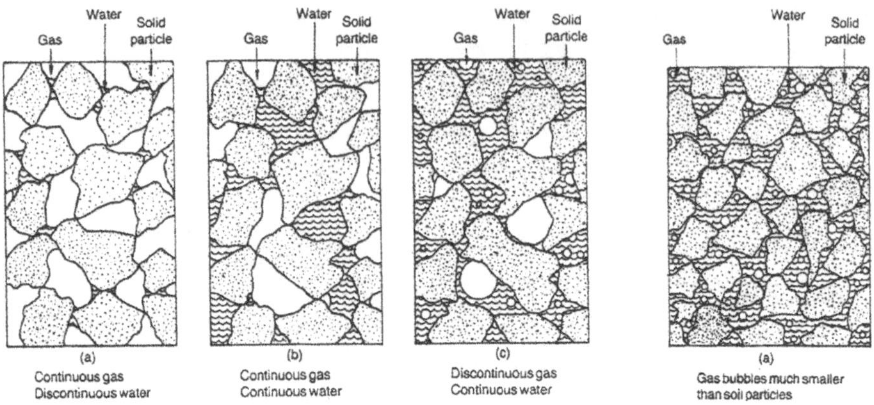


**Figure 21.** Classification of unsaturated soils into three groups (left) and the mostly studied case of gas bubbles much smaller than the soil particles; from Wheeler (1988).

It turned out that the basic structure (see Fig. 21) and the acoustic properties of unsaturated media are dependent on the *degree of saturation.* This is the proportion of the fluid volume to the entire volume of voids. Fig. 21 shows a classification of unsaturated soils into three groups and the mostly studied case where the gas bubbles are much smaller than the soil particles.

In contrast to the one-component fluid in the channels of a saturated porous medium the gas fluid mixture contains phase interfaces which affect both the stress state (surface tension) and the flow conditions (e.g. permeability). Additionally, the morphology (e.g. the degree of saturation) is changed permanently by phase transitions, appearance of big gas bubbles instead of some small bubbles etc. In soils, most of the time, the gas occurs in the form of discrete bubbles as shown on the right hand side of Fig. 21. This is the case if the degree of saturation lies above a certain critical degree of saturation which varies for different soils but is of order 85%. For fluid-gas-mixtures mostly it is assumed that the bubbles are spherical. Then the radius of curvature of the gas-fluid interface which determines the surface tension between gas and fluid and thus the difference of pressures in gas and fluid is equal to the radius of the bubble. Also for porous materials there exists an interrelation between size of bubbles, saturation and capillary pressure. An increase of saturation leads also to an increase of the capillary pressure. Hence, for a given radius of bubbles (which can be measured in experiments), there exists a macroscopical relation of the form

$$p_c = p_c(S), \tag{5.1}$$

where $p_c$ denotes the capillary pressure and $S$ the saturation. In the modeling of unsaturated porous materials the form of this relation is one of the most important questions. We do not go into any details here and remark solely that several researchers simply choose a relation from the micro scale. There are also attempts, not yet fully ripe, to create transitions from the micro to the macro scale (e.g. Pietruszczak and Pande (1996)). A lot of experiments have been performed on the acoustical behavior of un-

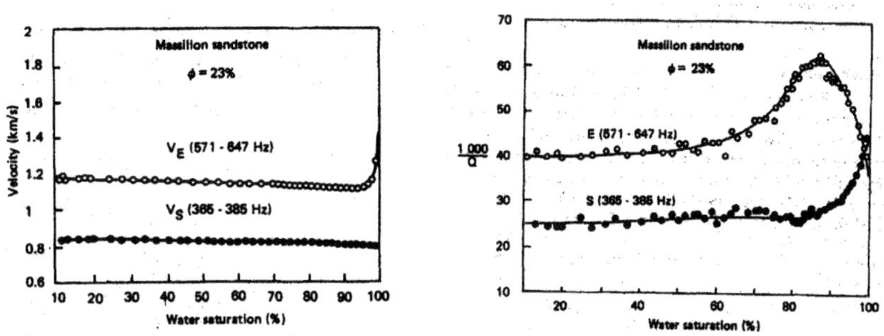


**Figure 22.** Velocity and attenuation $\left(\frac{1000}{Q}\right)$ of the P1-wave and the S-wave in sandstone in dependence on the water saturation, experiments by Murphy; from Bourbié et al. (1987).

saturated materials, especially of sediments (e.g. Anderson and Hampton (1980)). The experimenters claim that the occurrence of gas bubbles – also in small amounts – has high influence on the acoustic characteristics (propagation velocities, attenuations, reflection etc.). To show this we include here the results of some experiments on sandstone. The figures show the velocity and the attenuation in form of the quality factor of the $P1$ wave (denoted by E) and the shear wave (S) . For small amounts of bubbles (high water saturation) their behavior is particularly strong dependent on the saturation. Natural sediments contain usually a fraction of gas bubbles between zero and ten percent.

Helpful for the study of the influence of saturation in porous media may be early works treating the influence of gas bubbles on the sound velocity in pure water. Qualitatively wave velocities in unsaturated sediments should behave like the sound speed in a water-gas-mixture which is given in Fig. 23 (experiments performed by Wood in 1957). We see that there exists an extremely deep minimum in the sound velocity for a certain fraction of air in water.

The present article shows that both velocities and attenuations of surface waves (independent of the type of boundary) are highly dependent on the frequency. This is, of course, also true for unsaturated soils. In order to demonstrate this feature Fig. 24 is included which shows the attenuation of the $P1$ wave in dependence on the frequency and on the degree of saturation. We see that the attenuation depends very strong on both parameters.

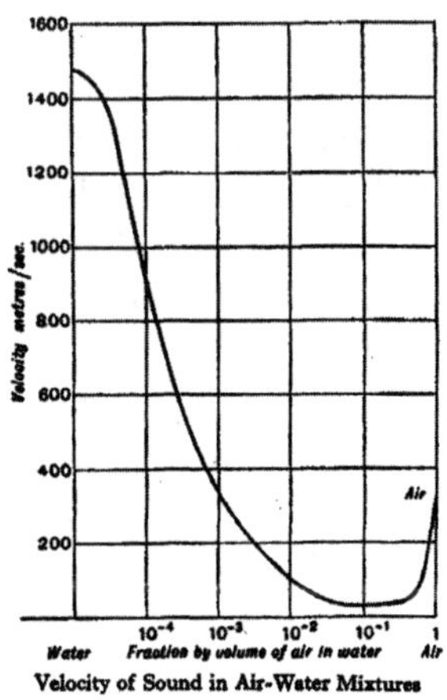


**Figure 23.** Sound speed in a water-air-mixture in dependence on the air fraction; from Wood (1957).

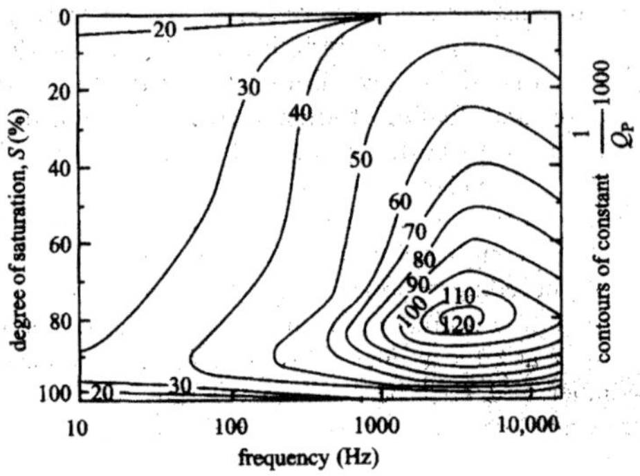


**Figure 24.** Influence of the frequency and the saturation on the attenuation of the $P1$-wave; from Santamarina (2001).

The above remarks show that even the propagation of bulk waves in unsaturated porous materials has not been investigated yet in this extent as for two-component systems. The investigation of surface waves for such materials does not exist at all. Due to its practical importance such a research direction seems to be a natural continuation of results presented in this article.

## Bibliography

K. Aki and P. Richards. *Quantitative seismology, theory and methods.* W. H. Freeman and Co., 1980.

B. Albers. Relaxation analysis and linear stability vs. adsorption in porous materials. *Contin. Mech. Thermodyn.*, 15(1):73–95, 2003. also WIAS-Preprint Nr. 721, 2002.

B. Albers and K. Wilmanski. Monochromatic surface waves on impermeable boundaries in two-component poroelastic media. Preprint 862, WIAS, 2003a. to appear in: Contin. Mech. Thermodyn.

B. Albers and K. Wilmanski. On modeling acoustic waves in saturated poroelastic media. Preprint 874, WIAS, 2003b. to appear in: J. Engn. Mech.

A. L. Anderson and L. D. Hampton. Acoustics of gas-bearing sediments. I. Background, II. Measurements and models. *J. Acoust. Soc. Am.*, 67(6):1865–1898, 1980.

J. H. Ansell. The roots of the stoneley wave equation for solid-liquid interfaces. *Pure Appl. Geophys.*, 94:172–188, 1972.

J. Bear. *Dynamics of Fluids in Porous Media.* Dover Publications, New York, 1988.

A. Bedford and M. Stern. A model for wave propagation in gassy sediments. *J. Acoust. Soc. Am.*, 73(2):409–417, 1983.

A. Ben-Menahem and S. J. Singh. *Seismic waves and sources.* Springer-Verlag, New York Heidelberg Berlin, 1981.

J. G. Berryman, L. Thigpen, and R. C. Y. Chin. Bulk elastic wave propagation in partially saturated porous solids. *J. Acoust. Soc. Am.*, 84(1):360–373, 1988.

T. Bourbié, O. Coussy, and B. Zinszner. *Acoustics of porous media.* Editions Technip, Paris, 1987.

G. Chao, D. M. J. Smeulders, and M. E. H. van Dongen. Shock induced borehole waves in porous formations: Theory and experiments. *J. Acoust. Soc. Am.*, 116(2), 2004.

H. Deresiewicz. The effect of boundaries on wave propagation in a liquid filled porous solid. IV. Surface waves in a half space. *Bull. Seismol. Soc. Am.*, 52:627–638, 1962.

H. Deresiewicz. The effect of boundaries on wave propagation in a liquid filled porous solid. VII. Surface waves in a half space in the presence of a liquid layer. *Bull. Seismol. Soc. Am.*, 54:425–430, 1964.

H. Deresiewicz and R. Skalak. On uniqueness in dynamic poroelasticity. *Bull. Seismol. Soc. Am.*, 53:783–788, 1963.

I. Edelman. Bifurcation of the biot slow wave in a porous medium. *J. Acoust. Soc. Am.*, 114(1):90–97, 2003.

I. Edelman and K. Wilmanski. Asymptotic analysis of surface waves at vacuum/porous medium and liquid/porous medium interfaces. *Contin. Mech. Thermodyn.*, 14(1): 25–44, 2002.

S. Feng and D. L. Johnson. High-frequency acoustic properties of a fluid/porous solid interface. I. New surface mode. *J. Acoust. Soc. Am.*, 74(3):906–914, 1983a.

S. Feng and D. L. Johnson. High-frequency acoustic properties of a fluid/porous solid interface. II. The 2D reflection Green's function. *J. Acoust. Soc. Am.*, 74(3):915–924, 1983b.

D. G. Fredlund and H. Rahardjo. *Soil Mechanics for Unsaturated Soils.* John Wiley & Sons, 1993.

P. Hess. Surface acousic waves in materials science. *Physics Today*, March:42–47, 2002.

H. Kutruff. *Ultrasonics, fundamentals and applications.* Elsevier, 1991.

C. Lai. *Simultaneous inversion of Rayleigh phase velocity and attenuation for near-surface site characterization.* PhD thesis, Georgia Institute of Technology, 1998.

N. G. H. Meyendorf, P. B. Nagy, and S. I. Rokhlin, editors. *Nondestructive Materials Characterization.* Springer, Berlin Heidelberg, 2004.

P. B. Nagy. Acoustics and ultrasonics. In Po-zen Wong, editor, *Methods in the Physics of Porous Media*, pages 161–221. Academic Press, 1999.

A. N. Norris. Stoneley-wave attenuation and dispersion in permeable formations. *Geophysics*, 54(3):330–341, 1989.

J. O. Owino and L. J. Jacobs. Attenuation measurements in cement-based materials using laser ultrasonics. *J. Engn. Mech.*, 125(6):637–647, 1999.

R. A. Phinney. Propagation of leaking interface waves. *Bull. Seismol. Soc. Am.*, 51(4): 527–555, 1961.

S. Pietruszczak and G. N. Pande. Constitutive relations for partially saturated soils containing gas inclusions. *J. Geotechnical Engrg.*, 122(1):50–59, 1996.

G. J. Rix, C. G. Lai, and S. Foti. Simultaneous measurement of surface wave dispersion and attenuation curves. *Geotechnical Testing Journal*, 24(4):350–358, 2001.

J. C. Santamarina. *Soils and Waves.* John Wiley & Sons, 2001.

B. A. Schrefler, L. Simoni, Li Xikui, and O. C. Zienkiewicz. Mechanics of partially saturated porous media. In C.S. Desai and G. Gioda, editors, *Numerical Methods and Constitutive Modelling in Geomechanics (Udine, 1989)*, volume 311 of *CISM Lecture Notes*, pages 169–209. Springer Verlag, Wien, 1990.

C. T. Schröder and W. R. Scott. On the complex conjugate roots of the Rayleigh equation: The leaky surface wave. *J. Acoust. Soc. Am.*, 110(6):2867–2877, 2001.

D. M. J. Smeulders, J. P. M. de la Rosette, and M. E. H. van Dongen. Waves in partially saturated porous media. *Transport in Porous Media*, 9:25–37, 1992.

E. Strick. III. The pseudo-Rayleigh wave. *Philos. Trans. Roy. Soc. (London), Ser. A*, 251:488–522, 1959.

I. Tolstoy, editor. *Acoustics, elasticity and thermodynamics of porous media: Twenty-one papers by M. A. Biot.* American Institute of Physics, 1992.

A. Udias. *Principles of Seismology.* Cambridge University Press, 1999.

I. A. Viktorov. *Rayleigh and Lamb waves. Physical theory and applications.* Plenum Press, New York, 1967.

S. J. Wheeler. A conceptual model for soils containing large gas bubbles. *Géotechnique*, 38:389–397, 1988.

J. E. White. *Underground sound. Application of seismic waves.* Elsevier, Amsterdam, 1983.

G. B. Whitham. *Linear and nonlinear waves.* John Wiley & Sons, New York, 1974.

K. Wilmanski. Waves in porous and granular materials. In K. Hutter and K. Wilmanski, editors, *Kinetic and continuum theories of granular and porous media*, number 400 in CISM Courses and Lectures, pages 131–186. Springer, Wien New York, 1999.

K. Wilmanski. Some questions on material objectivity arising in models of porous materials. In M. Brocato, editor, *Rational continua, classical and new*, pages 149–161. Springer-Verlag, Italia Srl, Milano, 2001.

K. Wilmanski. Propagation of sound and surface waves in porous materials. In B. Maruszewski, editor, *Structured media*, pages 312–326. Poznan University of Technology, Poznan, 2002.

K. Wilmanski. Tortuosity and objective relative acceleration in the theory of porous materials. Preprint 922, WIAS, 2004. to appear in: Proc. Royal Soc. (London), Ser. A.

K. Wilmanski. Elastic modelling of surface waves in single and multicomponent systems. In C. Lai and K. Wilmanski, editors, *Surface Waves in Geomechanics, Direct and Inverse Modelling for Soils and Rocks*, CISM Courses and Lectures. Springer Verlag, Wien-New York, 2005.

K. Wilmanski and B. Albers. Acoustic waves in porous solid-fluid mixtures. In K. Hutter and N. Kirchner, editors, *Dynamic response of granular and porous materials under large and catastrophic deformations*, volume 11 of *Lecture Notes in Applied and Computational Mechanics*, pages 285–314. Springer, Berlin, 2003.

K. W. Winkler, H.-L. Liu, and D. L. Johnson. Permeability and borehole Stoneley waves: Comparison between experiment and theory. *Geophysics*, 54(1):66–75, 1989.

C. J. Wisse, D. M. J. Smeulders, M. E. H. van Dongen, and G. Chao. Guided wave modes in porous cylinders: Experimental results. *J. Acoust. Soc. Am.*, 112(3):890–895, 2002.

A. B. Wood. *A Textbook of Sound.* G. Bell and Sons, London, 1957.

D. M. Wood. The behaviour of partly saturated soils: A review. *Cambridge University Report, Engineering Department*, 1979.

# Theory of Nonlinear Surface Waves and Solitons

Gérard A. Maugin

Université Pierre et Marie Curie, Laboratoire de Modélisation en Mécanique, UMR CNRS 7607, Paris, France

**Abstract.** In this contribution we explore the possibilities of existence of various types of nonlinear surface waves that may propagate in a half-space of elastic materials. Nonlinearities may be of physical and geometrical origins. Various types of surface waves and exemplary nonlinear wave equations are first presented. Then basic equations of nonlinear elasticity theory are introduced and the standard linear surface-wave problems are recalled, in particular the one governing Rayleigh waves. Then two large classes of problems are examined in the presence of nonlinearities, whether there exists no dispersion or in the presence of dispersion. In the first case, various methods leaning on asymptotics are presented ; these are the so-called iterative, projection, and Hamiltonian methods. This allows one to describe the evolution of nonlinear waves as they propagate, as also some exceptional cases of waves of permanent form. In the second case compensation between nonlinearity and dispersion may favor the existence of surface solitary waves of various types including surface SH envelope solitons guided by a thin film glued on the nonlinear substrate, as also solitary Rayleigh waves, with an emphasis on nonlocal effects. Generalizations including propagation along plates are given in the conclusion.

## 1 Introduction

Nonlinear wave propagation was traditionally attached to the dynamical behavior of fluids. Inaugurated by Georg Riemann's studies in the XIXth century and pursued by scientists such as Rankine, Hugoniot, Jouguet, Ernst Mach and others, it acquired its full splendor first with flow problems induced by developments in flight and aeronautics (first military and then civilian ; think of the sonic bang ; cf. Courant and Friedrichs, 1948) after WWII and then in plasmas and magnetohydrodynamics, due to the high-energy levels exhibited in some physical situations (astrophysics, controlled thermonuclear fusion research), see e.g., Jeffrey and Taniuti (1963) and Cabannes (1970). Explosions and the study of some armaments also drove some of the active research in this field. Natural phenomena related to water waves (bores, tsunamis, the " great wave " of Scott Russell) have also kindled the study of nonlinear localized waves such as solitary waves and solitons, while shock waves provided the main incentive for studies in nonlinear wave propagation before.

Comparatively, nonlinear wave propagation in solids is of much more recent extract. Some technological problems forced engineers to examine the problem of nonlinear wave propagation in plastic and viscoplastic materials (e.g., Rakhmatulin and Demianov, 1961 ; Mandel and Brun, 1975 ; Graham, 1993). Volume shock waves, sometimes initially formed from weaker discontinuity waves such as acceleration waves, were the only kind examined by these authors.

Obviously nonlinear phenomena had not escaped the attention of physicists dealing with wave propagation in solids, in particular in the domain of signal processing. Shock waves were not examined as the shock-formation distance (say, one meter) in such cases far exceeds the size of the devices considered, but the marked physical nonlinearity present in some crystals, in particular along certain crystallographic orientations, may cause some undesirable effects such as the production of harmonics, anisochronism, etc, in systems supposed to accurately keep a required frequency standard. We have considered all these effects in two books (Maugin, 1985, Maugin *et al*, 1992). If bulk (volume) waves are of fundamental interest in all cases, for they provide the basic scheme for the classification of possible modes of propagation, it must be emphasized that the very study of signal processing and the consideration of seismic waves lead to a necessary in depth study of so-called **surface waves.** In this case, the bulk of the energy is transported in the vicinity of the limiting surface that , therefore, serves as a waveguide for the mode of propagation. The combination of this notion of surface waves and the simultaneous presence of a relevant nonlinearity provides our subject of study in this contribution. The complexity of the resulting scheme is the price to be paid for advances in that field. Some of the arguments involved have been exposed in a book in the case of elastic crystals (Maugin, 1999). Also useful to the reader are lectures of another course given previously at the CISM in 1993 (Jeffrey and Engelbrecht , 1994).

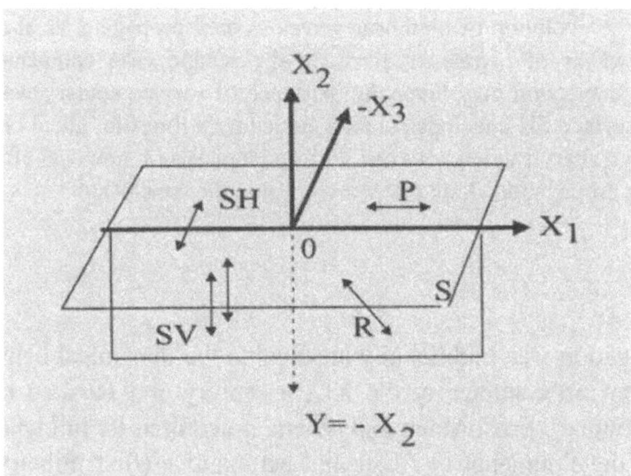


**Figure 1.** Various types of surface-wave polarization.

## 2 Types of Surface Waves on Solids

Types of surface waves of solids have been classified by geophysicists, hence a specific vocabulary ("vertical", "horizontal", "sagittal", etc). They are also classified according as their polarization (" longitudinal ", " transverse ", " quasi-longitudinal) and the mechanical degree of freedom they carry with them (" pressure ", " shear ", see Achenbach, 1973, Maugin, 1983). Accordingly, let the plane ($X_1$, $X_3$) be the limiting plane to the material body which is the half-space $X_2 < 0$. We consider waves propagating along direction $X_1$. The unit normal to the limiting plane is $\mathbf{n} = (0,1,0)$ in this coordinate system. Is called sagittal plane $S$ any plane parallel to the plane spanned by $X_3 = 0$ (see Figure 1). Any elastic wave whose elastic displacement is polarized along the propagation direction $X_1$ is called $P$ (for *pressure* or *primary*) wave or longitudinal wave. If the elastic polarization is parallel to $\mathbf{n}$, it is an $SV$ (shear-vertical) wave , and it is an $SH$ (shear-horizontal) wave if its polarization is along $X_3$ , hence perpendicular to $S$. Elastic waves whose polarization is simply parallel to $S$ but with any orientation in that plane are called $R$ (Rayleigh) waves. If the polarization is at any orientation compared to the orthogonal coordinate system introduced then it corresponds to a $GR$ (generalized Rayleigh) wave. With polarization close in some sense to those of $P$ or $SV$ and $SH$ waves , some waves are called quasi-longitudinal or quasi-transverse. As surface waves *per se*, all these waves are introduced as propagating sinusoidal signals of which the amplitude decreases rapidly (exponentially) with depth inward from the limiting plane. Pure $R$ waves , admixtures of $P$ and $SV$ waves, occur in *isotropic linear* homogeneous elasticity. Pure $SH$ waves also called *Love waves* occur in certain structures (e.g., linear elastic half space covered with another linear elastic body constituting a securely glued lid). In *anisotropic* linear elasticity (e.g., crystals), surface waves may be of the general $GR$ type , an oscillation in profile accompanying possibly the exponential decrease of amplitude with depth (cf. Dieulesaint and Royer, 1980). Because of the layered structure of the vicinity of the Earth surface, $SH$ waves have attracted much attention from geophysicists. In electronic equipment of mechanical-electrical energy conversion, it is Rayleigh waves that have been much exploited. Cutting a crystal along certain preferential planes favors the uncoupling between modes.

As noted, *linearity* is an asset for the uncoupling between various modes in elasticity. The situation is altogether different in *nonlinear elasticity*, even in the case of an isotropic body. Indeed, second order contributions may couple modes that were uncoupled in the linear approximation. This is true even for bulk waves. This is illustrated by the following example that concerns the generation of longitudinal waves by transverse waves.

**Example 1**. Consider a plane elastic wave propagating in the bulk (in the direction of coordinate $x = X_1$ ) of a nonlinear elastic *isotropic* body whose energy contains *cubic* terms in the strain. The body has Lamé coefficients $\lambda$ and $\mu$ of linear elasticity and material coefficients of nonlinear Murnaghan's elasticity $l$ and $m$ (corresponding to cubic terms in the energy) ; $\rho_0$ is the matter density in the reference configuration. Here right subscripts $t$ and $x$ will denote time and space partial derivatives, it being understood that we look for dynamical solutions that depend only on the pair $(t,x)$. We set $\alpha = \lambda + 2\mu$ , $\beta = 3\alpha + 2(l + 2m)$ and $\gamma = \alpha + m$, the elastic displacement $\mathbf{u}$ having Cartesian components $u$, $v$ , $w$ . In the absence of external body force we have the following three *coupled* equations of motions :

$$\begin{aligned} \rho_0 u_{tt} - \alpha u_{xx} &= \beta u_x u_{xx} + \gamma\left(v_x v_{xx} + w_x w_{xx}\right) , \\ \rho_0 v_{tt} - \mu v_{xx} &= \gamma\left(u_x v_{xx} + v_x u_{xx}\right) , \\ \rho_0 w_{tt} - \mu w_{xx} &= \gamma\left(u_x w_{xx} + w_x u_{xx}\right) . \end{aligned} \tag{2.1}$$

Because of isotropy the last two of equations (2.1) are symmetric in $v$ and $w$, and these two components play a similar source role in the first of eqns.(2.1) If we neglect quadratic contributions in eqns.(2.1) - which is not the same as setting $l = m = 0$, because there remains an effect of *geometrical* nonlinearity in coefficients $\beta$ and $\gamma$ - we obtain the uncoupled linear wave equations

$$\rho_0 u_{tt} - \alpha u_{xx} = 0 \ , \ \rho_0 v_{tt} - \mu v_{xx} = 0 \ ; \ \rho_0 w_{tt} - \mu w_{xx} = 0 \ . \tag{2.2}$$

These yield a longitudinal mode $u$ with squared speed of propagation $c_L^2 = \alpha / \rho_0$ and two transverse modes with equal squared speed $c_T^2 = \mu / \rho_0$. The question then arises whether we can have a purely transversely polarized motion such , say,

$$v = \bar{v}(x,t) \quad , \quad u = w = 0 \tag{2.3}$$

for the nonlinear system (2.1). This would directly yield the equations

$$0 = \gamma v_x v_{xx} \quad , \quad \rho_0 v_{tt} - \mu v_{xx} = 0 \ . \tag{2.4}$$

This requires that either $\gamma = 0$ , or essentially $v = 0$, in contradiction with the working hypothesis (2.3). Therefore, calling $f$ and $g$ functions of order one, and $A$ and $B$ orders of magnitude of the $v$ and $u$ components ($A$ is a small dimensionless parameter) , we should have considered a dynamical solution of the type

$$v(x,t) = A f(x,t) \quad , \quad u(x,t) = B g(x,t) \ , \quad B = 0(A^2). \tag{2.5}$$

Thus a pure transverse mode cannot exist without a second order longitudinal mode as companion, so that the system to be considered here is necessarily given by

$$\rho_0 u_{tt} - \alpha u_{xx} = \gamma v_x v_{xx} \quad , \quad \rho_0 v_{tt} - \mu v_{xx} = 0. \tag{2.6}$$

This is called *generation of longitudinal mode by a transverse mode.* This phenomenon holds true in the absence of limiting surface. The situation can only be worst, or more interesting, in the case of surface waves where boundary conditions will favor a cooperation of various modes.

## 3 A Gallery of Nonlinear wave Equations

Before attacking any nonlinear surface waves, it is necessary to identify and remember the essential types of nonlinear wave equations often met in the technical literature. These prototypical equations are paradigmatic of general situations or working hypotheses that will be emphasized. They are all given for a wave motion depending only on time $t$ and one space coordinate $x$. They are not always issued from solid mechanics but they may occur in that framework under appropriate circumstances as shown in the books by Maugin *et al* (1992) and Maugin (1999).

### 3.1. Linear wave equations

The standard wave equation known since d'Alembert is the paragon of *hyperbolic* systems (propagation at finite speed). It reads

$$u_{tt} - c_0^2 u_{xx} = \left(\frac{\partial}{\partial t} - c_0 \frac{\partial}{\partial x}\right)\left(\frac{\partial}{\partial t} + c_0 \frac{\partial}{\partial x}\right) u = 0. \tag{3.1}$$

The factorization shows that the general solution of $(3.1)_1$ is the sum of left-running and right-running plane waves. Introducing characteristic coordinates $\xi, \zeta = x \pm c_0 t$ , we can also write eqn.(3.1) as

$$\frac{\partial^2}{\partial \xi \partial \zeta} u = 0 \quad . \tag{3 .2}$$

Equations (3.1) or (3.2) should logically be called *two-directional* wave equations, since the simplest wave equations we can introduce clearly are *one-directional* wave equations such that

$$\frac{\partial}{\partial t} u \pm c_0 \frac{\partial}{\partial x} u = 0 \; or \; \frac{\partial}{\partial \xi} u = 0 \;\; and \frac{\partial}{\partial \zeta} u = 0 .$$

The last equations could be viewed as *evolution equations* for *u*, but here there is no evolution along the characteristic coordinates. This dualistic view between wave and evolution equations has some importance in the sequel.

## 3.2. Nonlinear hyperbolic equation

The simplest nonlinear generalization of equations $(3.3)_1$ is given by

$$u_t + c(u) u_x = 0 \quad . \tag{3.3}$$

This can be viewed as a *one-directional* wave equation whose velocity is no longer a datum of the problem (such as $c_0$) but it is a function of the solution *u*. Typically, that part of the wave signal which has a larger amplitude will travel faster that those with lower amplitude ; distorsion of the signal is bound to happen, with an eventual *breaking* of the wave and the occurrence of a nonunique solution saved by the emergence of a *shock* front. An example of wave equation such as (3.4) is

$$u_t + \left(\frac{u^2}{2}\right)_x = u_t + u u_x = 0 \; . \tag{3.3)'$$

For the two-directional wave equations a typical generalization of eqn.(3.1) is

$$u_{tt} - c_0^2 \left(1 + g(u_x)\right) u_{xx} = 0 , \tag{3.4}$$

where $g(u_x)$ is a polynomial. Such an equation appears in nonlinear elasticity for waves propagating in the bulk of a body (cf. Maugin *et al*, 1992).

### 3.3. Boussinesq and Korteweg-de Vries (KdV) equations

Let us consider the case where the polynomial in (3.4) reduces to the first order term. Then a standard equation of nonlinear wave theory is the so-called **crystal Boussinesq wave equation** given by

$$u_{tt} - \left(1 + \varepsilon u_x\right) u_{xx} - \varepsilon \delta^2 u_{xxxx} = 0 . \tag{3.5}$$

Here the material characteristic velocity is set equal to one, ε is a small parameter , and δ is a characteristic length. Equation (3.5) can be deduced from the elasticity with second displacement gradient (or gradient of strain), i.e., when some nonlocality is in action. It also follows from the continuum approximation of lattice equations in which the interactions between second neighbors are taken into account (cf.Maugin, 1999). The fourth order space derivative provides so-called dispersion. That is, if we imagine that the nonlinear contribution can be discarded in eqn.(3.5), then in the resulting linear equation, because of the nonhomogeneity of the polynomial of differentiation involved, signals with different frequencies (e.g., differing Fourier components) will travel at different speeds. Reinstituting nonlinearity, it may happen that the nonlinearity effect (steepening of a signal) and dispersion (broadening of the same signal) compensate exactly each other giving rise to the propagation at definite speeds of undisturbed strongly localized signals, i .e., signals that keep their shape while travelling, normally a property of linear systems only. This is the phenomenon of solitary waves, first observed for water waves by Scott Russell in the UK in the 19th century. A related remarkable equation is the so-called Korteweg-De Vries (KdV) equation which is the following one-directional nonlinear dispersive equation (1895):

$$q_{\tau} + 6q\, q_{\zeta} + \delta^2 q_{\tau\tau\tau} = 0 . \tag{3.6}$$

Here changes of function and variables have been performed, in particular $\zeta = x - t$ is a characteristic coordinate and $\tau = \varepsilon\, t/2$ is a slow time scale, and $q = \partial f/6\partial\zeta$ if $u(x,t) = f(\zeta, 2\tau) + 0(\varepsilon)$. This is the *reductive perturbation method*. Equation (3.6) admits spatially localized nonlinear solutions. It may be considered as the evolution equation or the one-directional wave equation associated with the Boussinesq equation (3.5). With this type of equation we leave the domain of hyperbolic equations *per se* to enter that of *nonlinear dispersive wave equations* (cf. Whitham, 1974), a practically unexplored domain of applied mathematics before the 1960s. Other examples of nonlinear dispersive wave equations are as follows

### 3.4 Sine-Gordon equation

This is the two-directional wave equation given by

$$u_{tt} - u_{xx} - \sin u = 0 . \tag{3.7}$$

In addition to a d'Alembert wave operator this equation presents *both* nonlinearity and dispersion through the *sine* function, hence the funny naming as it is a special case of Klein-Gordon equation $u_{tt} - u_{xx} - f(u) = 0$. The difference between eqn.(3.7) and eqn. (3.6) is that in the former case the speed is not related to the amplitude, although this is limited from above. Using linear characteristic coordinates, eqn.(3.7) also reads

$$\frac{\partial^2 u}{\partial \xi \partial \zeta} = \sin u \ . \tag{3.8}$$

This is *Enneper's equation* in the geometry of 2D surfaces of constant negative curvature, so that any known surface solution of (3.8) also provides a solution of eqn.(3.7). Equation (3.7) was first obtained in the theory of dislocations by Frenkel and Kontorova before WWII. But it is the theory of ferromagnets and superconducting devices (Josephson junctions) that really made it popular in the 1970s and 1980s.

## 3.5 Nonlinear Schroedinger equation :

This equation, often considered as a diffusive nonlinear evolution equation for a complex- valued field, reads

$$i u_t + p u_{xx} + q|u|^2 u = 0. \tag{3.9}$$

where the nonlinearity is cubic. This equation admits spatially localized solutions in the form of wave packets built from an envelope of the pulse type modulated by a harmonic carrier. The sign of the product *pq* determine the stability of the propagating solution (see so-called Zakharov criterion). Equation (3.9) is a canonical equation that will always obtain when studying the slowly varying envelope of small amplitude almost monochromatic signals in any weakly nonlinear dispersive wave system, but the expressions of *p* and *q* depend on the considered system, essentially on the curvature of the linear dispersion relation and on nonlinearities, repectively.

A remarkable property shared by eqns.(3.6), (3.7) and (3.9) is their so-called *exact integrability* by means of the *inverse scattering method* (ISM). This integrability results in the existence of *soliton* solutions, i.e. solitary wave solutions that do act as particles in the course of their interactions. The elements of these solutions cross each other without apparent disturbance if only for an alteration in phase. But their encounter does not generate spurious radiations. One can easily imagine the interest of such dynamical solutions for the transmission of information (e.g., pulse transmission along optical fibers governed by equations such as (3.9)).

## 3.6 Some generalizations

These generalizations of the just recalled equations will loose the exact integrability property and solitonic properties when the latter existed. First we have a well known generalization of eqn.(3.3) provided by Burgers' equation

$$u_t + u u_x = \nu u_{xx} . \tag{3.10}$$

If *u* is the velocity field of a fluid, then (3.10) is its equation of motion with kinematic viscosity $\nu$. Equation (3.10) my be viewed as a nonlinear evolution equation with diffusion. It is often considered as a " simple " model for turbulence.

**A. The Boussinesq paradigm.** Under this coinage we gather all equations that are direct generalizations of eqn.(3.5). They present additional nonlinear contributions and higher-order derivative terms . They may be deduced from crystal lattice physics at various levels of approximations . They are particularly documented in a book (Maugin, 1999). As a consequence their associated evolution equation, or one-directional wave equation, obtained by means of the reductive perturbation method will be more complex than the KdV equation (3.6), i.e., they will exhibit additional terms containing further nonlinearities and higher-order derivatives of the fields (e.g., fifth- order space derivative), such as in the *Kuramoto-Shivashinsky* equation that relates to water waves in fluids with surface tension.

Among the generalizationsof the KdV equations one is to be particularly emphasized . It is the *Benjamin-Ono equation*, which describes weakly nonlinear long internal gravity waves and exhibits a *nonlocal* contribution in the form of a Hilbert transform, e.g,

$$u_t + (1+u)u_x + aH[u_{xx}] = 0, \tag{3.11}$$

where a is a scalar coefficient and the *Hilbert transform* of $u$ is defined by ($PP$ = principal part in Cauchy's sense)

$$H[u(x)] = \frac{1}{\pi} PP \int_{-\infty}^{+\infty} \frac{u(x')}{x'-x} dx' . \tag{3.12}$$

**B. The double sine-Gordon equation.** This is the equations that reads (compare to eqn.(3.7))

$$u_{tt} - u_{xx} - \sin u + \gamma \sin 2u = 0 , \tag{3.13}$$

where $\gamma$ is a constant.

**C. The Zakharov system**. This is the system of two equations (a complex field a and a real-valued field n) such that

$$\begin{aligned} &ia_t + a_{xx} + 2an = 0 , \\ &n_{tt} - c_O^2 n_{xx} + \mu |a|_{xx}^2 = 0 . \end{aligned} \tag{3.14}$$

For a single travelling wave or if the inertial term (second-order time derivative) can be discarded in the second of (3.14) , then elimination of $n$ between the two resulting equations provides a single nonlinear (cubic) Schroedinger equation for $a$ as

$$ia_t + a_{xx} + \left(2\mu / c_0^2\right)|a|^2 a = 0. \tag{3.15}$$

The system (3.14) was first met by Zakharov in the study of coupled Langmuir ion-acoustic waves in plasmas, and generalizations were obtained by Maugin *et al* (1992) while studying elastic surface waves (see below).

# 4 Basic Equations of Elasticity

## 4.1 General equations

Let the smooth finite deformation of a body $B$ be described by the mapping

$$\mathbf{x} = \overline{\mathbf{x}}(\mathbf{X},t) \quad or \quad x_i = \overline{x}_i(X_K,t)\,;\; i = 1,2,3;\;\; K = 1,2,3 \tag{4.1}$$

from the reference configuration $K_R$ to the actual configuration $K_t$. The deformation gradient $\mathbf{F}$, its Jacobian determinant $J_F$, and the " physical " velocity $\mathbf{v}$ are given by

$$\mathbf{F} = \nabla_R \overline{\mathbf{x}} = \{\partial x_i / \partial X_K\} \;\;,\;\; J_F = det\,\mathbf{F} > 0 \;\;,\;\; \mathbf{v} = \left.\frac{\partial \overline{\mathbf{x}}}{\partial t}\right|_X . \tag{4.2}$$

Introducing the displacement $\mathbf{u}(\mathbf{x},t)$, we have

$$x_i = X_K \delta_i^K + u_i(\mathbf{X},t) \;\;,\;\; F_{iK} = \delta_{iK} + u_{i,K}\,. \tag{4.3}$$

If needed, the inverse motion and inverse deformation are defined by

$$\mathbf{X} = \overline{\mathbf{X}}(\mathbf{x},t) \;\;,\;\; \mathbf{F}^{-1} = \left.\frac{\partial \overline{\mathbf{X}}}{\partial \mathbf{x}}\right|_t \;\;. \tag{4.4}$$

At all regular material points $\mathbf{X}$ of a material body $B$, the balance equations of linear and angular momenta read as follows in the actual configuration :

$$\rho \frac{d\,\mathbf{v}}{d\,t} = div\,\sigma + \rho \mathbf{f} \;\;,\;\; \sigma = \sigma^T \;, \tag{4.5}$$

where $d/dt$ denotes the material time derivative, $\boldsymbol{\sigma}$ is the symmetric Cauchy stress (no spin, no body couple), and $\mathbf{f}$ is a body force per unit mass. On account of the following Piola transformation, identities, and mass conservation ($T$ = transpose)

$$\mathbf{T} = J_F\,\mathbf{F}^{-1}.\sigma \;\;,\;\; \nabla.(J_F^{-1}\mathbf{F}) \equiv \mathbf{0} \;\;,\;\; \nabla_R = \mathbf{F}^{\mathrm{T}}.\nabla \;\;,\;\; \rho_0 = \rho J_F \;, \tag{4.6}$$

where $\nabla = \partial / \partial \mathbf{x}$ and $\nabla_R = \partial / \partial \mathbf{X}$, $\mathbf{T}$ is the first Piola-Kirchhoff stress, and $\rho_0$ is the matter density at $K_R$.

The first of eqns.(4.5) can also be written in the following Piola-Kirchhoff form

$$\left.\frac{\partial}{\partial t}(\rho_0 \mathbf{v})\right|_X = div_R \mathbf{T} + \rho_0 \mathbf{f}\,. \tag{4.7}$$

If one further introduces the second Piola-Kirchhoff stress $\mathbf{S}$ by (T = transpose)

$$\mathbf{S} = \mathbf{T}.\left(\mathbf{F}^{-1}\right)^{T} = \mathbf{S}^{T}, \tag{4.8}$$

then eqn.(4.7) also reads in components

$$\frac{\partial}{\partial X_K}\left[S^{KL}\left(\delta_{iL} + u_{i,L}\right)\right] + \rho_0 f_i = \rho_0 \left.\frac{\partial^2 u_i}{\partial t^2}\right|_X . \tag{4.9}$$

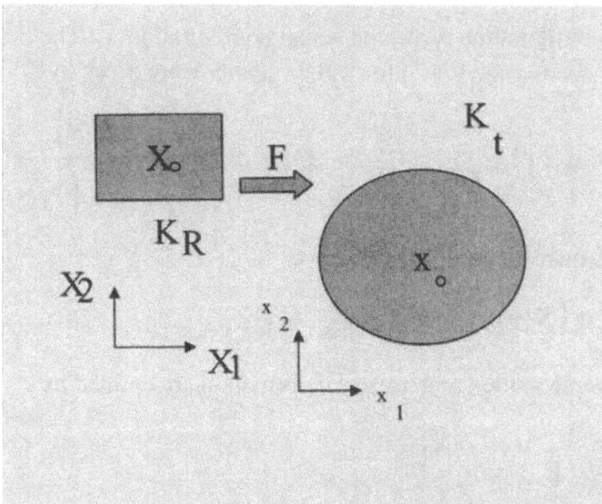


**Figure 2.** Finite-strain deformation of a continuous body

At *free* bounding surfaces of externally pointing unit normal **N**, the natural boundary condition associated with this equations is

$$N_K T_i^K = 0 . \tag{4.10}$$

At *perfect interfaces* (continuity of tractions and displacements) between two elastic bodies, we have the conditions

$$\mathbf{N}.[\mathbf{T}] = \mathbf{0} \quad , \quad [\mathbf{u}] = \mathbf{0}, \tag{4.11}$$

where the square brackets denote the jump at the interface.

In elasticity **S** is also called the energy stress as it is deduced from an energy density $W$ per unit volume of $K_R$ by the simple relation

$$\mathbf{S} = \frac{\partial W}{\partial \mathbf{E}} \quad , \; W = \overline{W}(\mathbf{E}) \quad , \; \mathbf{E} \equiv \frac{1}{2}\left(\mathbf{F}^T.\mathbf{F} - \mathbf{1}\right). \tag{4.12}$$

The latter quantity is the Lagrangian (material) finite strain tensor. If necessary we can also use the constitutive relation

$$\mathbf{T}=\frac{\partial W}{\partial \mathbf{F}} \quad , \quad W=\widetilde{W}(\mathbf{F}) \; , \tag{4.13}$$

the difference between the latter and eqn.(4.12) being that a rotational invariance condition (objectivity) has been used to introduce the dependency on **E**.

## 4.2 Particular cases of energy densities

For a *general anisotropic elastic solid* one usually considers an expansion of the function $\overline{W}$ in terms of the components of the tensor **E** with strains assumed to be sufficiently small, i .e., $\left(tr\mathbf{E}^2\right)^{1/2}=0(\varepsilon)<<1$ , e.g.,

$$\begin{aligned}\overline{W}\left(\mathbf{E}\right)=&\frac{1}{2}C^{KMMN}E_{KL}E_{MN}+\frac{1}{3!}C^{KLMNPQ}E_{KL}E_{MN}E_{PQ}+\\&+\frac{1}{4!}C^{KLMNPQRS}E_{KL}E_{MN}E_{PQ}E_{RS}+h.o.t\end{aligned} \tag{4.14}$$

where the tensorial coefficients $C$ of various orders are tensors of elasticity moduli of second, third, and fourth order, respectively. These possess obvious tensorial symmetries.

For *isotropic elastic bodies* in finite strains one rather considers the following expression that follows from Murnaghan (1951) -also Bland (1969) :

$$\overline{W}(\mathbf{E})=\frac{1}{2}\lambda\, I_1^2+\mu I_2+\alpha I_1^3+\overline{\beta}\, I_1 I_2+\gamma\, I_3+\xi I_1^4+\eta I_1^4 I_4 \quad , \tag{4.15}$$

where the invariants are defined by $I_\alpha = tr\,\mathbf{E}^\alpha$. The coefficients $\lambda$ and $\mu$ are the celebrated Lamé constants of linear isotropic elasticity for which both eqns. (4.14) and (4.15) reduce to

$$W(\mathbf{E})=\frac{1}{2}\lambda\left(tr\,\mathbf{E}\right)^2+\mu\; tr\mathbf{E}^2 \quad , \quad C^{KLMN}=\lambda\,\delta^{KL}\delta^{MN}+\mu\left(\delta^{KM}\delta^{LN}+\delta^{KN}\delta^{LM}\right). \tag{4.16}$$

## 4.3 The linear surface wave problem

In elasticity the small parameter is not a material coefficient, but the displacement solution itself, i.e., $\varepsilon=0(\nabla_R u)=0(\mathbf{E})$. Hence we can write

$$x_j=\delta_{jK}X_K+\varepsilon u_j(\mathbf{X},t), F_{iK}=\delta_{iK}+\varepsilon u_{i,K}\, , v_i=\varepsilon \dot{u}_i \; . \tag{4.17}$$

Accordingly,

$$T^K{}_i=\varepsilon\, C_{iKjL}u_{j,.L}+0\left(\varepsilon^2\right) , \tag{4.18}$$

and the following boundary value problem follows for a flat free surface $X_2 = 0$ and a field vanishing at infinity in the half plane $X_2 < 0$:

$$\rho_0 \ddot{u}_j = C_{j\alpha m\beta} u_{m,\alpha\beta} \quad in\, X_2 < 0, \tag{4.19}$$

$$C_{j2m\beta} u_{m,\beta} = 0 \quad at \quad X_2 = 0 \;, \tag{4.20}$$

$$u_j\left(X_\alpha, t\right) \to 0 \quad as \quad X_2 \to -\infty \,. \tag{4.21}$$

Here the convention that Greek subscripts take values 1 and 2 is adopted assuming that there is no dependence of the problem on the transverse coordinate $X_3$ . Wave solutions travelling at speed $c$ without distorsion are sought in the form

$$\mathbf{u} = \bar{\mathbf{u}}\left(\theta, X_2\right), \quad \theta = X_1 - ct \,. \tag{4.22}$$

From eqn.(4.22) there follows that

$$\left(C_{j\alpha m\beta} - \rho_0 c^2 \delta_{jm} \delta_{\alpha 1} \delta_{\beta 1}\right) u_{m,\alpha\beta} = 0, \tag{4.23}$$

where $u_{j,1} := \partial u_j / \partial \theta$ .

Equation (4.23) together with (4.20) and (4.21) provide an elliptic eigen-value problem for $c$, which is degenerate and possesses solutions which may be represented in terms of a general harmonic function in a half plane (cf. Parker 1985). However, we first examine *complex-valued periodic solutions* which are superpositions of partial waves such as

$$\bar{\mathbf{u}} = \mathbf{a}\, exp\, ik\left(\theta + sX_2\right) \;, \; k \; real \,. \tag{4.24}$$

Here $2\pi / |k|$ is the spatial *period* (wavelength) and $s$ is the *depth facctor*. On substituting from (4.24) into (4.23), we obtain an algebraic system in the form

$$L_{jm}(s) a_m \equiv \left\{ c_{j1m1} - \rho_0 c^2 \delta_{jm} + s\left(C_{j1m2} + C_{j2m1}\right) + s^2 C_{j2m2} \right\} a_m = 0 \,. \tag{4.25}$$

The compatibility condition for solving this is a *sextic* polynomial with real coefficients. We consider only the case where the six roots and corresponding eigenvectors occur in complex conjugate pairs. That is, we denote the solutions by (* indicates complex conjugacy)

$$\begin{aligned} & s^{(p)} = \alpha^{(p)} - i\beta^{(p)} \,; \mathbf{a}^{(p)} \\ & s^{(p)} = \alpha^{(p)} + i\beta^{(p)} \,; \mathbf{a}^{(p)} *, \; \beta^{(p)} > 0 \,, \; p = 1,2,3 \end{aligned} \tag{4.26}$$

Each partial wave is then written as

$$\mathbf{u} = \mathbf{a}^{(p)}\, exp\left(k\beta^{(p)} X_2\right) exp\left(ik\left(X_1 - ct + \alpha^{(p)} X_2\right)\right) \tag{4.27a}$$

and

$$\mathbf{u} = \mathbf{a}^{(p)} * exp\left(-k\beta^{(p)}X_2\right)exp\left(ik\left(X_1 - ct + \alpha^{(p)}X_2\right)\right) p = 1,2,3. \tag{4.27b}$$

The speed is $c$ and the constant phase is $X_1 + \alpha^{(p)}X_2$. Selecting the right variation in $X_2$ to guarantee the decay as $X_2 \to -\infty$, the complex surface displacement is given by

$$\mathbf{u} = \sum_{p=1}^{3} B^{(p)}\mathbf{a}^{(p)} \, exp\left(ik\left(\theta + s^{(p)}X_2\right)\right) \quad k > 0. \tag{4.28}$$

Substituting in the boundary condition this yields the following boundary matrix equation (cf.Parker, 1994)

$$M_{jp}B^{(p)} \equiv \sum_{p=1}^{3}\left(\left(C_{j2m1} + s^{(p)}C_{j2m2}\right)a_m^{(p)}B^{(p)}\right) = 0. \tag{4.29}$$

From this and the $c^2$ dependence of the $s^{(p)}$, it follows the compatibility condition for solving (4.29), i.e., $det\, M_{jp} = 0$ as an algebraic equation for $c^2$. To any root there will correspond a non-zero $B^{(p)}$ so that (4.8) will be written as

$$\begin{aligned} \mathbf{u} &= \mathbf{A}\left(kX_2\right)exp(ik\theta) \quad , \\ \mathbf{A}\left(kX_2\right) &= \sum_{p=1}^{3} B^{(p)}\mathbf{a}^{(p)} \, exp\left(iks^{(p)}X_2\right) \quad , \quad k > 0 \end{aligned} \tag{4.30}$$

But the solution

$$\mathbf{u}^* = \mathbf{A}^*\left(|k|X_2\right)exp(ik\theta), k < 0 \,, \tag{4.31}$$

is equivalent , so that the general expression for real surface wave solutions of period $2\pi / k$ and arbitrary complex constant $C$ will read

$$\mathbf{u} = \mathbf{A}(kX_2)\, exp\,(ik\theta) + C^*\mathbf{A}^*(kX_2)\, exp\,(-ik\theta) \quad , k > 0. \tag{4.32}$$

This was a reminder on the case of surface waves on an anisotropic linear elastic homogeneous substrate.

## 4.4 Rayleigh surface wave

The celebrated *Rayleigh surface wave* corresponds to the special case of the above for an *isotropic* body with elastic energy $(4.16)_1$. This is treated in many books (e.g., Achenbach, 1973) so that only the essential results will now follow. In the latter case the element

$$L_{3m} = L_{m3} = \left(\mu - \rho_0 c^2 + s^2 \mu\right)\delta_{3m} \tag{4.33}$$

of the matrix $L_{mj}$ is such that the eigenvector $\mathbf{a}^{(3)}$ and the corresponding pure imaginary root $s^{(3)} = -i\beta^{(3)}$ are given by

$$a_j^{(3)} = a_j^{(3)*} = \delta_{3j} \quad , \quad \beta^{(3)}\left(1 - \frac{c^2}{c_T^2}\right)^{1/2} \equiv \beta_2 \quad , \quad c_T^2 = \mu / \rho_0 , \tag{4.34}$$

while the other two roots $s^{(\alpha)} = -i\beta_\alpha$ are also purely imaginary and given by

$$s^{(1)} = -i\beta_1 \equiv -i\left(1 - \frac{c^2}{c_L^2}\right)^{1/2} \quad , \quad s^{(2)} = i\beta_2 = s^{(3)} \quad ; \quad c_L^2 = (\lambda + 2\mu) / \rho_0 , \tag{4.35}$$

with eigenvectors

$$a^{(1)} = (i, \beta_1, 0)^T \quad , \quad a^{(2)} = (i\beta_2, 1, 0)^T \quad , \quad a^{(3)} = (0,0,1)^T . \tag{4.36}$$

The traction vectors at $X_2 = 0$ for these solutions have the form $ik\, exp(ik\theta) T^{(p)}$ with (cf.eqn.(4.29))

$$T^{(p)} = \left(C_{j2m1} + s^{(p)} C_{j2m2}\right) a_m^{(p)} . \tag{4.37}$$

Accordingly, they are given by

$$\begin{aligned} T^{(1)} &= \left(2\mu\beta_1 \, , i\left(\lambda - \rho_0 c_L^2 \beta_1^2\right), 0\right)^T , \\ T^{(2)} &= \left(\mu\left(1 + \beta_2^2\right), -i2\mu\beta_2, 0\right)^T \quad , \\ T^{(3)} &= \left(0, 0, -i\mu\beta_2\right)^T \quad . \end{aligned} \tag{4.38}$$

The condition (4.29) provides a solution such as

$$B^{(1)} = B^{(2)} = 0, \quad \beta_2 = 0 \ , \ c^2 = c_T^2 . \tag{4.39}$$

This is a *shear horizontal* (*SH*) wave, but it is not a true surface wave since it is not accompanied by decay of its amplitude with depth. This is rather called a *surface skimming wave.*

True surface waves are obtained for

$$\frac{B^{(2)}}{B^{(1)}} = -\frac{2\beta_1}{1 + \beta_2^2} = \frac{\lambda\left(1 - \beta_1^2\right) - 2\mu\beta_1^2}{2\mu\beta_2} = -\frac{\left(1 + \beta_2^2\right)}{2\beta_2} , \tag{4.40}$$

$$B^{(3)} = 0. \tag{4.41}$$

The first and last expression provide the compatibility condition for $\beta_1$ and $\beta_2$ in the handy form

$$\left(1 + \beta_2^2\right)^2 = 4\beta_1\beta_2 \ . \tag{4.42}$$

But this is nothing but the celebrated Rayleigh equation written in full as

$$\left(\frac{c^2}{c_T^2}\right)^3 - 8\left(\frac{c^2}{c_T^2}\right)^2 + 8\frac{3\lambda + 4\mu}{\lambda + 2\mu}\frac{c^2}{c_T^2} - 16\frac{\lambda + \mu}{\lambda + 2\mu} = 0. \tag{4.43}$$

This, as we know (cf.Achenbach, 1973), provides one real subsonic (with respect to $c_T$) speed $c$. The corresponding surface wave is polarized parallel to the sagittal plane. We do not elaborate further on this classical case already introduced by other lecturers. However, it is salient to note that the just observed separation between a sagitally polarized Rayleigh wave and an *SH* surface skimming wave is characteristic of the isotropic half space limited by a *free flat* boundary. It was noticed elsewhere (Maugin, 1988) that, even in the isotropic material, an appropriate perturbation of the flat boundary (curvature, grating, grooves, roughness, mixed mechanical electrical condition, etc) which allows for a relaxation of the condition (4.20) may yield the existence of a true *SH* surface wave. This is the case of the so-called *Love waves* where the " big " perturbation there consists in the superposition of a perfectly glued " slow " elastic layer on the half space substrate. But while Rayleigh waves are non dispersive in the linear case , *SH* waves of the Love type are *dispersive* even in the linear case since the thickness of the superposed layer introduces a characteristic length in the system while the present elastic problem includes no such length. We shall return to this case later on.

# 5 Nonlinear Surface Waves in the Absence of Dispersion

## 5.1 Introductory remark

It should be remarked that in *nondispersive* systems the effects of nonlinearities on the propagation of surface (acoustic or elastic) waves are **weak**. As a practical picture, we note that the nonlinearities essentially influence the *evolution of wave profiles* over length scales much larger than the wave length of the input wave. Accordingly, for weak nonlinear effects the exploitation of mathematical methods used in the linear case (e.g., Fourier analysis) is appropriate together with asymptotic methods. It is simply observed (see below) that higher order harmonics couple resonantly with the fundamental wave so that the nonlinear solution can be sought as a superposition of surface acoustic waves of the linear problem of motion complemented by boundary conditions. As a result one has to solve a series of *inhomogeneous* (i.e., with source terms) linear problems when one considers an asymptotic expansion. The obvious result of this is the *slow* variation of the amplitudes at various wave numbers along the direction of propagation. New remarkable effects, such as the existence of nonlinear periodic surface waves of nondistorting form can be proved (Parker and Talbot, 1985).

It is difficult to list all contributors to this field without risk of missing some authors. Among the most active authors we note in particular Harvey and Tupholme (1991,1992), Hunter (1989), Kalyanasundaram (1981a,b), (1984), Kalyanasundaram *et al* (1982, 1990), Lardner (1983, 1986), Maugin (1985), Panoyotaros (2002), Parker (1988, 1990, 1994), Parker and David (1989a,b), Parker and Talbot (1985), Planat (&985), Teymur (1988), Tupholme and Harvey (1986), Vlannes and Bers (1982). Some of these works include a generalization to coupled electroelastic waves in piezoelectrics and electroelastic materials. Other works considered nonlinear magnetoelastic couplings (cf .works by Abd-Alla and Maugin - not cited here) . This long, incomplete, roster demonstrates an intense activity in the field in the 1980s essentially in relation to electro-magneto-mechanical devices of wave conversion.

In relation to the above introductory remark , if we pursue the idea of smallness already expressed in eqns.(4.17) and (4.18), then at the second order in $\varepsilon$ we observe that the displacement distribution considered in the foregoing section, $\bar{\mathbf{u}}(\theta, X_2)$ gives rise to perturbations that depend also on $X_1$. A straightforward expansion in $\varepsilon$, *à la* Poincaré, around the linear solution (4.32) shows that higher harmonics of the type $exp(\pm ink\theta)$ , $n$ =2,3,.. are successively introduced (see, e.g., Maugin, 1985). The resulting growth of higher harmonics involves annoying secular terms that are familiar in nonlinear oscillation theory. There are several techniques to deal with this difficulty, such as the introduction of *multiple scales* (cf.Planat, 1985 ; Parker, 1988), the application of the so-called *reductive perturbation method* (Jeffrey *et al*), and also the less satisfactory iterative methods. We shall present first the *iterative method* and then the so-called *projection method* favored by Reutov (1973), and Maradudin and co-workers (e.g., Maradudin and Mayer, 1991 ; Mayer, 1995, Parker *et al*, 1992), and finally the *Hamiltonian method* favored by Zalobotskaya and her co-workers (E.g., Hamilton *et al*, 1995a,b , 1999).

## 5.2 The iterative method

Here we follow Parker (1994) in exploiting the multiple scale technique , by introducing a so-called " slow " coordinate $X = \varepsilon X_1$ in the direction of propagation (i .e., the direction of evolution of the amplitudes) and start the analysis of the iterative method. Instead of (4.17) and (4.18) we shall have the following representations

$$u_{i,1} = \frac{\partial u_i}{\partial X_1} = \frac{\partial u_i}{\partial \theta} + \varepsilon \frac{\partial u_i}{\partial X} \quad , \quad u_{i,2} = \frac{\partial u_i}{\partial X_2} \quad , \quad u_{i,3} = 0, \tag{5.1}$$

$$v_i = -c \frac{\partial u_i}{\partial \theta}, \tag{5.2}$$

$$\frac{\partial}{\partial X^K} T_i^K - \rho_0 c^2 \bar{x}_{i,11} = -\varepsilon T_i^K{}_{,X} \quad in \ X_2 < 0, \tag{5.3}$$

$$T_j^2 = 0 \quad at \quad X_2 = 0, \tag{5.4}$$

with

$$T_i^K = T_{Ki} = \frac{\partial W}{\partial F^i{}_K}, \quad F^i{}_K = F_{iK} = \delta_{iK} + \varepsilon \hat{F}_{iK} = \delta_{iK} + \varepsilon u_{i,K} + \varepsilon^2 u_{i,X} \delta_{K1}, \tag{5.5}$$

or

$$T_{Ki} = \varepsilon C_{iKmL} \hat{F}_{mL} + \varepsilon^2 C_{iKmLpQ} \hat{F}_{mL} \hat{F}_{pQ} + 0(\varepsilon^3), \tag{5.6}$$

where

$$C_{iKmL}=\frac{\partial^2 W}{\partial F_{iK}\partial F_{mL}}\Big|_{\mathbf{F}=1}\quad,\quad C_{iKmLpQ}=\frac{\partial^3 W}{\partial F_{iK}\partial F_{mL}\,\partial F_{pQ}}\Big|_{\mathbf{F}=1}\ . \tag{5.7}$$

Here above in eqn.(5.3) on we have introduced the simplifying notation that $A_{,1}=\partial A/\partial\theta$. In general we shall obtain equations for elastic displacement in which elasticity tensorial coefficients of the $n$-th order involve both the thermodynamical elasticity coefficients of this order $n$ and elasticity coefficients of all orders less than $n$. This is due to the fact that nonlinearities of *both* physical and geometrical origins are involved. For instance, we shall have for (5.6)

$$T_{Ki}=\varepsilon\hat{C}_{iKmL}u_{m,L}+\frac{1}{2}\varepsilon^2\hat{C}_{iKmLpQ}u_{m,L}u_{p,Q}+0\left(\varepsilon^3\right), \tag{5.8}$$

wherein

$$\begin{aligned}&\hat{C}_{iKmL}=C_{ikmL}\ ,\\&\hat{C}_{iKmLpQ}=C_{iKmLpQ}+\left(\delta_{im}C_{LKpQ}+\delta_{ip}C_{QKmL}+\delta_{mp}C_{iKLQ}\right).\end{aligned} \tag{5.9}$$

Retaining terms $0(\varepsilon)$ at most in eqns.(5.3) and (5.4) yields

$$\begin{aligned}L_j(\mathbf{u}):&=\hat{C}_{j\alpha m\beta}u_{m,\alpha\beta}-\rho_0c^2u_{j,11}\\&=-\varepsilon\hat{C}_{j\alpha m\beta n\gamma}\left(u_{m,\beta}u_{n,\gamma}\right)_{,\alpha}-\varepsilon\left(C_{j1m\beta}+C_{j\beta m1}\right)u_{m,\beta X}+0\left(\varepsilon^2\right)\ \ in\ X_2<0\end{aligned} \tag{5.10}$$

and

$$\begin{aligned}B_j(\mathbf{u}):&=C_{j2m\beta}u_{m,\beta}\\&=-\varepsilon\hat{C}_{j2m\beta n\gamma}u_{m,\beta}u_{n,\gamma}-\varepsilon C_{j2m1}u_{m,X}+0\left(\varepsilon^2\right)\ \ at\ X_2=0\end{aligned}\quad, \tag{5.11}$$

while the condition (4.21) reads

$$u_j\left(\theta,X_2,X;\varepsilon\right)\to 0\quad as\,X_2\to-\infty\ . \tag{5.12}$$

No derivatives with rexpect to $X$ occuring in the problem of leading order, which problem is homogeneous , we have

$$\begin{aligned}&L_j(\mathbf{u})=0\quad in\ X_2<0\,,\\&B_j(\mathbf{u})=0\quad at\ X_2=0\,,\\&\mathbf{u}\to 0\qquad as\,X_2\to-\infty\ .\end{aligned} \tag{5.13}$$

Accordingly, the leading approximation is any solution of the problem (5.13) with $X$ entering as a *free* parameter. Obviously, the problem (5.13) is formally identical to the linear problem stated in (4.19)-(4.21). The most general solution of (5.13) therefore is a Fourier superposition of solutions of the type (4.30) , i.e.,

$$\mathbf{u} = \hat{\mathbf{u}}(\theta, X_2, X) = \int_{-\infty}^{+\infty} C(k, X)\mathbf{A}(kX_2) exp(ik\theta)\, dk\,, \tag{5.14}$$

where we must check that

$$C(-k, X) = C^*(k, X), \mathbf{A}(kX_2) = \mathbf{A}^*(|k|X_2)\, k<0\,, \tag{5.15}$$

so that $\hat{\mathbf{u}}$ will be real.

In the spirit of the asymptotic expansion the solution of the system (5.10)-(5.11) is sought in the form

$$\mathbf{u} = \hat{\mathbf{u}}(\theta, X_2, X) + \varepsilon\mathbf{w}(\theta, X_2, X; \varepsilon), \tag{5.16}$$

corresponding to the *inhomogeneous* system (5.10)-(5.11),i.e.,

$$\begin{aligned} L_j(\mathbf{u}) &= P_j(\theta, X_2; X, \varepsilon) \quad for \quad X_2 < 0\,, \\ B_j(\mathbf{u}) &= -Q_j(\theta, X_2; X, \varepsilon) \quad at \quad X_2 = 0\,, \end{aligned} \tag{5.17}$$

where we have set

$$\begin{aligned} P_j &:= -\hat{C}_{j\alpha m\beta n\gamma}\left(u_{m,\beta}u_{n,\gamma}\right)_{,\alpha} - \left(C_{j1m\beta} + C_{j\beta m1}\right)u_{m,\beta X} + 0(\varepsilon)\,, \\ Q_j &:= \hat{C}_{j2m\beta n\gamma}u_{m,\beta}u_{n,\gamma} + C_{j2m1}u_{m,X} + 0(\varepsilon)\quad . \end{aligned} \tag{5.18}$$

## 5.3 Evolution equation for quadratic nonlinearity

The reasoning of Parker (1994) goes as follows. At order *one* the expressions (5.18) involve only $\hat{\mathbf{u}}$ as given by eqn.(5.14). Therefore, $\mathbf{w}$ can be constructed either iteratively or by means of a perturbation expansion. Since $X$ enters $\hat{\mathbf{u}}$ only through $C(k,X)$, a formal inversion of the problem (5.17)-(5.18) would show that $X$ enters any approximation or iterate of $\mathbf{w}$ only through $C(k, X)$ . The associated *solvability* condition arising because problem (5.13) has many nontrivial solutions, imposes *integral* constraints on $P_j$ and $Q_j$ . This is an application of Fredholm's aternative. Accordingly, at each stage of the iterative procedure this yields an evolution equation for $C(k, X)$. Let us do this by means of Fourier transforms. We set

$$\begin{aligned} F\{f\} &= \bar{f}(k, X_2, X) = \frac{1}{2\pi}\int_{-\infty}^{+\infty} f(\theta, X_2, X) exp(-ik\theta)\, d\theta, \\ F^{-1}\{\bar{f}\} &= f(\theta, X_2, X) = \int_{-\infty}^{+\infty} \bar{f}(k, X_2, X) exp(ik\theta)\, dk \quad . \end{aligned} \tag{5.19}$$

The problem (5.17)-(5.18) expressed by means of Fourier transforms reads

$$\begin{aligned} &\overline{L}_{jm}\overline{w}_m = \overline{P}_j \quad , \; X_2 < 0 \; , \\ &\overline{B}_{jm}\overline{w}_n = -\overline{Q}_j \quad , \; X_2 = 0 \, , \\ &\overline{w}_m \to 0 \;\; as \;\; X_2 \to -\infty \; , \end{aligned} \tag{5.20}$$

where we have defined the following operators

$$\begin{aligned} &\overline{L}_{jm} := C_{j2mL}\frac{d^2}{d\,X_2^2} + ik\left(C_{j2m1} + C_{j1m2}\right)\frac{d}{d\,X_2} + \rho_0 k^2 c^2 \delta_{jm} - k^2 C_{j1m1} \; , \\ &\overline{B}_{jm} := C_{j2m2}\frac{d}{d\,X_2} + ik\,C_{j2m1} \; . \end{aligned} \tag{5.21}$$

The following Green identity is easily checked by integration by parts along the depth coordinate $X_2$ (with $D > 0$)

$$\int_{-D}^{0}\left(v_j\overline{L}_{jm}\overline{w}_m - \overline{w}_m\overline{L}^*_{mj}\,v_j\right)dX_2 = \left[v_j\overline{B}_{jm}\overline{w}_m - \overline{w}_m\overline{B}^*_{mj}\,v_j\right]_{-D}^{0} \; , \tag{5.22}$$

where $\overline{L}_{mj}$ (resp. $\overline{B}_{mj}$) is the complex conjugate of the transpose of $\overline{L}_{jm}$ (resp. $\overline{B}_{jm}$).

Accordingly, $\mathbf{v}$ is any solution of the homogeneous adjoint problem associated with (5.20), i.e.,

$$\begin{aligned} &\overline{L}^*_{mj}\,v_j = 0 \;\; for \;\; X_2 < 0 \, , \\ &\overline{B}^*_{mj}\,v_j = 0 \;\; at \; X_2 = 0 \quad , \\ &v_j \to 0 \;\; as \;\; X_2 \to -\infty \end{aligned} \tag{5.23}$$

One checks from (5.22) that

$$\int_{-\infty}^{0} v_j(k, X_2)\overline{P}_j\, dX_2 + v_j(k,0)\overline{Q}_j = 0 \; . \tag{5.24}$$

This is the *solvability condition*. But substitution of $u_j = A^*_j(kX_2)exp(-ik\theta)$ into the problem (5.13) shows that $v_j = A^*_j(kX_2)$ satisfies the problem (5.23). Accordingly, for any $k$, the most general relevant solution $\mathbf{v}$ is given by

$$\mathbf{v}(k, X_2) = \overline{\lambda}\,\mathbf{A}^*(kX_2) = \begin{cases} \overline{\lambda}\sum_{p=1}^{3} B^{(p)*}\mathbf{a}^{(p)*}exp\left(iks^{(p)*}X_2\right) & k > 0 \\ \overline{\lambda}\sum_{p=1}^{3} B^{(p)}\mathbf{a}^{(p)}exp\left(-iks^{(p)}X_2\right) & k < 0 \end{cases} \; . \tag{5.25}$$

Setting $\mathbf{A}_0 = \sum_{p=1}^{3} B^{(p)}\mathbf{a}^{(p)}$ the boundary value of $\mathbf{A}$ for $k > 0$, the condition (5.24) can be rewritten as

$$\int_{-\infty}^{0}\mathbf{A}^*(kX_2).\overline{\mathbf{P}}(k,X_2;X,\varepsilon)dX_2+\mathbf{A}_0{}^*.\overline{\mathbf{Q}}(k;X,\varepsilon)=0\ . \tag{5.26}$$

We can now start the iterative procedure by retaining terms 0(1) only in the expressions (5.18). We refer the reader to Parker (1994) for the detailed algebra consisting in obtaining first the Fourier transform of these expressions and then substitution of the latter into (5.26). The result is an evolution equation (in $X$) of the amplitude $C(k,X)$ such as

$$iJ\frac{\partial C(k,X)}{\partial X}+\int_{-\infty}^{+\infty}\Lambda(k-\kappa,\kappa)(k-\kappa)\kappa C(k-\kappa,X)C(\kappa,X)dX=0\ , \tag{5.27}$$

where the coefficient $J$ and the kernel $\Lambda$ are given by

$$J=\int_{-\infty}^{0}\left\{2C_{j1m1}A_m(\xi)A_j^*(\xi)+iC_{j2m1}\left[A_m(\xi)A_j^*(\xi)-A'_j(\xi)A_m^*(\xi)\right]\right\}d\xi, \tag{5.28}$$

and

$$\Lambda(k-\kappa,\kappa)=\int_{-\infty}^{0}\{iN_{j1}(X_2;k-\kappa)A_j^*(kX_2)-N_{j2}(X_2;k-\kappa,\kappa)A_j^*{}'(kX_2)\}kdX_2\ , \tag{5.29}$$

where $\mathbf{A}$ is the depth function , $\mathbf{A}'$ is its derivative with respect to its argument, and the $N_{j\alpha}$ , $\alpha$ =1, 2 , are given by

$$\begin{aligned}N_{j\alpha}(X_2;\nu,\kappa)=&C_{j2m2n2}A_m{}'(\nu X_2)A_j{}'(\kappa X_2)-C_{j\alpha m1n1}A_m(\nu X_2)A_n(\kappa X_2)\\&+i\left[C_{j\alpha m1n2}A_m(\nu X_2)A_n{}'(\kappa X_2)+C_{j\alpha m2n1}A_m^*{}'(\nu X_2)A_n(\kappa X_2)\right]\mathrm{I}\end{aligned} \tag{5.30}$$

The real coefficient $J$ measures the strain energy associated with the displacement field at order zero while $\Lambda$ exhibits the following symmetry and scaling properties :

$$\begin{aligned}&\Lambda(\nu,\kappa)=\Lambda(\kappa,\nu)=\Lambda^*(-\nu,-\kappa),\\&\Lambda(\alpha\nu,\alpha\kappa)=\Lambda(\nu,\kappa)\ \mathit{for\ any}\,\alpha>0.\end{aligned} \tag{5.31}$$

According to the last of these $\Lambda$ is only a function of the ratio $\kappa/k$ for $k>0$.

If we normalize the depth solution $\mathbf{A}(\xi)$ in such a way that $A_2(0)=1=A_2^*(0)=A_{02}$, then we find that (5.14) and (5.19) yield

$$\hat{u}_2(\theta,0,X)=\int_{-\infty}^{+\infty}C(k,X)exp(ik\theta)dk\ ,\ C(k,X)=\frac{1}{2\pi}\int_{-\infty}^{+\infty}\hat{u}_2(\theta,0,X_2)exp(-ik\theta)d\theta\,. \tag{5.32}$$

In plain words this means that (5.27) is an *evolution equation for the Fourier transform of the* ***surface elevation profile***. An equivalent equation was obtained by Lardner (1983,1986) by using a multiple-scale technique.

**Case of periodic wave forms.** We consider the special case for which u is given by

$$\mathbf{u}\left(\theta, X_2, X\right) = \sum_{n=-\infty}^{+\infty} C_n(X) A\left(nX_2\right) exp(in\theta). \tag{5.33}$$

With the normalization

$$A_2(0) = 1$$

and writing

$$C_n = C^*_{-n} = \frac{1}{2}\left(D_n + iE_n\right),\ n > 0,\ C_0 = 0$$

the surface elevation profile is given by

$$\hat{u}_2\left(\theta, 0, X\right) = \sum_{n=1}^{\infty}\left(D_n(X) cos\, n\theta - E_n(X) sin\, n\theta\right). \tag{5.34}$$

In that case the complicated equation (5.27) takes on the form of *coupled-amplitude equations*

$$J\frac{dC_n}{dX} = i\sum_{m=1}^{n-1} K(m/n)(n-m)m\, C_{n-m} C_n + 2i\sum_{m=n+1}^{\infty} K^*(n/m)n(m-n)\, C^*_{m-n} C_m. \tag{5.35}$$

This yields an equation for $E_n$ (with $D_n$=0) for materials which have reflectional symmetry in planes $X_1 = const$. In this case all $s^{(p)}$ are pure imaginary, the kernel $K(\nu)$ is real , and $C_n = iE_n / 2$. The evolution equation consequently reads

$$2\frac{dE_n}{dX} = -\sum_{m=1}^{n-1}\Gamma_{mn} m(n-m) E_m E_{n-m} + \sum_{m=1}^{\infty} 2\Gamma_{n,n+m} mnE_m E_{m+n}, \tag{5.36}$$

with $\Gamma_{mn} = J^{-1}K(m/n)$. The reader will find explicit expressions of $K(\nu)$ for isotropic materials in Parker (1988) (computations with 80 non-zero coefficients). A wave steepening characteristic of other nonlinear phenomena (water wave and nonlinear (fluid) acoustics) is observed; cf. Figure 3. It resembles a *shock formation.* Such coupled-amplitude equations as (5.29) have been proposed by different authors (Planat, 1985 ; Maugin, 1985 ; Kalyanasundaram *et al*, 1982 ; Vlannes and Bers, 1982).

## 5.4 Nondistorting wave forms

Parker and Talbot (1985) have shown that the above system, although not exhibiting any length scale, hence no dispersion, can still exhibit nondistorting waveform solutions (just like systems where dispersion would compensate for nonlinearity). These waves of " permanent form " are expected to travel at speed $\hat{c}$ slightly different from the speed $c$ of the linear problem. This means that we should look for solutions

$$\mathbf{u} = \tilde{\mathbf{u}}\left(\theta - \tilde{c}t, X_2\right), \tag{5.37}$$

where

$$\hat{c} = c / \left(1 - \varepsilon\tilde{c}\right) \cong c\left(1 + \varepsilon\tilde{c}\right). \tag{5.38}$$

Then through (5.14), we will have

$$C\left(k, X\right) = \tilde{C}\left(k\right) exp(-ik\tilde{c}X). \tag{5.39}$$

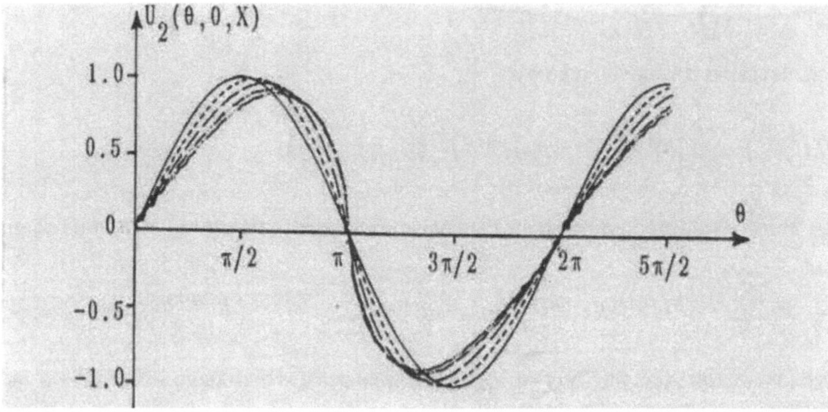


**Figure 3.** Profile deformation (wave breaking) for an initially sinusoidal surface acoustic wave on a"harmonic" isotropic material at succive ranges (distance of propagation)(after Parker, 1988).

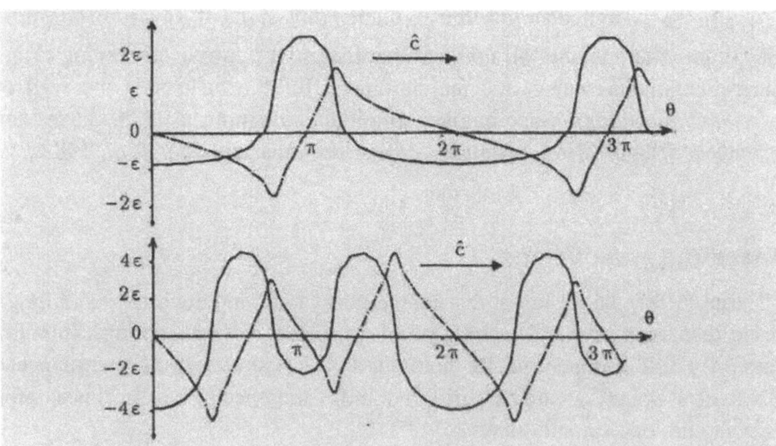


**Figure4.** Horizontal (interrupted line) and vertical (continuous line) displacement waveforms for one- and two-hump nondistorting waveforms on a "harmonic" isotropic material (after Parker and Talbot,1985)

Accordingly, such nondistorting waveforms correspond to solutions of the following integral equation [cf.eqn.(5.27)]

$$J\tilde{c}\,\tilde{C}(k)+\int_{-\infty}^{+\infty}\Lambda(k-\kappa,\kappa)(k-\kappa)\tilde{C}(k-\kappa)\kappa\, C(\kappa)d\kappa=0. \tag{5.40}$$

Parker and Talbot (1985) note that no analytic method of solution is known for such an equation. But they point out the following scaling properties. If we knew a solution $\tilde{C}=\tilde{C}_1(k)$ corresponding to the speed $\tilde{c}_1$, then $\tilde{C}=ab\tilde{C}_1(bk)$ would also be a solution but with speed $\tilde{c}$ given by $\tilde{c}=ab^{-1}\tilde{c}_1$ . It follows from this remark that if we know one nondistorting waveform solution

$$\tilde{\mathbf{u}}_1\left(\theta-\tilde{c}_1X,X_2\right)=\int_{-\infty}^{+\infty}\left[\tilde{C}(k)\mathbf{A}(kX_2)exp\,ik\left(\theta-\tilde{c}_1X\right)\right]dk, \tag{5.41}$$

then the following two-parameter family of solutions is also known :

$$\tilde{\mathbf{u}}=a\tilde{\mathbf{u}}_1\left(\left(\theta-ab^{-1}\tilde{c}_1X\right)/b,X_2/b\right). \tag{5.42}$$

For a *fixed* perturbation in speed $\tilde{c}$ , a/b=const., the peak amplitude and the wavelength are in fixed proportion according to (5.42), so that we can say that it is the *wave steepness* that characterizes the departure from the speed $c$ of the linear case. Parker and Talbot (1985) have produced computational results for waveforms $\tilde{\mathbf{u}}$ when the latter are *periodic*. In this case, with wavelength $2\pi$, the coefficients $C_n$ appearing in (5.33) are written $C_n=\tilde{C}_n exp(-in\tilde{c}X)$ and there follows an infinite set of algebraic equations for the quantities $P_n:=n\tilde{C}_n/\tilde{c}$ in the form (compare to (5.35))

$$JP_n+\sum_{m=1}^{n-1}K(m/n)P_mP_{n-m}+2\sum_{m=n+1}^{\infty}\frac{n}{m}K*(n/m)P_{m-n}*P_m=0 \tag{5.43}$$

***Example***. Considering the truncated expansion at $N=3$ (with $P_n=0$ *for* $n>N$ ) we obtain the following coupled-amplitude system :

$$\begin{aligned} &P_1+\Gamma_{12}^*P_1^*P_2+\frac{2}{3}\Gamma_{13}^*P_2^*P_3=0\ ,\\ &P_2+\Gamma_{12}P_1^2+\frac{4}{3}\Gamma_{13}^*P_1^*P_3=0\ ,\\ &P_3+2\Gamma_{13}P_1P_2=0\ , \end{aligned} \tag{5.44}$$

wherein $\Gamma_{mn}\equiv J^{-1}K(m/n)$. If we know $P_1$, then this sytem yields $P_2$ and $P_3$ as

$$P_2=-\frac{3\Gamma_{12}P_1^2}{3+8\Gamma_{13}\Gamma_{13}^*P_1^2}\ ,\quad P_3=-2\Gamma_{13}P_1P_2\ . \tag{5.45}$$

Examples of periodic nondistorting profiles are given by Parker and Talbot (1985) for special materials (isotropic nonlinear elastic in particular). The forms obtained are somewhat exotic, e.g., presenting one or two sharp humps per period, still others with nonsymmetric profile (cf. Figures 4).

## 5.5 The projection method

This method is based on the notion of projection of the nonlinear contribution solution at a certain order onto a solution of the linearized system and then two applications of Green's theorem. Let the linearized solution be

$$\bar{\mathbf{u}}(\mathbf{X},t) = \mathbf{w}\left(\mathbf{k}\middle|X_2\right) exp\left[i(\mathbf{k}.\mathbf{X}) - \omega(\mathbf{k})t\right], \tag{5.46}$$

with

$$\begin{aligned} &\mathbf{w} \to \mathbf{0} \;\; as \;\; X_2 \to -\infty \;\;, \\ &\omega(\mathbf{k}) = -\omega(-\mathbf{k}) \;\;, \\ &\mathbf{w}\left(-\mathbf{k}\middle|X_2\right) = \mathbf{w}^*\left(\mathbf{k}\middle|X_2\right). \end{aligned} \tag{5.47}$$

with $\mathbf{k} = (k,0,0)$ for a propagation along $X_1$. Note the change of notation to agree with the works of Maradudin *et al* (however, **w** reduces to a function $\mathbf{w}(kX_2)$ when there is no length scale - no dispersion - in the physical system).

In the *nonlinear case*, we shall look for an asymptotic solution in the form

$$\begin{aligned} \mathbf{u}(\mathbf{X},t) = &\,\varepsilon \mathbf{U}\left(\theta, X_2, X_1^{(1)}, X_3^{(1)}, T^{(1)}\right) \\ &+ \varepsilon^2 \mathbf{u}^{(2)}\left(\theta, X_2, X_1^{(1)}, X_3^{(1)}, T^{(1)}\right) + 0\left(\varepsilon^3\right) \;, \end{aligned} \tag{5.48}$$

wherein

$$\theta = X_1 - ct \;\;,\;\; X_1^{(1)} = \varepsilon X_1 \;\;,\;\; X_3^{(1)} = \varepsilon X_3 \;\;,\;\; T^{(1)} = \varepsilon t \tag{5.49}$$

define the characteristic coordinate and *stretched* coordinates in space-time. Here the dependency on $X_3^{(1)}$ is introduced for a possible finite lateral width of the wave profile. But substituting for (5.48) into the equations of the nonlinear problem (5.3)-(5.4), and collecting terms of the first order in $\varepsilon$, we see that **U** should be of the general form

$$\mathbf{U}\left(\theta, X_2, X_1^{(1)}, T^{(1)}\right) = \sum_k \mathbf{w}\left(k\middle|X_2\right) A\left(k, X_1^{(1)}, T^{(1)}\right) exp(ik\theta), \tag{5.50}$$

with $A$ real such that $A(-k) = A^*(k)$ and $k = 2\pi n / L_1$ (a more general representation would replace the sum by an integral). To second order in $\varepsilon$ eqn.(5.3) takes on the form $(\beta = 1,2; j = 1,2,3)$

$$
\begin{aligned}
-L_j\left(\mathbf{u}^{(2)}\right) := {} & \rho_0 c^2 u_{j,11} - \hat{C}_{jKmL} u^{(2)}_{m,KL} \\
= {} & \frac{1}{2}\frac{\partial}{\partial X_K}\hat{C}_{jKmLpQ} U_{m,L} U_{p,Q} + 2c\frac{\partial}{\partial T^{(1)}} U_{j,1} \\
& + \left(C_{jKm\beta} + C_{j\beta mK}\right)\frac{\partial}{\partial X^{(1)}_\beta} U_{m,K} \qquad in\, X_2 < 0,
\end{aligned}
\tag{5.51}
$$

while the boundary condition (5.4) yields

$$
\begin{aligned}
-B_j\left(\mathbf{u}^{(2)}\right) := {} & -C_{j2mK} u^{(2)}_{m,K} \\
= {} & \frac{1}{2}\hat{C}_{j2mKnL} U_{mK} U_{n,L} + C_{j2m\beta}\frac{\partial}{\partial X^{(1)}_\beta} U_m \quad at\ X_2 = 0.
\end{aligned}
\tag{5.52}
$$

Now we project (5.51) onto the solution $\overline{\mathbf{u}}$ of the linearized system and integrate in depth $X_2 < 0$ and over a wavelength $L_1$ in the direction of propagation , obtaining thus after application of the Green theorem the following expression where account is taken of the boundary conditions imposed on both linearized and actual problems :

$$
\begin{aligned}
\int_0^{L_1}\int_{-\infty}^{0}\Big\{ & \rho_0 c^2 \overline{u}^*_{j,1} u^{(2)}_{j,1} - C_{jKmL}\overline{u}^*_{j,K} u^{(2)}_{m,L} - \frac{1}{2}\hat{C}_{jKmLpQ}\overline{u}^*_{j,K} U_{m,L} U_{p,Q} + 2\rho_0 c\overline{u}^*_j \frac{\partial}{\partial T^{(1)}} U_{j,1} - \\
& - C_{jKm\beta}\overline{u}^*_{j,K}\frac{\partial}{\partial X^{(1)}_\beta} U_m + C_{j\beta mK}\overline{u}^*_j \frac{\partial}{\partial X^{(1)}_\beta} U_{m,K} \Big\} dX_2 dX_1 = 0 \quad .
\end{aligned}
\tag{5.53}
$$

Applying once more the Green theorem, the first two terms in this expression are shown to cancel out leaving an equation that no longer involves $\mathbf{u}^{(2)}$. This expresses the solvability condition which , by using the expressions (5.46) and (5.50), for $\overline{\mathbf{u}}$ and $\mathbf{U}$ respectively, can be rewritten in the form

$$
i\left\{\frac{\partial}{\partial T^{(1)}} + V_\beta \frac{\partial}{\partial X^{(1)}_\beta}\right\} A(k) = \sum_\kappa K_R(k,\kappa) A(\kappa) A(k-\kappa)
\tag{5.54}
$$

in which we have set (cf. Parker *et al* , 1992)

$$
\begin{aligned}
K_R(k,\kappa) = {} & N^{-1}\hat{C}_{jKmLpQ} \times \\
& \int_{-\infty}^{0}\left\{\left[\hat{Q}_K(-k) w_j\left(-k|X_2\right)\right]\left[\hat{Q}_L(\kappa) w_m\left(\kappa|X_2\right)\right]\left[\hat{Q}_Q(k-\kappa) w_p\left(k-\kappa|X_2\right)\right]\right\} dX_2 \\
= {} & k^2 \widetilde{K}_R(\kappa / \mathrm{k}) \quad ,
\end{aligned}
\tag{5.55}
$$

$$\hat{Q}_K(k) \equiv ik\delta_{K1} + \delta_{K2}\frac{\partial}{\partial X_2} \quad , \tag{5.56}$$

and **w** is normalized such that

$$N = 4\rho_0 ck \int_{-\infty}^{+\infty} w_j(-\mathbf{k}|X_2) w_j(\mathbf{k}|X_2) dX_2 \, , \tag{5.57}$$

while the vector of components $V_\beta = \{V_1, 0, V_3\}$ is the group velocity for generalized Rayleigh waves having wave vector k in the positive X1 direction. The kernel K can be evaluated from the linear generalized rayleigh wave results only. Moreover, in the limit of infinite period L1, the sums in both eqns.(5.50) and (5.54) can be replaced by integrals. Equation (5.54) or its limit with an integral in the right-hand side is the evolution equation sought for the solution of order $\varepsilon^2$. One should proceed to the third order in $\varepsilon$ in order to derive an evolution equation for the slow variation of A(*k*). This requires determining $\mathbf{u}^{(2)}$ and applying the projection technique again for $\mathbf{u}^{(3)}$. We refer to the works of Mayer, Maradudin and Parker for these lengthy developments.

## 5.6 The Hamiltonian technique

This is an alternative method for deriving the evolution equation for *weakly* nonlinear ssurface acoustic waves as favored , for instance, by Zabolotskaya, Hamilton (not the same Hamilton as the one of analytical mechanics fame) and their co-workers Knight, Meegam, and Il'insky (see reference list). In that view the exact solution of the linearized field equation and boundary condition is taken as a superposition of straight-crested surface waves

$$\mathbf{u}(X_1, X_2, t) = \frac{1}{2\pi}\int_{-\infty}^{+\infty} a(k,t)\tilde{\mathbf{w}}(k|X_2) exp(ikX_1) dk \, . \tag{5.58}$$

This is inserted in the Lagrangian from which (5.2) and (5.3) can be shown to follow, i.e., with possible smooth material inhomogeneities along the depth coordinate

$$\begin{aligned} L = \int_{-\infty}^{d} dX_2 \Bigg\{ & \int_{-\infty}^{+\infty} \frac{dk}{2\pi}\frac{1}{2}\rho_0 \tilde{w}_j(-k|X_2)\tilde{w}_j(k|X_2)\dot{a}(-k)\dot{a}(k) \\ & - \int_{-\infty}^{+\infty}\frac{dk}{2\pi}\frac{1}{2} C_{jKmL}(X_2)\left[D_K(-k)\tilde{w}_j(-k|X_2)\right]\left[D_L(k)\tilde{w}_m(k|X_2)\right] a(-k)a(k) \\ & - \int_{-\infty}^{+\infty}\frac{dk}{2\pi}\int_{-\infty}^{+\infty}\frac{dk'}{2\pi}\int_{-\infty}^{+\infty}\frac{dk''}{2\pi}\frac{1}{6}\hat{C}_{jKmLpQ}(X_2)\left[D_K(k')\tilde{w}_j(k'|X_2)\right]\left[D_L(k)\tilde{w}_m(k|X_2)\right] \times \\ & \times \left[D_Q(k'')\tilde{w}_p(k''|X_2)\right] 2\pi\delta(k+k'+k'')a(k)a(k')a(k'') \Bigg\} \end{aligned} \tag{5.59}$$

where the operators $D_K$ are defined just like the $\hat{Q}_K$ in (5.56). The second term in the right-hand side of (5.59) is integrated by parts taking into account that

$$C_{j2mL}\left(X_2 = d\right)\left[D_L(k)\widetilde{w}_m\left(k\middle|X_2\right)\right]_{X_2=d} = 0. \tag{5.60}$$

If we now perform the variation of the resulting action integral (5.59) with respect to the real and imaginary parts of the complex amplitudes $a(k)$, $k > 0$, or , equivalently to $a(-k)$, $k > 0$, we obtain the equation of motion for the amplitudes $a(k)$ as

$$\begin{aligned} 0 = \int_{-\infty}^{d} dX_2 \Big\{ & \rho_0 \widetilde{w}_j\left(-k\middle|X_2\right)\widetilde{w}_j\left(k\middle|X_2\right)\ddot{a}(k) - \widetilde{w}_j\left(-k\middle|X_2\right)D_K(k)C_{jKmL}\left(X_2\right)D_L(k)\widetilde{w}_m\left(k\middle|X_2\right)a(k) \\ & + \frac{1}{2}\hat{C}_{jKmLpQ}\left(X_2\right)\int_{-\infty}^{+\infty}\frac{d\kappa}{2\pi}\left[D_K(-k)\widetilde{w}_j\left(-k\middle|X_2\right)\right]\left[D_L(\kappa)\widetilde{w}_m\left(\kappa\middle|X_2\right)\right]\times \\ & \times\left[D_Q(k-\kappa)\widetilde{w}_p\left(k-\kappa\middle|X_2\right)\right]a(\kappa)a(k-\kappa) \Big\}. \end{aligned} \tag{5.61}$$

On decomposing now the complex amplitude $a(k,t)$ into one part that varies slowly in time and one that varies rapidly in time we write

$$a(k,t) = A(k,t)\,exp\left(-ic_R kt\right), \tag{5.62}$$

where $c_R$ is the Rayleigh speed for the homogeneous linear case, and neglecting the second time derivative of $A$ , the amplitude equation (5.61) leads to

$$\begin{aligned} \frac{\partial}{\partial t}A(k) = & \frac{1}{2kc_R}\left[\omega^2(k) - \left(c_R k\right)^2\right]A(k) \\ & + \int_{-\infty}^{+\infty}\frac{d\kappa}{2\pi}G(k,\kappa)A(\kappa)A(k-\kappa) \end{aligned}, \tag{5.63}$$

where $\omega(k)$ is the frequency of the linear surface wave with wave number $k$, and we have set

$$\begin{aligned} G(k,\kappa) = \frac{1}{N(k)}\int_{-\infty}^{+\infty} & dX_2 \hat{C}_{jKmLpQ}\left[D_K(-k)\widetilde{w}_j\left(-k\middle|X_2\right)\right]\times \\ & \times\left[D_L(\kappa)\widetilde{w}_m\left(k\middle|X_2\right)\right]\left[D_Q(k-\kappa)\widetilde{w}_p\left(k\middle|X_2\right)\right] \end{aligned} \tag{5.64}$$

and (compare (5.57)

$$N(k) = 4c_R|k|\int_{-\infty}^{d} dX_2 \rho_0\left(X_2\right)\widetilde{w}_j * \left(k\middle|X_2\right)\widetilde{w}_j\left(k\middle|X_2\right). \tag{5.65}$$

Here we have kept the possibility of the occurrence of smooth material inhomogeneities in both density and elasticity coefficients along the depth. For small dispersion

$\left|\omega^2(k)-c_R^2k^2\right| << \left(c_Rk\right)^2$, so that we can write $\omega^2(k)-\left(c_Rk\right)^2 \cong 2c_Rk^3\Delta k$, with $\Delta k \equiv \left[\omega(k)-\left(c_Rk\right)\right]/k^2$. Also, the integral over $X_2$ in (5.53)-5.65) may be extended from - ∞ to zero, and $\tilde{\mathbf{w}}$ may be replaced by **w**. If, furthermore, there are no material inhomogeneities, then (5.63) in fact recurs to the evolution equation of the projection method as proved by Eckl *et al* (2004). Such an equation was stated by Reutov (1973). We shall not pursue this line for which we refer the reader to the works of Hamilton *et al.*

# 6 Nonlinear Surface Waves in the Presence of Dispersion

## 6.1 Introductory remark

Some of the methods presented in Section 5 apply to inhomogeneous materials (dependence of properties on the depth coordinate $X_2$) and also to the case of dispersion. We repeatedly remind the reader that dispersion of waves occurs when a characteristic length is involved in the physical dynamical problem. In this case dispersion inhibits the growth of higher harmonics that resulted from nonlinearity. As a consequence solitary waves of various types may become possible (governed by one of the equations highlighted in Section 3) and nonlinear wave solutions described by a single sinusoidal carrier wave with slowly vaying amplitude may exist. Indeed, the equations recalled in Section 3 are *model equations* that one expects to hold in many circumstances. Accordingly, some may be tempted to describe apriori physical systems directly by such equations. A much harder task is to show that such equations can be deduced from general equations of continuum thermomechanics with due simplifications and assumptions. In early works mentioning the possibility of existence of dispersive nonlinear surface waves on solids, it is the first attitude mentioned that prevailed. This is true in particular of the works of Bataille and Lund (1982), Even *et al* (1981,1982), Cho and Miyagawa (1993), and Sakuma and Nishigushi (1990). Boussinesq and Korteweg-de Vries equations , the most familiar such equations thanks to fluid-mechanics studies, were a priori considered as model equations. It is only with the works of Maradudin (1988), Hadouaj and Maugin (1989- in greater detail , Maugin and Hadouaj, 1991) that a true derivation from basic principles of continuum mechanics was given corresponding to wave equations finally leading to a nonlinear Schroedinger equation (envelope solitons). Further works are those of Gorentsveig *et al* (1990,1991), Hadouaj, Malomed and Maugin (1991a,b), Kosevich (1991), Mayer and Maradudin (1991). It is only recently that a rather complete description of dispersive nonlinear Rayleigh waves was given (cf. Hamilton *et al*, 1995a,b ; Knight *et al* , 1997, Gusev *et al* , 1997) including the case of solitary and pulse-like Rayleigh waves (Eckl *et al*, 2001, 2002, 2004 ; Kovalev *et al* , 2002).

Without considering a material microstructure that would require a finer continuum mechanics description (for this, see Maugin, 1999), the simplest way to introduce a characteristic length in an already nonlinear surface wave problem (thus taking place in a nonlinear substrate $X_2$<0) is to glue an elastic layer on top of the half space, this layer having the desired elastic properties. In the linear case, this corresponds to the situation of *Love waves* . A short reminder on this is spot on (see also the contribution of Wilmanki in this volume).

## 6.2 Reminder on linear Love waves

These are *SH* polarized surface waves that exist when a homogeneous isotropic linear elastic layer of thickness $h$ is perfectly glued on top of an isotropic elastic half-space $X_2 < 0$, the former having a characteristic transverse elastic speed $\hat{c}_T$ smaller than that $\bar{c}_T$ of the half-space (so-called " slow " superposed layer). With the notation of Sections 4 and 5 this corresponds to the problem ($\alpha = 1,2$)

$$
\begin{aligned}
&\nabla_2^2 \hat{u} \equiv \hat{u}_{,\alpha\alpha} = \frac{1}{\hat{c}_T^2}\frac{\partial^2 \hat{u}}{\partial t^2} \quad in \;\; 0 < X_2 < h \quad , \\
&\nabla_2^2 \bar{u} = \frac{1}{\bar{c}_T^2}\frac{\partial^2 \bar{u}}{\partial t^2} \quad in \;\; X_2 < 0 \quad , \\
&\hat{u}_{,2} = 0 \quad at \quad X_2 = h \quad , \\
&[\mathrm{u}] = 0, \quad \left[\mu u_{,2}\right] = 0 \quad at \quad X_2 = 0 \quad ,
\end{aligned}
\tag{6.1}
$$

when both $\hat{\mathbf{u}}$ and $\bar{\mathbf{u}}$ are along the transverse direction $X_3$. A solution $u = A\left(k,h \middle| X_2\right) exp(ik\theta)$ exits only if $\hat{c}_T^2 < c^2 < \bar{c}_T^2$. The resulting wave has an exponential decay in amplitude with depth in $X_2 < 0$, and it presents a dispersion relation given by (cf. Achenbach, 1973 : Maugin, 1983)

$$
D_{Love}(k,h;\mathrm{c}) := tan\left\{\left(\frac{c^2}{\bar{c}_T^2} - 1\right)^{1/2} kh\right\} - \frac{\left(1 - \dfrac{c^2}{\bar{c}_T^2}\right)^{1/2}}{\left(\dfrac{c^2}{\hat{c}_T^2} - 1\right)^{1/2}} = 0 \quad , \hat{c}_T < c < \bar{c}_T \; . \tag{6.2}
$$

This yields the following sequence of reduced wave numbers :

$$
kh = \frac{1}{\left(\dfrac{c^2}{\bar{c}_T^2} - 1\right)^{1/2}} \left\{ tan^{-1} \frac{\left(1 - \dfrac{c^2}{\bar{c}_T^2}\right)^{1/2}}{\left(\dfrac{c^2}{\hat{c}_T^2} - 1\right)^{1/2}} + n\pi \right\} \quad ,n = 0,1,2,.. \tag{6.3}
$$

All modes $n \geq 1$ exhibit a **cutoff**, while the lowest mode $n = 0$ exists for all values of *kh*. As $h$ goes to zero with fixed *k*, *c* goes to $\bar{c}_T$ along the first mode. The variation of amplitude with depth is given by

$$A(k,h|X_2)=\begin{cases} cos\left[k(X_2-h)\left(\dfrac{c^2}{\hat{c}_T^2}-1\right)^{1/2}\right] & for\ 0<X_2\le h\ , \\ exp\left[\left(1-\dfrac{c^2}{\bar{c}_T^2}\right)kX_2\right] & for\ X_2<0\ . \end{cases} \tag{6.4}$$

The Love mode thus is exemplary. It illustrates many features that are typical of dispersive waves, i.e., dependence of speed and depth profile on wave number, the existence of multiple modes, and that of cutoff frequencies.

Obviously, Rayleigh-like waves also exist in the above structure considering the general case of two anisotropic linear elastic materials. These can be analyzed by superposing partial waves.

## 6.3 The notion of nonlinear dispersion law

We are examining here the *influence of nonlinearity* on the dispersive wave such as described in the superposed layer (of finite thickness) problem of the foregoing paragraph. This was first dealt with for Love waves by Kalyanasundaram (1981). To look into this problem we consider at first a wave of any polarization and seek a solution with linearized approximation in the form

$$\mathbf{u}=\mathbf{U}=a\mathbf{A}(k|X_2)\ exp\left[i(kX_1-\omega(k)t)\right]+c.c, \tag{6.5}$$

where a length normalization is such that $h=1$, and $a$ is a free parameter. Calling $N_j(\mathbf{u})$ and $M_j(\mathbf{u})$ the right-hand sides defined in eqns.(5.10) and (5.11), we have to examine the following nonlinear *and* dispersive wave problem :

$$\begin{aligned} &L_j(\bar{\mathbf{u}})=N_j(\bar{\mathbf{u}})\ in\quad X_2<0\ , \\ &L_j(\hat{\mathbf{u}})=N_j(\hat{\mathbf{u}})\ in\ 0<X_2<1\ , \\ &B_j(\hat{\mathbf{u}})=M_j(\hat{\mathbf{u}})\ at\ X_2=1\ , \\ &\left[B_j(\mathbf{u})-M_j(\mathbf{u}\right]=0\ ,\ [\mathbf{u}]=0\ at\ X_2=0\ , \\ &\bar{\mathbf{u}}\to\mathbf{0}\quad \text{as}\quad X_2\to-\infty \end{aligned} \tag{6.6}$$

being cautious to select the correct material coefficients in each region of space. Although not written explicitly, we will also use the already used notation

$$\theta=X_1-ct\ ,\ X_2,\ X=\varepsilon X_1\ ,\ c=\omega(k)/k\,.$$

We look for an expansion $\mathbf{u}=\mathbf{U}+\varepsilon\mathbf{v}+\varepsilon^2\,\mathbf{w}+\ldots$ Equations (6.6) then provide the following inhomogeneous system for $\mathbf{v}$ :

$$
\begin{aligned}
&L_j(\overline{\mathbf{v}}) = -\overline{C}_{jKmLpQ}\left(\overline{U}_{m,L}\overline{U}_{p,Q}\right)_{,K} \quad , \; X_2 < 0 \quad , \\
&L_j(\hat{\mathbf{v}}) = -\hat{C}_{jKmLpQ}\left(\hat{U}_{m,L}\hat{U}_{p,Q}\right)_{,K} \quad , 0 < X_2 < 1 , \\
&B_j(\hat{\mathbf{v}}) = -\hat{C}_{j2mLpQ}\hat{U}_{m,L}\hat{U}_{p,Q} \quad , \quad X_2 = h = 1 , \\
&B_j(\overline{\mathbf{v}}) - B_j(\hat{\mathbf{v}}) = \hat{C}_{j2mLpQ}\hat{U}_{m,L}\hat{U}_{p,Q} - \overline{C}_{j2mLpQ}\overline{U}_{m,L}\overline{U}_{p,Q} \; , \; X_2 = 0 \; ,\mathrm{I} \\
&\overline{\mathbf{v}} = \hat{\mathbf{v}} \quad at \;\; X_2 = 0 \quad , \quad \overline{\mathbf{v}} \to \mathbf{0} \quad as \;\; X_2 \to -\infty \, .
\end{aligned}
\tag{6.7}
$$

with an obvious notation for the material coefficients. The right-hand sides of this system involve the characteristic coordinate $X$ only through factors $exp(\pm 2ik\theta)$, so that this system will be automatically solved for $\hat{\mathbf{v}}$ and $\overline{\mathbf{v}}$ unless we meet the rare circumstance where $\omega(2k) = 2\omega(k)$. Apart from this very special case, $\mathbf{v}$ will therefore have the form

$$
\mathbf{v} = aa^* \mathbf{v}^{(0)}\left(k\middle|X_2\right) + a^2\mathbf{v}^{(2)}\left(k\middle|X_2\right) exp(2ik\theta) + c.c \, . \tag{6.8}
$$

Nonlinear effects of evolution will arise only when the system (6.6) is solved for the next order solution **w.** But for this we need to account for cubic terms in the stress constitutive equation, hence to account for fourth-order elasticity coefficients - quantities which are more than rarely known (cf. Maugin, 1985) - leading to a replacement of eqns. (6.7) including cubic terms in the gradient of **u,** and source terms in $exp(\pm ik\theta)$ and $exp(\pm 3ik\theta)$. In the end the solvability condition will be expressed as an orthogonality condition of the type of (5.26) as (along with the corresponding complex conjugate)

$$
\begin{aligned}
&\int_{-\infty}^{0} \mathbf{A}^*\left(k\middle|X_2\right).\overline{\mathbf{P}}^1 dX_2 + \int_{0}^{1} \mathbf{A}^*\left(k\middle|X_2\right).\hat{\mathbf{P}}^1 dX_2 + \\
&\qquad + \mathbf{A}^*(k|0).\overline{\mathbf{Q}}^1 + \mathbf{A}^*(k|1).\hat{\mathbf{Q}}^1 = 0 \quad .
\end{aligned}
\tag{6.9}
$$

This equation will finally be of the form

$$
iJ\frac{da}{dY} + Na^2a^* = 0 , \tag{6.10}
$$

where $Y = \varepsilon^2 X_1$ , $J$ is a generalization of (5.28) with an integration of properties over the $X_2$ interval $\left(-\infty,1\right]$, and $N$ is a complicated expression (cf. Kalyanasundaram, 1981) that involves determining the depth behaviors of $\mathbf{v}^{(0)}$ and $\mathbf{v}^{(2)}$ in the expansion (6.8). The solution of (6.10) is

$$
a = a_0 \; exp\left(i\nu\left|a_0\right|^2 Y\right) \; , \quad \nu \equiv N / J \quad , \tag{6.11}
$$

so that (6.5) can be rewritten as

$$
\mathbf{u} = a_0\mathbf{A}\left(k\middle|X_2\right) exp\, i\left\{\left(k + \nu\varepsilon^2\left|a_0\right|^2\right)X_1 - \omega(k)t\right\} + c.c \, . \tag{6.12}
$$

This clearly shows that the primary influence of nonlinearity on dispersive surface waves in the present case is to perturb the wave speed of the guided mode. If we can invert the dispersion relation so that $k = \tilde{k}(\omega)$ , then according to (6.12) the *nonlinear dispersion* behavior is given by

$$k = \tilde{k}(\omega) + v|\varepsilon a_0|^2 , \tag{6.13}$$

hence a perturbation in the wave speed that is quadratic in the amplitude $|\varepsilon a_0|$ of a harmonic wave train, a common feature of guided waves. For a pure Love wave $\mathbf{A}(k|X_2) = \tilde{A}(X_2)\mathbf{e}_3$, but $\mathbf{v}$ is sagitally polarized as it is checked that $\mathbf{v}.\mathbf{e}_3 = 0$. For a Rayleigh wave, one verifies that the nonlinear wave remains sagittally polarized , i.e., $\mathbf{u}.\mathbf{e}_3 = 0$, $\mathbf{v}.\mathbf{e}_3 = 0$, and this also holds for the next order, $\mathbf{w}.\mathbf{e}_3 = 0$.

## 6.4 Thin-film Approximation

If in Section 6.2 the thickness $h$ of the superposed layer is much smaller than a typical wavelength $L = 2\pi / k$ , then the dispersion will be small and, following a procedure due to Tiersten (1969), one may eliminate the displacement field $\hat{\mathbf{u}}$ in the layer in an approximation in powers of $kh$ in order to obtain an effective boundary condition at $X_2 = 0$. This is called the *thin-film approximation*. Instead of the classical boundary condition (5.4) expressing the absence of surface traction , we will have then the following condition (we assume no dependence on $X_3$ , and $\alpha$ =1,3; cf. Figure 5)

$$\begin{aligned} T^2_{.\alpha}(X_1, X_2 = 0_+, t) = -h\{ & \hat{\rho}_0 \ddot{\hat{u}}_\alpha(X_1, 0_+, t) - G^{(1)}_{\alpha\beta} \hat{u}_{\beta,11}(X_1, 0_+, t) \\ & - G^{(2)}_{\alpha\beta\gamma} \hat{u}_{\beta,11} \hat{u}_{\gamma,11} \} \\ & - h^2 \{ G^{(3)}_{\alpha\beta} \hat{\rho}_0 \ddot{\hat{u}}_{\beta,1} + G^{(4)}_{\alpha\beta} \hat{u}_{\beta,111} \} \\ & + 0(u^2 h^2, u^3 h, uh^3) \end{aligned} , \tag{6.14}$$

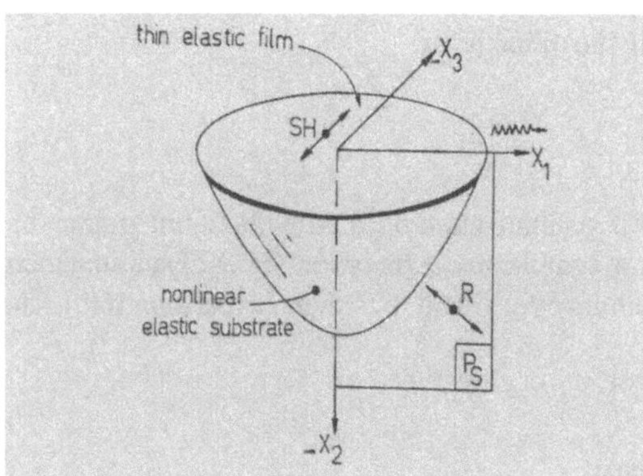


**Figure 5.** Linear elastic thin-film perfectly glued on a nonlinerar elastic halfspace

where $\hat{\rho}_0$ is the matter density of the thin film and the various $G^{(1)}$, $G^{(3)}$ and $G^{(4)}$ depend on second-order elasticity coefficients of the thin film and $G^{(2)}$ depends on both its second and third order elasticity coefficients, having assumed for the sake of generality a nonlinear elastic thin film. When the latter is linear elastic , and considering the first order approximation of (6.14) the latter reduces to

$$T^2_{.\alpha} = -h\left(\hat{\rho}_0\ddot{\hat{u}}_\alpha - G^{(1)}_{\alpha\beta}\hat{u}_{\beta,11}\right), \tag{6.15}$$

where the left-hand side would be given by the stress contribution from the half-space, i.e., by an expression of the type given by eqn.(5.11). In this situation, which is the easiest to conceive and to realize practically, *nonlinearity* is due to the substrate and *dispersion* to the linear elastic coating. With some luck the conditions for the propagation of solitary waves of permanent form can be achieved thanks to a strict compensation between the two effects. As a matter of fact we shall have the following fully *dynamical* boundary condition at $X_2$=0:

$$\begin{aligned} B_j(\mathbf{u}) &:= \overline{C}_{j2m\beta}u_{m,\beta} + \overline{C}_{j2m\beta n\gamma}u_{m,\beta}u_{n,\gamma} + \overline{C}_{j2m\beta n\gamma p\nu}u_{m,\beta}u_{n,\gamma}u_{p,\nu} \\ &= -(\hat{\rho}_0 h)\ddot{u}_j + \left(hG^{(1)}_{jm}\right)u_{m,11} + 0\left(\varepsilon^2, h^2, uh\right) \ , \end{aligned} \tag{6.16}$$

where the $\overline{C}$'s are effective tensorial elasticity coefficients of the half-space up to and including fourth-order elasticity (see below for this inclusion), and only one single **u** appears since $[\mathbf{u}] = 0$ at $X_2$ =0.

For a purely *SH* polarized wave , only $u_3 \neq 0$ , and this yields essentially $(\beta,\gamma,\nu = 1,2)$

$$\begin{aligned} B_3(\mathbf{u}) &:= \overline{C}_{323\beta}u_{3,\beta} + \overline{C}_{323\beta3\gamma}u_{3,\beta}u_{3,\gamma} + \overline{C}_{323\beta3\gamma3\nu}u_{3,\beta}u_{3,\gamma}u_{3,\nu} \\ &= -(\hat{\rho}_0 h)\ddot{u}_3 + \left(hG^{(1)}_{33}\right)u_{3,11} \ . \end{aligned} \tag{6.17}$$

Maugin and Hadouaj (1991) have proposed to approach the same problem by considering that the superposed film is an *interface* made of an elastic film of vanishing thickness that possesses both inertia (i.e., a mass density $\tilde{\rho}_0$ per unit surface) and isotropic membrane elasticity of Lamé coefficients $\hat{\lambda}$ and $\hat{\mu}$ , while the nonlinear substrate is a nonlinear isotropic elastic body with energy density given by (4.15). Their deduction is based on the fact that when $X_2 = 0$ is an interface , the bulk equation (4.7) - with $\mathbf{f} = \mathbf{0}$ - couples with the " interface equation of motion " via the jump of the classical stress at $X_2$. This interface equation reads (cf . Maugin, 1999)

$$\frac{\hat{\partial}}{\partial t}\left(\tilde{\rho}_0\dot{\hat{\mathbf{u}}}\right) = A^{-1}\mathbf{P}.\hat{div}_R\hat{\mathbf{T}} + \mathbf{N}.[\mathbf{T}], \tag{6.18}$$

where $A$ is a geometrical quantity, $\mathbf{P} := \mathbf{1} - \mathbf{N}\otimes\mathbf{N}$ is the projector onto the surface, $\hat{\mathbf{T}}$ is the surface Piola-Kirchhoff stress (an essentially two-dimensional geometrical object), $\hat{div}_R$ is the surface divergence, $\hat{\partial}/\partial t$ is the material time derivative in following the surface, and **[T]** is the jump of the bulk stress tensor across the boundary. For a purely *SH* polarized elastic signal, accounting for an isotropic linear behavior in the surface plane , eqn.(6.18) is shown (Hadouaj and Maugin, 1989) to reduce to the simple equation

$$\tilde{\rho}_0 \ddot{\hat{u}}_3 - \hat{\mu}\hat{u}_{3,11} = -\left[T^2_{.3}\right]. \tag{6.19}$$

The right-hand side contribution here is none other than the negative of the quantity $B_3(\mathbf{u})$ defined in (6.17) for a general anisotoric substrate. But for the case of energy (4.15) and wave motion $u_3$ only, it is shown that eqn.(4.9) in the bulk $X_2 < 0$, with $\mathbf{f}=\mathbf{0}$, takes on the following form ($\beta = 1,2$)

$$\rho_0 \ddot{\bar{u}}_3 = \mu \bar{u}_{3,\beta\beta} + \delta_{eff}\left(\bar{u}_{3,K}\,\bar{u}_{3,\beta}\,\bar{u}_{3,\beta}\right)_{,K}. \tag{6.20}$$

Here $\mu$ the bulk shear modulus and $\delta_{eff}$ is an effective elasticity coefficient of the *fourth* order defined by

$$\delta_{eff} = \delta + \left(\bar{\beta} + \frac{3}{2}\gamma\right) + \left(\mu + \frac{1}{2}\lambda\right) \tag{6.21}$$

in terms of the coefficients of the expression (4.15). We must also satisfy the constraint (film perfectly glued to the substrate)

$$\bar{\mathbf{u}} = \hat{\mathbf{u}} \quad at \quad X_2 = 0. \tag{6.22}$$

We have

$$\left[T^2_{.3}\right] = -\bar{T}^2_{.3}\left(X_2 = 0\right) = -\left(\mu + \delta_{eff}\bar{u}_{3,\beta}\,\bar{u}_{3,\beta}\right)\bar{u}_{3,2}\ . \tag{6.23}$$

For the assumed isotropy of both the substrate and the thin film, the nonlinearity of the substrate, and the a priori chosen polarization for **u**, we immediately check that (6.19), together with (6.2), agrees with (6.17), if we set in a natural way $\tilde{\rho}_O = \hat{\rho}_0 h$ and $\mu = h\,G^{(1)}_{33}$. There then remains only linear and cubic terms in the expression of $B_3$. Note that this situation together with the special expression (6.20), is very peculiar. It is a consequence of the combined isotropy and special choice of polarization of the wave. Had we considered a generalized Rayleigh-like wave, quadratic terms would have remained in both equations (6.20) and (6.22) and going to fourth order elasticity would probably have been supererogatory. For the time being, however, eqns.(6.19), (6.22) and (6 .23) form the basis of our analysis that we shall expand following an original mathematical approach to surface envelope solitons. The reason for this is that if the primary signal entered in the system (say, through a transducer, or via a natural cause (earthquake)), is of the *SH* type and is $0(\varepsilon)$, then the Rayleigh subsystem will develop an $0(\varepsilon^2)$ component (cf. eqn.(2.6)). Thus, in a first approach , the nonlinear mutual coupling can be discarded resulting in the study of a pure *SH*, both *nonlinear* and *dispersive*, mode.

## 6.5 Envelope SH solitons

Here we follow the orignal work of Hadouaj and Maugin (1989) - for much greater detail see Maugin and Hadouaj (1991). On substituting (6.23) into (6.19) and effecting the appropriate nondimensionalization, we arrive at the following system :

$$W_B(U):=\beta^2 U_{,TT}-\left(U_{,XX}+U_{,YY}\right)=\beta^2\Delta T_B^{NL}(U) \quad , \quad Y>0 \ ,$$
$$W_F\left(\hat{U}\right):=\hat{U}_{,TT}-\hat{U}_{YY}=U_{,Y}+\beta^2\Delta\, T_F^{NL} \quad at \ Y=0 \quad , \tag{6.24}$$
$$U=\hat{U} \quad at \ Y=0 \ , \quad U\left(X,Y\to\infty,T\right)=0 \quad ,$$

where we have set

$$T_B^{NL}=\left(U_{,X}\Phi\right)_{,X}+\left(U_{,Y}\Phi\right)_{,Y} \quad , \quad T_F^{NL}=U_{,Y}\Phi \quad , \quad \Phi\equiv U_{,X}^2+U_{,Y}^2 \ , \tag{6.25}$$

$$\Delta=c^2\Delta_{eff}=\delta_{eff}\,/\,\mu \quad , \quad \beta=c^{-1}=c_F\,/\,c_T \ , \tag{6.26}$$

and

$$c_F^2=\hat{\mu}\,/\,\tilde{\rho}_0 \quad , \quad c_T^2=\mu\,/\,\rho_0 \quad , X=k_a X_1 \quad , \quad Y=-k_a X_2 \quad , \ U=k_a u_3 \tag{6.27}$$

together with $k_a\equiv\mu\,/\,\hat{\mu}$. The $W$'s in (6.24) are linear wave operators (so-called d'Alembertian operators). Time is normalized with respect to a characteristic frequency $\omega_a$ such that $\omega_a^2=c_F^2k_a^2$. Note the change of sign in the definition of the depth coordinate (the fourth of (6.27)). Also, $\Phi$ in fact is an invariant for isotropic elasticity in the present problem so that the elastic energy is none other than a function $W=\overline{W}(\Phi)$. The parameter $\Delta$ - which may be of any sign depending on the nonlinear material - is the nonlinearity parameter of the nondimensional problem. Finally, $\beta<1$ the ratio of shear linear wave speed in the film and in the bulk, is the relevant dispersion parameter. The condition imposed on $\beta$ guarantees the existence of the linear Love-type solution ; cf. the last of eqns.(6.2). As a matter of fact, we may first consider the *linear* harmonic approximation of the above-stated system with frequency $\omega$ , wavenumber $k$, and depth parameter $\chi$ in the form

$$U\left(X,Y,T\right)=A_0\,exp\left(-\chi Y\right)exp\left\{i(kX-\omega T)\right\}. \tag{6.28}$$

Then the linearized version of the problem (6.24) provides Murdoch's (1976) surface-wave solution for linear elastic waves on a half-space with surface tension. This has a dispersion relation obtained by eliminating the depth coefficient $\chi$ between the linear " bulk " and "thin film" " dispersion relations " deduced by substituting (6.28) in $(6.24)_{1,2}$ :

$$D_B\left(k,\omega;\chi\right):=\beta^2\omega^2-k^2+\chi^2=0 \ ,$$
$$D_F\left(k,\omega;\chi\right):=\omega^2-k^2-\chi=0 \tag{6.29}$$

The result is the dispersion relation (the subscript $M$ stands for "Murdoch")

$$D_L\left(\omega,k\right)\equiv D_M\left(\omega,k\right):=\beta^2\omega^2-k^2+\left(\omega^2-k^2\right)^2=0\,. \tag{6.30}$$

This is a function of the parameter $s_0=c_F\,/\,c_T$. Existence of the wave requires that that

$$s_0<s<1,\text{or } \beta<1\,. \tag{6.31}$$

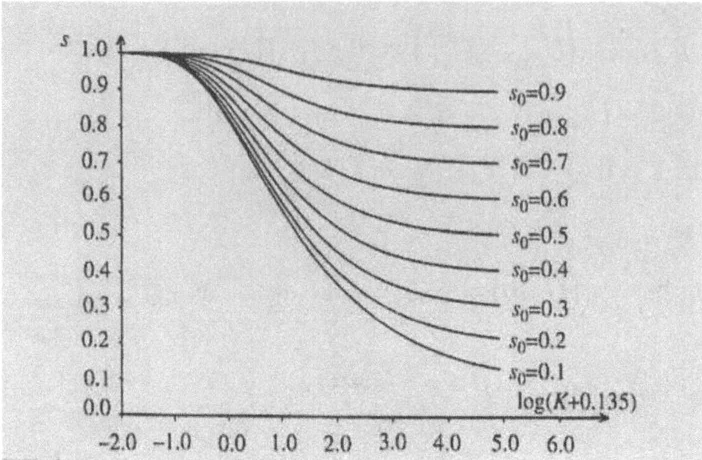


**Figure 6.** Dimensionless speed of Murdoch's SH waves vs dimensionless wavenumber (logarithmic scale) as a function of $s_0$ (after Maugin and Hadouaj, 1991)

The glued thin film must be " slower " than the substrate (cf. Figure 6). Had we neglected the superposed thin film but kept the nonlinearity of the substrate, we would have obtained no *SH* mode in the linearized analysis but the possibility of propagation of such a mode in the nonlinear analysis as was proved by Mozhaev (1989). This is an example of " perturbation " of the boundary condition favoring the existence of Love waves. But such waves are not as rich as those that we describe now.

**Solitary-wave solution for envelope signals.** We now consider the nonlinear regime but still for small amplitudes, with solutions of the type

$$U = A\,exp\left(-\chi Y\right)cos\,\theta + \ldots, \qquad \theta = kX - \omega T\,. \tag{6.32}$$

As the nonlinearities/fourth-order elasticities act as third-harmonic generators, (6.32) implies that the nonlinear contributions in eqns.$(6.24)_{1,2}$ have the form

$$\begin{aligned} T_B^{NL} &\cong \left(9\chi^4 - 3k^4 + 2k^2\chi^2\right)\frac{A^3}{4}exp\left(-3\chi Y\right)cos\,\theta + terms\ in\ \ cos\,3\theta\ , \\ T_F^{NL} &\cong -\left(\chi k^2 + 3\chi^3\right)\frac{A^3}{4}cos\,\theta + terms\ in\ \ cos\,3\theta \end{aligned} \tag{6.33}$$

For $A$ now depending on $X$, $Y$ and $T$, we can look for the " *nonlinear* " bulk and film dispersion relations in the form

$$D_B^{NL}(k,\omega,\chi,A) \equiv D_B(k,\omega;\chi)+\frac{\Delta\beta^2}{4}\left(9\chi^4-3k^4+2k^2\chi^2\right)A^2\, exp(-2\chi Y) \\ = \varepsilon\lambda^{(1)}+\varepsilon^2\lambda^{(2)}+.... \tag{6.34}$$

and

$$D_F^{NL}(k,\omega,\chi;A) \equiv D_F(k,\omega;\chi)-\frac{\Delta\beta^2}{4}\left(\chi k^2+3\chi^3\right)A^2 , \\ = \varepsilon m^{(1)}+\varepsilon^2 m^{(2)}+..... \tag{6.35}$$

where the right-hand sides represent perturbations to be determined. This is indeed achieved by seeking an asymptotic expansion with

$$x=\varepsilon X \ , \ y=\varepsilon Y \ , \ t=\varepsilon T \ , \\ U(X,Y,T)=f(\theta,A)+\varepsilon U^{(1)}+\varepsilon^2 U^{(2)}+... \tag{6.36}$$

where $\theta$ is a general phase such that

$$\theta_{,X}=k \ , \ \theta_{,T}=-\omega \ , \ \theta_{,Y}=i\chi \ . \tag{6.37}$$

Here we are following the technique of Whitham and Newell (cf.Newell, 1985 ; Maugin , 1999, Appendix A.6). On evaluating the derivatives $\partial/\partial T$, $\partial/\partial X$, $etc$, we find at the first order that $U_0$ is given by the linear solution (6.25) while at the second order we have

$$U^{(1)}=0 \ , \ \lambda^{(1)}=0 \ , \ m^{(1)}=0, \tag{6.38}$$

and we simultaneously obtain equations that have for meaning that of a *conservation of wave action* :

$$\frac{\partial}{\partial t}\left(\beta^2\omega A^2\right)+\frac{\partial}{\partial x}\left(kA^2\right)+\frac{\partial}{\partial y}\left(i\chi A^2\right)=0 \quad y>0 \ , \\ \frac{\partial}{\partial t}\left(\omega A^2\right)+\frac{\partial}{\partial x}\left(kA^2\right)-\frac{i}{2}\frac{\partial}{\partial y}\left(A^2\right)=0 \quad , \quad y=0 \ . \tag{6.39}$$

At the next order we shall obtain the secularity condition as (6.34) and (6.35) - i.e., the contributions in $\lambda^{(2)}$ and $m^{(2)}$ - in the form

$$D_B^{NL}(\omega,k,\chi,A)-\frac{\varepsilon^2}{A}W_B A=0 \quad , \quad y>0 \ , \\ D_F^{NL}(\omega,k,\chi,A)-\frac{\varepsilon^2}{A}W_F A=0 \quad , \quad y=0 \ . \tag{6.40}$$

These may be called *nonlinear " dispersive " dispersion relations*. They result from a double expansion in which $\varepsilon$ and $A$ are of the same order. They can also be viewed as wave equations for the amplitude $A$. They have to be complemented by the following equations (that follow from eqns.(6.37)) :

$$\frac{\partial k}{\partial t}+\frac{\partial \omega}{\partial x}=0 \ , \quad \frac{\partial \chi}{\partial t}-i\frac{\partial \omega}{\partial y}=0 \ . \tag{6.41}$$

Once $\chi$ has been eliminated from (6.39) and (6.40), these equations and (6.41) form a system that governs $k$, $\omega$, and $A$. The solution of this system would be a tremendous task in a general case. That is why we consider the so-called **almost monochromatic limit** in which the solution is assumed to be very close to the harmonic regime $(\omega_0, k_0, \chi_0)$ which satisfies the dispersion law (6.29) or (6.30). Thus we can write

$$k = k_0 + \varepsilon\phi_{,X} \ , \quad \omega = \omega_0 - \varepsilon\phi_{,t} \ , \quad \chi = \chi_0 - i\varepsilon\phi_{,y} \, . \tag{6.42}$$

We want to determine the *slowly varying* quantities $A$ and $\phi$. To that purpose we introduce

$$\xi = x - \omega_0' t + i\chi_0' y \, , \tau = \varepsilon t \, , A \to \varepsilon A \, . \tag{6.43}$$

Here the primes indicate a differentiation with respect to $k$ at $k_0$. Now we use eqns.(6.39) and (6.40) and the first two differentials of eqn.(6.30) evaluated at the $(\omega_0, k_0)$ point and introduce the complex amplitude $a = A exp(i\phi)$, to obtain two nonlinear Schroedinger equations from which, by eliminating the second derivative of $\chi_0$ with respect to k, we deduce a unique nonlinear Schroedinger equation for the amplitude $a$, in the canonical form

$$ia_{,\tau} + pa_{,\xi\xi} + q|a|^2 a = 0 \quad at \ y=0. \tag{6.44}$$

In the last equation we have set

$$\tau = \varepsilon t \ , \ \xi = x - \omega_0' t \ , \ y = 0 \ , \tag{6.45}$$

and

$$p = \frac{\omega_0''}{2} \ , \quad q = \frac{3}{8}\Delta\beta^2 \frac{\beta^2\omega_0^2\left(\beta^2\omega_0^2 - 2k_0^2\right)}{\omega_0\left[\beta^2 + 2\left(\omega_0^2 - k_0^2\right)\right]} \, . \tag{6.46}$$

According to a criterion obtained by Zakharov, if the product $pq > 0$, then (6.44) provides stable propagating solutions of the so-called *bright-envelope soliton type* (optical jargon !) such that finally at the interface we have the displacement solution (cf. Figure 7)

$$\begin{aligned}\hat{u}(X,t) = \varepsilon\eta\left|\frac{\omega_0''}{q}\right|^{1/2} exp\{i(k_0 X - \omega_0 T + 2\varepsilon^2\omega_0'' T\eta^2 + \phi_0)\}\times \\ \times sech\{2\varepsilon\eta(X_0 - c_0 T - X_0)\}\end{aligned} \ , \tag{6.47}$$

while for $pq < 0$ , we would have a stable " *dark-envelope soliton* " solution of the form

$$\hat{u}(X,t) = \varepsilon|\omega_0''|^{1/2}\, tanh\{\varepsilon(X - c_0T - X_0)\}exp\{i(k_oX - \omega_0T)\}. \tag{6.48}$$

The solution (6.47) contains three parameters $\eta$, $X_0$ , and $\phi_0$ and the condition $pq>0$ translates into the condition $1 > \beta^2 > 1/2$. This is satisfied by a lithium niobate $LiNbO_3$ substrate for which $\Delta > 0$ and an aluminium film. The solution (6.48) contains two parameters $\varepsilon$ and $X_0$ , and the existence condition $pq<0$ translates into $0 < \beta^2 < 1/2$ and would be realized by gold or silver on top of lithium niobate. Obviously, these are examples from signal-processing devices. Finally, at this degree of approximation the dependence on depth for both solutions (6.47) and (6.48) is simply obtained by multiplying by the factor $exp(-\chi_0 Y)$ with $\chi_0 \cong \omega_0^2 - k_0^2$ as in the linear solution.

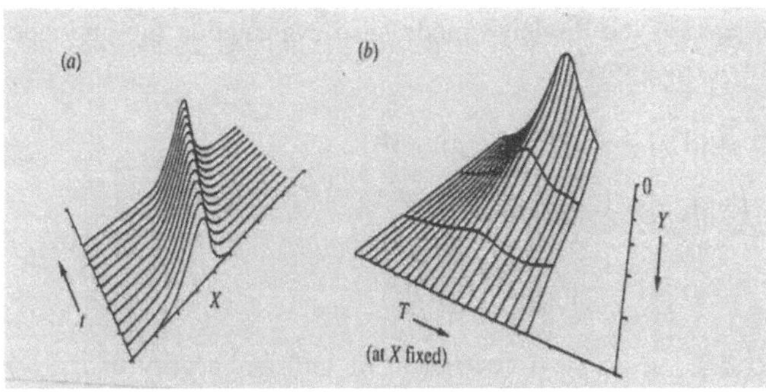


**Figure 7.** Numerical simulation of a bright-envelope surface soliton: (a) square of the envelope at the interface, (b) checking the ex ponential decrease with depth during propagation (simulation on the discrete system, after Hadouaj and Maugin, 1992).

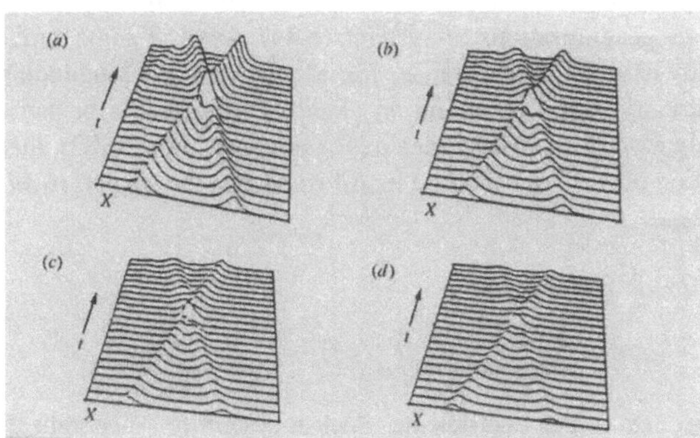


**Figure 8.** Head-on collision of unequal bright-envelope surface elastic solitons in system (6.24); simulation on the discrete system of exact equations: (a) through (b), respectively at the interface, at the 15[th] atomic layer below, at thc 25[th] layer below, and at the 35[th] layer below (after Hahouaj and Maugin, 1992).

Equation (6.44) is endowed with the property of *exact integrability* in soliton theory (see Newell, 1985), so that two bright-envelope solitons behave just like particles in a head-on collision, recovering their shape and other properties, save for their phase, once they have encountered each other. This behavior of the physical system itself has been checked directly numerically on the original system of equations (6.24) by Hadouaj and Maugin (1992), almost exactly (see figure 8) , so that we can say that the system (6.24) is *nearly* exactly integrable.

**The generalized Zakharov system.** If we now consider the inevitable coupling generated by the nonlinearities between the SH mode and Rayleigh modes, their amplitude being $0(\varepsilon)$ and $0(\varepsilon^2)$ , respectively, then the same method followed to construct the above-reduced system with a proper rescaling of the nonlinear dispersion relation is used. The result of this lengthy procedure (cf. Maugin, Hadouaj, and Malomed ,1992) is that the complex amplitude a of the SH component, at the interface, and the real x-gradients $n_1 = v_{,x}$ and $n_2 = w_{,x}$ of the longitudinal (P) and shear vertical (SV) components of the Rayleigh mode, also evaluated at the interface, jointly satisfiy the following system of equations :

$$\begin{aligned} & ia_{,t} + a_{,xx} \pm \lambda |a|^2 a + 2a(n_1 + n_2) = 0 \ , \\ & (n_1)_{,tt} - c_L^2 (n_1)_{,xx} = -\mu_L (|a|^2)_{,xx} \ , \\ & (n_2)_{,tt} - c_T^2 (n_2)_{,xx} = -\mu_T (|a|^2)_{,xx} \ . \end{aligned} \tag{6.49}$$

Here $\lambda = \lambda(\Delta, \beta, \omega_0, k_0)$ is a coefficient of self-nonlinearity of the *a-mode* , and $\mu_L$ and $\mu_T$ are the mutual interaction coefficients of which we do not give the explicit expressions. We have called systems of the type (6.49) *generalized Zakharov systems* (GZS) as if we discard one of the *n*'s and $\lambda = 0$, the system (6.49) reduces to (3.14) , which is equivalent to a cubic Schroedinger equation when we further neglect the inertia of the *n*-mode and consider only one single soliton. Systems of the type (6.49) present extremely interesting properties that have been thoroughly studied by Hadouaj, Malomed and Maugin (1991a,b), to whom we refer the reader. Furthermore, some generalizations of systems (6.49) useful in some surface wave propagation problems are easily devised . For instance, considering only the longitudinal component $n_1 = n$, its nonlinear contribution, and assuming an effect of discreteness or some stratification in the direction of propagation (this was the idea of Bataille and Lund, 1982), then it is not difficult to see that the equation governing *n* will be modified so that the system (6.49) will be replaced by the following one :

$$\begin{aligned} & ia_{,t} + a_{,xx} \pm 2\lambda |a|^2 a + 2an = 0 \, , \\ & n_{,tt} - c_0^2 n_{,xx} + \left( \mu |a|^2 + \alpha n^2 + \beta n_{,xx} \right)_{,xx} = 0 , \end{aligned} \tag{6.50}$$

where $\alpha$ and $\beta$ are real scalar coefficients. Such a system is a generalization of some systems appearing in plasma physics. They are difficult to deal with as (6.50) couples a cubic Schroedinger equation to a Boussinesq equation. So much for essentially *SH* surface-wave motions.

To conclude this paragraph we should note that instead of considering the dominant nonlinearity in the substrate it was shown by some authors (e.g., Kivshar,1991 ; Porubov and

Samsonov, 1995) that a dominant nonlinearity in the superimposed layer is also conceivable, leading to the problem of a nonlinear film covering a linear substrate. So much for essentially *SH* surface-wave motions.

## 6.6 Solitary Rayleigh waves and nonlocality

The general dispersive systems with coupling between modes obtained with a solid substrate and a coating film have been the subject of thorough investigations. In reason of the obviously cumbersome algebra, here we shall content ourselves with giving a few characteristic results obtained when the coating film presents the dominant nonlinearity (cf. Kovalev et al, 2002). The resulting system is somewhat easier to handle than in the case of dominant nonlinearity in the substrate as the substrate being linear, the displacement field in it can be eliminated and the solution is practically directly sought in the form of a nonlinear wave solution for the displacement in the film at $X_2 = 0$. Once the displacement field is known at the surface, then the displacement at any point in the linear substrate can be computed simply by using Green's function. The presence of the thin film is still taken care of by exploiting the effective boundary condition , but this time with the latter including the second order nonlinearity of the film and higher-order linear dispersion. Then a travelling-wave ansatz is introduced for the sagittal component of the displacement field. It is shown thus that a single *integrodifferential* equation for one scalar variable only results therefrom . This equation in fact contains terms with Hilbert transforms as in the Benjamin-Ono equation recalled in eqn.(3.11). Just like for Rayleigh waves in nonlinear substrates (Parker *et al*, 1992), an evolution equation can be obtained for Rayleigh waves in the system under consideration by using a projection method already illustrated in Paragraph 5.5 above. What we want to emphasize here are the *nonlocal* aspects of the problem (cf. the title of the contribution by Parker, 2004). For instance, already in the linear problem (with isotropic materials), it is noted Kovalev *et al*, 2002) that any of the longitudinal or transverse (SV) component of the Rayleigh elastic displacement has derivatives at $X_2 = 0$ that are related by ($Y \equiv -X_2$)

$$\left.\frac{\partial u}{\partial Y}\right|_{X_2=0} = \left(1-\frac{c^2}{c_S^2}\right)^{1/2} H\left[\frac{\partial u}{\partial X_1}\right]_{X_2=0}, \tag{6.51}$$

where $c_S = c_L$ for the longitudinal components, and $c_S = c_T$ for the transverse ones, and $H$ denotes the Hilbert transform of the quantity within square brackets. Also, with a rescaling of coordinates and the field variable $u_2$ as $\eta \propto X_1 - ct$, and $U$ the dimensionless $u_2$ , one shows that the alluded to integrodifferential equation at $X_2 = 0$ reads

$$U - \beta_1 H\left[U_{,\eta}\right] - \beta_2 U_{,\eta\eta} - \alpha_3 H\left[U U_{,\eta}\right] = 0 \ , \tag{6.52}$$

where the coefficients are also dimensionless but expressible in terms of physical coefficients of the model. This is to be compared to the case of the traditional Benjamin-Ono equation (3.1) where we can write

$$U - U^2 - H\left[U_{,\eta}\right] = 0. \tag{6.53}$$

or to the KdV case for which both dispersion and nonlinearity are local, and we could write

$$U - U^2 - U_{,\eta\eta} = 0. \tag{6.54}$$

by using a travelling-wave ansatz in both cases.

Equation (6.52) belongs in a general class of equations. Such equations can exhibit periodic pulse train solutions and also solitary-wave solutions, some of them having the form of a " Mexican hat " (cf.Figure). Such solutions exhibit an algebraic decay both in the substrate and along the surface. They are thus true surface waves. We refer the reader to the original work of Kovalev *et al* (2002) for technical details which are clearly out of scope here.

## 7 Conclusions

In this contribution we have tried to exhibit the possibilities of existence of special classes of nonlinear waves of the surface wave type travelling over a substrate. In the absence of coating on the limiting surface we have witnessed the essential effects of *nonlinearity* on the evolution of propagating signals, sometines highlighting the possibility of the propagation of permanent wave forms. With the addition of a coating (superposed layer or thin film) on top of the susbtrate the problem becomes much richer in that a natural *dispersion* comes into the picture. Solitary waves and nearly soliton solutions are exhibited whether of the Love (*SH*) type or true Rayleigh nonlinear waves when a compensation between nonlinearity and dispersion occurs. Finally, in the end of section 6, we have pinpointed the *nonlocality* of some of these surface wave problems. The quest for such relevant problems in both signal-processing devices or in geophysics is not ended. Much more work is needed, but the mathematical technical difficulties must not be overlooked. We have tried to keep these to a strict minimum, but it is clear that the study of surface waves in such solid structures presents challenges to the applied mathematicians. What we have not treated here is the lateral evolution of some of the considered waves. A more or less standard equation representing such effects of limited lateral extension of non linear waves is the so-called *Kadomtsev-Petiashvili* equation associated with a quasi-two dimensional propagation in a plane :

$$\left(4u_{,t} + 12uu_{,x} + u_{,xxx}\right)_{,x} = 3u_{,zz}, \tag{7.1}$$

where z denotes the lateral coordinate , and the left-hand side appears to be the x-derivative of the left-hand side of a Korteweg-de Vries equation. This equation is exactly integrable in the sense of soliton theory (cf. Newell, 1985). In our case, it is possible that depending on the sign and values of some material coefficients the wave will suffer laterally a focusing or a defocusing as the wave travels. It is easily conceived that this focusing or defocusing phenomenon can happen when one studies the nonlinear longitudinal propagation of elastic waves in a nonlinear elastic slab. We have not reported here such studies but it is also obvious that the nonlinear wave propagation in such structures as elastic plates and shells is close to our present concern. For these we refer the reader to the studies by Collet and Pouget (1996, 1998, 2001), , and to a recent work of Porubov , Maugin and Mareev (2004). The latter work reports the interesting phenomenon of a considerable increase in the amplitude of a two-dimensional nonlinear strain wave propagating in an elastic plate influenced by an external medium, that is found during the formation of a two-dimensional localized strain wave from an arbitrary initial pulse (cf. Figure 9). In these cases the existence of localized bell-shaped strain waves of permanent form (solitary waves) again results from the balance between nonlinearity and dispersion. But dispersion may be intrinsic, such as caused by the internal structure of the material, or caused by the finite width of the wave guide, while nonlinearity is provided by both the elastic features of the material (physical nonlinearity on which we focused in these lecture notes) and by the nonlinear relationship between displacement and strains (geometrical nonlinearities). An external medium in interaction with the elastic structure may affect the nonlinear and dispersive features of the wave guide through its lateral surface. This situation may be of interest in some dynamical geophysical problems (cf. Engelbrecht and Peipman, 1992).

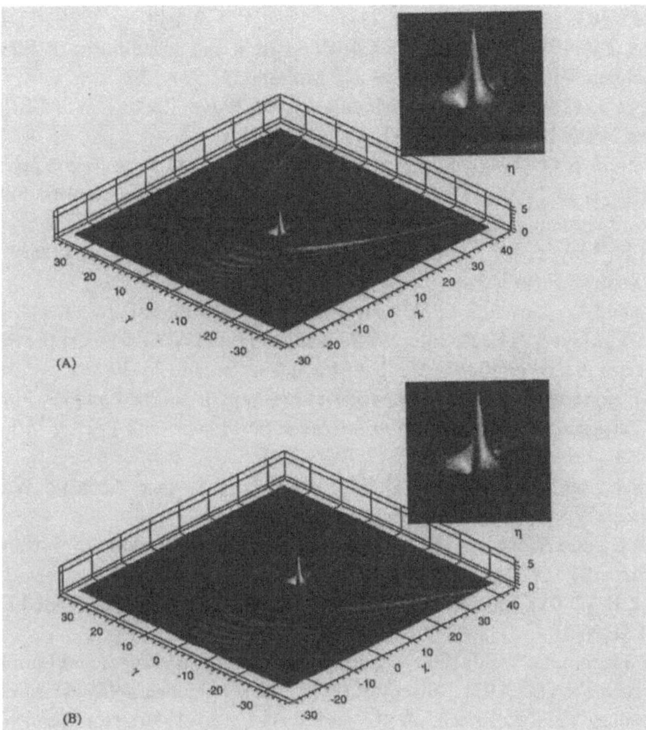


**Figure 9.** Formation of a lump from a Gaussian input at (A) $t$= 1.5; (B) $t$=2.5, for two-dimensiosnal long nonlinear strain waves in an elastic plate (after Porubov, Maugin and Mareev, 2004)

***Acknowledgement*:** The author benefits from a Max-Planck Award for International Co-operation (2002-2005). The present notes would not exist without the achievements and kind collaboration of my co-close workers H.Hadouaj, B.A.Malomed, C.I.Christov, A.S.Kovalev and A.P.Mayer, among others.

## References

Achenbach, J.D. (1973). Wave Propagation in Elastic Solids. Amsterdam : North-Holland.
Bataille, K. , and Lund F. (1982). Nonlinear Waves in Elastic Media. Physica 6D : 95-104.
Bland, D.R. (1969). Nonlinear Dynamical Elasticity. Waltham, Mass :Blaisdell.
Cabannes, H. (1970). Fluiddynamics and Magnetohydrodynamics. NewYork : Wiley-Interscience.
Cho, Y., and Miyagawa N. (1993). Surface Acoustic Wave Solitons Propagating on the Metallic Grating Waveguide. Appl.Phys.Lett. 63 :1188-1190.
Collet, B., and Pouget J. (1996). Nonlinear Propagation of Wave Packets on an Elastic Substrate Coated with a Thin Film. In Markov, K.. Ed., Continuum Models and Discrete Systems, Singapore : World Scientific Publ. 395-403..
Collet, B., and Pouget J. (1998). Two-dimensional Modulation and Instabilities of Flexural Waves of a Thin Plate on Nonlinear Elastic Foundation. Wave Motion 27 : 521-534.
Collet, B., and Pouget J. (2001). Nonlinear Modulation of Wave Packets in a Shallow-Shell on An Elastic Foundation. Wave Motion 34 : 63-81.
Courant, R., and Friedrichs K.O. (1948). Supersonic Flows and Shock Waves. New York : J.Wiley.
Dieulessaint, E., and Royer D. (1980). Elastic Waves in Solids (Translation from the French), Chichester, U.K : J.Wiley.
Eckl, C., Kovalev, A.S., Mayer, A.P., Lomonosov, A.M., and Hess P. (2004), Solitary Surface Acoustic Waves. Physical Review E (in press)
Eckl, C., Mayer, A.P., and Kovalev A.S. (2002). Phys.Rev.Lett. 89 : 095501.
Eckl, C., Schöllmann, J., Mayer, A.P., Kovalev, A.S., and Maugin G.A. (2001), On the Stability of Surface Acoustic Pulse Trains in Coated Elastic Medbia. Wave Motion 34 : 35-49.
Engelbrecht, J. (1983). Nonlinear Wave Processes of Deformation in Solids. London : Pitman.
Engelbrecht, J., and Peipman, P. (1992). Nonlinear Waves in a Layer with Energy Influx. Wave Motion, 16 : 173-181.
Ewen, J., Gunshor, R.L., and Weston, V.H. (1981). Solitons in Surface Acoustic Waves Media . In : Proc.1981 Ultrasonics Symposium. New York : I.E.E.E 295-298.
Ewen, J., Gunshor, R.L., and Weston, V.H. (1982). An Analysis of Solitons in Surface Acoustic Wave Devices. J.Appl.Phys. 53 : 5682.
Fu, Y.B., and Hill, S.L.B. (2001). Propagation of Steady Nonlinear Waves in a Coated Elastic Half-space. Wave Motion 34 : 109-29.
Goldstein, R.V. , and Maugin, G.A. Eds.(2004). Surface Waves in Anisotropic and Laminated Bodies and Defects Detection (NATO ARW,Moscow, 2002), NATO Science Series, Dordrecht : Kluwer.
Gorentsveig, V.I., Kivshar, Yu.S, Kosevich, A.M., and Syrkin, E.S. (1990). Nonlinear Surface Modes in Crystals. Phys.Lett. 144 : 479-486.
Gorentsveig, V.I., Kivshar, Yu.S, Kosevich, A.M. and Syrkin, E.S., (1991). Nonlinear Surface Modes in Crystals. Int.J.Engng.Sci. 29 : 271-279.
Graham, R.A., (1993). Solids under High Pressure Shock Compression. Berlin : Springer.
Gusev, V., Lauriks, W. and Thoen, J. (1997). Theory for the Time Evolution of the Nonlinear Rayleigh Waves in an isotropic Solid. Phys.Rev. B55 : 9344-9347.
Hadouaj, H. and Maugin, G.A. (1989). Une onde solitaire se propageant sur un substrat élastique recouvert d'un film mince. C.R.Acad.Sci.Paris II-309 : 1877-1881.
Hadouaj, H. and Maugin, G.A. (1992). Surface Solitons on Elastic Structures : numerics. Wave Motion, 16 : 115-125.

Hadouaj, H., Malomed, B.A. and Maugin, G.A. (1991a). Dynamics of a Soliton in the Generalized Zakharov System. Phys.Rev. A44 : 3925-3931.

Hadouaj, H., Malomed, B.A. and Maugin G.A. (1991b)., Soliton-soliton Collisions in the Generalized Zakharov System. Phys.Rev. A44 : 3932-3940.

Hamilton, M.F., Il'inskii, Y.A., and Zabolotskaya, E.A. (1995a). Local and Nonlocal Nonlinearity in Rayleigh Waves. J.Acoust.Soc.Amer. 97 : 882-890.

Hamilton, M.F., Il'inskii, Y.A., and Zabolotskaya, E.A. (1995b). Evolution Equation for Nonlinear Rayleigh Waves. J.Acoust.Soc.Amer. 97 : 891-897.

Hamilton, M.F., Il'inskii, Y.A., and Zabolotskaya, E.A. (1999). Nonlinear Surface Acoustic Waves in Crystals. J.Acoust.Soc.Amer. 105 : 639-651.

Harvey, A.P., and Tupholme, G.E. (1991). Nonlinear Mode Coupling of Two co-directional Surface Acoustic Waves on a Piezoelectric Solid. Int.J.Engng.Sci. 29 : 987-998.

Harvey, A.P., and Tupholme, G.E. (1992). Propagation of Anisotropic Elastic and Piezoelectric Nonlinear Surface waves. Wave Motion 16 :125-136.

Hunter, J.K. (1989).Nonlinear Surface Waves.In : Lindquist, W.B., Current Progress in Hyperbolic Problems and Computations. Providence : New York. American Mathematical Society. 185-202,

Kalyanasundaram, N. (1981a). Nonlinear Surface Acoustic Waves on an Isotropic Solid. Int.J.Engng.Sci., 19 : 279-286.

Kalyanasundaram, N., (1981b). Finite-amplitude Love Waves on an Isotropic Layered Half Space. Int.J.Engng.Sci. 19 :287-293.

Kalyanasundaram, N. (1984). Nonlinear Propagation Characteristics of Bleustein-Gulayev Waves. J.Sound Vibr. 96 : 411-420.

Kalyanasundaram, N., Parker, D.F., and David E.A. (1990). The Spreading of Nonlinear Elastic Surface Waves. J.Elasticity 24 : 79-103.

Kalyanasundaram, N., Randam, R., and Prasad, T. (1982). Coupled Amplitude Theory of Nonlinear Surface Acoustic Waves. J.Acoust.Soc.Amer. 72 : 488-492.

Kivshar, Yu.S. (1991). Solitons in a Nonlinear Elastic Medium. Phys.Rev. B43 : 3493-3499.

Knight, E.Yu., Hamilton, M.F., Il'inskii, Y.A., and Zabolotskaya, E.A. (1997). General Theory for the Spectral Evolution of Nonlinear Rayleigh Waves. J.Acoust.Soc.Amer. 102 : 1402-1417.

Kosevich, Yu.A. (1991). Nonlinear Shear Surface Wave at the Interfaces and Planar defects of Crystals. Int.J.Engng.Sci. 29 : 327-330.

Kovalev, A.S., Mayer, A.P., Eckl, C. and Maugin, G.A., (2002), Solitary Rayleigh Waves in the Presence of Surface Nonlinearities.Phys.Rev. E66 :036615/1-15

Kulikowsky, A., and Sveshnikova, L.M. (1995). Nonlinear Waves in Elastic Media. Boca-Raton, Florida : C.R.C. Press.

Jeffrey, A., and Engelbrecht, J., Eds. (1994). Nonlinear Waves in Solids (CISM Lecture Notes, 1993). Wien : Springer.

Jeffrey, A., and Taniuti T. (1963). Nonlinear Wave Propagation with Applications to Physics and Magnetohydrodynamics. New York : Academic Press.

Lardner, R.W. (1983). Nonlinear Surface Waves on an Elastic Solid. Int.J.Engng.Sci. 21: 1331-1342.

Lardner, R.W. (1985). Waveform Distorsion and Shock Development in Nonlinear Rayleigh Waves. Int.J.Engng.Sci. 23 : 113-118.

Lomonosov, A.M., Mayer, A.P. and Hess, P. (2001). In : Levy, M., Bass, H.E., and Stren, E. Eds. Experimental Methods in the Physical Sciences. San Diego : Academic Press. Vol.3, 65.

Maradudin, A.A. (1988). Nonlinear Surface Acoustic Waves and Their Associated Solitons. In : Parker, D.F, and Maugin, G.A., Eds. (1988), pp.62-71.

Maradudin, A.A., and Mayer A.P. (1991). Surface Acoustic Waves on Nonlinear Susbtrates . In : Boardman, A., Bertolotti, M., and Twardowski, T., Eds. Nonlinear Waves in Solid State Physics. New York : Plenum. 13-161.

Mandel, J., and Brun, L., Eds. (1975). Mechanical Waves in Solids (CISM Lecture Notes). Wien : Springer.
Maugin, G.A. (1983). Surface Elastic Waves with Transverse Horizontal Polarization. In : Hutchinson, J.W., Ed. Advances in Applied Mechanics. New York : Academic Press. Vol .3, 373-434. .
Maugin, G.A. (1985)., Nonlinear Electromechanical Effects, A series of Lectures. Singapore : World Scientific.
Maugin, G.A. (1988). Shear Horizontal Surface Acoustic Waves in Solids. In : Parker, D.F, and Maugin, G.A., Eds., (1988), pp.258-27.
Maugin, G.A. (1994). Physical and Mathematical Models of Nonlinear Waves. in : Jeffrey, A., and Engelbrecht, J. Eds (1994), pp.109-233.
Maugin, G.A. (1999). Nonlinear Waves in Elastic Crystals. Oxford, U.K. :Oxford University Press.,
Maugin, G.A., and Hadouaj, H. (1991). Solitary Surface Waves on an Elastic Substrate Coated with a Thin Film. Phys.Rev. B44 : 1266-1280.
Maugin, G.A., Hadouaj, H., and Malomed, B.A. (1992). Nonlinear Coupling Between SH Surface Solitons and Rayleigh Modes on Elastic Structures. Phys.Rev. B45 : 9688-9694.
Maugin, G.A., Pouget, J., Drouot, R. and Collet, B. (1992). Nonlinear Electromechanical Couplings. New York : J.Wiley.
Mayer, A.P. (1995). Surface Acoustic Waves in Nonlinear Elastic Media. Physics Reports 256 : 237-366.
Mayer, A.P., and Kovalev, A.S. (2003). Nonlinear Waves Guided a Liquid-solid Interface, Proc.Estonian Acad.Sci.Math.Phys. 52 : 43-48.
Mayer, A.P., and Maradudin, A.A. (1991). Effects of Nonlinearity and Dispersion on the Propagation of Surface Acoustic Waves. In : Maugin, G.A. Ed. Continuum Models and Discrete Systems, London : Longman. Vol.2, 306-315.
Meegan, G.F., Hamilton, M.F., Il'inskii, Y.A., and Zabolotskaya, E.A. (1999). Nonlinear Rayleigh and Scholte Waves. J.Acoust.Soc.Amer. 106 : 1712-1723.
Mozhaev, V.G. (1989). A New Type of Acoustic Waves in Solids Due to Nonlinearity. Phys.Lett. A139 : 333-337.
Murdoch, A.I. (1976). The Propagation of Surface Waves in bodies with Material Boundaries, J.Mech.Phys.Solids 24 : 137-146.
Murnaghan, F.D. (1951). Finite Deformations of an Elastic Solids. New York : J.Wiley.
Nayanov, V.I. (1986). Surface Acoustic Cnoïdal Waves and Solitons in a LiNbO3-(SiO film)., JETP Lett. 44 : 314-317.
Newell, A.C. (1985). Solitons in Mathematics and Physics. Philadelphia : S.I.A.M.
Panayotaros, P. (2002). An Expansion Method for Nonlinear Rayleigh Waves., Wave Motion 36 : 1-21.
Parker, D.F. (1988).Wave Form Evolution for Nonlinear Surface Acoustic Waves, Int.J.Engng.Sci. 26 : 59- 75.
Parker D.F. (1990). In : Boardman, A.D., Twardowski, T., and Bertolotti, M. Eds. Nonlinear Waves in Solid State Physics. New York : Plenum. 163-183,
Parker, D.F. (1994). Nonlinear Surface Acoustic Waves and Waves on Stratified media. In : Jeffrey, A., and Engelbrecht, J. Eds (1994), pp.289-347.
Parker, D.F. (2004). Nonlinearity in Elastic Surfaces Waves acts nonlocally. In : Goldstein, R.V., and Maugin, G.A. Eds (2004) . Dordrecht : Kluwer (in press).
Parker, D.F., and David, E.A. (1989a). Nonlinear Piezoelectric Surface Waves. Int.J.Engng.Sci. 27 : 565-581.
Parker, D.F., and David, E.A. (1989b). Spreading of Nonlinear Surface Waves on Piezoelectric Solids. In : McCarthy, M.F., and Hayes, M.A., Eds. Elastic Wave Propagation. Amsterdam : Elsevier. 125-131.
Parker, D.F., Mayer, A.P., and Maradudin, A.A. (1992). The Projection Method for Nonlinear Surface Acoustic Waves. Wave Motion 16 : 151-162.

Parker, D.F., and Maugin G.A. Eds (1988). Recent Developments in Surface Acoustic Waves. Berlin : Springer.

Parker, D.F., and Talbot, F.M. (1985). Analysis and Computation for Nonlinear Elastic Surface Waves of Permanent Form. J.Elasticity 15 : 389-426.

Planat, M. (1985). Multiple-scale Analysis of the Nonlinear Acoustic Wave Propagation in Anisotropic Crystals. J.Appl.Phys. 57 : 4911-4925.

Planat, M. (1991). Observation of Acoustic Envelope Solitary Waves Generated by Metallic Interdigitated Transducers in a Quartz Crystal. In : Maugin, G.A. Ed., Continuum Models and discrete Systems. London : Longman. Vol.2, 339-349.

Porubov, A.V., and Samsonov A.M. (1995). Long Nonlinear Strain Waves in Layered Elastic Half-space. Int.J.Non-linear Mech. 30 : 861-877.

Rakhmatulin, Kh.A., and Demianov Y. (1961). Resistance to Intense Loads of Short Duration (in Russian). Moscow : Fizmatgiz.

Reutov, V.P. (1973). Izvestiya Vysshikh Uchebnykh Zavedenii, Radiofizika 16 : 1690 [English translation : Radiophys.Quant.Electron. 16 : 1307]

Sakuma, T., and Nishigushi N. (1990). Theory of the Surface Acoustic Soliton, V. Approximate Soliton Solutions of the Korteweg-de Vries Type. Phys.Rev. B41 : 12117-12125.

Teymur, M. (1988). Nonlinear Modulation of Love Waves on an Isotropic Hyperelastic Layered Half-space. Int.J.Engng.Sci. 26 : 907-927.

Tiersten, H.F. (1969). Elastic Surface Waves Guided by Thin Film. J.Appl.Phys. 40 : 770-789.

Tupholme, G.E., and Harvey, A.P. (1988). Nonlinear Surface Acoustic Waves on a Piezolecetric Solid. Int.J.Engng.Sci. 26 : 1161-1166.

Zabolotskaya, E.A. (1992). Nonlinear Propagation of Plane and Circular Rayleigh Waves in Isotropic Solids. J.Acoust.Soc.Amer. 91 : 2569-2575.

Vlannes, N.P. and Bers, A. (1982). Nonlinear Coupled-Mode Theory for Surface Acoustic Waves. J.Appl.Physics 53 : 372-378.

Whitham, G.B. (1974). Linear and Nonlinear Waves. New York : J.Wiley-Interscience.

Wright, T.W., Ed. (1986). Nonlinear Wave Propagation in Solids, Vol.AMD-77. New York : Society of Mechanical Engineers.

# Experimental Soil Behaviour: its Testing by Waves and Engineering Applications

Renato Lancellotta

Dept. of Structural and Geotechnical Eng.,
Politecnico di Torino, corso Duca degli Abruzzi 24, 10129 Torino, Italy
renato.lancellotta@polito.it

## 1 Introduction

This paper is aimed at presenting features of soil behaviour, which can be of relevance when using tests, based on wave propagation, for its characterization. In particular, examples are provided to **motivate** both in situ and laboratory tests, and some basic aspects related to **particulate nature** of soils and its **anisotropy** are also addressed.

To motivate the use of seismic methods for site characterization, let start by recalling that coarse grained materials are difficult to sample soils, so that the current practice in Geotechnical Engineering still relies on empirical correlations, mainly related to penetration tests (see Jamiolkowski et al., 1985; Lancellotta, 1995; for a detailed presentation of this subject).

To outline shortcomings of these procedures, we consider a practical example concerned with the mechanical characterization of the "Messina Gravel" formation (Crova et al., 1992).

The subsoil of Messina Strait consists of gravel and sand deposits of holocene and pleistocene epoch, superimposed to soft rocks of pliocene and miocene epoch. On the Sicilian shore (see Figure 1), sand and gravel deposits extend to a depth of about 200 m below the actual G.L.

In particular, at the location of the anchor block[1], the soil is composed of sand and gravel of middle pleistocene (500.000 to 600.000 years old), whereas at the location of the bridge tower the soil is composed of sand and gravel of holocene epoch (6000 to 15.000 years).

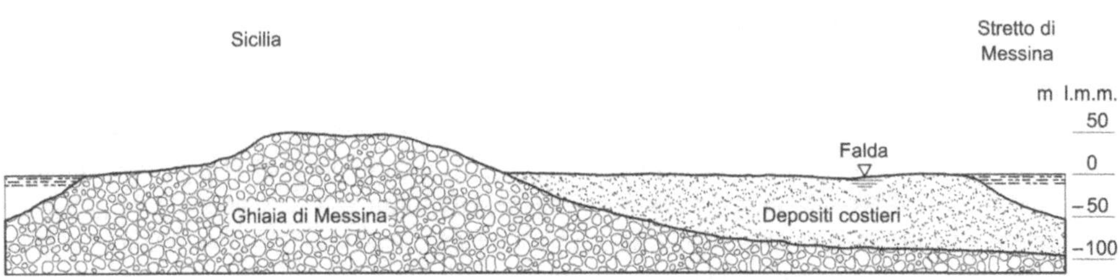


**Figure 1:** Messina Gravel deposit (Crova et al., 1992).

[1] Reference is here made to the design of the Messina Strait bridge.

The circumstance, that both these formations have been deposited in similar environment and exhibit similar grading and mineralogical composition, offered the possibility to explore the influence of the different age on the results of penetration tests and seismic tests. As expected, whereas penetration tests in both formations gave similar results, a clear difference merged when referring to the shear modulus, inferred from shear wave velocity measurements (see Figure 2), performed through cross-hole tests. The non invasive character of this tests preserves the initial structure of soil deposits and the major influence of all diagenetic phoenomena contributing to a stiffer mechanical response.

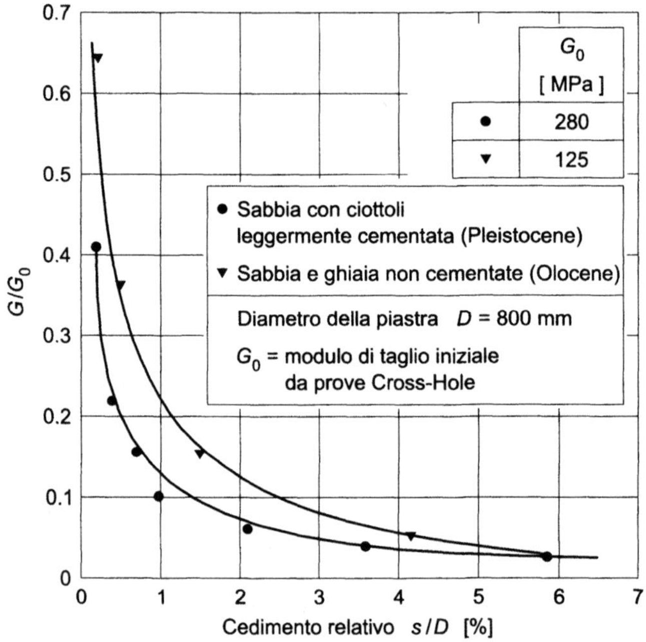


**Figure 2:** Shear modulus from cross-hole tests and plate loading tests (Ghionna et al., 1993).

Diagenetic processes begin immediately after deposition and continue until metamorphism takes over (when reactions occur at elevated temperature and/or pressure). Diagenetic processes include dissolution of grains at points of contact, producing sutured contacts if grains are of similar solubility, or concavo-convex contacts if one grain dissolves preferentially. In addition, when silica cement (deriving from pressure dissolution, as well as from dissolution of feldspars or biogenic silica) is precipitated around quarz grains in optical continuity there is a syntaxial overgrowth of quarz grains. Calcite cements are also common in grain supported structures. Finally the precipitation of clay minerals (a process known as authigenesis) also has a great effect on permeability and porosity. All these processes, whose path depends on depositional environment and climate, sediment composition and texture, strongly affect the mechanical response of coarse grained soils at low strains, so that it is mandatory to rely on tests preserving the soil structure, a possibility offered by seismic measurements[2].

## 2 A Reminder of Wave Propagation Theory

In continuum mechanics, a wave can be considered a singular surface for some fields, propagating with a finite speed depending on the properties of the porous medium.

Let be $\sigma(t)$ denote an orientable surface whose equation in the current configuration is:

$$f(\mathbf{x},t)=0 \quad . \tag{1}$$

If **n** denotes the unit normal vector to the above defined surfaces in the current configuration, it turns out:

$$\mathbf{n}=\frac{\nabla f}{|\nabla f|}, \qquad \nabla f=\left(\frac{\partial f}{\partial x^i}\right) . \tag{2}$$

The speed of the moving surface $\sigma(t)$, with respect to the chosen frame of reference (so, it is independent of the motion of the continuum body), is defined *speed of displacement* and is given by:

$$v_n=\mathbf{v}\cdot\mathbf{n}=-\frac{(\partial f/\partial t)}{|\nabla f|} \quad . \tag{3}$$

The velocity of the surface with respect to the material points, instantaneously lying upon it, is called *local speed of propagation*:

$$c=\left(v_n-\dot{\mathbf{x}}\cdot\mathbf{n}\right) \quad . \tag{4}$$

Let now assume the three-dimensional region $C(t)$ be divided by the surface $\sigma(t)$ into two regions $C^-(t)$, $C^+(t)$, and consider the function:

$$\varphi(\mathbf{x},t)=\begin{cases}\varphi_1(\mathbf{x},t) & \forall\mathbf{x}\in C^-(t)\\ \varphi_2(\mathbf{x},t) & \forall\mathbf{x}\in C^+(t)\end{cases} . \tag{5}$$

If, $\forall\mathbf{r}\in\sigma(t)$, the function approaches finite limits:

[2] Foti et al. (2002) and Foti and Lancellotta (2004) have also shown the capability of seismic tests to detect soil porosity.

$$\lim_{\mathbf{x}\to\mathbf{r},\mathbf{x}\in C^-} \varphi_1(\mathbf{x},t) = \varphi^-(\mathbf{r},t), \qquad \lim_{\mathbf{x}\to\mathbf{r},\mathbf{x}\in C^+} \varphi_2(\mathbf{x},t) = \varphi^+(\mathbf{r},t), \tag{6}$$

than we can define on $\sigma(t)$ the jump of the function as:

$$[\![\varphi]\!] = \varphi^+ - \varphi^-. \tag{7}$$

A basic aspect is now the following: *the derivatives of a function wich is continuous through a singular surface are not independent from each other*. This is due to the fact that the jumps of the considered field are linked to the geometric character of the surface through a single parameter. To prove this, if $\varphi(\mathbf{x},t)$ is continuous across the singular surface, being:

$$d\varphi = \varphi_{,i}dx^i + \varphi_{,t}dt\ , \tag{8}$$

the continuity assumption requires that:

$$[\![\varphi_{,i}]\!]dx^i + [\![\varphi_{,t}]\!]dt = 0, \tag{9}$$

for any $dx^i$ *and* $dt$ satisfying:

$$dx^i n^i - v_n dt = 0. \tag{10}$$

Because (9) and (10) are not independent, certainly a quantity $g$ exists such that

$$\begin{aligned} &[\![\varphi_{,i}]\!] = gn^i \\ &[\![\varphi_{,t}]\!] = -gv_n \\ &[\![\varphi_{,i}]\!]v_n + [\![\varphi_{,t}]\!]n^i = 0 \end{aligned} \qquad . \tag{11}$$

Equations (11) correspond to Hadamard's geometric, kinematic and mixed compatibility conditions. In addition, by considering the material derivative:

$$\dot{\varphi} = \varphi_{,t} + \varphi_{,k}v^k \quad , \tag{12}$$

the previous relations allow to obtain:

$$[\![\dot{\varphi}]\!] = -gc\,. \tag{13}$$

If the derivative of the field under consideration are continuous across the surface up to the order $(n-1)$, while certain derivatives of order *n* are discontinuous, then the surface is considered a singular surface of order *n*.

As an example, if the field $\varphi$ is the motion and is discontinuous, the surface is said to be *singular of order zero*, as it is the case of fractures in solids.

A singular surface of order 1 is a surface where some first derivative of the motion fails to be continuous and is called *shock wave*.

Singular surfaces of order 2 are defined *acceleration* or *ordinary waves*, since the motion and the velocity are continuous across them, but this does not hold for second derivative of the motion.

If we consider a second order discontinuity, the following relations hold:

$$\begin{aligned} [\![\varphi_{,ik}]\!] &= [\![(\varphi_{,i})_{,k}]\!] = g_i n_k \\ [\![\varphi_{,ki}]\!] &= [\![(\varphi_{,k})_{,i}]\!] = g_k n_i \end{aligned} \quad , \tag{14}$$

and being:

$$\frac{g_i}{n_i} = \frac{g_k}{n_k} = \gamma \quad ,$$

we can write:

$$\begin{aligned} [\![\varphi_{,ik}]\!] &= \gamma n_i n_k \\ [\![\ddot{\varphi}]\!] &= \gamma c^2 \end{aligned} \quad . \tag{15}$$

By substituting now the constitutive equation

$$\sigma_{ik} = C_{iklm}\varepsilon_{lm} \tag{16}$$

into the equation of the motion

$$\rho\ddot{u}_i = \rho b_i + \sigma_{ik,k} \quad , \tag{17}$$

and by applying the jump operator, taking into account the continuity of $\rho, u_i, u_{i,j}, b_i$, we obtain:

$$\rho c^2 a_i = C_{iklm} n_k n_l a_m \quad . \tag{18}$$

that can be written in the form:

$$\left(C_{ihlm} n_k n_l - \rho c^2 \delta_{im}\right) a_m = 0 \quad . \tag{19}$$

Equation (19) shows that the squared speed of local propagation are the eingenvalues of the acoustic tensor:

$$A_{im} = C_{iklm} n_k n_l \ . \tag{20}$$

The above results will now be exploited in the context of soil anisotropy.

## 3 Soil Anisotropy

Soils, during depositional processes, usually experience one-dimensional deformation, so that particles move vertically without any tendency to move laterally. The so-called initial anisotropy reflects this depositional history and a rather realistic model is, in this case, the cross-anisotropic body: the soil response is different if the loading direction changes from vertical to horizontal, but it is the same when changes occur in the horizontal plane.

To analyse the geometrical character of wave propagation, let indicate the non vanishing components of the $C_{ijhk}$ as

$$\begin{aligned}
C_{1111} &= C_{2222} = A \\
C_{3333} &= C \\
C_{3311} &= C_{3322} = F \\
C_{2323} &= C_{1313} = L \\
C_{1212} &= N \\
C_{1122} &= C_{2211} = A - 2N
\end{aligned} \tag{21}$$

When considering wave propagation in the $x_3$ direction (assumed to be the vertical direction, see Figure 3.a), equations (19) write

$$\begin{bmatrix} \frac{L}{\rho} - c^2 & 0 & 0 \\ 0 & \frac{L}{\rho} - c^2 & 0 \\ 0 & 0 & \frac{C}{\rho} - c^2 \end{bmatrix} \begin{Bmatrix} a_1 \\ a_2 \\ a_3 \end{Bmatrix} = 0 \ , \tag{22}$$

so that a compressional P wave, characterized by displacement components $(0,0,a_3)$, propagates with a velocity $c=\sqrt{C/\rho}$, whereas a shear S wave, with transverse displacement components $(a_1,a_2,0)$, propagates with $c=\sqrt{L/\rho}$.

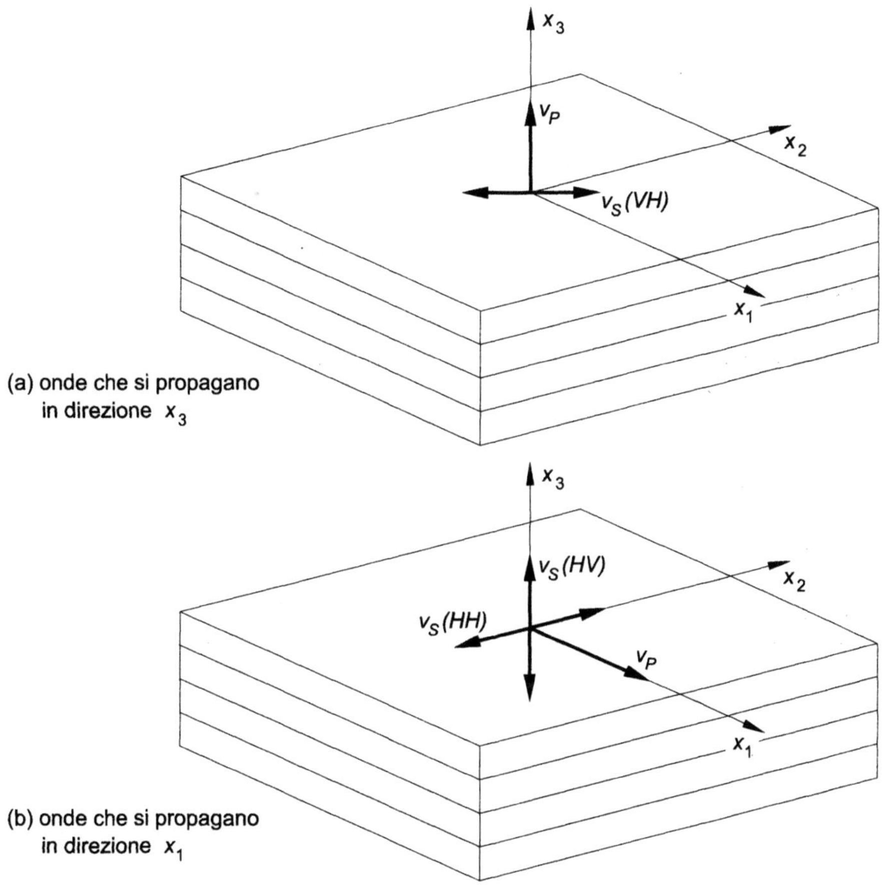


**Figure 3:** Wave propagation in a cross-anisotropic medium

By considering a wave propagating in the $x_1$ (see Figure 3.b), equations (19) write

$$\begin{bmatrix} \frac{A}{\rho}-c^2 & 0 & 0 \\ 0 & \frac{N}{\rho}-c^2 & 0 \\ 0 & 0 & \frac{L}{\rho}-c^2 \end{bmatrix} \begin{Bmatrix} a_1 \\ a_2 \\ a_3 \end{Bmatrix} = 0 \quad , \tag{23}$$

and in this case we deduce 3 different speed of propagation.

The displacement vector $(a_1,0,0)$ corresponds to a P wave with propagation speed $c_P = \sqrt{A/\rho}$; the displacement vector $(0,a_2,0)$ corresponds to a S wave with $c_S = \sqrt{N/\rho}$ and the displacement vector $(0,0,a_3)$ to a S wave with $c_S = \sqrt{L/\rho}$. Being the speed of propagation of S waves different from each other, the phoenomenon is indicated as **wave splitting**.

We still observe that the above deduced geometrical aspects are inherent to the fact that, from the properties of the costitutive tensor $C_{iklm}$, it follows that the acoustic tensor $A_{im}$ is symmetric and positive definite, so that its eingenvalues are real and positive and their corresponding vectors are orthogonal. But, contrary to the isotropic case, when considering any general direction of propagation (i.e. different from those previously considered), displacements are neither truly longitudinal nor truly transverse, so that the motion cannot be decoupled anymore into P and S waves. Referring to experimental measurements, the above arguments lead to the concusion that data from cross-hole tests and down-hole tests cannot be directly compared (at least theoretically) and the interpretation of down-hole tests is far from being straightforward.

In anisotropic media, the directions along which the propagating plane wave are purely longitudinal or transverse are called pure mode axes. These axes are defined in terms of symmetry properties (Waterman, 1959): an axis is a pure mode axis if it is orthogonal to a plane of reflection symmetry, or if it is orthogonal to an axis of sixfold rotational symmetry, or if it is itself an axis of twofold or greater rotational symmetry.

In addition to the initial anisotropy, any subsequent loading history gives the material an induced anisotropy, so that properties have a directional dependence, linked to stress components.
In order to have a physical insights of this aspect, it is convenient to make reference to DEM[3] analyses, which take into account the particulate nature of soils.

Figures 4 and 5 (see Dobry and Ng, 1989) show force chains, deduced by inspection of interparticle contact forces. In particular, they refer to the direction of larger contact forces, whereas minor contact forces (i.e. lesser than 25% of the maximum value), not shown in Figures, provide support to stiffer columns. It is relevant to note that, in presence of isotropic loading, there are many force paths in different directions, whereas shearing gives rise to a limited number of parallel stiff chains.

[3] Discrete Element Method (see Cundall, 1976; Cundall and Strack, 1979).

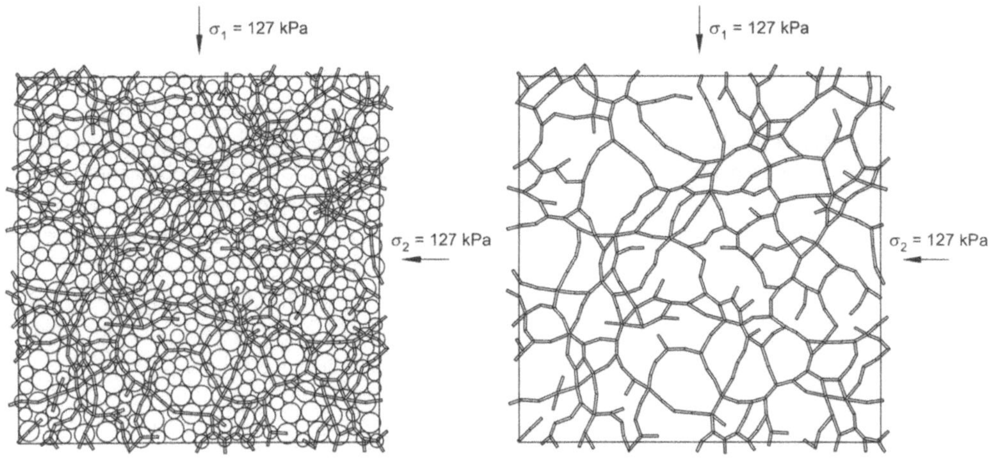


**Figure 4:** Force chains due to isotropic stress state (Dobry and Ng, 1989).

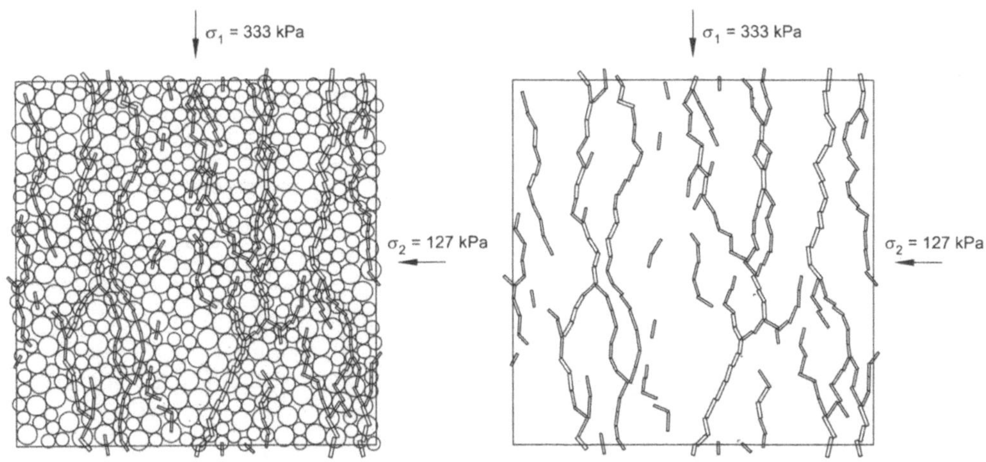


**Figure 5:** Force chains due to anisotropic stress state (Dobry and Ng, 1989).

This pattern explains the experimental evidence (Stokoe et al., 1980, Kuvano and Jardine, 2002) that the P wave velocity mainly depends on the stress component in the direction of propagation; on the contrary, the shear wave velocity is controlled by stress components in the direction of propagation and particle motion. In general, the $S_{VH}$ and the $S_{HV}$ velocity measured on sand samples depend more on normal effective stress acting in the direction of wave propagation that on that acting in the direction of polarization.

## 4 Barotropy and Pycnotropy

Experimental evidences reported in Figures 6 and 7 prove that soil behaviour is both density and pressure dependent. For this reason, elastic modulii (as well as other parameters) cannot be considered material properties, but they are parameters dependent on the current state (as represented by the current specific volume and mean effective pressure). The implication is that soil parameters, measured in free field conditions by using seismic waves, must be corrected when used in engineering practice, according to the relevant current state, strictly linked to the problem being analysed.

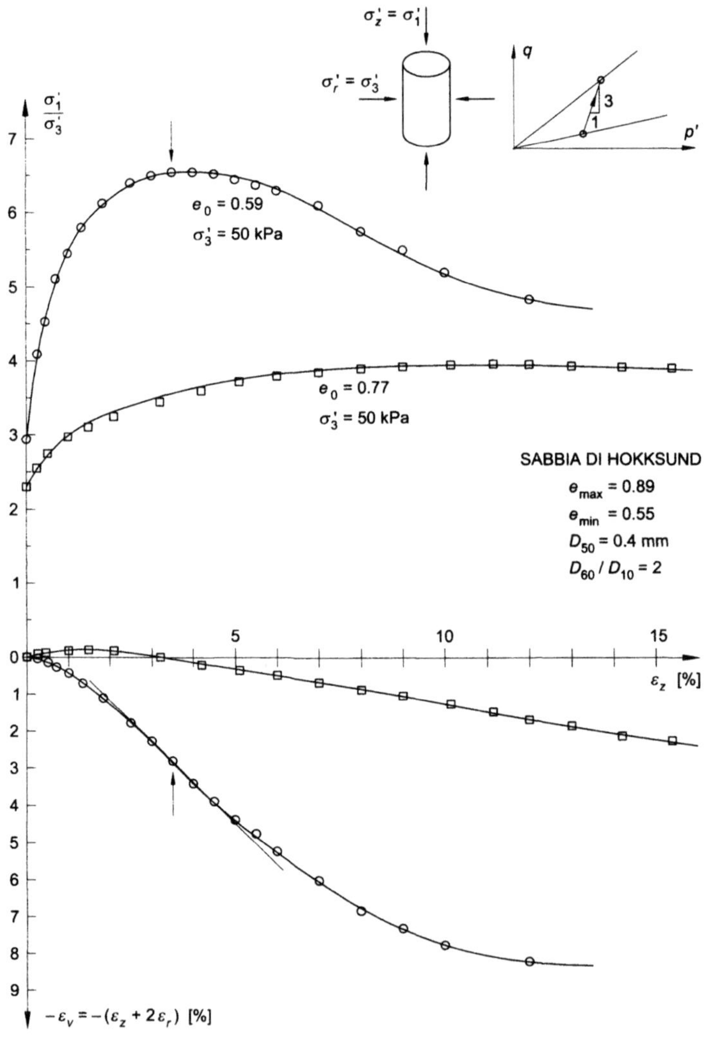


**Figure 6:** Influence of void ratio on sand behaviour (Lancellotta, 2004).

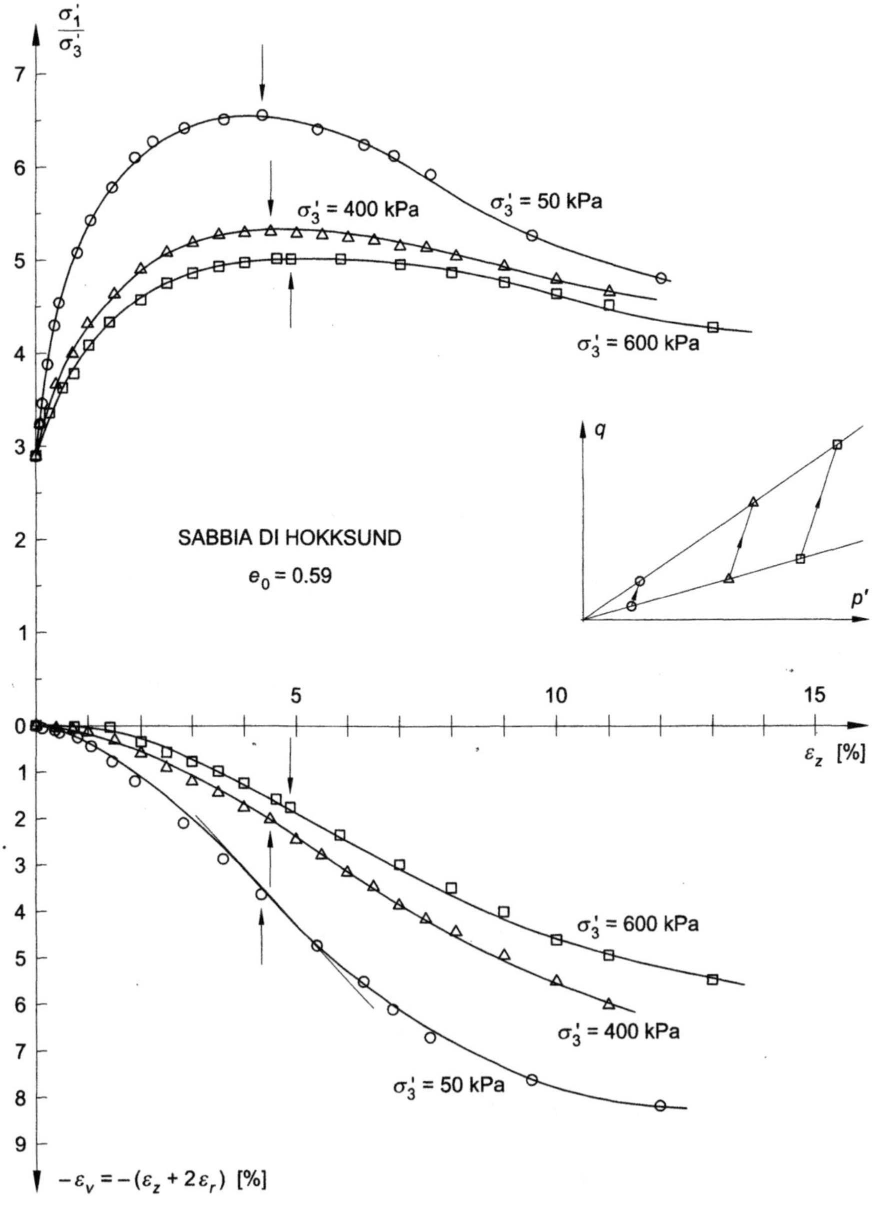


**Figure 7:** Influence of confining stress on sand behaviour (Lancellotta, 2004).

In order to take into account the pressure dependence, the starting point is to refer to Hertz's theory, so that it is common in geotechnical engineering to assume that the bulk modulus $K$ depends on pressure according to the power function

$$\frac{K}{p_a} = m_K \left( \frac{p'}{p_a} \right)^n \quad , \tag{24}$$

where $p_a$ is a reference pressure, $m_K$ $e$ $n$ are dimensionless constants, related to the nature of the tested material and $p' = \frac{1}{3} tr(\sigma'_{ij})$. Usually, the exponent $n$ ranges from 0.3 to 0.5.

Despite the considerations regarding soil anisotropy, in engineering practice isotropy is a convenient hypotesis, since only two soil parameters are required, and a common assumption is that the Poisson's ratio is constant. It follows that the shear modulus $G$ is also a power function of the mean effective pressure (with the same exponent)

$$\frac{G}{p_a} = m_G \left( \frac{p'}{p_a} \right)^n \quad . \tag{25}$$

The shortcoming of the above assumptions is that the obtained elastic model is not anymore a conservative one (Zytynski et al., 1978), i.e., when considering a closed path energy can be dissipated or produced, as it is shown in the sequel.
Referring for sake of simplicity to conventional triaxial tests, the incremental work input is given by

$$dW = p' d\varepsilon_V + q d\varepsilon_S \quad , \tag{26}$$

where $q = \sigma'_a - \sigma'_r$ is the deviator stress and $\varepsilon_V = tr(\varepsilon_{ij})$ and $\varepsilon_S = \frac{2}{3}(\varepsilon_a - \varepsilon_r)$.

By taking into account the constitutive law, the incremental work input can be written

$$dW = \frac{p'}{K} dp' + \frac{q}{3G} dq \quad , \tag{27}$$

and, since a necessary condition for $dW$ to be path independent is[4]

$$\frac{\partial}{\partial q}\left( \frac{p'}{K} \right) = \frac{\partial}{\partial p'}\left( \frac{q}{3G} \right) \quad , \tag{28}$$

it follows that this condition cannot be satisfied, being

[4]Reference is made to Schwarz's theorem on partial derivatives. The above condition is also a sufficient one for a simply connected domain.

$$\frac{\partial}{\partial p'}\left(\frac{q}{3G}\right)=-\frac{q}{3G^2}\frac{\partial G}{\partial p'}\neq 0 \quad ,$$

whereas the l.h.s. vanishes.

On the contrary, by using the same argument it can be proved that, by assuming the bulk modulus depending on both the effective mean pressure $p'$ and the deviator stress $q$, the required structure for the bulk modulus should be

$$\frac{K}{p_a}=m_K\left[\left(\frac{p'}{p_a}\right)^2+\frac{2}{9}\frac{1+\upsilon}{1-2\upsilon}\left(\frac{q}{p_a}\right)^2\right]^{n/2} , \qquad (29)$$

in order to assure that the elastic model is conservative.

**References**


Crova R., Jamiolkowski M., Lancellotta R., Lo Presti D. (1992): Geotechnical characterization of gravelly soils at Messina site: selected topics, Proc. Wroth Memorial Symposium, Oxford, Thomas Telford, 199-218.

Cundall P.A. (1976): Explicit Finite-Difference Methods in Geomechanics, Proc. 6th Conf. on Numerical Methods in Geomechanics, 132-150.

Cundall P.A., Strack O.D.L. (1979): A discrete numerical model for granular assemblies, Géotechnique, 29, 1, 47-65.

Dobry R., Ng T.T. (1989): Discrete modelling of stress-strain behaviour of granular media at small and large strains, Proc. 1st US Conf. On Discrete Element Methods.

Foti S., Lai C., Lancellotta R. (2002): Porosity of fluid-saturated porous media from measured seismic wave velocity, Géotechnique, 52, 5, 359-373.

Foti S., Lancellotta R. (2004): Soil porosity from seismic velocities, Géotechnique, Technical Note, 54, 8, 551-554.

Ghionna V., Jamiolkowski M., Lancellotta R., Pedroni S. (1993): Based capacity of bored piles in sands from in situ tests, Proc. 2nd Int. Seminar on Deep foundations on bored and auger piles, Balkema, 67-75.

Jamiolkowski M., Ladd C.C., Germain J.T., Lancellotta R. (1985): "New developments in field and laboratory testing of soils". Theme Lecture, 11th ICSMFE, S.Francisco, 1, 57-152.

Kuwano R., Jardine R.J.: On the applicability of cross-anisotropic elasticity to granular materials at very small strains, Géotechnique, 52, 10, 727-749.

Lancellotta R. (1995): "Geotechnical Engineering", Balkema, pp.436.

Lancellotta R. (2004): Geotecnica, 3 ed., Zanichelli, 481 pp.

Stokoe K.H., Roesset J.M., Knox D.P., Kopperman S.E., Sudhiprakarn C. (1980): Development of a large scale triaxial testing device for wave propagation studies, Geotechnical Engineering Report GR 80-10, Un. Texas at Austin.

Waterman P.G. (1959): Orientation dependence of elatic waves in single crystals, Phys. Rev., 113, 1240-1253.

Zytynski M., Randolph M.F., Nova R., Wroth C.P. (1978): On modelling the unloading- reloading behaviour of soils, Int. Journal For Numerical and Analytical Methods in Geomechanics, 2, 87-94.

GPSR Compliance

*The European Union's (EU) General Product Safety Regulation (GPSR) is a set of rules that requires consumer products to be safe and our obligations to ensure this.*

*If you have any concerns about our products, you can contact us on ProductSafety@springernature.com*

In case Publisher is established outside the EU, the EU authorized representative is:

Springer Nature Customer Service Center GmbH
Europaplatz 3
69115 Heidelberg, Germany

**Batch number: 10584556**

Printed by Printforce, the Netherlands